Fundamentals of Condensed Matter Physics

Based on an established course and covering the fundamentals, central areas, and contemporary topics of this diverse field, *Fundamentals of Condensed Matter Physics* is a much-needed textbook for graduate students.

The book begins with an introduction to the modern conceptual models of a solid from the points of view of interacting atoms and elementary excitations. It then provides students with a thorough grounding in electronic structure and many-body interactions as a starting point to understanding many properties of condensed matter systems – electronic, structural, vibrational, thermal, optical, transport, magnetic, and superconducting – and the methods used to calculate them.

Taking readers through the concepts and techniques, the text gives both theoretically and experimentally inclined students the knowledge needed for research and teaching careers in this field. It features over 246 illustrations, 9 tables and 100 homework problems, as well as numerous worked examples, for students to test their understanding.

Marvin L. Cohen is University Professor of Physics at the University of California at Berkeley and Senior Faculty Scientist at the Lawrence Berkeley National Laboratory. His research covers a broad spectrum of subjects in theoretical condensed matter physics. He is an elected member of the National Academy of Sciences, the American Academy of Arts & Sciences, and the American Philosophical Society. He has received numerous awards, including the US National Medal of Science, the Buckley Prize and the Lilienfeld Prize of the American Physical Society, the von Hippel Award of the Materials Research Society, and the Dickson Prize in Science. He has received honorary degrees from the University of Montreal, the Hong Kong University of Science & Technology, and the Weizmann Institute of Science. He is a former President of the American Physical Society.

Steven G. Louie is Professor of Physics at the University of California at Berkeley and Senior Faculty Scientist at the Lawrence Berkeley National Laboratory. His research spans a broad spectrum of topics in theoretical condensed matter physics and nanoscience. He is an elected member of the National Academy of Sciences, the American Academy of Arts & Sciences, and an academician of the Academia Sinica of the Republic of China (Taiwan). He has won numerous awards and prizes for his work, including the Rahman Prize and the Davisson–Germer Prize of the American Physical Society, the Materials Theory Award of the Materials Research Society, and the Sustained Outstanding Research in Solid State Physics Award of the US Department of Energy. He and Cohen shared the Foresight Institute Richard P. Feynman Prize in Nanotechnology (Theory).

Fundamentals of Condensed Matter Physics

MARVIN L. COHEN

University of California, Berkeley

STEVEN G. LOUIE

University of California, Berkeley

CAMBRIDGE
UNIVERSITY PRESS

CAMBRIDGE
UNIVERSITY PRESS

Shaftesbury Road, Cambridge CB2 8EA, United Kingdom

One Liberty Plaza, 20th Floor, New York, NY 10006, USA

477 Williamstown Road, Port Melbourne, VIC 3207, Australia

314–321, 3rd Floor, Plot 3, Splendor Forum, Jasola District Centre, New Delhi – 110025, India

103 Penang Road, #05–06/07, Visioncrest Commercial, Singapore 238467

Cambridge University Press is part of Cambridge University Press & Assessment,
a department of the University of Cambridge.

We share the University's mission to contribute to society through the pursuit of
education, learning and research at the highest international levels of excellence.

www.cambridge.org
Information on this title: www.cambridge.org/9780521513319

First published 2016
Reprinted 2017

A catalogue record for this publication is available from the British Library

Library of Congress Cataloging-in-Publication data
Names: Cohen, Marvin L., author. | Louie, Steven G., 1949– author.
Title: Fundamentals of condensed matter physics / Marvin L. Cohen and Steven
G. Louie, Department of Physics, University of California at Berkeley,
Materials Sciences Division, Lawrence Berkeley National Laboratory.
Description: Cambridge, United Kingdom : Cambridge University Press, 2016. |
2016 | Includes bibliographical references and index.
Identifiers: LCCN 2016003013| ISBN 9780521513319 (hardback) |
ISBN 0521513316 (hardback)
Subjects: LCSH: Condensed matter.
Classification: LCC QC173.454.C64 2016 | DDC 530.4/1–dc23
LC record available at http://lccn.loc.gov/2016003013

ISBN   978-0-521-51331-9   Hardback

Additional resources for this publication at www.cambridge.org/cohenlouie

To Suzy and Jane

Contents

Part III Optical and transport phenomena

Part IV Many-body effects, superconductivity, magnetism, and lower-dimensional systems

Preface

The field of condensed matter physics is the largest branch of physics worldwide and probably the most diverse. Undergraduate courses in this area are ubiquitous and most research universities offer graduate courses. Over the past 50 years, the undergraduate course has been open to physicists, chemists, materials scientists, engineers, and, to a smaller extent, biologists. The graduate course slowly evolved in many institutions from a course for theorists to one that welcomed students interested in a career in experimental condensed matter physics and materials research. In recent years, the proportion of chemists, materials scientists, engineers, and researchers in nanoscience has increase significantly in graduate courses in condensed matter physics.

There are numerous undergraduate texts. The prime example is *Introduction to Solid State Physics (ISSP)* authored by C. Kittel. At the graduate level, no single text has emerged as the canonical choice. N. Ashcroft and N. D. Mermin's book *Solid State Physics* is sometimes chosen since it contains advanced topics going beyond Kittel's *ISSP*, although Ashcroft & Mermin is often used as an undergraduate text. J. Ziman's *Principles of the Theory of Solids* is at roughly the same level as Ashcroft & Mermin, with excellent physical examples and discussions of concepts. C. Kittel's *Quantum Theory of Solids,* which was written for the graduate course at the University of California, Berkeley, at a time when students taking the course were predominantly theorists, is somewhat limited in scope and generally considered difficult by graduate students not intent on a career in theoretical condensed matter physics. Other texts, such as those by J. Callaway, O. Madelung, M. Marder, and J. Patterson and B. Bailey, are considered to be at the right level and suitable. Many others are in the recommended, but not required, category and are useful when specific subjects are considered. Examples include those authored by E. Kaxiras, R. Martin, M. Balkanski and R. Wallis, P. Yu and M. Cardona, and M. Cohen and J. Chelikowsky for electronic and optical properties of solids; M. Tinkham, P. de Gennes, and R. Schrieffer for superconductivity; G. Mahan for many-particle physics; and P. Chailken and T. Lubensky for phase transitions and soft matter systems. There are also many other excellent texts on specific subjects in this field.

The present text, *Fundamentals of Condensed Matter Physics*, is intended to cover the "mainstream" subjects in this field at the graduate level. It is probably impossible to produce a book that fills the complete bill for a course, as J. D. Jackson's E&M text has done for electricity and magnetism, because the range of topics is so broad. To cover the whole field would require many volumes. Hence the intent here is to write a text that covers the central topics on a level that will prepare a student to enter research, and that can serve as a higher-level source to sit alongside undergraduate texts for researchers in this field.

This text is based on lectures given as part of the condensed matter physics graduate course at Berkeley since 1965. The course, called Physics 240A and B, is a two-semester (or three-quarter) course covering 90 hours of lectures. In addition, there is a weekly discussion section for going over problem sets. Over the decades, this course was taught by one of the authors, either Marvin L. Cohen or Steven G. Louie, with guest lecturers from time to time. Student evaluations of the course have been high. The class size has typically been 20–30 students plus auditors. Of the more than one thousand students who took this course, a fair number have taken academic positions and have reported back that they have used their course notes successfully to teach similar courses in various departments including physics, chemistry, materials science, and engineering.

The book is divided into four main parts. Part I is devoted to the development of basic concepts. It begins with an introduction to the modern conceptual models of a solid from the points of view of (i) interacting atoms and (ii) elementary excitations, and then develops a thorough grounding on the basic elements needed to understand many of the properties of solids and the methods used to calculate them. Part II concerns the fundamentals of electron interactions, electron dynamics, and response functions, that control and exhibit the properties of and phenomena in condensed matter. Parts III and IV focus on the different properties and phenomena that are central to modern condensed matter and materials research. These include vibrational, thermal, optical, and transport properties, superconductivity, magnetism, and lower-dimensional systems, with emphasis on developing a physical understanding of real material systems. A range of theoretical techniques is developed as needed. The mathematical level varies, as does the degree of detail, in a manner similar to what one would experience in the world of research. Topics and techniques, such as band structure methods, pseudopotentials, density functional theory, effective Hamiltonians, electron dynamics, dielectric functions, electron–electron and electron–hole interactions, Berry's phase, Boltzmann transport theory, optical response, cooperative phenomena, many-body Green's functions, and diagrammatic and quasiparticle approaches, are explored and motivated by "real problems" associated with understanding and calculating material properties. Experimental techniques are described but not in detail.

Because of the breadth of the field and the limitation to one volume, some subjects are not treated in depth and others are left out completely. However, the success of the course at Berkeley, and hopefully this text, is that it takes a student through the concepts and techniques for many central areas of condensed matter physics, and establishes the level needed to start current research. Hence, the intent is to take students with a good knowledge of graduate quantum mechanics and undergraduate condensed matter physics to a level where they can do cutting-edge research. The book is suitable for a one-semester course (covering most of Parts I and II and some selected topics in Parts III and IV) or a two-semester course (covering essentially all of Parts I–IV with the option of omitting some topics as desired by the instructor).

This book would not have been possible without help from many people. We would especially like to thank Ms. Katherine de Raadt for her help with editorial matters and for producing the manuscript. We would like to acknowledge Cheol-Hwan Park for some of his class notes from the Berkeley course, David Penn for suggestions and critical readings of part of the text, Felipe Jornada and Sangkook Choi for producing most of the figures and for useful suggestions, and Meng Wu for his contributions in producing the problem sections. We profited from helpful corrections of the various chapters of the text by Gabriel Antonius, Brad Barker, Ting Cao, Sinisa Coh, Zhenglu Li, Jamal Mustafa, Chin-Shen Ong, Diana Qiu, Liang Tan, and Derek Vigil-Fowler. Finally, we thank Simon Capelin of Cambridge University Press for his guidance and patience.

Marvin L. Cohen
Steven G. Louie
Berkeley, CA, USA
Spring, 2016

PART I

BASIC CONCEPTS: ELECTRONS AND PHONONS

1 Concept of a solid: qualitative introduction and overview

1.1 Classification of solids

Condensed matter physics and solid state physics usually refer to the same area of physics, but in principle the former title is broader. Condensed matter is meant to include solids, liquids, liquid crystals, and some plasmas in or near solids. This is the largest branch of physics at this time, and it covers a broad scope of physical phenomena. Topics range from studies of the most fundamental aspects of physics to applied problems related to technology.

The focus of this book will be primarily on the quantum theory of solids. To begin, it is useful to start with the concept of a solid and then describe the two commonly used models that form the basis for modern research in this area. The word "solid" evokes a familiar visual picture well described by the definition in the Oxford Dictionary: "Of stable shape, not liquid or fluid, having some rigidity." It is the property of rigidity that is basic to the early studies of solids. These studies focused on the mechanical properties of solids. As a result, until the nineteenth century the most common classification of solids involved their rigidity or mechanical properties. The Mohs hardness scale (talc – 1; calcite – 3; quartz – 7; diamond – 10) is a typical example. This is a useful but limited approach for classifying solids.

The advent of atomic theory brought more microscopic concepts about solids. Solids were viewed as collections of more or less strongly interacting atoms. From the point of view of atomic theory, a gas is described in terms of a collection of almost independent atoms, while a liquid is formed by atoms that are weakly interacting. This picture leads to a description of the formation of solids, under pressure or by freezing, in which the distances between atoms are reduced and, in turn, this causes them to interact more strongly. Molecular solids are formed by condensing molecular gases.

Hence, the development of atomic physics and chemical analysis led to a more detailed classification of solids according to chemical composition. Although for most studies of solids it is necessary to establish the identity of the constituent atoms, such a scheme provides limited insight into many of the basic concepts of condensed matter physics.

Solids can also be classified according to their crystalline structure. This monumental feat has occupied crystallographers and applied mathematicians for over a hundred years. The discovery of X-ray crystallography added an important component to the atomic model. The chemical view of a crystalline solid as a collection of strongly interacting atoms can be expanded by adding that the atoms arrange themselves in a periodic structure. This

Table 1.1 Classification of solids using resistivity.		
Class	Typical resistivity (Ω cm)	Example
metal	10^{-6}	copper
semimetal	10^{-3}	bismuth
semiconductor	10^{-2}–10^9	silicon
insulator	10^{14}–10^{22}	diamond

would not be true for non-crystalline solids such as window glass. For most modern studies of solids, it is usually assumed that the structure and chemical composition are known and that information of this kind is the starting point for most investigations. When new materials are discovered, chemical and structural analyses are often the first steps in the characterization process.

Properties other than mechanical, chemical, and structural can be used to classify solids. Electromagnetic and thermal characteristics are commonly used. In particular, the resistivity ρ is the most used property, since it involves a single scalar quantity or a symmetric tensor with a range of values for different substances that varies at room temperature by about 28 orders of magnitude. For many materials ρ can be measured with great precision. Even though this approach focuses on a macroscopic property, with some theoretical analysis, it provides considerable insight into the microscopic nature of solids. Table 1.1 lists the classification according to resistivity of the four classes of solids: metals, semimetals, semiconductors, and insulators. Typical resistivities and examples for each class are also given. Although the divisions between classes are approximate, this classification of solids is extremely useful for a wide variety of studies and applications.

The chemical and structural classifications lead naturally to a model of crystalline solids based on interacting atoms. This is a fruitful approach and suggests that it may be possible to explain and predict the properties of solids through an adjustment or perturbation of the properties of atoms. The use of the resistivity for the classification of types of solids suggests another view, where the responses of a solid to external probes are used to classify solids. Electromagnetic, thermal, or mechanical probes can give rise to responses that may arise from the collective nature of the interacting particles in solids. However, viewing the solid as perturbed atoms may not lead naturally to an interpretation of its properties in terms of collective or cooperative effects. The interacting atoms approach is convenient for describing ground-state properties, whereas collective effects are best explained in terms of a model based on the excitations of the solid. Both models are discussed in more detail below.

1.2 A first model of a solid: interacting atoms

Consider aluminum metal. It is a relatively soft solid with hardness between 2 and 2.9 on the Mohs scale and it is composed of a single element. It crystallizes in the face-centered

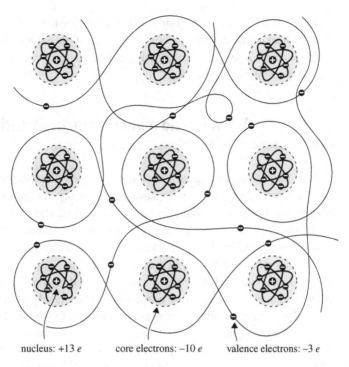

nucleus: +13 e core electrons: $-10\,e$ valence electrons: $-3\,e$

Figure 1.1 Schematic drawing of a solid consisting of atoms with atomic number $z = 13$ containing cores and valence electrons. Here e denotes the charge of a proton.

cubic structure with cubic lattice constant $a = 4.5$ Å, it is a reasonably good conductor ($\rho = 2.8 \times 10^{-6}$ Ω cm at 20°C), and it becomes a superconductor at $T_c = 1.19$ K.

What model can be used as a basis for explaining these properties? Why is aluminum different from copper, solid argon, solid oxygen, silicon, sodium chloride, or anthracene?

The most straightforward approach is to start from the chemical and atomic descriptions and to consider aluminum metal as a collection of interacting aluminum atoms. Each aluminum atom is composed of a nucleus with atomic number $Z = 13$. Hence in the neutral atom, there are 13 protons and 13 electrons. The electronic configuration can be easily divided into 10 core electrons, denoted by principal and angular momentum quantum labels as $[(1s)^2(2s)^2(2p)^6]$, and three outer valence electrons $[(3s)^2(3p)^1]$.

One picture of solid aluminum is a collection of cores with each core composed of a nucleus plus the tightly bound core electrons. Because of the cancellation of charge of 10 protons by the core electrons, the core has an effective charge $Z_{\text{eff}} = 3$. Moving around the cores and between them is a collection of itinerant, nearly-free electrons; there are three per core. A schematic drawing is given in Fig. 1.1. Cores and valence electrons move in a potential set by their mutual interactions. The laws of motion are well known and are described by quantum mechanics. The Schrödinger equation is sufficient in most cases; however, the Dirac equation may be necessary if relativistic effects are to be incorporated. The forces are also known; only electromagnetism plays a role. Gravity and weak interactions are too feeble, and strong interactions are too short-ranged. However, the problem is not an easily solvable one since there are of the order of 10^{23} interacting particles in solids per cm^3. Approximate methods must be used and simpler models developed. These will be discussed in Chapter 2.

1.3 A second model: elementary excitations

A drastically different approach to studying condensed matter systems is also in common use. In quantum mechanics and quantum field theory, it is often convenient to change the description of a system when discussing its excited states. This picture is based on the excitations that can emerge when a system is not in its ground state. A standard example is a harmonic oscillator of mass m moving in a quadratic potential $\frac{1}{2}m\omega^2 x^2$. This system has an energy spectrum defined by a quantum number n, $E_n = (n+\frac{1}{2})\hbar\omega$, which can be viewed as a ground state having energy $E_0 = \frac{1}{2}\hbar\omega$, and higher-energy states in which quanta of excitation energy $\hbar\omega$ can be created and subsequently destroyed in any non-negative integral number n. Another example is the electromagnetic field, which can can be viewed as a collection of quantized particle-like excitations, the photons, each one characterized by a wavevector $\mathbf{k}$, a polarization direction $\hat{\epsilon}$, and energy $\hbar\omega = \hbar c|\mathbf{k}|$, where c is the velocity of light.

The examples above give an operational definition of a quantum system that is described by its excitations, each one defined by its energy and other specific physical characteristics. Once the excitations are identified, the next task is to study the interactions between various excitations and the manner in which these excitations appear, disappear, or are modified when the external conditions of the quantum system change. In the end, the quantum system will be characterized by its elementary excitations when probes are used to study its properties.

The elementary excitations of a solid can often be divided into two classes: quasiparticles and collective excitations. Quasiparticles are usually fermions and resemble well-defined excited states of the non-interacting real particles of the solid. Collective excitations are usually bosons and do not resemble their constituent real particles. In the majority of cases, collective excitations are associated with macroscopic collective motions of the system, which in turn are described by quanta of generalized harmonic oscillators that can be created or destroyed in an integral number n. Each quantum provides an excitation energy $\hbar\omega$.

The language of second quantization is the natural one to use for this model. Since elementary excitations can be created and destroyed, and since symmetric (Bose–Einstein) or antisymmetric (Fermi–Dirac) conditions must be satisfied, creation and destruction operators, with their attendant commutation and anticommutation rules, are among the fundamental descriptive tools of this approach.

To explain the properties of a solid, it is advantageous to define the elementary excitations of the solid, to describe their properties and characteristics, to evaluate their interactions, and to determine how they respond to external probes.

1.4 Elementary excitations associated with solids and liquids

Quasielectrons. Quasielectrons (or electrons for short) are quasiparticles that behave like non-interacting electrons in low-lying excited states. They are fermions that are characterized by energy and quantum numbers, such as their wavevectors and spin orientations. Their properties include the effects of the environment in which they move. An important example is that of an electron experiencing the interaction with other electrons that can change its free electron mass m to an effective mass $m^* > m$. The resultant quasielectron can be described by one-particle states of spin $\frac{1}{2}$ and charge $-|e|$. Typical excitation energies for quasielectrons are of the order of the Coulomb interaction energy between two electrons separated by a crystal lattice parameter a: $\frac{e^2}{a} \cong 5$ eV. Typical velocities associated with quasielectrons are $v \cong (\frac{e^2}{am})^{1/2} \cong 10^8$ cm/s.

Hole. The removal of an electron from an orbital that is normally occupied in the ground state is called a hole. The analogy is made with the Dirac theory of positrons. A hole is a quasiparticle that has charge $+|e|$, spin $\frac{1}{2}$, and energies and velocities similar to quasielectrons. When electrons are injected into or removed from a solid in processes such as those associated with quantum tunneling phenomena and electron emission, they are often treated as single particles without any reference to the holes left behind. The holes are usually studied separately.

Phonon. A phonon is a collective excitation (boson) associated with lattice vibrations or sound waves. It is defined by a wavevector $\mathbf{q}$, a branch or polarization mode index α, and an energy $\hbar\omega$. Typical energies are of the order of $k_B T_D$, where k_B is Boltzmann's constant and T_D is the Debye temperature. Since T_D is of the order of room temperature ($\sim$300 K), a typical phonon energy is $\hbar\omega \cong 0.025$ eV.

Plasmon. A plasmon is a collective excitation (boson) associated with the collective motion of the electronic charge density. It is characterized by a wavevector $\mathbf{q}$ and an energy of the order of the classical plasma energy (in three dimensions)

$$\hbar\omega_p = \hbar\left(\frac{4\pi ne^2}{m}\right)^{\frac{1}{2}}, \tag{1.1}$$

where n is the density of valence electrons per unit volume. For typical solids, $\hbar\omega_p \cong 10$ eV. This value can be smaller by an order of magnitude or more in low-density electron or hole systems such as those found in semimetals and degenerate semiconductors.

Magnon. A magnon is the collective excitation (boson) associated with spin waves or spin excitations, resulting from the occurrence of spin reversals in an ordered magnetic system. Typical energies are of the order of the ordering temperatures (Curie or Néel). These can be as high as 10^{-1} eV, but are usually much lower ($\sim 4 \times 10^{-5}$ eV).

Polaron. A polaron is a special type of quasielectron that exists in crystals. It can be regarded as an electron or a hole moving through a crystal and carrying a lattice deformation or strain with it. If the strain is expressed in terms of excitations of phonons, this leads to the view of a polaron as an electron accompanied by a cloud of phonons. The terms "polaron" and "polaron effects" are often used to describe changes in electron properties arising from electron–phonon interactions in general.

Exciton. An exciton is a bound or quasibound state of a quasielectron and a hole. This elementary excitation is similar to positronium, and it often behaves as a boson which can be decomposed into its two component fermions or radiatively annihilated. Excitons are usually observed in insulators and semiconductors. Typical binding energies are $\sim$0.025 eV in three-dimensional systems.

Superconducting quasiparticles. Superconducting quasiparticles are fermions that describe the electronic excited states of a superconductor. They are sometimes called Cooper particles (not Cooper pairs) or Bogoliubons. Because of the physics inherent in the description of the superconducting ground state, these quasiparticles are viewed as a linear combination of quasielectrons and holes. Typical energies are of the order of the superconducting transition temperature between 10^{-5} eV and 10^{-2} eV, depending on the material.

Roton. A roton can be viewed as a special phonon associated with a local energy minimum in the dispersion relation at a finite wavevector. This description usually applies to liquid He^4. Typical roton energies in helium are of the order of 10^{-3} eV.

1.5 External probes

Knowledge about the properties of solids is obtained from measurements done under well-defined conditions. These are either equilibrium situations, in which the temperature and externally applied static electric and magnetic fields are predetermined, or dynamical situations in which energy, momentum, angular momentum, and other dynamical quantities are exchanged with the environment. In the latter case, the agents that effect this exchange are also microscopic quanta, the so-called test or probe particles. Some of these are listed below.

Photons. Electromagnetic probes are the most commonly used probes in solids, for example, they are used in absorption spectra, reflectivity spectra, and photoemission spectra. The energy range of the useful photons spans the available electromagnetic spectrum, from radio frequency studies of metals ($\hbar\omega \cong 2 \times 10^{-8}$ eV) to γ-radiation studies of the Mössbauer effect ($\hbar\omega \cong 2 \times 10^6$ eV).

Electrons. Electrons are used to probe solids in a variety of ways. They are injected and extracted through electrical contacts and tunneling junctions, or used as scattering particles in electron beams. Typical energies vary with the experiment: $\sim$1 meV for superconductive tunneling; $\sim$1 eV for semiconductor tunneling; $\sim 10^{-2}$ to 2 eV for scanning tunneling microscope investigations of surfaces; $\sim$ 10 to 100 eV for low-energy electron diffraction at solid surfaces; and $\sim$100 keV to 1 MeV for high-energy electron microscopy.

Positrons. Positron annihilation in solids arising from electron–positron interactions provides useful information for investigating the electronic properties of solids – primarily metals. Photons are emitted when the annihilation occurs and studies of the emitted radiation give information about the electronic structure.

Neutrons. Neutron scattering has become the standard and preferred technique to study magnetic structural properties and properties of phonons, magnons, and other collective excitations in solids, owing to the fact that a neutron is a neutral particle with a magnetic moment.

Muons and pions. Muons and pions are used as probes, but they are not as commonly used as most of the other particles discussed here. Muons are more versatile than pions because of their similarity to heavy electrons. Their magnetic moments and decay modes can give information not obtained with other methods.

Protons. Protons can be used to study the structures of crystalline solids, primarily by examining their trajectories in solids.

Atoms. Light atoms and ions are usually employed to study surfaces; sometimes they can be used to probe deeper into solids. Their role is often similar to that of electrons in scattering experiments.

1.6 Dispersion curves

All the probe particles can be characterized by a momentum (or wavevector or wavelength) and an energy (or frequency). The particles are assumed to exist in free space or vacuum. For massive particles with mass m, the dispersion curve, which is the functional relation between the energy E and the momentum $\mathbf{p}$ or wavevector $\mathbf{k}$, is given by

$$E = \frac{\mathbf{p}^2}{2m} = \frac{\hbar^2 \mathbf{k}^2}{2m}, \tag{1.2}$$

or, in the relativistic limit, by

$$E = (\hbar^2 \mathbf{k}^2 c^2 + m^2 c^4)^{1/2} - mc^2, \tag{1.3}$$

where c is the speed of light. The probing photon in free space (a massless particle) is described by the dispersion curve, connecting the frequency ω and wavevector $\mathbf{k}$, $\omega = c\mathbf{k}$. These dispersion curves are illustrated in Fig. 1.2. In a solid or liquid, most elementary excitations are also defined by a wavevector and an energy or frequency. The functional dependence of the energy on the wavevector, that is, the dispersion curve, constitutes one of the most fundamental properties of the excitations to be determined.

The quasielectron dispersion curve can be used to distinguish the various types of solids: metals – no gap in the spectrum and the existence of a Fermi surface; semiconductors and insulators – an electronic energy gap of order 0.1 to 10 eV, caused by the ion core potential and uniquely related to specific $\mathbf{k}$-space locations; and superconductors – an energy gap for creating a quasielectron and quasihole pair of order $2\Delta < 10^{-1}$ eV caused by dynamic interactions of the electrons.

Some examples of dispersion curves for quasiparticles are shown in Fig. 1.3. The quasielectrons and holes in metals (Fig. 1.3(a)) have excitation energies which start from zero. Excitations with zero energy define a surface in $\mathbf{k}$-space known as the Fermi surface. For

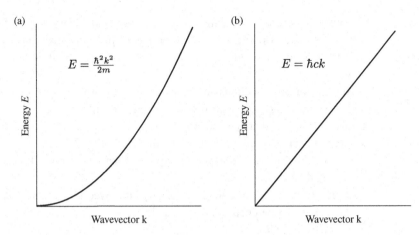

Figure 1.2   Dispersion curves for probe particles. (a) Free particle of mass m. (b) Photons in vacuum.

semiconductors, the excitation spectrum has a different starting value for electrons and holes, which depends on the chemical potential μ (or E_F) of the system. Silicon is used as an example in Fig. 1.3(b). The lowest-energy electron excitation is at a well-defined k-vector, $\mathbf{k} = \mathbf{k}_{cb}$, and the lowest-energy hole state is at $\mathbf{k} = \mathbf{k}_{vb}$. For the silicon example, $\mathbf{k}_{vb} = 0$, while $\mathbf{k}_{cb}$ has a finite value along the (100) direction of $\mathbf{k}$-space, and the minimum energy required to create a quasielectron and a hole is 1.1 eV. This is the minimum bandgap E_g for silicon.

Superconducting quasiparticle spectra also exhibit an energy gap. The minimum energy for creating both a superconducting quasielectron and quasihole requires an energy of 2Δ. This is usually a much smaller gap than in semiconductors ($2\Delta = 0.3$ meV in aluminum), and it appears at what, in the normal state, was the Fermi surface of the metal. The semiconducting and superconducting gaps are of very different natures; their characteristics are discussed in subsequent chapters.

It is often convenient to choose a specific form for the dispersion curve near local minima or maxima and in other regions where quadratic approximations are appropriate. For example, in the case of an electron or hole at a band minimum with wavevector $\mathbf{k}_0$, it is convenient to write

$$E(\mathbf{k}) = \tfrac{1}{2}\hbar^2(\mathbf{k} - \mathbf{k}_0) \cdot A \cdot (\mathbf{k} - \mathbf{k}_0), \tag{1.4}$$

where the tensor $A_{ij} \equiv 1/m^*_{ij}$ can be interpreted as an inverse effective mass tensor. For a free electron, the relation (Eq. (1.4)) holds exactly with

$$A_{ij} = m^{-1}\delta_{ij}, \tag{1.5}$$

where m is the free electron mass (see Eq. (1.2) and Fig. 1.2). In solids, effective masses may differ substantially from the free electron mass. For example, $m^* \cong 0.01m$ for quasielectrons near the conduction band minimum in InSb. This small m^* value is caused by the static crystal potential, which is also responsible for the existence of the energy gap. In sodium metal, where the Fermi surface is, to a very good approximation, a sphere of radius

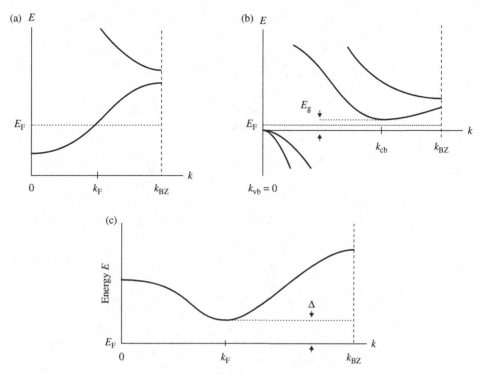

Dispersion curves. (a) Quasielectrons and holes in a simple metal, for example, an alkali metal. Quasielectrons are excited into states $E > E_F, k > k_F$, where E_F and k_F are the Fermi energy and wavevector and k_{BZ} is the wavevector for the Brillouin zone edge. Hole states are by convention often plotted by the dispersion curve for $E < E_F$ and $k < k_F$. However, since it costs energy to remove an electron to create a hole, the hole branch of the dispersion curve should be inverted and the energies measured positively with hole energies between zero and E_F. (b) Quasielectrons and holes in a semiconductor. For the case illustrated, the lowest-energy quasielectron wavevector is centered at k_{cb} at the conduction band minimum, while the lowest-energy hole wavevector is at the valence band maximum centered near $k_{vb} = 0$. These two kinds of states are separated by energy gap E_g. As in (a), the hole branch of the dispersion curve should be inverted, with the energies measured positively. (c) Quasiparticles in a superconductor. Here the quasielectron and hole branches of (a) are combined. The superconducting gap Δ for creating a single quasiparticle appears at k_F.

k_F, the quasiparticle dispersion curve is given by

$$E(\mathbf{k}) = \frac{\hbar^2}{2m^*}|k^2 - k_F^2|. \tag{1.6}$$

Equation (1.6) is valid for both quasielectrons ($|\mathbf{k}| > k_F$) and holes ($|\mathbf{k}| < k_F$) with an effective mass of $m^* \cong 1.25m$. This enhancement of 25% over the free electron value is not caused by the static crystal potential. It arises from the dynamic interactions between electrons and phonons and interactions among the electrons.

In Fig. 1.4, examples of dispersion curves for collective excitations are given. The specific properties related to features of particular curves are discussed in later chapters.

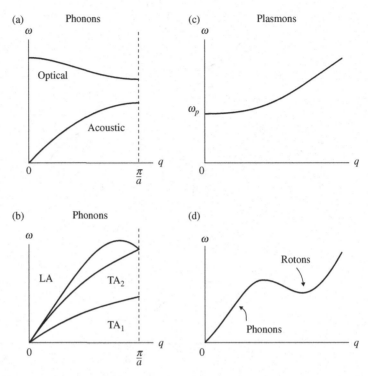

Dispersion curves (schematic) for collective excitations. (a) Phonons in a one-dimensional solid with lattice constant a and two masses per unit cell. (b) Phonons in a three-dimensional solid with one atom/cell. LA, TA_1, and TA_2 refer to the longitudinal acoustical mode and the two transverse acoustical modes, respectively. (c) Plasmons in a three-dimensional metal where ω_p is the classical plasma frequency. (d) Phonons and rotons in liquid He-4.

In restricted regions of given spectra, useful analytic approximations are commonly made. For instance, for each mode α of the acoustic phonon dispersion relation near $\mathbf{q} = 0$, it is possible to write

$$\hbar\omega_\alpha = \hbar v_\alpha |\mathbf{q}|, \tag{1.7}$$

where v_α is the speed of sound propagation for mode α.

The determination and interpretation of the dispersion curves for the elementary excitations can be complex. However, a scheme or investigational approach can be outlined. This approach consists of: (1) defining the bare elementary excitation by means of a Hamiltonian formalism; (2) solving the equations of motion to determine the dispersion curves for the "bare" elementary excitations; (3) solving for the "final" spectrum of the excitations after including the necessary interactions among the excitations; (4) including the effects of the external probes and their interactions with the excitations; and (5) solving the new coupled equations to determine the response functions for the condensed matter system.

In the above scheme, steps (1) and (2) are usually handled by ordinary quantum mechanical methods, using the Schrödinger equation appropriate for one body in an external potential. Steps (3), (4), and (5) are concerned with the many-body aspects that can be

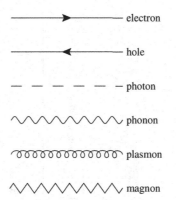

Figure 1.5   Diagrammatic representation of some elementary excitations and probe particles. Spin and polarization indices (not shown) can be added to the wavevector to specify a state.

evaluated using second quantization and Green's function techniques, which will be discussed later. In these steps, it is convenient to have a convention to express the physical processes in graphical form. A standard and very convenient choice is the use of Feynman diagrams.

1.7 Graphical representation of elementary excitations and probe particles

Elementary excitations, probe particles, and their interactions can be represented graphically (Feynman diagrams). When describing physical processes, it is often assumed that time develops from the left to the right, or from the bottom to the top of the diagram used to describe the process. Each elementary excitation or probe particle is depicted by a line (a different type of line for each type of excitation) and a label that characterizes its quantum numbers. Some graphic representations are shown in Fig. 1.5.

1.8 Interactions among particles

In order to accomplish the last three parts of the scheme designed to explain the properties of condensed matter systems using elementary excitations, it is necessary to describe how the elementary excitations interact with one another and with the probe particles. The interactions of interest are given in Table 1.2. It is helpful to develop interactive Hamiltonians that can be analyzed by means of diagrams and use these to give a physical picture of the processes involved.

Table 1.2 Important interactions between excitations and probe particles.

	Elementary Excitations									Probe Particles			
	Electron	Hole	Photon	Plasmon	Magnon	Polaron	Exciton	Superconducting quasiparticle	Roton	Photon	Electron	Positron	Neutron
Electron	x	x	x	x	x			x		x	x	x	
Hole	x	x	x	x	x			x		x	x	x	
Photon	x	x	x	x	x			x		x	x		x
Plasmon	x	x	x	x	x				x	x	x		x
Magnon	x	x	x	x	x					x			x
Polaron			x			x				x			
Exciton	x	x	x				x			x	x		
Superconducting quasiparticle	x	x	x					x	x	x	x		
Roton			x						x	x			x

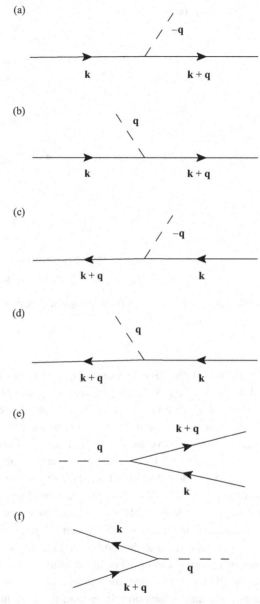

Figure 1.6 Diagrams representing some electron–photon and hole–photon processes. The individual processes are discussed in the text. The electron is represented by the wavevector **k** and a right-pointing arrow, and a hole is represented by a left-pointing arrow and its wavevector is given by the negative of the electron wavevector **k** or **k** + **q** used in this figure.

1.8.1 Quasiparticle–boson interactions

Many physical processes in condensed matter systems involve the interactions of quasiparticles with bosons. The bosons may correspond to the collective excitations of the system

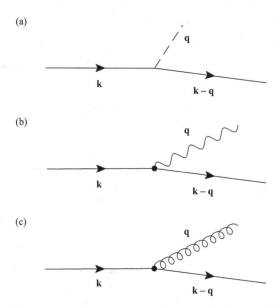

Figure 1.7 Some possible quasielectron–boson interactions. (a) Electron–photon. (b) Electron–phonon. (c) Electron–plasmon.

or to the probe particles. Some examples are given in Fig. 1.6, where electrons and holes are used as the prototypes for the quasiparticles, and photons represent bosons. The following cases are given graphically in Fig. 1.6: (a) an electron having wavevector **k** emits a photon of wavevector $-\mathbf{q}$ and is scattered into a state described by wavevection $\mathbf{k} + \mathbf{q}$; (b) an electron with wavevector **k** absorbs a photon of wavevector **q** and is scattered into a state $\mathbf{k} + \mathbf{q}$; (c) a hole with wavevector $-\mathbf{k} - \mathbf{q}$ emits a photon at wavevector $-\mathbf{q}$ and is scattered into the state $-\mathbf{k}$; (d) a hole of wavevector $-\mathbf{k} - \mathbf{q}$ absorbs a photon of wavevector **q** and is scattered into the state $-\mathbf{k}$; (e) the creation of an electron of wavevector $\mathbf{k} + \mathbf{q}$ and a hole of wavevector $-\mathbf{k}$ by a photon of wavevector **q**; and (f) the annihilation of an electron of wavevector $\mathbf{k}+\mathbf{q}$ and a hole of wavevector $-\mathbf{k}$ to produce a photon of wavevector **q**.

The photon representing the boson in Fig. 1.6 can be replaced by other bosons. Examples of common electron–boson interactions are shown in Fig. 1.7. These include the electron–photon interaction and others, such as the electron–phonon interaction and the electron–plasmon interaction. In each of these examples an electron with wavevector **k** emits a boson of wavevector **q** and is scattered into a state described by wavevector $\mathbf{k} - \mathbf{q}$.

The examples above are the fundamental interactions for a large number of physical phenomena. For example, the electron–photon interaction is basic to the processes involved in optical absorption and emission, electron photoemission, X-ray scattering, Compton scattering, Raman scattering, cyclotron resonance, and many other physical phenomena. The electron–phonon interaction is responsible for electrical and thermal resistance, ultrasonic attenuation, polaron formation, and superconductivity. Electron–plasmon interactions dominate electron energy loss in many systems when fast or high-energy electrons

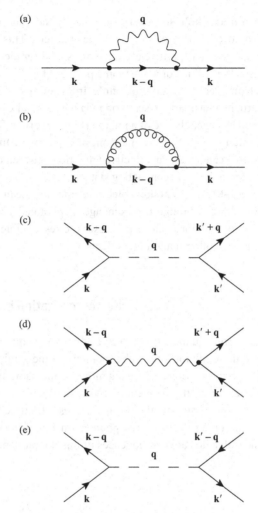

Figure 1.8 (a) Electron self-energy diagram involving phonon emission and absorption by the same electron. (b) Electron self-energy diagram involving plasmon emission and absorption. (c) Electron–electron Coulomb interaction or photon exchange. (d) Electron–electron interaction via phonon exchange. (e) Electron–hole Coulomb interaction.

pass through solids. Finally, electron-magnon interactions determine magnetic resonance phenomena and transport properties of magnetic materials.

1.8.2 Quasiparticle–quasiparticle interactions

Because the quasiparticle concept is only an approximation, all quasiparticles interact directly, in most cases weakly, with one another via residual interactions, or indirectly via the exchange of a collective excitation. These processes can be described using Feynman diagrams, as shown in Fig. 1.8, and are discussed below.

Electron and hole self-energies. The emission and reabsorption of a boson "dresses" what would otherwise be a "bare" quasiparticle. This process produces changes in the energy and the quasiparticle effective mass, and for electrons can lead to the formation of a polaron (Fig. 1.8(a)) or a plasmonic polaron (Fig. 1.8(b)).

Electron–electron and hole–hole interactions. This interaction leads to ordinary quasiparticle–quasiparticle scattering via the residual Coulomb interaction, which can also be viewed as the exchange of a photon (Fig. 1.8(c)) or, for appropriate cases, the exchange of a phonon (Fig. 1.8(d)), a magnon, or a plasmon. In particular, the phonon exchange (Fig. 1.8(d)) can lead to an effective attraction between quasielectrons, which can result in a transition to the superconducting state.

Electron–hole interaction. Electrons and holes can interact via the Coulomb interaction (Fig. 1.8(e)) through the exchange of photons. They can also interact via phonons, plasmons, or magnons. These processes are responsible for the formation of excitons, for screening, and other many-body effects.

1.8.3 Collective excitation interactions

Because second quantization can be used to view collective excitations in terms of quanta of harmonic oscillators, any anharmonicity in the original Hamiltonian leads to direct interactions between collective excitations of the same type. This anharmonicity causes a renormalization of the frequencies ("dressing") of the excitations as well as the decay of the modes into a multiplicity of other modes ("lifetime" effects). A typical example of an anharmonic effect is shown for phonons in Fig. 1.9(a). Ferroelectricity and the thermal expansion of solids are consequences of this interaction.

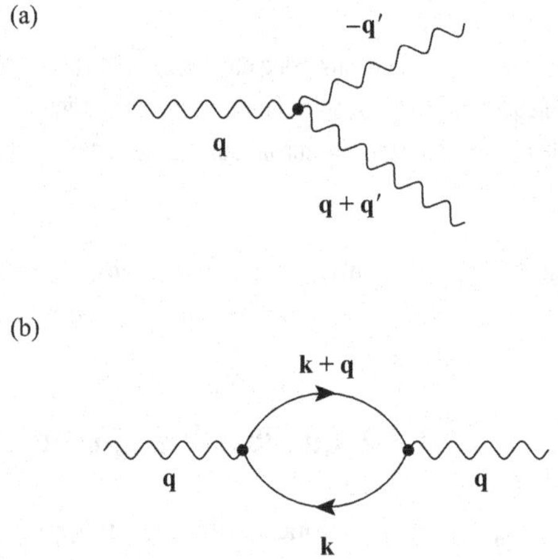

Figure 1.9   Examples of interactions among collective excitations. (a) Phonon–phonon scattering. (b) Phonon self energy caused by electron–hole pair creation and destruction.

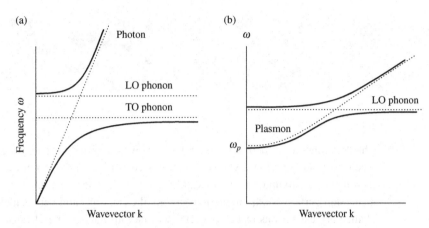

Figure 1.10 Coupled modes (solid line) formed from bare collective excitations (dashed lines). (a) An optical phonon–photon complex (also called a polariton). The longitudinal optical (LO) and transverse optical (TO) phonon dispersion curves are assumed to be flat near $\mathbf{k} = 0$. (b) A longitudinal optical phonon–plasmon complex.

Collective excitations can also be "dressed" by the creation of virtual quasielectron–hole pairs, as shown in Fig. 1.9(b). This process leads to the softening of the frequency of the modes, which can also be responsible for phase transitions.

The interaction between two different types of collective excitations is a good example of mode coupling. We can have optical phonon–photon coupled excitations (polaritons), optical phonons–plasmon coupled excitations, exciton–photon coupled excitations, magnon–phonon coupled excitations, magnon–photon coupled excitations, and so forth.

If the dispersion curves for any two of these modes intersect, that is, there is a point of common excitation energy and wavevector, the interaction becomes paramount in the neighborhood of the crossover point. There the two modes can be mixed in approximately equal amounts, and each of the original collective excitations loses its identity. Examples for the polariton and the longitudinal optical phonon–plasmon complex are shown in Fig. 1.10.

2 Electrons in crystals

The "scheme" described in Chapter 1 will now be applied to determine the properties of electrons (and holes) in crystals. The first step is to use an appropriate Hamiltonian and then make approximations that exhibit the quasiparticles and allow calculations of their dispersion curves. Next, interactions among the electrons and between electrons and other elementary excitations or test particles will be introduced. The physical properties of the solid, such as the optical properties, can then be described in terms of response functions, such as the dielectric function, that measure the responses of the solid to external probes.

2.1 General Hamiltonian

The starting point is based on the first model of a solid described in Chapter 1: a collection of ion cores and itinerant valence electrons. We assume that the electronic core states are atomic-like in character, undeformable, and tightly bound to the nuclei. Therefore, for many applications, each core, which consists of the nucleus and core electrons, can be treated as a single particle. There is a Coulomb attraction between the positive cores and the valence electrons. Near a nucleus, a negative test charge would experience an attractive potential corresponding to all the protons in the nucleus. However, outside the core, the effective charge of the core is reduced by the number of core electrons. For example, in silicon the charge near the nucleus is $+14|e|$, while the core charge is $+4|e|$, where e is the charge of an electron.

The total Hamiltonian H_T for the system of cores and valence electrons is

$$H_T = \sum_i \frac{\mathbf{p}_i^2}{2m} + \sum_n \frac{\mathbf{p}_n^2}{2M_n} + \frac{1}{2}\sum_{ij}' \frac{e^2}{|\mathbf{r}_i - \mathbf{r}_j|} + \frac{1}{2}\sum_{nn'}' \frac{Z_n Z_{n'} e^2}{|\mathbf{R}_n - \mathbf{R}_{n'}|} + \sum_{n,i} V_n(\mathbf{r}_i - \mathbf{R}_n) + H_R,$$

(2.1)

where the first right-hand term is the kinetic energy of the valence electrons, the second term is the kinetic energy of the cores, the third term is the Coulomb interaction between electrons, the fourth term is the Coulomb interaction between cores, the fifth term is the interaction between electrons and cores, and the final term represents the relativistic corrections, including spin–orbit coupling.

In Eq. (2.1), the primed summation indicates either $i \neq j$ or $n \neq n'$; $\mathbf{r}_i$, $\mathbf{p}_i$, e, and m are the electron position, momentum, charge, and mass, while $\mathbf{R}_n$, $\mathbf{p}_n$, Z_n, and M_n are the core position, momentum, charge, and mass. A straightforward approach for solving Eq. (2.1) is not practical since an exact solution would imply the handling of $\sim 10^{23}$ quantum numbers to describe the system. Even if available, these quantum numbers would be too numerous for practical applications, and most of them are likely to have obscure or non-useful physical meaning. Approximations are a necessity and must be introduced from the start. Three basic approximations are described below.

2.2 The Born–Oppenheimer adiabatic approximation

The cores are much heavier than the electrons; for example, $M \sim 5 \times 10^4 m$ for aluminum. At densities corresponding to typical solids, the cores behave like classical particles, but the electrons form a degenerate electron gas, leading to a difference of orders of magnitude in the kinetic energy of the two species. This difference in mass and kinetic energy implies that the cores are significantly more sluggish than the electrons; therefore the electrons can react almost instantaneously to any motion of the cores.

For most systems, it can be assumed that the electrons feel the instantaneous potential produced by the cores in their "frozen" or "fixed" positions. Therefore, we define an electronic Hamiltonian for fixed cores,

$$H_e = \sum_i \left[\frac{\mathbf{p}_i^2}{2m} + \sum_n V_n(\mathbf{r}_i - \mathbf{R}_n) \right] + \frac{1}{2} {\sum_{ij}}' \frac{e^2}{|\mathbf{r}_i - \mathbf{r}_j|} + H_R, \qquad (2.2)$$

and determine its energy spectrum $E_e^l(\{\mathbf{R}_n\})$, where l labels the ith excited state which depends implicitly on the core positions $\{\mathbf{R}_n\}$. In particular, with the electronic system in the ground state, the energy $E_e^0(\{\mathbf{R}\})$ defines the potential $V_{ec}(\{\mathbf{R}\})$ within which the cores move. Remaining contributions of Eq. (2.1) can be assigned to the core part of the Hamiltonian,

$$H_c = \sum_n \frac{\mathbf{p}_n^2}{2M_n} + \frac{1}{2} {\sum_{nn'}}' \frac{Z_n Z_{n'} e^2}{|\mathbf{R}_n - \mathbf{R}_{n'}|} + V_{ec}(\{\mathbf{R}_n\}). \qquad (2.3)$$

The term $V_{ec}(\{\mathbf{R}_n\})$, which is $E_e^0(\{\mathbf{R}_n\})$, represents an electron–core term which can be evaluated once Eq. (2.2) is solved and then used in Eq. (2.3) to determine the energy specturm for the cores.

The procedure described above is the Born–Oppenheimer[1] adiabatic approximation. It provides the important step of separating the electronic and core degrees of freedom. The electronic part (Eq. (2.2)) leads primarily to the determination of the properties of the electrons, holes, excitons, plasmons, and magnons. The core part (Eq. (2.3)) is used to describe the core motions and phonons. When the phonons are coupled to the electrons, including terms in Eq. (2.1) that go beyond Eqs. (2.2) and (2.3), one can examine polarons, superconductivity, resistivity, and other properties of solids. At this point, we restrict ourselves to the electronic Hamiltonian (Eq. (2.2)).

---

[1] M. Born and J. R. Oppenheimer, "Zur Quantentheorie der Molekeln," *Ann. Physik* 84(1927), 457.

2.3 The mean-field approximation

Even though the Hamiltonian (Eq. (2.2)) does not include the cores as dynamical variables, it still contains a very large number of particles, namely, all the valence electrons in the solid. Another approximation is needed. The simplest commonly employed one is the Hartree[2] mean-field approximation, which assumes that each electron moves in the average or mean field created by the cores together with all the other electrons. In the Hartree approach, one assumes that the electronic wavefunction of all the valence electrons is approximated by a product of one-electron wavefunctions, and each one is characterized by some one-electron spatial and spin quantum numbers. The effects of the Pauli exclusion principle are taken into account by requiring that no pairs of one-electron wavefunctions (orbitals) in the product have an identical set of quantum numbers. As discussed in standard quantum textbooks, demanding that the ground state has the lowest energy for the electronic system results in a set of self-consistent Euler–Lagrange equations (the Hartree equations) for the one-electron orbitals and energies with potential $V(\mathbf{r}, \{\mathbf{R}_n\})$.

The Hartree mean-field approximation accomplishes the important task of separating the Hamiltonian (Eq. (2.2)) into a sum of one-electron Hamiltonians

$$H_e = \sum_i H(\mathbf{r}_i, \{\mathbf{R}_n\}), \tag{2.4}$$

where

$$H(\mathbf{r}, \{\mathbf{R}_n\}) = \frac{p^2}{2m} + V(\mathbf{r}, \{\mathbf{R}_n\}). \tag{2.5}$$

Another approach involves the Hartree–Fock approximations, which is also a mean-field approach for the electronic ground state. The Hartree–Fock method approximates the wavefunction by a determinant of one-electron orbitals, and thus automatically satisfies all the symmetry requirements of the Pauli principle. However, although more accurate, this approach has several complicating features which often make it inconvenient for many applications. Further discussion of this point appears in Chapter 6.

2.4 The periodic potential approximation

The potential $V(\mathbf{r}, \{\mathbf{R}_n\})$ of Eq. (2.5) is a one-electron potential that describes the forces acting on the electron labeled by the coordinate $\mathbf{r}$. However, it depends on the one-electron states occupied by all the other electrons, a consequence of the Hartree approximation; it also depends parametrically on the "fixed" position of all the cores $\{\mathbf{R}_n\}$.

Even though the problem has been reduced to solving for the wavefunction of one electron at a time, this calculation must be accomplished self-consistently, since each electron state depends on the states of all the other electrons for an arbitrary core configuration.

[2] D. R. Hartree, *The Calculation of Atomic Structures* (New York: Wiley, 1957).

Again, a straightforward calculation is extremely complicated and another approximation is needed.

At this point, we rely on experiment which tells us that, in a crystalline solid, the cores form, to a very good approximation, an ordered periodic array. We thus take as a starting point a perfect crystal with core position $\{\mathbf{R}_n\}$ determined by X-ray crystallography or other means. We also assume at this stage that there are no defects or surfaces, and we assume, temporarily, an infinite number of atoms in the crystal.

With $\{\mathbf{R}_n\}$ fixed, we may then write

$$V(\mathbf{r}, \{\mathbf{R}_n\}) = V(\mathbf{r}), \tag{2.6}$$

where $V(\mathbf{r})$ has the symmetry of the periodic crystal arrangement. Because of the periodicity, an electron at point $\mathbf{r}$ experiences the same potential at all points $\mathbf{r}'$ which are related to $\mathbf{r}$ through the periodicity and symmetry of the crystal. This potential $V(\mathbf{r})$ now contains an enormous amount of information: the chemical composition of the crystal, the crystal structure and all its relevant parameters, the equilibrium positions of the cores, and the averaged potential acting on an electron produced by all the others.

We have now arrived at the one-electron periodic potential model. The use of this model enables us to determine the spectrum of electrons and, with it, to answer approximately an amazingly large number of questions about the properties and behavior of electrons in solids at a mean-field level.

2.5 Translational symmetry, periodicity, and lattices

The crystal potential $V(\mathbf{r})$ describes the ideal crystal, that is, a collection of electrons and static cores in a perfectly periodic arrangement. Given an arbitrary point $\mathbf{r}$ in the crystal, there is an infinite array of points $\{\mathbf{r} + \mathbf{R}_n\}$ such that, without change of orientation, the crystal appears exactly the same regardless of which point the arrangement is viewed from. This means that

$$V(\mathbf{r} + \mathbf{R}_n) = V(\mathbf{r}). \tag{2.7}$$

This collection of all vectors $\{\mathbf{R}\}$ constitutes a set of discrete points which is called a Bravais lattice. In three dimensions, any periodic lattice $\{\mathbf{R}\}$ can be written as

$$\mathbf{R}_n = \mathbf{R}_{n_1 n_2 n_3} = n_1 \mathbf{a}_1 + n_2 \mathbf{a}_2 + n_3 \mathbf{a}_3, \tag{2.8}$$

where the three vectors $\mathbf{a}_i$ are not coplanar,

$$(\mathbf{a}_1 \times \mathbf{a}_2) \cdot \mathbf{a}_3 \neq 0, \tag{2.9}$$

and where $n_1, n_2, n_3 = 0, \pm 1, \pm 2, \pm 3, \ldots$.

Lattice vectors. The vectors $\mathbf{R}_n$ are called lattice vectors or translation vectors; the three vectors $\mathbf{a}_i$ are called primitive lattice vectors or primitive translation vectors. The parallelepiped defined by the three primitive lattice vectors $\mathbf{a}_1, \mathbf{a}_2$, and $\mathbf{a}_3$ has a volume Ω_p given by

$$\Omega_p = |(\mathbf{a}_1 \times \mathbf{a}_2) \cdot \mathbf{a}_3|, \tag{2.10}$$

and the lattice points in a Bravais lattice have a density $(1/\Omega_p)$. Several important concepts follow.

Primitive cell. A primitive cell is a region of space that, when translated by "all" the vectors in a Bravais lattice, exactly fills all of the space of the crystal without either voids or overlaps. Each primitive cell contains only one lattice point. The volume of a primitive cell is Ω_p, and the parallelepiped defined above is one possible kind of primitive cell. Although the volume of a primitive cell is constant and equal to Ω_p, its shape is largely arbitrary, and, in principle, it can consist of several disconnected parts.

Unit cell. A unit cell is a region of space which, when translated by a subset of vectors of the Bravais lattice, exactly fills the space of the crystal with no voids or overlaps. Hence the volume of any unit cell Ω_u satisfies the condition $\Omega_u \geq \Omega_p$. In fact,

$$\Omega_u = v\Omega_p, \quad v = 1, 2, 3, \ldots \tag{2.11}$$

A primitive cell can now be redefined as a unit cell of minimum volume.

Basis. A basis is the set of atoms and their coordinates within the unit cell which form the crystal. The basis for a primitive cell is smaller than the basis for a unit cell, and it is a minimal basis when related to a primitive cell. A basis must contain an integral number of the formula units that characterize the crystal. In more mathematical terms, a basis is given by a finite (and usually small) set of vectors $\{\boldsymbol{\tau}_\mu\}, \mu = 1, 2, \ldots, N_b$, where N_b is the number of atoms in the basis.

Because the origin of a unit cell is arbitrary, two bases, which differ by a constant displacement, $\{\boldsymbol{\tau}_\mu\}$ and $\{\boldsymbol{\tau}_\mu + \boldsymbol{\tau}_0\}$ for all μ, define the same crystal. Also, the change of any basis vector $\boldsymbol{\tau}_\mu$ by a translation vector $(\boldsymbol{\tau}_\mu + \mathbf{R}_n)$ yields a different basis and a different unit cell, but it still describes the same crystal.

The most convenient way to define a crystal is to give its Bravais lattice and one primitive-cell basis. In some cases, to illustrate symmetries that are not obvious, the crystal is defined by a nonprimitive-cell basis and the corresponding subset of translation vectors. A two-dimensional example for a triangular crystal of formula AB_2 is given in Fig. 2.1. Shown there are the Bravais lattice and a basis of three atoms.

Wigner–Seitz cell. The Wigner–Seitz cell is a special primitive cell. It is built by drawing a line from the origin of the Bravais lattice to nearest-neighbor lattice points, and then constructing all the planes that bisect and are perpendicular to the lines, as illustrated in Fig. 2.2 for one and two dimensions. The smallest polyhedron that remains at the origin is the Wigner–Seitz cell. It is the most compact of all primitive cells and exhibits the maximum possible symmetry of the lattice. In the examples of Wigner–Seitz cells given in Fig. 2.2, (a) is a one-dimensional lattice, (b) is a two-dimensional square lattice, and (c) shows a two-dimensional hexagonal Wigner–Seitz cell for a triangular lattice.

Reciprocal lattice. The reciprocal lattice is a mathematical construction of great importance in the study of the quantum mechanical properties of solids. Given a crystal and its

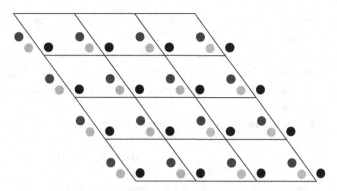

Figure 2.1 A two-dimensional lattice with a basis. The shaded circles represent different atoms.

(a)

(b)   (c)

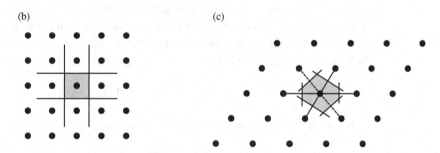

Figure 2.2 (a) One-dimensional Wigner–Seitz cell. (b) and (c) Two-dimensional Wigner–Seitz cells.

Bravais lattice (Eq. (2.8)) defined by the three primitive translation vectors $\mathbf{a}_1, \mathbf{a}_2$, and $\mathbf{a}_3$, we define a reciprocal lattice

$$\mathbf{G}_m \equiv m_1 \mathbf{g}_1 + m_2 \mathbf{g}_1 + m_3 \mathbf{g}_3 \quad (m_1, m_2, m_3 = \text{integers}), \tag{2.12}$$

such that, for any translation vector $\mathbf{R}_n$,

$$\mathbf{G}_m \cdot \mathbf{R}_n = 2\pi v \quad (v = \text{integer}) \tag{2.13}$$

for any m_1, m_2, m_3.

It can easily be shown that

$$\mathbf{g}_i = 2\pi \frac{\mathbf{a}_j \times \mathbf{a}_k}{(\mathbf{a}_1 \times \mathbf{a}_2) \cdot \mathbf{a}_3}, \tag{2.14}$$

where $(i \rightarrow j \rightarrow k) = (1 \rightarrow 2 \rightarrow 3)$, in cyclic order, satisfies the requirements with $v = n_1 m_1 + n_2 m_2 + n_3 m_3$.

Given the fact that the magnitude of $\mathbf{a}_i$ is measured in units of length, typically Å or Bohr radii a_0, the magnitudes of the $\mathbf{g}_i$, $|\mathbf{g}_i|$, are measured in units of inverse length such as Å^{-1} or a_0^{-1}. This space of "inverse length" where we define the reciprocal lattice is called "reciprocal space."

Periodic functions. Any function $u(\mathbf{r})$ that has the periodicity of the lattice

$$u(\mathbf{r} + \mathbf{R}_n) = u(\mathbf{r}) \tag{2.15}$$

can be expressed in a "discrete" Fourier expansion

$$u(\mathbf{r}) = \sum_m b_m e^{i\mathbf{G}_m \cdot \mathbf{r}}, \tag{2.16}$$

where $\mathbf{G}_m$ are reciprocal lattice vectors as given by Eq. (2.12).

Brillouin zone. Since the reciprocal lattice is a lattice, we may define unit cells and primitive cells for it in a manner similar to that was done for the real-space lattice. In particular, if we build a primitive cell following the same construction described for the Wigner–Seitz cell in the Bravais lattice, we obtain a polyhedron in reciprocal space. This important cell is called a Brillouin zone, and it is used extensively. Figure 2.2 can equally be used to represent one-dimensional and two-dimensional Brillouin zones, but the units are in inverse lengths, as described previously. For three dimensions, the Brillouin zone volume is usually measured in units of Å^{-3} or a_0^{-3}, and is given by

$$\Omega_{\text{BZ}} = |(\mathbf{g}_1 \times \mathbf{g}_2) \cdot \mathbf{g}_3|. \tag{2.17}$$

The use of Eq. (2.14), some simple vector algebra, and Eq. (2.10) yields

$$\Omega_{\text{BZ}} = \frac{8\pi^3}{|(\mathbf{a}_1 \times \mathbf{a}_2) \cdot \mathbf{a}_3|} = \frac{8\pi^3}{\Omega_{\text{p}}}. \tag{2.18}$$

A second and higher Brillouin zone can be defined by regions in space between the bisecting planes. These regions are mapped back into the first Brillouin zone by mapping reciprocal lattice vectors. The mapping is

$$\mathbf{k} \rightarrow \mathbf{k} - \mathbf{G}_m. \tag{2.19}$$

This is shown for the square lattice and the construction of the higher Brillouin zones in Fig. 2.3.

Periodic boundary conditions. All crystals, even the ideal, perfect ones, are finite. Their finiteness is essential for defining a crystal volume Ω_x, and the finiteness has important physical consequences. The presence of a crystal surface or boundary destroys the periodicity of the crystal, and many of the concepts defined above lose their meaning.

This unsatisfactory situation is resolved with the aid of a mathematical device that in some sense makes the crystal finite and periodic. This is the concept of "periodic boundary

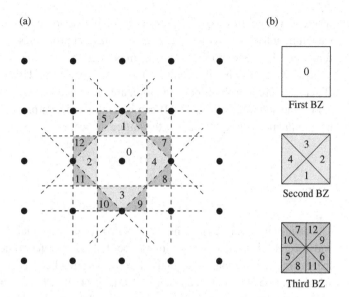

Figure 2.3 (a) Construction of the two-dimensional Brillouin zones. (b) The first, second, and third Brillouin zones in two dimensions.

conditions." These conditions state that the crystal closes onto itself and that the last atom on the right is followed, without a loss of periodicity, by the first atom on the left. Similar conditions apply to other dimensions for an n-dimenional crystal. (This corresponds to replacing a line by a circle, a plane by a torus, and the whole crystal by a "super torus.")

The details of this approach will be dealt with later. For now, we eliminate finite crystal effects and consider a crystalline solid of volume Ω_x containing N primitive cells. Hence, for N_i primitive cells in the $\mathbf{a}_i$-direction ($i = 1, 2, 3$ in three dimensions), any translation of the form $\mathbf{R}_{v_1 N_1, v_2 N_2, v_3 N_3}$, where v_i = integer, takes the crystal back into itself and leaves the crystal invariant. Therefore, there are

$$N \equiv N_1 \cdot N_2 \cdot N_3 \qquad (2.20)$$

primitive cells in the crystal, and the crystal volume is

$$\Omega_x = N\Omega_\mathrm{p}. \qquad (2.21)$$

Bloch's theorem. We now return to the one-electron-in-a-periodic-potential problem. The Hamiltonian (Eq. (2.5)) in the periodic potential approximation yields the Schrödinger equation

$$H(\mathbf{r})\psi(\mathbf{r}) \equiv \left\{ -\frac{\hbar}{2m}\nabla^2 + V(\mathbf{r}) \right\} \psi(\mathbf{r}) = E\psi(\mathbf{r}), \qquad (2.22)$$

where $V(\mathbf{r})$ satisfies Eq. (2.7). Because of the periodic symmetry of the potential, the wavefunctions that satisfy Eq. (2.22) have distinct properties and are called Bloch[3] wavefunctions. The theorem that states or illustrates the properties of Bloch wavefunctions is referred to as Bloch's theorem or the Bloch condition. Three standard ways of stating Bloch's theorem or the Bloch condition are described below.

(1) The solutions of Eq. (2.22) can be expressed as a planewave $e^{i\mathbf{k}\cdot\mathbf{r}}$ times a function having the periodic symmetry of the lattice; hence,

$$\Psi_{\mathbf{k}}(\mathbf{r}) = u_{\mathbf{k}}(\mathbf{r})e^{i\mathbf{k}\cdot\mathbf{r}}, \tag{2.23}$$

where

$$u_{\mathbf{k}}(\mathbf{r} + \mathbf{R}_n) = u_{\mathbf{k}}(\mathbf{r}). \tag{2.24}$$

The condition (Eq. (2.24)) is the same as that for the potential (Eq. (2.7)). At this point, the wavevector $\mathbf{k}$ is used as a quantum number to label the electronic state. Later we will show that energy levels group into bands. This is expected because of the existence of unit cells which confine electrons in the sense of Eq. (2.24). The bands will be indexed by the letter n. Hence, a specific state will be assigned the label $(n, \mathbf{k})$, where n indicates the band and $\mathbf{k}$ specifies a particular state in that band. We suppress the band index n for now.

(2) Bloch functions satisfy the periodic translational symmetry condition

$$\Psi_{\mathbf{k}}(\mathbf{r} + \mathbf{R}_n) = e^{i\mathbf{k}\cdot\mathbf{R}_n}\Psi_{\mathbf{k}}(\mathbf{r}). \tag{2.25}$$

This condition, which relates a lattice translation to a wavefunction phase factor, is very useful for illustrating the consequences of lattice periodicity on the wavefunction and matrix elements involving Bloch functions.

(3) Since functions having the periodicity of the lattice can be expressed in a discrete Fourier sum (Eq. (2.16)), the periodic part of the Bloch function can be written as

$$u_{\mathbf{k}}(\mathbf{r}) = \sum_{\mathbf{G}} b_{\mathbf{k}}(\mathbf{G})e^{i\mathbf{G}\cdot\mathbf{r}}. \tag{2.26}$$

This results in a useful and common form for the Bloch wavefunction

$$\psi_{\mathbf{k}}(\mathbf{r}) = \sum_{\mathbf{G}} b_{\mathbf{k}}(\mathbf{G})e^{i(\mathbf{k}+\mathbf{G})\cdot\mathbf{r}}. \tag{2.27}$$

It is possible to use any of the three descriptions above to obtain the benefits of periodic translational symmetry via Bloch's theorem. The above three statements describe the same physical properties. One-dimensional proofs of Bloch's theorem, also called Floquet's theorem, appear in most introductory texts on solids. Below we give a proof of Eq. (2.25).

We begin by defining a set of translation operators $\{T_n\}$ such that

$$T_n f(\mathbf{r}) = f(\mathbf{r} + \mathbf{R}_n), \tag{2.28}$$

---

[3] F. Bloch, "Uber die Quantenmechanik der Elektronen in Kristallgittern," Z. Phys. 52(1928), 555.

where $f(\mathbf{r})$ is an arbitrary function and $n = (n_1, n_2, n_3)$ is a set of integers (here we consider a three-dimensional crystal) where we have dropped the vector sign on n. Hence, $\mathbf{R}_n$ is defined as

$$\mathbf{R}_n = n_1\mathbf{a}_1 + n_2\mathbf{a}_2 + n_3\mathbf{a}_3. \tag{2.29}$$

We can describe T_n as an operator that takes $\mathbf{r} \to \mathbf{r}' = \mathbf{r} - \mathbf{R}_n$ and then relabel $\mathbf{r}' = \mathbf{r}$, where $\mathbf{r} = (x, y, z)$ and $\mathbf{r}' = (x', y', z')$.

If the operator T_n acts on the Hamiltonian $H(\mathbf{r})$ given by Eq. (2.22),

$$H(\mathbf{r}) = -\frac{\hbar^2}{2m}\nabla^2 + V(\mathbf{r}), \tag{2.30}$$

it leaves $H(\mathbf{r})$ invariant. This is obvious for the second term because of the periodic nature of the potential. To examine the action of T_n on the kinetic energy term, we first look at the x component of $\mathbf{r}$,

$$T_n\left[\frac{\partial^2}{\partial x^2}f(\mathbf{r})\right] = \frac{\partial^2}{\partial x'^2}f(\mathbf{r}') = \frac{\partial^2}{\partial x^2}f(\mathbf{r} + \mathbf{R}_n) = \frac{\partial^2}{\partial x^2}T_nf(\mathbf{r}), \tag{2.31}$$

and similarly for $\partial^2/\partial y^2$ and $\partial^2/\partial z^2$. Therefore, T_n commutes with ∇^2 and thus with $H(\mathbf{r})$:

$$[T_n, H(\mathbf{r})] = 0. \tag{2.32}$$

The action of two translation operators T_n and T_ℓ on $f(\mathbf{r})$ yields

$$T_nT_\ell f(\mathbf{r}) = f(\mathbf{r} + \mathbf{R}_n + \mathbf{R}_\ell) = T_\ell T_n f(\mathbf{r}) = T_{\ell+n}f(\mathbf{r}). \tag{2.33}$$

Therefore, $\{T_n\}$ is a set of commuting operators that commute with $H(\mathbf{r})$. Hence, we can diagonalize all T_n and $H(\mathbf{r})$ simultaneously.

Choosing θ_n to represent the eigenvalues corresponding to the action of the translation operator on an eigenfunction of Eq. (2.30), $\psi_n(\mathbf{r})$, we have

$$T_n\psi_n(\mathbf{r}) = \theta_n\psi_n(\mathbf{r}). \tag{2.34}$$

Using Eq. (2.33), we have

$$\theta_n\theta_\ell = \theta_{n+\ell}. \tag{2.35}$$

We now break up the translation operator into components:

$$T_n \equiv T_{(n_1n_2n_3)} = \left[T_{a_1}^{n_1}\right]\left[T_{a_2}^{n_2}\right]\left[T_{a_3}^{n_3}\right] \tag{2.36}$$

and

$$\theta_{(n_1n_2n_3)} = \theta_{a_1}^{n_1}\theta_{a_2}^{n_2}\theta_{a_3}^{n_3}. \tag{2.37}$$

Since T is a translation operator and the electron wavefunction is normalized, θ can only be a complex number with unity absolute value. Letting

$$\theta_{a_1} = e^{i\alpha_1}; \quad \theta_{a_2} = e^{i\alpha_2}; \quad \theta_{a_3} = e^{i\alpha_3} \tag{2.38}$$

yields

$$\theta_n = \theta_{(n_1 n_2 n_3)} = e^{i(n_1 \alpha_1 + n_2 \alpha_2 + n_3 \alpha_3)}. \tag{2.39}$$

Since, from Eqs. (2.12)–(2.14),

$$\mathbf{g}_i \cdot \mathbf{R}_n = 2\pi n_i, \tag{2.40}$$

we may write

$$\theta_n = e^{i\mathbf{k} \cdot \mathbf{R}_n}, \tag{2.41}$$

with

$$\mathbf{k} \equiv \left(\frac{\alpha_1}{2\pi}\right) \mathbf{g}_1 + \left(\frac{\alpha_2}{2\pi}\right) \mathbf{g}_2 + \left(\frac{\alpha_3}{2\pi}\right) \mathbf{g}_3. \tag{2.42}$$

We note here, since the α_i's are defined only in the range $0 \leq \alpha \leq 2\pi$, that $\mathbf{k}$ is a vector in reciprocal space and may be taken to be confined to within the first Brillouin zone. We then can use $\mathbf{k}$ as a quantum number, labeling the eigenstates that simultaneously diagonalize $\mathbf{H}$ and all the T_n's. Equation (2.34) becomes

$$T_n \psi_{\mathbf{k}}(\mathbf{r}) = e^{i\mathbf{k} \cdot \mathbf{R}_n} \psi_{\mathbf{k}}(\mathbf{r}), \tag{2.43}$$

which gives one of the standard forms of Bloch's theorem:

$$\psi(\mathbf{r} + \mathbf{R}_n) = e^{i\mathbf{k} \cdot \mathbf{R}_n} \psi_{\mathbf{k}}(\mathbf{r}). \tag{2.44}$$

Arbitrariness of the k-vector and reduction to the Brillouin zone. Because of the properties of Bloch functions, we can write

$$\psi_{\mathbf{k}}(\mathbf{r}) = e^{i\mathbf{k} \cdot \mathbf{r}} u_{\mathbf{k}}(\mathbf{r}) = e^{i\mathbf{k} \cdot \mathbf{r}} \left[\sum_{\mathbf{G}} b_{\mathbf{k}}(\mathbf{G}) e^{i\mathbf{G} \cdot \mathbf{r}} \right] = e^{i(\mathbf{k} - \mathbf{G}_0) \cdot \mathbf{r}} \left[\sum_{\mathbf{G}} b_{\mathbf{k}}(\mathbf{G}) e^{i(\mathbf{G} + \mathbf{G}_0) \cdot \mathbf{r}} \right]$$

$$= e^{i(\mathbf{k} - \mathbf{G}_0) \cdot \mathbf{r}} \left[\sum_{\mathbf{G}'} b_{\mathbf{k}}(\mathbf{G}' - \mathbf{G}_0) e^{i\mathbf{G}' \cdot \mathbf{r}} \right] = e^{i(\mathbf{k} - \mathbf{G}_0) \cdot \mathbf{r}} u_{\mathbf{k} - \mathbf{G}_0}(\mathbf{r}) = \psi_{\mathbf{k} - \mathbf{G}_0}(\mathbf{r}), \tag{2.45}$$

where $\mathbf{G}_0$ is any specific reciprocal lattice vector, $\mathbf{G}' = \mathbf{G} + \mathbf{G}_0$, and $u_{\mathbf{k} - \mathbf{G}_0}$ is a periodic function with properties similar to those of $u_{\mathbf{k}}(\mathbf{r})$. Equation (2.45) states that both $\mathbf{k}$ and $\mathbf{k} - \mathbf{G}_0$ are equally acceptable labels for describing the translational properties of ψ. In other words, any vector $\mathbf{k}'$ in reciprocal space that differs from $\mathbf{k}$ by a reciprocal lattice vector $\mathbf{G}$ (Eqs. (2.12)–(2.14)) can be chosen to represent the "planewave" part of a Bloch function; the choice is arbitrary. Mathematically, this is because, in Eq. (2.44), $(\mathbf{k} - \mathbf{G}_0) \cdot \mathbf{R}_n = \mathbf{k} \cdot \mathbf{R}_n - 2\pi \nu$, with ν an integer, and therefore the $\mathbf{G}_0$ term has no physical consequences.

This indeterminacy is normally resolved by imposing extra conditions on $\mathbf{k}$. The most common one is to require that the magnitude $|\mathbf{k}|$ of the $\mathbf{k}$-vector is the smallest possible one. This condition restricts $\mathbf{k}$ uniquely to the interior of the Brillouin zone, and it leaves at most a small number of possible choices if $\mathbf{k}$ is on the Brillouin zone boundary. In this last case, the choice is left to the individual. For example, in one dimension where $G_n = (2\pi/a)n$, the label k is restricted to $-(\pi/a) < k \leq (\pi/a)$.

3 Electronic energy bands

In Chapter 2, the complex problem of calculating the properties of electrons in perfect crystals is reduced (through a series of approximations) to an investigation of a one-electron Hamiltonian, which we now attempt to solve. Bloch's theorem and symmetry considerations will be used to simplify the computation. At this point, the construction of the potential itself will not be described, and it will be assumed that a periodic potential consisting of electron–electron and electron–core contributions has been formulated.

We begin by considering a free electron model to introduce the approach and to develop a framework for the discussion. After a description of some of the consequences of symmetry principles, a potential will be considered to demonstrate the existence of energy bands and energy bandgaps. Initially, the examples chosen will be one-dimensional. Most of the conclusions based on the one-dimensional cases are general, or at least suggestive of what can be expected for two and three dimensions.

3.1 Free electron model

We first consider the one-electron Schrödinger equation

$$\left[\frac{\mathbf{p}^2}{2m} + V(\mathbf{r})\right]\Psi = E\Psi, \tag{3.1}$$

with $V(\mathbf{r}) = V(\mathbf{r}+\mathbf{R})$, and explore the case where $V(\mathbf{r})$ is much less than the kinetic energy $p^2/2m$. The translational boundary conditions on the electronic wavefunction are enforced; that is, we still have Bloch waves, but beyond this, it is assumed that the potential has only a slight effect on the electrons.

Let us begin by examining electrons in a one-dimensional box of length L, and as a first approximation to a weak potential, we set $V(x) = 0$. This is the free electron model or free electron gas. The Schrödinger equation is

$$\frac{p^2}{2m}\Psi_k^0 = E_0(k)\Psi_k^0, \tag{3.2}$$

and its solution is straightforward, consisting of planewaves Ψ_k^0 and energy $E_0(k)$ for a given k, where

$$\Psi_k^0 = \frac{1}{\sqrt{L}}e^{ikx} \tag{3.3}$$

and

$$E_0(k) = \frac{\hbar^2 k^2}{2m}. \tag{3.4}$$

The wavevector k is not restricted, and k goes from $-\infty$ to $+\infty$, that is, no potential and no Brillouin zone. If we plot $E_0(k)$, it has the form shown in Fig. 3.1(a). Plotting $E_0(k)$ with no restriction on k is referred to as using the "extended zone scheme." When the $E_0(k)$ curve is mapped back into the first Brillouin zone, this is called the "reduced zone scheme." The extended zone scheme is appropriate for free electrons, since the Brillouin zone has no real significance here because the lattice periodicity does not exist for this case.

In the reduced zone scheme, the $E_0(k)$ of the free electron model is mapped into the first Brillouin zone in anticipation of turning on the potential. The prescription for the mapping, based on the result of Eq. (2.45), is

$$k' = k - \frac{2n\pi}{a}, \tag{3.5}$$

with n equal to an integer, or

$$k' = k - G. \tag{3.6}$$

This gives a mapping for Ψ_k^0 and $E_0(k)$ as

$$E_0(nk') = E_0\left(k' + \frac{2n\pi}{a}\right). \tag{3.7}$$

There is now another quantum number n in addition to k' that is restricted to the first Brillouin zone.

Figure 3.1(b) contains the dispersion curve or band structure (because of the appearance of "bands" of energy) for the free electron model when it is mapped into the first Brillouin zone. It will be shown presently that a periodic potential can introduce energy gaps between energy bands that are degenerate at the k-points, $k = 0$ and $k = \pm\pi/a$. Before demonstrating this feature, we will assume that it is plausible and discuss the influence of some symmetry principles on the form of $E_0(nk)$.

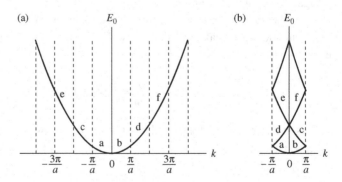

Figure 3.1 Energy bands for a free electron model. (a) Extended zone scheme. (b) Reduced zone scheme.

3.2 Symmetries and energy bands

We begin by discussing some general features of symmetry that are applicable in one, two, or three dimensions, and then focus on a few features of one-dimensional energy bands. Translational symmetry led to Bloch functions and Bloch's theorem. If the Bloch form is assumed for the wavefunction (Eq. (2.23)) and inserted into the one-electron Hamiltonian (Eq. (2.22)), we arrive at the equation for the $u_{\mathbf{k}}(\mathbf{r})$ part of the Bloch wavefunction

$$\left[\frac{1}{2m}\left(\mathbf{p}^2 + 2\hbar\mathbf{k}\cdot\mathbf{p} + \hbar^2\mathbf{k}^2\right) + V(\mathbf{r})\right]u_{n,\mathbf{k}}(\mathbf{r}) = E_n(\mathbf{k})u_{n,\mathbf{k}}(\mathbf{r}). \tag{3.8}$$

Since this is an eigenvalue problem with periodic boundary conditions, the eigenvalues take on discrete values labeled by n. Once the above equation is solved, the band structure $E_n(\mathbf{k})$, giving the electronic energy for each state labeled by the band index n and wavevector $\mathbf{k}$, is obtained. The full Bloch wavefunction $\Psi_{n\mathbf{k}}$ can be constructed using the periodic part obtained from the solution of Eq. (3.8) using Eq. (2.23). The label n is the band index described earlier.

A lattice can have other symmetries in addition to periodic translations. For example, it can be invariant to rotations through a specific angle. The group of rotations, reflections, and combinations of the two is called the point group. These are the symmetries obtained by keeping one point fixed, and like periodic translational symmetries, the point group symmetries can be exploited. (The point group operations together with translation symmetry operations form the space group of the crystal. For some crystals, there are also operations that involve point group operations plus a translation that is not a translation vector, leaving the system invariant. These are called nonsymmorphic operations.) In particular, it can be shown that the energy bands given by $E_n(\mathbf{k})$ have the same symmetry as the crystal. To describe what this implies, consider the operator θ, which can be a rotation such that

$$\mathbf{r}' = \theta\mathbf{r}. \tag{3.9}$$

In order to show that

$$E_n(\mathbf{k}) = E_n(\mathbf{k}'), \tag{3.10}$$

where $\mathbf{k}' = \theta\mathbf{k}$, we apply θ to the wave equation for $u_{\mathbf{k}}(\mathbf{r})$ (Eq. (3.8)). Since scalar products are invariant under rotation,

$$\mathbf{p}'^2 = (\theta\mathbf{p}\cdot\theta\mathbf{p}) = \mathbf{p}^2, \quad \mathbf{k}'^2 = \mathbf{k}^2, \quad \mathbf{k}\cdot\mathbf{p} = \mathbf{k}'\cdot\mathbf{p}'. \tag{3.11}$$

Therefore,

$$\left[\frac{1}{2m}\left(\mathbf{p}^2 + 2\hbar\mathbf{k}\cdot\mathbf{p} + \hbar^2\mathbf{k}^2\right) + V(\mathbf{r})\right]u_{n,\mathbf{k}'}(\theta\mathbf{r}) = E_n(\mathbf{k}')u_{n,\mathbf{k}'}(\theta\mathbf{r}), \tag{3.12}$$

and $u_{nk'}(\theta\mathbf{r})$ satisfies the same equation as $u_{nk}(\mathbf{r})$. Hence, a situation similar to Eq. (2.44) exists, and a band index n can be chosen such that

$$E_n(\mathbf{k}') = E_n(\mathbf{k}), \quad \text{where } \mathbf{k}' = \theta\mathbf{k}. \tag{3.13}$$

Schematically, $\theta E_n(\mathbf{k}) = E_n(\mathbf{k}')$, and $E_n(\mathbf{k})$ possesses the full rotational symmetry of the crystal structure. Operationally, the translational symmetry is assured by the Brillouin zone scheme, and beyond this, $E_n(\mathbf{k})$ is taken to have the space group symmetry.

We consider inversion separately because its consequences are related to those of time-reversal symmetry. All Bravais lattices (i.e. structures) without a basis have inversion symmetry. In some cases, inversion symmetry remains even after the inclusion of a basis. Using the above arguments, since $V(\mathbf{r}) = V(-\mathbf{r})$, then $E_n(\mathbf{k}) = E_n(-\mathbf{k})$. However, even without inversion symmetry, if the system being considered contains no magnetic fields and the forces are time-reversal invariant, such as the Coulomb force, it can be shown that $E_n(\uparrow \mathbf{k}) = E_n(\downarrow -\mathbf{k})$, where $\uparrow$ and $\downarrow$ denote spin configurations that are related by time reversal symmetry.

The proof for the case without spin–orbit interaction comes from taking complex conjugates of both the wave equation and wavefunctions, and this is essentially equivalent to a time-reversal operation. Applying the complex conjugate operator to a Bloch function

$$\Psi_{n,\mathbf{k}}^*(\mathbf{r}) = e^{i(-\mathbf{k})\cdot\mathbf{r}} u_{n,\mathbf{k}}^*(\mathbf{r}), \tag{3.14}$$

the wave equation for Eq. (3.8) becomes

$$\left[\frac{1}{2m}\left(\mathbf{p}^2 - 2\hbar\mathbf{k}\cdot\mathbf{p} + \hbar^2\mathbf{k}^2\right) + V(\mathbf{r})\right]u_{nk}^*(\mathbf{r}) = E_n(\mathbf{k})u_{nk}^*(\mathbf{r}). \tag{3.15}$$

So $u_{n,\mathbf{k}}^*(\mathbf{r})$ and $u_{n,\mathbf{k}}(\mathbf{r})$ satisfy the same equation if $\mathbf{k}$ is replaced by $-\mathbf{k}$, and we can choose

$$u_{n,-\mathbf{k}}^* = u_{n,\mathbf{k}} \tag{3.16}$$

and

$$E_n(\uparrow \mathbf{k}) = E_n(\downarrow -\mathbf{k})$$

by the same approach used before. This degeneracy is called the Kramers degeneracy.

When discussing states and bands in the first Brillouin zone, there are some definitions that are commonly used. Degeneracy implies more than one state with the same $\mathbf{k}$, and E. In cases where we consider $\Psi_{n,\mathbf{k}}$ and $\Psi_{n,\mathbf{k}'}$ such that $\mathbf{k}' = \theta\mathbf{k}$ and in the cases where we consider $\Psi_{n,\mathbf{k}}$ and $\Psi_{n,\mathbf{k}+\mathbf{G}}$, we do not call these states degenerate. In the latter case, the states are considered identical. Most degeneracies result from symmetries, while others are called accidental.

The symmetry considerations discussed here can save considerable effort in computing energy bands. For example, if $E_n(\mathbf{k})$ is needed throughout the Brillouin zone for a face-centered cubic lattice, the calculation need only be done in 1/48 of the Brillouin zone because of space group symmetry and the condition $E_n(\mathbf{k}) = E_n(-\mathbf{k})$. Other states in the full Brillouin zone are identical in energy to those in the 1/48 section and can be obtained by a mapping.

3.2.1 Symmetries and energy bands in one dimension

Several features of $E(k)$ with k confined to the Brillouin zone defined by $-\pi/a < k \leq \pi/a$ are easily deduced from symmetry principles for a strictly one-dimensional crystal with lattice constant a. It is possible to show that (1) $E(k)$ is non-degenerate; (2) $E(k = 0)$ is a maximum or a minimum; (3) $E(k)$ approaches $k = \pm\pi/a$ with zero slope; and (4) $E(k)$ has a maximum or a minimum only at $k = 0$ and $\pm\pi/a$ for a lattice with a center of symmetry.

$E(k)$ is non-degenerate. For one dimension, the one-electron Schrödinger equation (3.1) is just an ordinary homogeneous linear differential equation, so there are at most two independent solutions $\Psi(x)$ for a given energy E, which are labeled by the pair of quantum numbers nk. For a system without spin, time reversal and inversion both require $E(k) = E(-k)$, but we exclude the k and $-k$ pair of states from our definition of degeneracy. One-dimensional bands are therefore, non-degenerate and do not "overlap."

$E(k = 0)$ is a maximum or a minimum. Since $E(k) = E(-k)$, and assuming the derivative can be defined, then

$$\left.\frac{dE(k)}{dk}\right|_{k=0} = \lim_{\delta k \to 0} \frac{E(\delta k) - E(-\delta k)}{2\delta k} = 0. \tag{3.17}$$

Therefore, $E(k)$ is a maximum or minimum at $k = 0$.

$E(k)$ approaches $k = \pm\frac{\pi}{a}$ with zero slope. Translational symmetry requires that

$$E\left(-\frac{\pi}{a} + \delta k\right) = E\left(-\frac{\pi}{a} + \delta k + \frac{2\pi}{a}\right) = E\left(\frac{\pi}{a} + \delta k\right). \tag{3.18}$$

Since from time-reversal symmetry,

$$E\left(\frac{\pi}{a} + \delta k\right) = E\left(-\frac{\pi}{a} - \delta k\right),$$

then

$$\left.\frac{dE(k)}{dk}\right|_{k=\pm\pi/a} = \lim_{\delta k \to 0} \frac{E\left(\pm\frac{\pi}{a} + \delta k\right) - E\left(\pm\frac{\pi}{a} - \delta k\right)}{2\delta k} = 0. \tag{3.19}$$

So, $E(k)$ is a maximum or minimum at $k = \pm\pi/a$.

$E(k)$ has a maximum or minimum only at $k = 0$, $\pm\frac{\pi}{a}$. This means that the $E(k)$ curve can have the form shown in Fig. 3.2(a) but not Fig. 3.2(b). If another maximum or minimum existed, then, as the figures show, we would have a situation where $E(k_1) = E(k_2)$ with $k_1 \neq k_2$. This condition would exceed the maximum number of degenerate solutions that the differential equation (3.1) could have. Because of the above restriction, one-dimensional conductors have hole or electron states only at $k = 0$ and $k = \pm\pi/a$.

The proofs given neglect the spin degree of freedom and assume inversion or time-reversal symmetry, and that the system is not a one-dimensional structure embedded in three dimensions with internal symmetries. For example, if our system consists of an array of atoms with special symmetry properties, these would have to be considered.

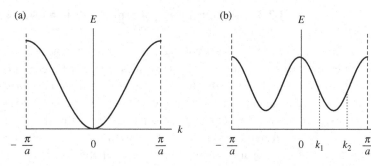

Figure 3.2 (a) Possible one-dimensional $E(k)$ with maxima and minima at $k = 0, \pm\pi/a$. (b) One-dimensional $E(k)$ with minima not at $k = 0, \pm\pi/a$. This case is not allowed.

3.2.2 Energy bands and gaps: the Kronig–Penney model

In the previous discussion, we anticipated the existence of separated energy bands and bandgaps. The gaps are regions of energy where electronic states do not exist; these gaps separate the states into bands. It is the interaction of the electrons with the periodic crystalline potential that gives rise to the bandgaps if we neglect strong electron correlation effects. There is a one-dimensional model of a periodic potential that is exactly solvable and demonstrates the existence of separated energy bands and bandgaps. This Kronig–Penney[1] model can be solved easily, and it provides a good example of the consequences of a periodic potential. It is also a useful starting point for many other calculations; hence, it is often used instructionally and as a research tool. We begin by considering an infinite periodic array of δ-function potentials of integrated strength aV separated by a distance a. The potential can be written as

$$V(x) = aV \sum_{n=-\infty}^{\infty} \delta(x - na). \qquad (3.20)$$

This potential represents the simplest periodic, one-dimensional model.

Consider two adjacent regions (labeled 1 and 2) on either side of a δ-function that is placed at $x = 0$. In these regions, $V = 0$ and the eigenfunctions are linear combinations of planewaves with momentum $\hbar k_0 = \sqrt{2mE}$,

$$\Psi_{1 \text{ or } 2} = A_{1 \text{ or } 2}\, e^{ik_0 x} + B_{1 \text{ or } 2}\, e^{-ik_0 x}, \qquad (3.21)$$

and the energy of the electron E is composed of the kinetic energy term only. Continuity of the wavefunction yields

$$\Psi_1(0^-) = \Psi_2(0^+), \qquad (3.22)$$

$$A_1 + B_1 = A_2 + B_2. \qquad (3.23)$$

[1] R. D. Kronig and W. G. Penney, "Quantum mechanics of electrons in crystal lattices", *Proc. R. Soc. Lond. A* 130(1931), 499.

Near $x = 0$, by integration of the Schrödinger equation,

$$-\frac{\hbar^2}{2m}\left[\Psi'(0^+) - \Psi'(0^-)\right] + aV\Psi(0) = 0. \tag{3.24}$$

This equation gives a discontinuous slope for the wavefunction that results from the singularities in the potential and yields the following condition:

$$-\frac{i\hbar^2 k_0}{2m}(A_1 - B_1 - A_2 + B_2) + aV(A_1 + B_1) = 0. \tag{3.25}$$

Also, Bloch's theorem requires

$$\Psi_2(a) = e^{ika}\Psi_1(0); \quad \Psi_2'(a) = e^{ika}\Psi_1'(0). \tag{3.26}$$

Hence,

$$A_2 e^{ik_0 a} + B_2 e^{-ik_0 a} = e^{ika}(A_1 + B_1) \tag{3.27}$$

and

$$A_2 e^{ik_0 a} - B_2 e^{-ik_0 a} = e^{ika}(A_1 - B_1). \tag{3.28}$$

Simultaneous solution of Eqs. (3.23), (3.25), (3.27), and (3.28) gives

$$\cos ka = \cos k_0 a + \frac{Vma^2}{\hbar^2}\frac{\sin k_0 a}{k_0 a}. \tag{3.29}$$

The solution of Eq. (3.29) yields $E(k)$, but since $E(k)$ is not in an explicit form, the equation must be solved numerically or graphically. Before considering the solution, we will change notation. Let

$$P = \frac{Vma^2}{\hbar^2}, \quad K = ka, \text{ and } \alpha^2 = k_0^2 a^2 = \frac{2mEa^2}{\hbar^2}. \tag{3.30}$$

The right- and left-hand sides of Eq. (3.29) become, respectively,

$$f(\alpha) \equiv \cos\alpha + P\frac{\sin\alpha}{\alpha}$$

and

$$g(K) \equiv \cos K.$$

Equation (3.29) is equivalent to

$$f(\alpha) = g(K). \tag{3.31}$$

We begin an analysis of Eq. (3.31) by considering a repulsive potential. The attractive potential case will be discussed later. For the repulsive case, $P > 0$, which means that the energy $E > 0$ and $\alpha^2 > 0$. So α is real. We take $\alpha > 0$ and solve Eq. (3.31) graphically.

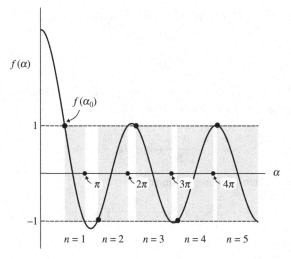

Figure 3.3   Graphical solution of a Kronig–Penney model of δ-functions.

Since $|g(K)| \leq 1$, the only values of α that satisfy Eq. (3.31) are those consistent with $|f(\alpha)| \leq 1$. Graphically, this can be solved as shown in Fig. 3.3. The acceptable α's are given by the shaded regions; the rest are forbidden. For a given k, there exists an infinite set of solutions for $f(\alpha)$ given by the intersection of the value $\cos K$ and $f(\alpha)$. As $\alpha \to \infty$, the forbidden regions get smaller, then $f(\alpha) \to \cos \alpha$, $K \to \alpha$, and $k \to k_0$. This means that, as expected, the potential has less of an effect on the particle as the energy of the particle becomes very large.

The regions containing acceptable values for α can be labeled by a band index n. Using the expression $\alpha^2 = 2mEa^2/\hbar^2$, the acceptable values for $\alpha(k)$ yield separate acceptable values for $E(k)$, and hence we have demonstrated the existence of energy bands. Symmetry properties are preserved. Since $\cos K = \cos(-K)$, then $\alpha(k) = \alpha(-k)$ and $E_n(k) = E_n(-k)$; also, the solution for $\cos(K + 2n\pi)$ is the same as the solution for $\cos K$, and we need only take $-\pi < K \leq \pi$ or $\pi/a < k \leq \pi/a$. This is just the first Brillouin zone. If we solved for $\alpha(k)$ or $E(k)$, we would get energy bands that are consistent with the restrictions discussed earlier for a one-dimensional $E(k)$.

A schematic $E(k)$ curve for the first three bands of the Kronig–Penney model is given in Fig. 3.4. The first band has its lowest energy at $E(k = 0) = E_0$, a constant that depends on the value of the potential. At the Brillouin zone edge, $ka = \pi$, $\alpha = \pi$, and $E(k) = \hbar^2 \pi^2/2ma^2$, which is the free electron value for this value of k. In fact, for this model, the Kronig–Penney energy reduces to the free electron value at $k = n\pi$ at the top of a band, since the potential term in Eq. (3.31) is inoperative at these points.

The nth gap, therefore, starts at $ka = n\pi$ and $f(n\pi) = (-1)^n$. Using this fact, it is possible to find the size of the gap for large n when the potential energy is small compared to the kinetic energy, that is, for a "nearly-free electron" system. Suppose the nth gap starts at $\alpha = n\pi$ and ends at $\alpha = n\pi + \Delta\alpha$, where $\Delta\alpha$ is small. Using Eq. (3.31) and the fact

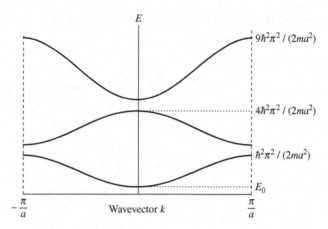

Figure 3.4   Schematic $E(k)$ for the Kronig–Penney model.

that f does not change much with the small variation in α, we have

$$(-1)^n = \cos(n\pi + \Delta\alpha) + P\frac{\sin(n\pi + \Delta\alpha)}{(n\pi + \Delta\alpha)} \ . \tag{3.32}$$

Expanding, we get

$$\Delta\alpha = \frac{2P}{n\pi}, \tag{3.33}$$

and the energy gap becomes

$$\Delta E = E(n\pi + \Delta\alpha) - E(n\pi) = \frac{2\hbar^2 P}{ma^2} = 2V. \tag{3.34}$$

This is a well-known result for the bandgap E_g for a nearly-free electron; however, it is usually obtained using perturbation theory on a free electron model, as will be shown next.

3.3 Nearly-free electron model

The model of electrons interacting with a weak potential is an excellent one for representing the electronic structure of many solids. In many metals, the electrons can be considered as almost free, so their energies and wavefunctions are not expected to differ much from the results obtained for a free electron model (FEM). The standard approach is to use perturbation theory on the FEM to obtain results for the nearly-free electron model (NFEM) rather than using the Kronig–Penney method. Because of the appropriateness of the model for real systems, the model also illustrates many of the general features of band structures.

Since the extension of the FEM and the NFEM to the three-dimensional case is straightforward with no fundamental complication, we proceed directly to three dimensions.

Letting $V(\mathbf{r}) \neq 0$ (but weak), we can obtain $E(\mathbf{k})$ via standard perturbation theory. This gives to second order

$$E(\mathbf{k}) = E_0(\mathbf{k}) + V_{\mathbf{kk}} + \sum_{\mathbf{k} \neq \mathbf{k}'} \frac{|V_{\mathbf{k'k}}|^2}{E_0(\mathbf{k}) - E_0(\mathbf{k}')}, \tag{3.35}$$

with

$$V_{\mathbf{k'k}} = \frac{1}{\Omega_x} \int e^{i(\mathbf{k}-\mathbf{k}')\cdot\mathbf{r}} V(\mathbf{r}) d\mathbf{r} \tag{3.36}$$

and

$$V_{\mathbf{kk}} = \frac{1}{\Omega_x} \int V(\mathbf{r}) d\mathbf{r} = V_0, \tag{3.37}$$

where $E_0(\mathbf{k})$ is the FEM energy, Ω_x is the crystal volume, which for convenience can be set to unity, V_0 is the average potential, and the $V_{\mathbf{kk}'}$ are matrix elements of the potential between FEM states $\Psi_{\mathbf{k}}^0$ and $\Psi_{\mathbf{k}'}^0$, where $\Psi_{\mathbf{k}}^0(\mathbf{r}) = \frac{e^{i\mathbf{k}\cdot\mathbf{r}}}{\sqrt{\Omega_x}}$. The constant V_0 shifts all the energies irrespective of their $\mathbf{k}$-values; hence, it is used just to set the zero of energy.

Since $V(\mathbf{r})$ is periodic with lattice periodicity, it can be expanded in the reciprocal lattice

$$V(\mathbf{r}) = \sum_{\mathbf{G}} V(\mathbf{G}) e^{i\mathbf{G}\cdot\mathbf{r}}, \tag{3.38}$$

where (with $\Omega_x = 1$)

$$V(\mathbf{G}) = \int e^{-i\mathbf{G}\cdot\mathbf{r}} V(\mathbf{r}) d\mathbf{r}. \tag{3.39}$$

This expansion simplifies the expression for $V_{\mathbf{kk}'}$, that is,

$$V_{\mathbf{k'k}} = \sum_{\mathbf{G}} V(\mathbf{G}) \int e^{i(\mathbf{k}-\mathbf{k}'+\mathbf{G})\cdot\mathbf{r}} d\mathbf{r} = \sum_{\mathbf{G}} V(\mathbf{G}) \delta_{\mathbf{k'},\mathbf{k}+\mathbf{G}}. \tag{3.40}$$

The energy becomes

$$E(\mathbf{k}) = E_0(\mathbf{k}) + V_0 + \sum_{\mathbf{G} \neq 0} \frac{|V(\mathbf{G})|^2}{E_0(\mathbf{k}) - E_0(\mathbf{k}+\mathbf{G})}. \tag{3.41}$$

The "second-order" term is usually small, since for most $\mathbf{k}$'s it is essentially the square of the potential energy divided by the kinetic energy difference. The whole scheme breaks down when $E_0(\mathbf{k}) = E_0(\mathbf{k}+\mathbf{G})$. At these $\mathbf{k}$-points, we need degenerate perturbation theory. We note that at these $\mathbf{k}$-points,

$$2\mathbf{k} \cdot \mathbf{G} + \mathbf{G}^2 = 0, \tag{3.42}$$

which is the famous Bragg condition used in X-ray scattering from crystals. It is also the condition that determines the Brillouin zone boundary. For example, in one dimension,

$$2k\frac{\pm 2\pi}{a} + \left(\frac{2\pi}{a}\right)^2 = 0.$$

Hence,

$$k = \frac{\pm \pi}{a}, \tag{3.43}$$

which are the edges of the Brillouin zone.

Hence, the unperturbed energies are degenerate at the Brillouin zone boundaries, and degenerate perturbation theory is required to solve the Schrödinger equation for **k**-points near these boundaries. This is done by considering the wavefunction to be a linear combination of the zero-order degenerate states. In one dimension, only one G is involved. We therefore have

$$\Psi_k = a\Psi_k^0 + b\Psi_{k+G}^0$$

and

$$H\Psi_k = \left[\frac{p^2}{2m} + V(r)\right]\Psi_k = E\Psi_k. \tag{3.44}$$

Taking matrix elements of the Hamiltonian between Ψ_k^0 and Ψ_{k+G}^0, we get

$$aH_{kk} + bH_{k,k+G} = Ea \tag{3.45}$$

and

$$aH_{k+G,k} + bH_{k+G,k+G} = Eb,$$

where

$$H_{kk'} = [E_0(k) + V_0]\delta_{kk'} + V_{kk'}. \tag{3.46}$$

Since

$$V_{k,k+G} = V_{k+G,k}^* = V(-G)$$

and

$$E_0(k) = E_0(k+G), \tag{3.47}$$

solving for Eq. (3.44), we have

$$E(k) = E_0(k) + V_0 \pm \sqrt{|V(G)|^2}. \tag{3.48}$$

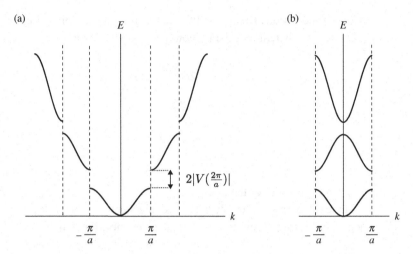

Figure 3.5 Nearly-free electron model. (a) Extended zone scheme. (b) Reduced zone scheme.

Setting $V_0 = 0$, we have

$$E(k) = E_0(k) \pm |V(G)|. \tag{3.49}$$

The degenerate levels are split by $2|V(G)|$. A similar result was obtained earlier for the higher gaps in the Kronig–Penney model.

For one dimension in the extended zone scheme, the $E(k)$ curve is essentially a parabola with gaps at $k = \pm n\pi/a$, as shown in Fig. 3.5(a). In the reduced zone, we have the situation shown in Fig. 3.5(b).

If we calculate the wavefunctions near one of these gaps, the existence of the gap can be understood on physical grounds. The solutions for a and b (assuming an attractive potential) are for the lower state

$$a = b = \frac{1}{\sqrt{2}}$$

and for the upper state

$$a = -b = \frac{1}{\sqrt{2}}, \tag{3.50}$$

which give

$$\Psi_1 \propto \cos\frac{1}{2}Gx$$

and

$$\Psi_2 \propto \sin\frac{1}{2}Gx. \tag{3.51}$$

The electronic density $|\Psi_1|^2$ is peaked at the lattice points, while $|\Psi_2|^2$ is peaked between the lattice points. For an attractive potential, the $|\Psi_1|^2$ charge distribution lies lower in energy than that corresponding to $|\Psi_2|^2$, and the energy difference is $2|V(G)|$. Behavior of this kind is expected near bandgaps.

In summary, the NFEM yields energy bands, which are almost identical to those of the FEM, except near zone edges. In the NFEM, energy gaps appear, and these can be extremely important in determining the electronic properties of the system.

3.4 Tight-binding model

At the other limit relative to the NFEM is the tight-binding model (TBM). This approximation is most appropriate for solids in which the constituent atoms resemble slightly perturbed free atoms. Whereas the NFEM starts with completely free electrons perturbed by a weak potential, the TBM begins with atomic-like states and perturbs them with a potential due to interaction with nearby atoms. The NFEM is appropriate for close-packed solids in which the overlap between electrons from adjacent atoms is large. Hence, metals are particularly suited to this type of analysis. The TBM is more appropriate for wide gap insulators where the overlap between adjacent atoms is small and atomic separations are large. Semiconductors tend to fall in between, and for these cases, both methods have been used.

It is possible to begin with the Kronig–Penney model to gain insight into the TBM. For example, consider a situation in which the above Kronig–Penney potential is appropriate for atomic-like systems where the starting point is a set of bound states. For the repulsive potential case considered earlier, it is not possible to get a bound state. This can be seen by examining the Kronig–Penney equation, assuming a bound state with $E < 0$. The condition $E < 0$ implies that α is imaginary. Assuming $\alpha = i\phi$, we have

$$\cos K = \cosh \phi + \frac{P \sinh \phi}{\phi}. \tag{3.52}$$

This can be solved only if we consider an attractive potential.

Now P (Eq. (3.30)) is negative, and one has the possibility that E can be negative. Equation (3.52) can be written as

$$\cos K = \cosh \phi - |P| \frac{\sinh \phi}{\phi}. \tag{3.53}$$

This is easily solvable if $|\phi| \gg 1$. Expanding, we find

$$\cos K = \frac{e^\phi}{2} \left(1 - \frac{|P|}{\phi} \right). \tag{3.54}$$

Since $|\cos K| \leq 1$ and e^ϕ is large, we need $\left(1 - \frac{|P|}{\phi} \right) \sim 0$ (or $\phi = |P| + \delta\phi$). Equation (3.54) becomes (to lowest order)

$$\cos K = \frac{1}{2} e^{|P|} \frac{\delta\phi}{\phi} = \frac{\delta\phi}{2|P|} e^{|P|} \tag{3.55}$$

and

$$\delta\phi = 2|P|e^{-|P|}\cos K. \tag{3.56}$$

We can find the energy by using

$$E = -\frac{\hbar^2\phi^2}{2ma^2} = -\frac{\hbar^2|P|^2}{2ma^2}\left(1 + 4e^{-|P|}\cos K + \cdots\right) \tag{3.57}$$

or

$$E = -\frac{mV^2a^2}{2\hbar^2}\left[1 + 4\exp\left(\frac{-|V|ma^2}{\hbar^2}\right)\cos ka + \cdots\right]. \tag{3.58}$$

The first term is negative, and it represents the energy of a bound-state in a one-dimensional well; hence, we have bound-state solutions. For large a, the potential wells (negative delta functions in our case) are separated, and N independent potentials give N independent bound states. As a gets smaller, the wells interact, and the level spreads. This spread is given by the second term in Eq. (3.58). The total spread ΔE has the form

$$\Delta E = \frac{4mV^2a^2}{\hbar^2}\exp\left(\frac{-|V|ma^2}{\hbar^2}\right) = \frac{4mV^2a^2}{\hbar^2}e^{-a/\ell}, \tag{3.59}$$

and the spread is small until $a \sim \ell$, where $\ell \equiv \frac{\hbar^2}{|V|ma}$ is the radius of the bound-state electron.

The above calculation is similar to a "tight-binding-type calculation." If atoms are isolated, the energy levels are sharp. As the distance between the atoms gets smaller, the tails of the exponential bound-state wavefunctions "feel" each other and interact. This causes the energy levels to spread out into bands of levels.

Summarizing, in the Kronig–Penney model, we are able to find a bound state for a negative potential, and, for N isolated delta functions, this state is N-fold degenerate. As the delta functions are brought closer together, they begin to interact, and the states spread in energy (Fig. 3.6) according to the expression

$$E = E_0 + 4E_0e^{-a/\ell}\cos ka + \cdots \tag{3.60}$$

The delta function potentials interact when $a \sim \ell$, that is, when the distance between the potentials is of the order of the radius of the bound-state wavefunction.

If we now extend this idea to a collection of atoms, we expect the free atom levels to spread as the distance between atoms decreases (Fig. 3.7). The spreading can be so large that the bands corresponding to different levels can overlap. For the case where the overlap is small, the spreading is small, and we expect the wavefunctions to be similar to those in the atomic case. It is then reasonable to expect atomic states to form a good basis set for representing the wavefunctions describing the solid. It is also reasonable to expect the crystal potential to look like a superposition of atomic potentials $V_a(\mathbf{r} - \mathbf{R}_\ell - \boldsymbol{\tau}_a)$ about

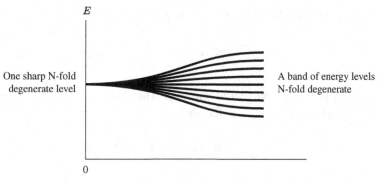

Figure 3.6 Spreading of one N-fold degenerate δ-function level into a band as a function of the inverse of the lattice constant using the Kronig–Penney model.

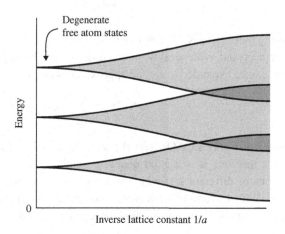

Figure 3.7 Spreading of atomic levels into bands as the lattice constant a decreases.

the lattice points $\mathbf{R}_\ell$ and basis vector $\boldsymbol{\tau}_a$, i.e.

$$V(\mathbf{r}) = \sum_{\ell,a} V_a\,(\mathbf{r} - \mathbf{R}_l - \boldsymbol{\tau}_a), \qquad (3.61)$$

where a labels the basis atoms.

The atomic wavefunctions satisfy the equation

$$\left[\frac{\mathbf{p}^2}{2m} + V_a(\mathbf{r})\right]\phi_{ta} = E_{ta}\phi_{ta}, \qquad (3.62)$$

with t labeling the atomic states. We may now form the so-called linear combination of atomic orbitals (LCAOs) to serve as basis functions to expand the electronic states in a crystal. An LCAO $\phi_{\mathbf{k}}^{ta}(\mathbf{r})$ is defined as

$$\phi_{\mathbf{k}}^{ta}(\mathbf{r}) \equiv \frac{1}{\sqrt{N}} \sum_n e^{i\mathbf{k}\cdot\mathbf{R}_n} \phi_{ta}(\mathbf{r} - \mathbf{R}_n - \boldsymbol{\tau}_a), \qquad (3.63)$$

where N is the number of unit cells. The phase factor $e^{i\mathbf{k}\cdot\mathbf{R}_n}$ is to ensure that $\phi_{\mathbf{k}}^{ta}$ is of the Bloch form. Let $\mathbf{r} \to \mathbf{r}+\mathbf{R}_m$, then

$$\phi_{\mathbf{k}}^{ta}(\mathbf{r} + \mathbf{R}_m) = \frac{1}{\sqrt{N}} \sum_n e^{i\mathbf{k}\cdot\mathbf{R}_n} \phi_{ta}\left[\mathbf{r} - (\mathbf{R}_n - \mathbf{R}_m) - \boldsymbol{\tau}_a\right]. \qquad (3.64)$$

Relabeling $\mathbf{R}_\ell = \mathbf{R}_n - \mathbf{R}_m$,

$$\begin{aligned}
\phi_{\mathbf{k}}^{ta}(\mathbf{r} + \mathbf{R}_m) &= \frac{1}{\sqrt{N}} \sum_\ell e^{i\mathbf{k}\cdot\mathbf{R}_\ell} e^{i\mathbf{k}\cdot\mathbf{R}_m} \phi_{ta}(\mathbf{r} - \mathbf{R}_\ell - \boldsymbol{\tau}_a) \\
&= e^{i\mathbf{k}\cdot\mathbf{R}_m} \phi_{\mathbf{k}}^{ta}(\mathbf{r}),
\end{aligned} \qquad (3.65)$$

showing that $\phi_{\mathbf{k}}^{ta}(\mathbf{r})$ is a Bloch function.

For a crystal with weakly interacting atoms, the crystalline electron wavefunctions $\Psi_{\mathbf{k}}(\mathbf{r})$ can be expanded to a good approximation in terms of the LCAOs. That is,

$$\Psi_{\mathbf{k}}(\mathbf{r}) = \sum_{ta} \alpha_{ta}(\mathbf{k}) \phi_{\mathbf{k}}^{ta}(\mathbf{r}). \qquad (3.66)$$

We note that only LCAOs with the same $\mathbf{k}$ enter into Eq. (3.66). This is because, from Bloch's theorem, $\mathbf{k}$ is a good quantum number and electronic states do not mix Bloch functions of different $\mathbf{k}$'s, which is a basic consequence of translational symmetry of a crystal. With the form of the wavefunction in Eq. (3.66), the Schrödinger equation ($H\Psi_{\mathbf{k}} = E(\mathbf{k})\Psi_{\mathbf{k}}$) may be written in the form of a generalized matrix eigenvalue equation:

$$\left(H_{ij}(\mathbf{k})\right) \begin{pmatrix} \alpha_1(\mathbf{k}) \\ \cdot \\ \cdot \\ \cdot \\ \alpha_M(\mathbf{k}) \end{pmatrix} = E(\mathbf{k}) S_{ij} \begin{pmatrix} \alpha_1(\mathbf{k}) \\ \cdot \\ \cdot \\ \cdot \\ \alpha_M(\mathbf{k}) \end{pmatrix}, \qquad (3.67)$$

where (i,j) run over the indices (ta). That is, M in Eq. (3.67), which is equal to the dimension of the matrices H and S, is equal to the sum of the number of relevant atomic states on all of the atoms in the unit cell. The elements of the Hamiltonian matrix H and the overlap matrix S are given by

$$H_{ij}(\mathbf{k}) = \langle \phi_{\mathbf{k}}^i | H | \phi_{\mathbf{k}}^j \rangle \qquad (3.68)$$

and

$$S_{ij}(\mathbf{k}) = \left\langle \phi_{\mathbf{k}}^i \,\middle|\, \phi_{\mathbf{k}}^j \right\rangle. \qquad (3.69)$$

Solutions of Eq. (3.67) lead to a band structure with M bands within this formalism.

As a simplest possible illustration, let us consider the case of a crystal with one atom per unit cell; on each atom there is only one relevant atomic state of interest. In this case, $M = 1$ with

$$
\begin{aligned}
H_{11}(\mathbf{k}) &= \left\langle \phi_{\mathbf{k}}^1(\mathbf{r}) \left| H \right| \phi_{\mathbf{k}}^1(\mathbf{r}) \right\rangle \\
&= \frac{1}{N} \sum_{n,m} e^{i\mathbf{k}\cdot(\mathbf{R}_n - \mathbf{R}_m)} \int_{\Omega_x} \phi_1^*(\mathbf{r} - \mathbf{R}_m) H(\mathbf{r}) \phi_1(\mathbf{r} - \mathbf{R}_n) d^3 r.
\end{aligned}
\tag{3.70}
$$

Using the factor that $H(\mathbf{r})$ is invariant with respect to a translation vector, we may rewrite Eq. (3.70) (by defining $\mathbf{R}_p = \mathbf{R}_n - \mathbf{R}_m$) as

$$
\begin{aligned}
H_{11} &= \frac{1}{N} \sum_{p,m} e^{i\mathbf{k}\cdot\mathbf{R}_p} \int_{\Omega_x} \phi_1^*(\mathbf{r}) H(\mathbf{r}) \phi_1(\mathbf{r} - \mathbf{R}_p) d^3 r \\
&= \sum_p e^{i\mathbf{k}\cdot\mathbf{R}_p} h_{11}(\mathbf{R}_p).
\end{aligned}
\tag{3.71}
$$

Here we have denoted

$$
h_{11}(\mathbf{R}_p) = \int_{\Omega_x} \phi_1^*(\mathbf{r}) H(\mathbf{r}) \phi_1(\mathbf{r} - \mathbf{R}_p) d^3 r,
\tag{3.72}
$$

which is often called the hopping integral, and we use the fact that the summation over m gives rise to a factor of N. Similarly,

$$
\begin{aligned}
S_{11}(\mathbf{k}) &= \sum_p e^{i\mathbf{k}\cdot\mathbf{R}_p} \int_{\Omega_x} \phi_1^*(\mathbf{r}) \phi(\mathbf{r} - \mathbf{R}_p) \, d^3 r \\
&= \sum_p e^{i\mathbf{k}\cdot\mathbf{R}_p} s_{11}(\mathbf{R}_p).
\end{aligned}
\tag{3.73}
$$

The solution to the 1×1 matrix eigenvalue problem in Eq. (3.67) becomes (for this simple case)

$$
\begin{aligned}
E(\mathbf{k}) &= \frac{H_{11}(\mathbf{k})}{S_{11}(\mathbf{k})} \\
&= \frac{\sum_p e^{i\mathbf{k}\cdot\mathbf{R}_p} h_{11}(\mathbf{R}_p)}{\sum_p e^{i\mathbf{k}\cdot\mathbf{R}_p} s_{11}(\mathbf{R}_p)}.
\end{aligned}
\tag{3.74}
$$

This solution may be transformed to a more intuitive form by dividing H into the atomic part H_a plus the part due to the rest of the crystal $H^1 = H - H_a$. Then Eq. (3.74) becomes

$$
E(\mathbf{k}) = E_a + \frac{\sum_p e^{i\mathbf{k}\cdot\mathbf{R}_p} h_{11}^1(\mathbf{R}_p)}{\sum_p e^{i\mathbf{k}\cdot\mathbf{R}_p} s_{11}(\mathbf{R}_p)}.
\tag{3.75}
$$

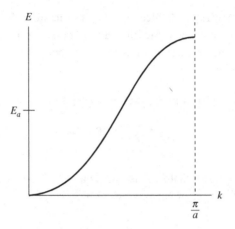

Schematic $E(\mathbf{k})$ in the k_x direction for a tight-binding band for a simple cubic lattice with a single s-orbital on each site.

The first term is the energy of the atomic state, and the second term contains $h_{11}^{1}(\mathbf{R}_p)$, which is negligibly small for $p = 0$.

As an example, consider a simple cubic lattice with only nearest-neighbor interactions. The lattice vectors are $(a, 0, 0)$, $(0, a, 0)$, and $(0, 0, a)$. Assuming the same matrix elements for the six nearest-neighbor sites (which is appropriate for s-like orbitals),

$$E(\mathbf{k}) = E_a + 2h(a, 0, 0)\left[\cos k_x a + \cos k_y a + \cos k_z a\right]. \qquad (3.76)$$

This result is similar to the Kronig–Penney result in Eq. (3.60). If $h(a, 0, 0)$ is negative, the minimum energy for the simple cubic lattice is $E_a + 6h(a, 0, 0)$, and the width of the band is $12|h(a, 0, 0)|$. (See Fig. 3.8 for a schematic example.)

The tight-binding method is useful to illustrate how atomic states broaden into bands. It can be used for accurate band calculations in cases where the overlap between neighboring atoms is small and the bands are narrow. This method becomes more accurate as the number of atomic orbits on each atom and further-neighbor interactions are included. The TBM has been exploited with success in calculations on molecules and solids.

3.5 Electron (or hole) velocity in a band and the f-sum rule

Once the band structure for a crystal has been computed and the $E_n(\mathbf{k})$ curve is known, it is possible to evaluate the velocity of an electron (or hole) and the effective mass m^* characterizing the electron or hole in a band state. In addition, if the energies of the states are known at one $\mathbf{k}$-point, the band structure at neighboring points can be obtained. In the following discussion, these features are derived, as is the f-sum rule which connects the effective mass, bandgaps, and momentum matrix elements.

For simplicity, let us consider non-degenerate energy levels. We know the energy levels at a particular $\mathbf{k}$-point; that is, we know $E_n(\mathbf{k}_0)$ for all n. (The discussion may be generalized to degenerate states.) From this information, we can determine the energy at a neighboring $\mathbf{k}$-point $\mathbf{k}'$, where

$$\mathbf{k}' = \mathbf{k}_0 + \mathbf{k}. \tag{3.77}$$

Consider the wave equation for the periodic part of the Bloch function at $\mathbf{k}'$:

$$\left[\frac{(\mathbf{p} + \hbar \mathbf{k}')^2}{2m} + V(\mathbf{r}) \right] u_{n\mathbf{k}'}(\mathbf{r}) = E_n(\mathbf{k}') u_{n\mathbf{k}'}(\mathbf{r}), \tag{3.78}$$

then

$$\left[\frac{(\mathbf{p} + \hbar \mathbf{k}_0)^2}{2m} + V(\mathbf{r}) + \frac{\hbar \mathbf{k} \cdot (\mathbf{p} + \hbar \mathbf{k}_0)}{m} + \frac{\hbar^2 \mathbf{k}^2}{2m} \right] u_{n\mathbf{k}'}(\mathbf{r}) = E_n(\mathbf{k}') u_{n\mathbf{k}'}(\mathbf{r}). \tag{3.79}$$

The first two terms on the left-hand side give the energy levels at $\mathbf{k}_0$, and the last term is a number that is second order in $\mathbf{k}$. The third term, which we shall call H', can be used as a perturbation if $\mathbf{k}$ is small. When this term is used in this way, the method is called "$\mathbf{k} \cdot \mathbf{p}$ perturbation theory." To second order in $\mathbf{k}$, we have

$$E_n(\mathbf{k}_0 + \mathbf{k}) = \frac{\hbar^2 \mathbf{k}^2}{2m} + E_n(\mathbf{k}_0) + \langle u_{n\mathbf{k}_0} | H' | u_{n\mathbf{k}_0} \rangle + \sum_{n \neq n'} \frac{|\langle u_{n'\mathbf{k}_0} | H' | u_{n\mathbf{k}_0} \rangle|^2}{E_n(\mathbf{k}_0) - E_{n'}(\mathbf{k}_0)}, \tag{3.80}$$

where the u's are orthonormal since they are solutions to $H(\mathbf{k}_0)$.

Another way to obtain $E_n(\mathbf{k}')$ to second order is to expand around $\mathbf{k}_0$ in a Taylor series,

$$E_n(\mathbf{k}_0 + \mathbf{k}) = E_n(\mathbf{k}_0) + \sum_{i=1}^{3} \frac{\partial E_n(\mathbf{k}_0)}{\partial k_{oi}} k_i + \frac{1}{2} \sum_{i,j=1}^{3} \frac{\partial^2 E_n(\mathbf{k}_0)}{\partial k_{oi} \partial k_{oj}} k_i k_j. \tag{3.81}$$

By equating terms of the same order in Eqs. (3.80) and (3.81), we obtain the desired relations. The first-order terms yield

$$\begin{aligned} \frac{\partial E_n(\mathbf{k}_0)}{\partial k_{oi}} &= \frac{\hbar}{m} \langle u_{n\mathbf{k}_0} | p_i + \hbar k_{oi} | u_{n\mathbf{k}_0} \rangle \\ &= \frac{\hbar}{m} \langle n\mathbf{k}_0 | p_i | n\mathbf{k}_0 \rangle, \end{aligned} \tag{3.82}$$

where we have used the Dirac notation for the full Bloch state and the relation

$$\langle n\mathbf{k}_0 | p_i | n\mathbf{k}_0 \rangle = \langle u_{n\mathbf{k}_0} | p_i + \hbar k_{oi} | u_{n\mathbf{k}_0} \rangle. \tag{3.83}$$

This result allows a calculation of the velocity in the direction i for an electron in the state $(n, \mathbf{k}_0)$, using

$$v_{i,n,\mathbf{k}_0} = \frac{1}{\hbar} \frac{\partial E_n(\mathbf{k}_o)}{\partial k_{oi}}. \tag{3.84}$$

Equating second-order terms yields the f-sum rule

$$\frac{m}{(m_n)_{ij}} = \delta_{ij} + \sum_{n' \neq n} f_{nn'}^{ij}, \tag{3.85}$$

where the effective mass is defined as

$$\frac{1}{(m_n)_{ij}} \equiv \frac{1}{\hbar^2} \frac{\partial^2 E_n(\mathbf{k})}{\partial k_i \partial k_j}, \tag{3.86}$$

and $f_{nn'}^{ij}$ is the oscillator strength which is related to matrix elements between Bloch states of the momentum operator

$$f_{nn'}^{ij} = \frac{1}{m} \frac{\langle n'\mathbf{k}_0 | p_i | n\mathbf{k}_0 \rangle \langle n\mathbf{k}_0 | p_j | n'\mathbf{k}_0 \rangle + \langle n'\mathbf{k}_0 | p_j | n\mathbf{k}_0 \rangle \langle n\mathbf{k}_0 | p_i | n'\mathbf{k}_0 \rangle}{E_n(\mathbf{k}_0) - E_{n'}(\mathbf{k}_0)}. \tag{3.87}$$

We shall further discuss the meaning of the effective mass in Chapter 5.

The main features of the f-sum rule can be illustrated by considering a model of a semi-conductor with just two bands separated by a gap E_g. For simplicity, we assume constant momentum matrix elements $|p|$ and spherical bands, and investigate how the effective mass of each band depends on the gap by considering Eq. (3.85), using these approximations to obtain

$$\frac{m}{m_n^*} = 1 + \frac{2}{m} \sum_{n' \neq n}^{2} \frac{|p|^2}{E_n - E_{n'}}. \tag{3.88}$$

For band 2, which would be the conduction band of the semiconductor (as seen in Fig. 3.9),

$$\frac{m}{m_2^*} = 1 + \frac{2}{m} \frac{|p|^2}{E_2 - E_1} = 1 + \frac{2}{m} \frac{|p|^2}{E_g}. \tag{3.89}$$

Hence, $m_2^* > 0$ and, if E_g is small, $m_2^* \ll m$. For band 1, the valence band,

$$\frac{m}{m_1^*} = 1 + \frac{2}{m} \frac{|p|^2}{E_1 - E_2} = 1 - \frac{2}{m} \frac{|p|^2}{E_g}, \tag{3.90}$$

and $m_1^* < 0$, since the absolute value of the second term is usually greater than unity.

If E_g is very small, such that the second term on the right sides of Eqs. (3.89) and (3.90) dominates, then $|m_1^*| = |m_2^*|$. This model gives the main dependence of the effective masses on bandgaps, i.e. small gaps imply small masses. There are applications of the f-sum rule to interpret optical properties, since the momentum matrix elements can be related to optical transition probabilities.

Next, we return to the example of the Kronig–Penney model and examine the velocities and effective masses. Consider the repulsive potential case and the calculations of the electron effective mass and velocity near $k = 0$ and $k = \pi/a$. For the first band (Fig. 3.4), $E(k = 0) \equiv E_0$, which is a constant depending on the potential. At $k = \frac{\pi}{a}$,

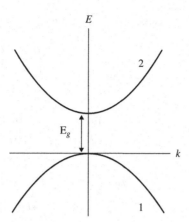

E

E_g

k

2

1

Figure 3.9 Two bands separated by a gap E_g (the two-band model).

$E_1 = \pi^2 \hbar^2 / 2ma^2$, which is independent of the potential. Let us begin with $k = 0$, and letting $\alpha_0 \equiv \alpha(k = 0)$, we expand Eq. (3.31) about this point in a Taylor series

$$1 - \frac{1}{2}K^2 = f(\alpha_0) + f'(\alpha_0)(\alpha - \alpha_0). \tag{3.91}$$

Since $f(\alpha_0) = 1$, this gives

$$\alpha = \alpha_0 - \frac{1}{2}\frac{K^2}{f'(\alpha_0)}. \tag{3.92}$$

The energy E can now be expressed in terms of α:

$$
\begin{aligned}
E &= \frac{\hbar^2 \alpha^2}{2ma^2} = \frac{\hbar^2}{2ma^2}\left[\alpha_0 - \frac{1}{2}\frac{K^2}{f'(\alpha_0)}\right]^2 \\
&= \frac{\hbar^2 \alpha_0^2}{2ma^2} - \frac{\hbar^2 \alpha_0}{2mf'(\alpha_0)}k^2.
\end{aligned} \tag{3.93}
$$

If we now express E using an effective mass, we have

$$E = E_0 + \frac{\hbar^2}{2m^*}k^2 \tag{3.94}$$

and

$$\left.\frac{m^*}{m}\right|_{k \to 0} = -\frac{f'(\alpha_0)}{\alpha_0}. \tag{3.95}$$

As shown in Fig. 3.3, the slope of $f(\alpha_0), f'(\alpha_0)$ is negative. This means that $m^* > 0$. As P increases, $f(0) = 1 + P$ gets larger, and the slope at $f(\alpha_0)$ is larger. Hence, large P means that m^* is large, i.e. when the potential is stronger, the band is flatter.

We can also compute $v(k = 0)$ using

$$v = \frac{1}{\hbar} \frac{\partial E}{\partial \alpha} \frac{\partial \alpha}{\partial K} \frac{\partial K}{\partial k}. \tag{3.96}$$

Now,

$$\left. \frac{\partial \alpha}{\partial K} \right|_{\alpha \to \alpha_0} = -\frac{\sin K}{f'(\alpha_0)} = \frac{-K}{f'(\alpha_0)} \tag{3.97}$$

for small K. Since $E = \frac{\hbar^2 \alpha^2}{2ma^2}$,

$$v = \left. \frac{-\hbar \alpha_0 k}{mf'(\alpha_0)} \right|_{k \to 0}. \tag{3.98}$$

We obtain this same result if we use Eq. (3.95) and the expression for the velocity $v = \hbar k/m^*$. Similarly, for the top of the first band,

$$\left. \frac{m^*}{m} \right|_{k \to \pi/a} = \frac{f'(\pi)}{\pi} = \frac{-P}{\pi^2}. \tag{3.99}$$

Since $f'(\pi)$ is negative, m^* is negative. If P is large, again $f'(\pi)$ is large and m^* is large. If P is small, the gap to the next band is small, and $f(\alpha)$ must have a smaller slope; that is, the turning point of $f(\alpha)$ is closer to the -1 line. This is consistent with the statement that near a small bandgap the masses are small.

The velocity at $k = \pi/a$ can also be computed. Near $K = \pi - \Delta K, f(\alpha) = \cos(\pi - \Delta K)$, where $\Delta K \to 0$ and

$$\frac{\partial \alpha}{\partial K} = \frac{\Delta K}{f'(\pi)}, \tag{3.100}$$

yielding

$$v = \left. \frac{\hbar \pi^2 \Delta K}{mPa} \right|_{k = \pi/a - \Delta K/a}. \tag{3.101}$$

3.6 Periodic boundary conditions and summing over band states

In the previous sections, $E_n(\mathbf{k})$ was examined in general and for specific models in various limits. At this point, we assume that $E_n(\mathbf{k})$ is known and explore the problem of using this function in evaluating properties such as response functions, densities of states, etc. Calculations of this type can involve sums or integrals over states. Because crystals have a finite size, the states associated with the electronic or vibrational properties (to be discussed in the next chapter) do not have a continuous range but are characterized by discrete wavevectors. As discussed in Chapter 2, periodic boundary conditions (PBCs) can be imposed as

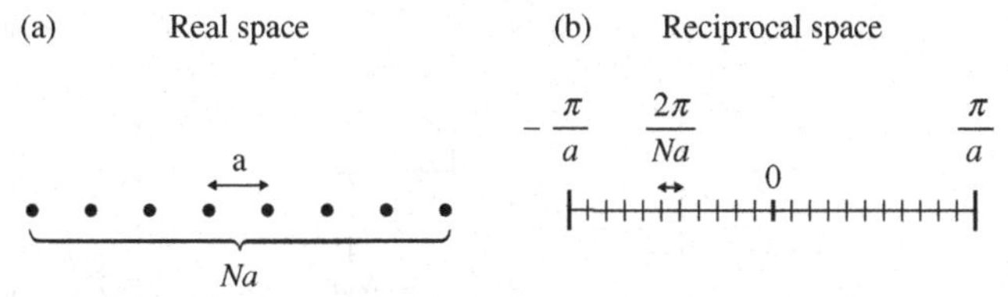

(a)   Real space   (b)   Reciprocal space

Figure 3.10 (a) One-dimensional crystal of length Na. (b) Representative cell in k-space arising from finite sample size through PBCs.

a convenient method to account for finite crystal size. This approach leads to a straightforward method for counting states and for doing sums and integrals over **k**-states in the Brillouin zone.

In one dimension, PBCs imply that for a crystal of N cells of lattice constant a,

$$\Psi(x + Na) = \Psi(x). \tag{3.102}$$

It is as if the one-dimensional line were joined into a circle. If we next use the Bloch condition, then

$$\Psi(x + Na) = e^{ikNa}\Psi(x),$$

and therefore,

$$e^{ikNa} = 1. \tag{3.103}$$

This gives the following condition for the allowed k-states:

$$k = \frac{2\pi}{Na}n \quad (n = 0, \pm 1, \pm 2, \ldots). \tag{3.104}$$

For one dimension, the Brillouin zone is defined by

$$-\frac{G}{2} < k \le \frac{G}{2}, \tag{3.105}$$

where G is the reciprocal lattice vector $2\pi/a$ and the allowed k-states are discrete, with n constrained by

$$-\frac{N}{2} < n \le \frac{N}{2}. \tag{3.106}$$

Therefore, there are N k-states in the Brillouin zone spaced $\frac{2\pi}{Na}$ apart, as shown in Fig. 3.10. For a long crystal, N is large, and there are many states that are close together in k-space. A similar plot is shown for two dimensions in Fig 3.11.

In three dimensions, we have

$$\Psi(\mathbf{r} + N_i\mathbf{a}_i) = \Psi(\mathbf{r}), \quad i = 1, 2, 3, \tag{3.107}$$

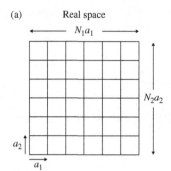
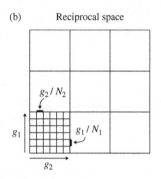

Figure 3.11 (a) Two-dimensional crystal of size $N_1 a_1$ and $N_2 a_2$. (b) Reciprocal lattice with **k**-space mesh arising from finite sample size.

where N_i is the number of cells in the direction $\hat{a}_i$ that form the crystal. Using the Bloch condition, we have

$$\exp(i\mathbf{k} \cdot N_i \mathbf{a}_i) = 1,$$

and

$$\mathbf{k} \cdot N_i \mathbf{a}_i = 2\pi m_i, \tag{3.108}$$

where $m_i = 0, \pm 1, \pm 2, \ldots$

If we express **k** in terms of primitive basis vectors in the reciprocal lattice, then

$$\mathbf{k} = \ell_1 \mathbf{g}_1 + \ell_2 \mathbf{g}_2 + \ell_3 \mathbf{g}_3, \tag{3.109}$$

where the ℓ_i's are not integers, and using

$$\mathbf{g}_i \cdot \mathbf{a}_j = 2\pi \delta_{ij} \tag{3.110}$$

we find $\mathbf{k} \cdot N_i \mathbf{a}_i = 2\pi N_i \ell_i = 2\pi m_i$. For a large crystal, N_i is large and the values of **k** are closely spaced, that is,

$$\ell_i = \frac{m_i}{N_i}, \tag{3.111}$$

such that

$$\mathbf{k} = \sum_{i=1}^{3} \ell_i \mathbf{g}_i = \sum_{i=1}^{3} m_i \left(\frac{\mathbf{g}_i}{N_i} \right). \tag{3.112}$$

Since $|\mathbf{g}| \sim \frac{1}{|a|}$, the spacing is of the order of the inverse of the crystal size. The mesh for **k** is very fine (Fig. 3.12) and depends only on the number of unit cells in the sample we are working with. Effects arising from finite size effects are dominant in small-particle research. (See Chapter 16.)

Using the above description for three-dimensional systems, we can count the number of **k**-states in a Brillouin zone. This number is the volume of the Brillouin zone divided by

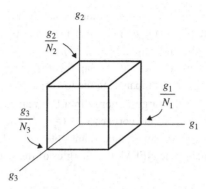

Figure 3.12 Three-dimensional cube illustrating the fine mesh for **k**-states in the Brillouin zone arising from finite sample size.

the volume of a cell containing only one state, that is,

$$\frac{\mathbf{g}_1 \cdot (\mathbf{g}_2 \times \mathbf{g}_3)}{\frac{\mathbf{g}_1}{N_1} \cdot \left(\frac{\mathbf{g}_2}{N_2} \times \frac{\mathbf{g}_3}{N_3}\right)} = N_1 N_2 N_3 = N. \tag{3.113}$$

Hence, the total number of **k**-states in the Brillouin zone equals the number of unit cells in the crystal N.

A common problem in solid state physics consists in summing a function of wavevector $f(\mathbf{k})$ over **k**-states in the Brillouin zone. If the samples are large, the **k**-states are close together, and the sum over states can be taken as an integral, that is,

$$\sum_{\mathbf{k}} f(\mathbf{k}) = \frac{\Omega_x}{(2\pi)^3} \int f(\mathbf{k}) d^3 k, \tag{3.114}$$

where Ω_x is the crystal volume and $\Omega_x = N\Omega_c$, where Ω_c is the cell volume. If we again ask how many states are contained in the Brillouin zone, the above expression can be used with $f(\mathbf{k}) = 1$ to obtain

$$\frac{\Omega_x}{(2\pi)^3} \int_{BZ} d^3 k = \frac{\Omega_x \Omega_{BZ}}{(2\pi)^3} = \frac{\Omega_x}{\Omega_c} = N. \tag{3.115}$$

3.7 Energy bands for materials

Using the formalism obtained thus far in this chapter, it is possible to give an explanation of the results shown in Table 1.1 concerning the variation in the resistance of materials and their classification as metals, semiconductors and/or insulators, and semimetals. One more aspect of energy band structure is needed, and this is the concept of overlapping bands. For one dimension, without accidental degeneracies, this is not possible, but it is allowed in two and three dimensions, as shown next.

Consider a two-dimensional square lattice with $E(\mathbf{k})$ having gaps at the Brillouin zone boundary. To show that the first band can have a higher energy than the second, or can "overlap" it, consider the situation shown in Fig. 3.13. There will be gaps between the bands at the boundary of the zone. For example, along k_x, a gap will appear at $(\frac{\pi}{a}, 0)$ between band 1 and band 2 (Fig. 3.13).

To decide about possible overlap for this example, we can consider the highest energy state in band 1 and compare it to the lowest energy state in band 2. For an FEM, the largest **k** will produce the largest energy in a band since $E_0(\mathbf{k}) = \frac{\hbar^2 k^2}{2m}$, and the largest **k** in the Brillouin zones is at the corner $\mathbf{k}_1 = (\frac{\pi}{a}, \frac{\pi}{a})$. The lowest energy in band 2 would then be at $\mathbf{k}_2 = (\frac{\pi}{a}, 0)$. Assuming a constant potential value V for the Fourier components in Eq. (3.38) and using the NFEM, the energies of the two states $E_n(\mathbf{k})$ for bands $n = 1, 2$ are

$$E_1(\mathbf{k}_1) = E_0(\mathbf{k}_1) + V_0 - |V| = \frac{\hbar^2}{2m} 2\left(\frac{\pi}{a}\right)^2 + V_0 - |V| \qquad (3.116)$$

and

$$E_2(\mathbf{k}_2) = E_0(\mathbf{k}_2) + V_0 + |V| = \frac{\hbar^2}{2m}\left(\frac{\pi}{a}\right)^2 + V_0 + |V|, \qquad (3.117)$$

where $|V|$ is one half of the gap between the bands. Overlap occurs if $E_2 < E_1$, or if

$$\frac{\hbar^2}{2m}\left(\frac{\pi}{a}\right)^2 > 2|V|. \qquad (3.118)$$

This situation occurs if the potential V is weak compared to the kinetic energy. This is a particularly easy condition to satisfy for elongated Brillouin zone shapes as the kinetic energy difference term becomes larger. Since we can demonstrate the possibility of overlap for two-dimensional cases, it is easy to extend this argument to the possibility of overlap in three dimensions.

We now have enough information to examine the periodic table, in order to look at elements and compounds that crystallize in distinct structures in an attempt to give a "broad-brush" argument for classifying materials in terms of their band structure, and

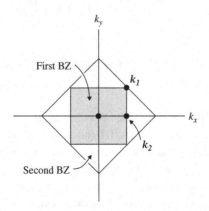

Figure 3.13   Two-dimensional Brillouin zones, showing that overlapping bands can exist in two dimensions.

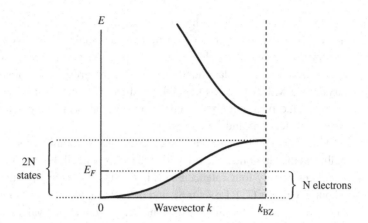

Figure 3.14 Schematic band structure for bcc Na, and illustration of the counting of states in a band.

to understand the classification of materials as metals, semiconductors, insulators, and semimetals. Essentially, it is the concepts of energy bands, the number of states in a band, bandgaps, and band overlap that allow us to explain the electronic properties of many solids. We shall for now ignore the effects of many-electron interaction, which can play an important role in highly correlated electron systems.

The basic argument is that if a band is partially filled, then there are states close by in energy into which an electron can scatter with negligible energy. If a band is full and a bandgap exists, then it takes a non-negligible energy to excite the electron into the next band. If bands overlap, then the situation is closer to that of a partially filled band. Hence, we expect full bands with bandgaps above them to lead to semiconductors and insulators, while partially filled bands and overlapping bands result in metals and semimetals.

We have shown that there are N **k**-states in a Brillouin zone for a specific band, where N is the number of cells in the crystal. Since electrons are fermions, we can put two electrons in each state because of the spin quantum number. Hence, two electrons per cell can be accommodated by each band. For valence 1 systems, we have one electron per atom, and, if there is one atom in a cell, then a band will be half-filled.

Using Na, we can illustrate this as an example of a valence 1 system. For Na in a bcc lattice, there is one atom per unit cell. Each atom has 10 core electrons ($1s^2 2s^2 2p^6$) and one valence electron $3s^1$. Hence, there are N valence electrons in the solid, which can be accommodated by 1/2 of a band, as shown in Fig. 3.14. The valence electrons fill the band to the Fermi level E_{F}. The next unoccupied state that can accommodate an electron, if the system is excited, is $\approx E_{\mathrm{F}}/N$ in energy above the last occupied state. For Na and other typical metals of macroscopic size, this is an infinitesimally small energy and lower than any achievable temperature. Therefore, alkali metals that have one valence electron conduct easily unless the systems are extremely small. The same argument can be made for noble metals like Cu, Ag, and Au in the fcc structure and similarly for valence 3 elements like Al in the fcc structure. In the latter cases, 3/2 bands will be filled.

For valence 2 metals, such as Be, Mg, Zn, and Cd, in the hcp structure with two atoms per cell, one expects four electrons per cell to completely fill two bands and produce semiconductors or insulators. However, overlap causes partial occupancy of bands and metallic characteristics. A similar situation exists for the group V elements As, Sb, and Bi, which crystallize with two atoms per cell, yielding 10 electrons in a cell. Here, a slight overlap between the fifth and sixth bands allows partial occupancy and semimetallic properties with free electrons and free holes.

The group IV elements are of special interest, particularly the first four, which can crystallize in the diamond structure with two atoms per cell. For this structure, the eight valence electrons occupy four valence bands in C, Si, Ge, and Sn, and there is a minimum gap of about 5.5 eV for C, 1.1 eV for Si, and 0.76 eV for Ge. For Sn, called αSn in this structure, the gap is zero. To thermally excite an electron out of a covalent bond (or occupied states) in C, Si, or Ge requires a Boltzmann factor of $e^{\frac{-E_g}{k_B T}}$. At room temperature, which is about 0.025 eV, this inhibits conduction, unlike the case for metals or semimetals. Materials with gaps of order 5 eV, such as diamond, conduct very little and are insulators, whereas gaps of order 1 eV yield semiconductors. Systems like αSn and graphene have no gap and have been called ideal semimetals or zero-gap semiconductors. The details of the transport of electrons and holes in these materials are complex. Semiconductors and some insulators can be doped with excess carriers to increase their conductivity. In some cases, the doping can be so high that these systems resemble metals with carriers obeying Fermi–Dirac statistics. These are called degenerate semiconductors.

For compounds, the above rules are generally applicable. For example, GaAs and ZnSe can crystallize in the zincblende structure, which, like the diamond structure, has two atoms per cell. Since the total number of valence electrons is eight, just as in Ge, there will be four filled valence bands and a gap to the next unoccupied state that is the bottom of the conduction band.

The group IV, III–V, and II–VI semiconductors provide a good example of how band structure calculations can be done for real materials. To illustrate this, we choose the series Ge, GaAs, and ZnSe. These materials have approximately the same lattice constant, and, as mentioned before, they have similar structures with eight valence electrons per unit cell. If we take a nearly-free electron type approach and assume we can obtain the electron–core potential from experiment, the picture of the solid is similar to what is shown in Fig. 1.1, except that the valence electrons concentrate between atoms and form bonds.

Constructing an appropriate potential for valence electron–core interactions is possible. Since the core electrons are left out of the problem, this is called the pseudopotential approach. When the pseudopotential is calculated from first principles, the resulting potential is called an *ab initio* pseudopotential. This will be discussed in later chapters. The empirical pseudopotential method (EPM)[2] involves extraction of the potential from experimental data, such as those from measuring the optical properties. This is the approach we will discuss here.

[2] M. L. Cohen and T. K. Bergstresser, "Band structures and pseudopotential form factors for fourteen semiconductors of the diamond and zincblende structures," *Phys. Rev.* 141(1966), 789.

We begin by assuming that the potential in the solid is composed of a linear combination of atomic potentials,

$$V_p(\mathbf{r}) = \sum_{\mathbf{R},\tau} V_a(\mathbf{r} - \mathbf{R} - \boldsymbol{\tau}), \tag{3.119}$$

where $V_a(\mathbf{r} - \mathbf{R} - \boldsymbol{\tau})$ is now the appropriate atomic pseudopotential, to be obtained by fitting experimental data. Hence,

$$V_p = \sum_{\mathbf{R},\tau} V_p^a(\mathbf{r} - \mathbf{R} - \boldsymbol{\tau}_\ell). \tag{3.120}$$

Because of its lattice periodicity, the potential can be expanded in reciprocal space as

$$V_p(\mathbf{r}) = \sum_{\mathbf{G}} V_p(\mathbf{G}) e^{i\mathbf{G}\cdot\mathbf{r}}, \tag{3.121}$$

where

$$V_p(\mathbf{G}) = \frac{1}{\Omega_x} \int V_p(\mathbf{r}) e^{-i\mathbf{G}\cdot\mathbf{r}} d\mathbf{r} = \frac{1}{\Omega_x} \sum_{\mathbf{R},\tau} \int V_p^a(\mathbf{r} - \mathbf{R} - \boldsymbol{\tau}) e^{-i\mathbf{G}\cdot\mathbf{r}} d\mathbf{r}, \tag{3.122}$$

where Ω_x is the crystal volume. Hence,

$$V_p(\mathbf{G}) = \frac{1}{\Omega_x} N \sum_{\tau} e^{-i\mathbf{G}\cdot\tau} \int V_p^a(\mathbf{r}) e^{-i\mathbf{G}\cdot\mathbf{r}} \, d\mathbf{r}. \tag{3.123}$$

For w atoms per cell, using Ω_a as the volume occupied by one atom on average,

$$V_p(\mathbf{G}) = \frac{1}{w\Omega_a} \sum_{\tau} e^{-i\mathbf{G}\cdot\tau} \int V_p^\tau(\mathbf{r}) e^{-i\mathbf{G}\cdot\mathbf{r}} d\mathbf{r}. \tag{3.124}$$

Since the atoms in the basis can be different, we here use the notation of $V_p^\tau(\mathbf{r})$ to denote the potential of the atom on site $\boldsymbol{\tau}$. The Fourier transform $\frac{1}{\Omega_a} \int V_p^\tau(\mathbf{r}) e^{-\mathbf{G}\cdot\mathbf{r}} d\mathbf{r} = V_\tau(\mathbf{G})$ is radially symmetric, so we choose $V_\tau(|\mathbf{G}|)$ as our form factor for the atom

$$V_p(\mathbf{G}) = \frac{1}{w} \sum_{\tau} e^{-i\mathbf{G}\cdot\tau} V_\tau(|\mathbf{G}|). \tag{3.125}$$

We can also separate the atoms into the different types ℓ and write

$$V_p(\mathbf{G}) = \sum_{\ell} S_\ell(\mathbf{G}) V_\ell(|\mathbf{G}|), \tag{3.126}$$

with

$$S_\ell(\mathbf{G}) = \frac{1}{w_\ell} \sum_{\tau_\ell} e^{-i\mathbf{G}\cdot\tau_\ell}, \tag{3.127}$$

where w_ℓ is the number of atoms of the ℓth type in the unit cell. For example, for Ge, in the diamond structure with two atoms per cell,

$$S(\mathbf{G}) = \tfrac{1}{2}\left(e^{-i\mathbf{G}\cdot\boldsymbol{\tau}} + e^{i\mathbf{G}\cdot\boldsymbol{\tau}}\right) = \cos(\mathbf{G}\cdot\boldsymbol{\tau}). \qquad (3.128)$$

Solving for $E_n(\mathbf{k})$, we expand the states in planewaves

$$\phi_{\mathbf{k}} = \sum_{\mathbf{G}} \alpha_{\mathbf{k}}(\mathbf{G}) e^{i(\mathbf{k}+\mathbf{G})\cdot\mathbf{r}} \qquad (3.129)$$

and solve ($H_p = \frac{\mathbf{p}^2}{2m} + V_p(\mathbf{r})$),

$$H_p\,|\phi\rangle = E\,|\phi\rangle, \qquad (3.130)$$

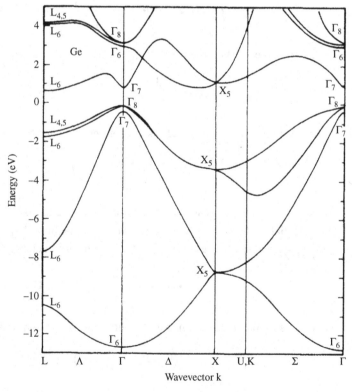

Figure 3.15 Band structure of germanium calculated using the EPM. The zero-energy reference is at the valence band maximum. The labels denote **k**-points in the fcc Brillouin zone corresponding to the high symmetry points $\Gamma(0,0,0)$, $L(\tfrac{1}{2},\tfrac{1}{2},\tfrac{1}{2})$, $X(0,1,0)$, $U(\tfrac{1}{4},1,\tfrac{1}{4})$, and $K(\tfrac{3}{4},\tfrac{3}{4},0)$, (in units of $\tfrac{2\pi}{a}$) and the symmetry lines Λ, Δ, and Σ connect the symmetry points.

which, when expressed in matrix form, is

$$
\left(H_{\mathbf{GG'}}(\mathbf{k}) \right)
\begin{pmatrix} \alpha_{\mathbf{k}}(\mathbf{G}_1) \\ \cdot \\ \cdot \\ \cdot \\ \alpha_{\mathbf{k}}(\mathbf{G}_N) \end{pmatrix}
= E
\begin{pmatrix} \alpha_{\mathbf{k}}(\mathbf{G}_1) \\ \cdot \\ \cdot \\ \cdot \\ \alpha_{\mathbf{k}}(\mathbf{G}_N) \end{pmatrix}.
\tag{3.131}
$$

The Hamiltonian matrix (in atomic units, $e = \hbar = m = 1$) becomes

$$
H_{\mathbf{GG'}}(\mathbf{k}) = \tfrac{1}{2}(\mathbf{k} + \mathbf{G})^2 \delta_{\mathbf{GG'}} + V_p(\mathbf{G} - \mathbf{G'}).
\tag{3.132}
$$

For the diamond structure, we consider the lowest $\mathbf{G}$'s corresponding to (in units of $2\pi/a$)

$$
G^2 = 0, 3, 4, 8, 11, 12, \ldots
\tag{3.133}
$$

Since $V_p(G^2 = 0) = $ constant and $S(\mathbf{G}) = 0$ for $G^2 = 4$ and 12, the first three non-trivial $V_p(G^2)$'s are $V_p(3)$, $V_p(8)$, and $V_p(11)$. Fitting to experiments for Ge yields $V_p(3) = -0.23$ Ry, $V_p(8) = 0.01$ Ry, and $V_p(11) = 0.06$ Ry. Only these three form factors are needed to compute the band structure of Ge, if we neglect the higher components.

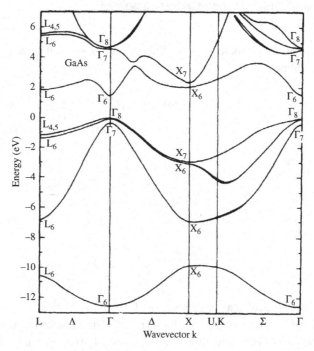

Figure 3.16   Band structure of gallium arsenide calculated using the EPM. The valence band maximum is taken to be the zero of energy.

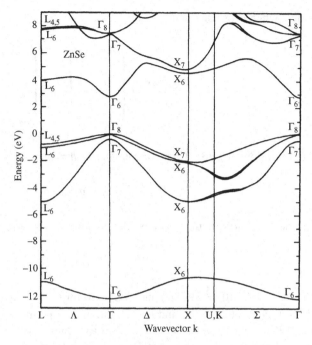

Figure 3.17 Band structure for zinc selenide calculated using the EPM. The energy zero is taken to be the top of the valence band. The Zn $3d$ level is not explicitly included in the pseudopotential calculation. Experimentally, this level occurs at approximately 9 eV below the valence band maximum.

For GaAs, there are two different atoms per cell (τ and $-\tau$ point to different atoms) yielding

$$V(\mathbf{G}) = \tfrac{1}{2}\left(e^{i\mathbf{G}\cdot\boldsymbol{\tau}} V_{\mathrm{Ga}} + e^{-i\mathbf{G}\cdot\boldsymbol{\tau}} V_{\mathrm{As}}\right)$$

$$= V_S \cos \mathbf{G}\cdot\boldsymbol{\tau} + iV_A \sin \mathbf{G}\cdot\boldsymbol{\tau}, \qquad (3.134)$$

where

$$V_S = \frac{V_{\mathrm{Ga}} + V_{\mathrm{As}}}{2} \qquad (3.135)$$

and

$$V_A = \frac{V_{\mathrm{Ga}} - V_{\mathrm{As}}}{2}. \qquad (3.136)$$

The same calculation can be done for ZnSe by letting $V_{\mathrm{Ga}} \to V_{\mathrm{Zn}}$ and $V_{\mathrm{As}} \to V_{\mathrm{Se}}$. Excellent band structures using the EPM for GaAs and ZnSe can be obtained by assuming $V_S = V_{\mathrm{Ge}}$ for both GaAs and ZnSe, and choosing $V_A(G^2 = 3), V_A(G^2 = 4), V_A(G^2 = 8)$, and $V_A(G^2 = 11)$ to be $(0.07, 0.05, 0, 0.01)$ Ry for GaAs and $(0.18, 0.12, 0, 0.03)$ Ry for ZnSe. The resulting band structures, using the EPM, are shown in Figs. 3.15, 3.16, and 3.17.

4 Lattice vibrations and phonons

4.1 Lattice vibrations

Until now, we have considered model systems made up of atomic cores at fixed equilibrium positions given by

$$\mathbf{R}_{\ell,j}^0 = \mathbf{R}_\ell^0 + \boldsymbol{\tau}_j^0, \tag{4.1}$$

where $\mathbf{R}_\ell^0$ locates the lattice points, and $\boldsymbol{\tau}_j^0$ is the basis vector giving the position of the jth atom in the basis.

Lattices vibrate because of temperature and external fields. Even at zero temperature, zero-point motion will move the cores from their equilibrium positions. To account for the motion, we denote the displacement of a core by $\boldsymbol{\xi}_{\ell j}(t)$, and then the core position at time t is

$$\mathbf{R}_{\ell,j}(t) = \mathbf{R}_{\ell,j}^0 + \boldsymbol{\xi}_{\ell,j}(t). \tag{4.2}$$

For most excitations, the $|\boldsymbol{\xi}|$ displacements are much less than a lattice constant, but this assumption breaks down near the melting point or for very light atoms like hydrogen and helium even at lower temperature.

We now consider the energy of a vibrating lattice. The energy can be written as the sum of the kinetic energy and potential energy for the cores, $H = T + U$, where

$$T = \sum_n \frac{\mathbf{P}_n^2}{2M_n} \tag{4.3}$$

and

$$U = U(\{\mathbf{R}_{\ell,j}\}). \tag{4.4}$$

To assure crystal stability, $U_0 = U(\{\mathbf{R}_{\ell,j}^0\})$ is a minimum, i.e. at the equilibrium structure,

$$\nabla U = 0, \tag{4.5}$$

$$\nabla^2 U > 0. \tag{4.6}$$

Using a Taylor series expansion about the equilibrium positions,

$$U = U_0 + \frac{1}{2} \sum_{\substack{n,\alpha \\ n',\alpha'}} \frac{\partial^2 U}{\partial \zeta_n^\alpha \partial \zeta_{n'}^{\alpha'}} \zeta_n^\alpha \zeta_{n'}^{\alpha'} + \cdots , \tag{4.7}$$

where $\alpha, \alpha' = x, y, z$, and $n, n' = (\ell, j)$ is the site index. If we let $U_0 = 0$ and truncate the series after second order in the so-called harmonic approximation, then

$$U = \frac{1}{2} \sum_{\substack{n,\alpha \\ n',\alpha'}} A_{nn'}^{\alpha\alpha'} \zeta_n^\alpha \zeta_{n'}^{\alpha'} . \tag{4.8}$$

From the equation of motion,

$$\dot{\mathbf{P}}_n = -\nabla_{\xi_n} U, \tag{4.9}$$

we have

$$M_n \ddot{\zeta}_n^\alpha = -\sum_{n'\alpha'} A_{nn'}^{\alpha\alpha'} \zeta_{n'}^{\alpha'} . \tag{4.10}$$

Hence, $A_{nn'}^{\alpha\alpha'}$ is the atomic force constant in the α direction on the nth atom when the n'th atom is displaced in the α' direction.

In the harmonic approximation (i.e. terminating Eq. (4.7) in second order of the displacements), we may think of the crystal as a collection of atoms connected by springs and the A's as the spring constants. The atomic force constant matrix A has some useful symmetry properties. It can be shown that A is a symmetric matrix, it depends only on $\mathbf{R}_\ell - \mathbf{R}_{\ell'}$ owing to crystalline periodicity, and it also has the group symmetry properties of the crystal.

We can now explore the normal modes of vibration. In a mechanical system, the number of degrees of freedom $= d \times r \times N$, where d is the number of dimensions, r is the number of basis atoms, and N is the number of unit cells. To solve for the normal modes, we look for solutions in which each atom of mass M_j is moving periodically with the same frequency ω,

$$\zeta_{\ell j}^\alpha(t) = C_{\ell j}^\alpha \frac{1}{\sqrt{M_j}} e^{-i\omega t}, \tag{4.11}$$

where C is the amplitude of vibration. Using the equation of motion (Eq. (4.10)), we obtain

$$\omega^2 C_{\ell j}^\alpha = \sum_{\ell' j' \alpha'} D_{jj'}^{\alpha\alpha'} (\mathbf{R}_\ell - \mathbf{R}_{\ell'}) C_{\ell' j'}^{\alpha'}, \tag{4.12}$$

where

$$D_{jj'}^{\alpha\alpha'} (\mathbf{R}_\ell - \mathbf{R}_{\ell'}) = \frac{A_{jj'}^{\alpha\alpha'} (\mathbf{R}_\ell - \mathbf{R}_{\ell'})}{\sqrt{M_j M_{j'}}} \tag{4.13}$$

is a real $drN \times drN$ symmetric matrix. Thus, there exist drN real eigenvalues ω_i^2. Physically, ω^2 should be positive. However, if for some reason a particular mode has $\omega_i \to 0$, it would suggest a possible structural phase transition, since this indicates a vanishing restoring face for distortions given by this mode.

Analogous to the electron case, using periodic translation symmetry, we seek solutions in Bloch form. For a given $\mathbf{q}$,

$$C_{\ell j}^{\alpha} = C_j^{\alpha} \, e^{i\mathbf{q} \cdot (\mathbf{R}_\ell^0 + \tau_j)}. \tag{4.14}$$

Equation (4.12) becomes

$$\omega^2 C_j^{\alpha} = \sum_{j'\alpha'} \underbrace{\left[e^{-i\mathbf{q} \cdot (\tau_j - \tau_{j'})} \sum_{\ell} D_{jj'}^{\alpha\alpha'}(\mathbf{R}_\ell - \mathbf{R}_{\ell'}) e^{-i\mathbf{q} \cdot (\mathbf{R}_\ell^0 - \mathbf{R}_{\ell'}^0)} \right]}_{D_{jj'}^{\alpha\alpha'}(\mathbf{q})} C_{j'}^{\alpha'}. \tag{4.15}$$

Equation (4.15) is a set of $d \times r$ coupled equations. The matrix $D_{jj'}^{\alpha\alpha'}(\mathbf{q})$ is called the dynamical matrix. We arrive at

$$\omega^2 C_j^{\alpha} = \sum_{j'\alpha'} D_{jj'}^{\alpha\alpha'}(\mathbf{q}) C_{j'}^{\alpha'}. \tag{4.16}$$

This is an $rd \times rd$ matrix eigenvalue problem. Hence, lattice periodicity reduces our initial system of drN equations to a system of rd equations. We arrive at the solutions $\omega_\lambda(\mathbf{q})$, where $\lambda = 1, \ldots, rd$ (the number of phonon branches) with $\mathbf{q}$ in the first Brillouin zone, where there are N $\mathbf{q}$-points allowed, as discussed in the previous chapter.

As an example, let us reexamine the familiar model of a one-dimensional chain of atoms with a one-atom basis (Fig. 4.1) and nearest-neighbor connecting springs with force constant γ.

The solution to this model[1] for a simple spring model derivation is

$$\omega^2(q) = \frac{4\gamma}{M} \sin^2\left(\frac{qa}{2}\right) = \frac{2\gamma}{M}(1 - \cos qa), \tag{4.17}$$

---

[1] C. Kittel, *Introduction to Solid State Physics*, 8th ed. (New York: Wiley, 2005).

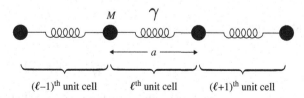

$$(\ell-1)^{\text{th}} \text{ unit cell} \qquad \ell^{\text{th}} \text{ unit cell} \qquad (\ell+1)^{\text{th}} \text{ unit cell}$$

Figure 4.1   Monatomic chain of atoms connected by springs.

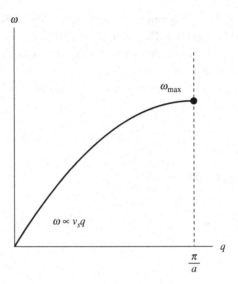

Figure 4.2   Acoustic phonon band structure for a monatomic chain of atoms.

which is shown in Fig. 4.2. We can explore this example using the dynamical matrix

$$D_{\ell\ell'} = \frac{A_{\ell\ell'}}{M} = \frac{\gamma}{M}\left\{-\delta_{\ell'-\ell,-1} - \delta_{\ell'-\ell,1} + 2\delta_{\ell-\ell',0}\right\}, \tag{4.18}$$

where δ's are Kronecker delta functions and, as in Eq. (4.13),

$$D_{\ell\ell'} = D(\ell - \ell'). \tag{4.19}$$

The Fourier transform of Eq. (4.18) gives

$$
\begin{aligned}
D(q) &= \sum_{\ell'} D(\ell - \ell')e^{-iq\cdot(R_\ell - R_{\ell'})} \\
&= \sum_{\ell'}\left\{-\frac{\gamma}{M}(\delta_{\ell'-\ell,-1} + \delta_{\ell'-\ell,1} - 2\delta_{\ell-\ell',0})\right\} \times e^{-iqa(\ell-\ell')} \\
&= -\frac{\gamma}{M}(e^{-iqa} + e^{iqa} - 2) = \frac{2\gamma}{M}(1 - \cos qa).
\end{aligned}
\tag{4.20}
$$

Then, from Eq. (4.16),

$$\omega^2 C = D(q)C, \tag{4.21}$$

which yields

$$\omega^2 = \frac{2\gamma}{M}(1 - \cos qa). \tag{4.22}$$

Hence, we obtain the same result as is found for the simple solution to the spring model using Newton's law.

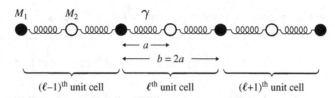

Figure 4.3 Diatomic chain of atoms connected by springs.

We note that $\omega = \omega_{\text{max}}$ when $qa = \pi$, i.e. $q = \pi/a$, which is the Brillouin zone boundary, and therefore $\omega_{\text{max}} = 2\sqrt{\gamma/M}$. The saturation of the dispersion curve results from the discreteness of the atomic arrangement in the solid, and q is only meaningful inside the first Brillouin zone. If we examine the long wavelength limit, then

$$\lim_{q \to 0} \omega^2 = \frac{2\gamma}{M} \frac{1}{2} (qa)^2. \tag{4.23}$$

Therefore,

$$\omega(q \to 0) = \sqrt{\frac{\gamma}{M}} aq = v_s q, \tag{4.24}$$

where

$$v_s = \sqrt{\frac{\gamma}{M}} a \tag{4.25}$$

represents the sound velocity.

Another example that is useful to explore is a one-dimensional diatomic chain (Fig. 4.3). The dynamical matrix $D_{jj'}(q)$ is a 2×2 matrix in this case. From Eq. (4.13), we find that

$$D_{11}(\ell - \ell') = \frac{2\gamma}{M_1} \delta_{\ell'-\ell,0}, \tag{4.26}$$

$$D_{22}(\ell - \ell') = \frac{2\gamma}{M_2} \delta_{\ell'-\ell,0}, \tag{4.27}$$

and

$$D_{12}(\ell - \ell') = \frac{-\gamma}{\sqrt{M_1 M_2}} (\delta_{\ell'-\ell,0} + \delta_{\ell'-\ell,-1}). \tag{4.28}$$

Thus, the symmetric dynamical matrix is

$$D(q) = \begin{pmatrix} \dfrac{2\gamma}{M_1} & -\dfrac{2\gamma}{\sqrt{M_1 M_2}} \cos qa \\ -\dfrac{2\gamma}{\sqrt{M_1 M_2}} \cos qa & \dfrac{2\gamma}{M_2} \end{pmatrix}, \tag{4.29}$$

and solving the eigenvalue problem $\omega^2 C = DC$ yields

$$\omega^2 = \gamma \left(\frac{1}{M_1} + \frac{1}{M_2} \right) \pm \gamma \sqrt{\left(\frac{1}{M_1} + \frac{1}{M_2} \right)^2 + \frac{-4\sin^2(qa)}{M_1 M_2}}. \qquad (4.30)$$

As expected, there are two solutions, and thus two vibrational branches (one is called acoustic and the other optical). This result can also be found using elementary methods. If we again examine the limiting cases, as $q \to 0$,

$$\omega_+^2 \to 2\gamma \left(\frac{1}{M_1} + \frac{1}{M_2} \right), \qquad (4.31)$$

$$\omega_-^2 \to \frac{2\gamma}{M_1 + M_2}(qa)^2. \qquad (4.32)$$

If $M_1 = M_2$,

$$\omega_- \to \sqrt{\frac{\gamma}{M}}qa, \qquad (4.33)$$

$$\omega_+ \to 2\sqrt{\frac{\gamma}{M}}, \qquad (4.34)$$

and the ω_- branch becomes that of the monatomic chain result (see Fig. 4.4).

At the zone boundary where $q = G/2 = \pi/b = \pi/2a$ (see Fig. 4.4), we find a gap. If we assume $M_2 > M_1$, then

$$\omega_+^2 = \frac{2\gamma}{M_1} \qquad (4.35)$$

and

$$\omega_-^2 = \frac{2\gamma}{M_2}. \qquad (4.36)$$

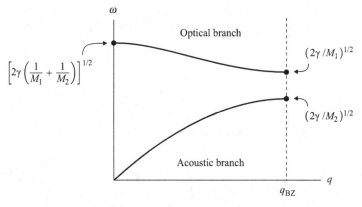

Figure 4.4　Lattice vibration dispersion curves for a diatomic chain of atoms with $M_2 > M_1$.

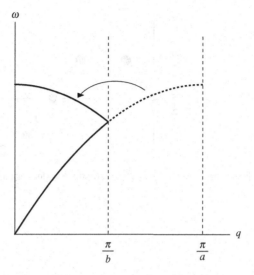

Figure 4.5 Mapping of the lattice vibration dispersion curves for a diatomic chain of equal masses from that of a monatomic chain.

However, if we set $M_1 = M_2$, then at the zone boundary the gap goes to zero.

When the two masses are equal, the periodicity of the system is half of the initial unit cell. So, if we consider a unit cell with half the initial periodicity, we would obtain a G that is twice the initial G. However, the dispersion curve will have only one acoustic branch. This leads to a dispersion curve where the upper half of the acoustic mode is folded back into an optical mode in the smaller Brillouin zone (Fig. 4.5).

We now examine the motion of the atoms of a one-dimensional diatomic chain. The normal mode equation has the form

$$\omega^2 \begin{pmatrix} C_1 \\ C_2 \end{pmatrix} = \begin{pmatrix} D_{11} & D_{12} \\ D_{21} & D_{22} \end{pmatrix} \begin{pmatrix} C_1 \\ C_2 \end{pmatrix}. \tag{4.37}$$

By solving the above eigenvalue problem, we obtain

$$\frac{C_1}{C_2} = \frac{D_{12}(q)}{\omega^2 - D_{11}(q)}. \tag{4.38}$$

The displacement of the atoms in terms of C_j is

$$\xi_{\ell j} = \frac{C_j}{\sqrt{M_j}} e^{iq \cdot (R_\ell + \tau_j) - i\omega t}. \tag{4.39}$$

Therefore, after using Eq. (4.38), the ratio of the displacement of the atoms is

$$\frac{\xi_1(q)}{\xi_2(q)} = \frac{\cos qa}{1 - \frac{\omega^2(q)}{2\gamma} M_1}. \tag{4.40}$$

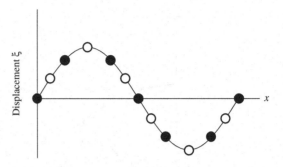

Figure 4.6   Atomic displacement versus distance *x* for an acoustic mode for a diatomic chain.

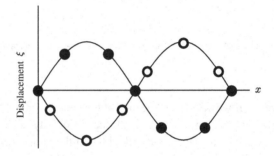

Figure 4.7   Atomic displacement versus distance *x* for an optical mode for a diatomic chain.

For the acoustic mode, as $q \rightarrow 0$, $\omega \rightarrow 0$, and thus $\xi_1/\xi_2 \rightarrow 1$. Hence, the displacements of the atoms in one unit cell are in phase and of similar magnitude as expected for the acoustic mode, as shown in Fig. 4.6.

For the optical mode, as $q \rightarrow 0$, $\omega_+^2 \rightarrow 2\gamma \left(1/M_2 + 1/M_1 \right)$, and thus $\xi_1/\xi_2 \rightarrow -M_2/M_1$. Hence, the oscillations are out of phase, and the lighter mass has a larger amplitude, as expected of two oppositely charged atoms in an EM wave where the two atoms move in opposite directions; hence the name optical mode (Fig. 4.7).

If we extend our study of the one-dimensional chain to higher dimensions, we can illustrate the procedure for counting branches. For a crystal of d dimensions with r atoms per unit cell, the total number of branches in the phonon dispersion curve is $r \times d$. Of these $r \times d$ branches, there are d acoustic modes and $d \times (r - 1)$ optical modes. For example, consider KBr in the rock salt structure. Since KBr has two atoms per unit cell, there are six branches in the KBr phonon band structure, three of which are acoustic and three of which are optical (Fig. 4.8). The physical origin of having d acoustic modes arises from translational invariance of space, i.e. displacing all atoms by the same amount along any direction in d-space results in no restore force and hence there are d branches with $\omega(q \rightarrow 0) = 0$.

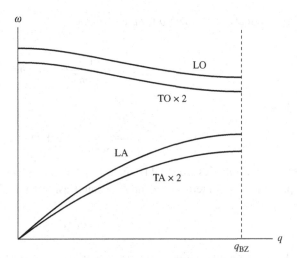

Figure 4.8   Lattice vibration dispersion curve for KBr. The transverse branches are doubly degenerate.

4.2 Second quantization and phonons

We now turn to an elementary excitation view of lattice vibrational properties by associating phonons with the vibrational modes. Phonons are viewed as being particle-like and wave-like in analogy with electromagnetic radiation. We begin by considering a simple one-dimensional harmonic oscillator, satisfying the equation

$$m\ddot{x} = -\gamma x. \tag{4.41}$$

The solution is $x(t) = x_0 e^{-i\omega t}$, where $\omega = \sqrt{\gamma/m}$ for spring constant γ and mass m. If we take a quantum view starting with the Hamiltonian

$$H = \frac{p^2}{2m} + \frac{1}{2}\gamma x^2, \tag{4.42}$$

then the energy levels are

$$E_n = \hbar\omega\left(n + \frac{1}{2}\right) \quad (n = 0, 1, 2, \dots). \tag{4.43}$$

It is convenient to take a second quantized view of this problem using creation and annihilation operators. The standard approach is to form a linear combination of the coordinate and momentum variables, x and p, and form new variables, a and $a^\dagger$, such that these variables satisfy the commutation relation

$$[a, a^\dagger] = 1. \tag{4.44}$$

The combinations of x and p for a and $a^\dagger$ are

$$a = \frac{1}{\sqrt{2\hbar\omega m}}(p - im\omega x) \tag{4.45}$$

and

$$a^\dagger = \frac{1}{\sqrt{2\hbar\omega m}}(p + im\omega x). \tag{4.46}$$

This choice satisfies Eq. (4.44) since $[x, p] = i\hbar$. Replacing x and p in Eq. (4.42) by expressing them in terms of a and $a^\dagger$, using Eqs. (4.45) and (4.46), yields

$$H = \frac{1}{2}\hbar\omega(aa^\dagger + a^\dagger a). \tag{4.47}$$

A matrix representation for a and $a^\dagger$ can be introduced based on labeling states by the index n. We have the following eigenvalue equations for the harmonic oscillator

$$a\,|n\rangle = \sqrt{n}\,|n-1\rangle \tag{4.48}$$

and

$$a^\dagger\,|n\rangle = \sqrt{n+1}\,|n+1\rangle, \tag{4.49}$$

which are consistent with Eq. (4.44). In addition, Eqs. (4.48) and (4.49) give

$$a^\dagger a\,|n\rangle = n\,|n\rangle, \tag{4.50}$$

and then, using Eq. (4.47), we have

$$H\,|n\rangle = \left(n + \frac{1}{2}\right)\hbar\omega\,|n\rangle = E_n\,|n\rangle. \tag{4.51}$$

From these essential features of the simple one-dimensional harmonic oscillator, we can propose a picture of excitations involving the creation and destruction of quanta in various states and the view of representing a state using the occupation number corresponding to specific quanta. This concept involves the setting up of a many-body state by operating on a vacuum state with creation and destruction operators to increase or decrease the number of particles in a given state. The state vector $\Psi_{n_1,n_2,\ldots,n_i\ldots}$ represents a state with n_1 particles in state 1, n_2 particles in state 2, and so on. The creation operators $a_i^\dagger$ repeatedly acting on the vacuum state $|0\rangle$ can produce this state vector. Also, when the destruction operator a_i operates on the state vector, it destroys a particle in the ith state. There is an overall factor such that

$$a_i\Psi_{n_1,n_2,\ldots,n_i\ldots} = \sqrt{n_i}\Psi_{n_1,n_2,\ldots,n_i-1\ldots}, \tag{4.52}$$

$$a_i^\dagger\Psi_{n_1,n_2,\ldots,n_i\ldots} = \sqrt{n_i+1}\Psi_{n_1,n_2,\ldots,n_i+1\ldots}, \tag{4.53}$$

$$a_i^\dagger a_i\Psi_{n_1,n_2,\ldots,n_i\ldots} = n_i\Psi_{n_1,n_2,\ldots,n_i\ldots}. \tag{4.54}$$

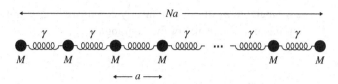

Figure 4.9 Finite linear chain with nearest-neighbor interactions.

Therefore, the operator $\hat{n}_i = a_i^{\dagger} a_i$ measures the number of particles in the ith state, and it is called the number operator.

Using this short description of second quantization and the occupation number representation for the state of a system, we can go beyond the simple harmonic oscillator and treat the model of interacting atoms, which we considered before, to represent lattice vibrations in a one-dimensional solid. This model uses a linear chain of N atoms of length $L = Na$ with nearest-neighbor interactions (Fig. 4.9). Unlike the harmonic oscillator case, the "extension of the springs" depends on the relative displacements of the atoms. Hence, a simple real-space Hamiltonian composed of a combination of the squares of position and momentum is not possible. However, a decoupling occurs if we use a Fourier transformation of the original Hamiltonian, and this will lead us to the form we need.

We begin with the Hamiltonian containing the kinetic and potential energies expressed in terms of real-space momenta p_n and coordinates ξ_n,

$$H = \frac{1}{2M} \sum_n p_n^2 + \frac{1}{2} \gamma \sum_n (\xi_{n+1} - \xi_n)^2, \tag{4.55}$$

where

$$\left[\xi_n, p_{n'} \right] = i\hbar \delta_{n,n'}. \tag{4.56}$$

As stated above, the desired form for the Hamiltonian is $H = \sum_k H_k$.

To accomplish this, we Fourier-transform the coordinate

$$\xi_n = \frac{1}{\sqrt{N}} \sum_k \xi_k e^{ikna}, \tag{4.57}$$

and

$$\xi_k = \frac{1}{\sqrt{N}} \sum_n \xi_n e^{-ikna}. \tag{4.58}$$

Using PBCs,

$$k = \frac{2\pi \ell}{Na} = \frac{2\pi \ell}{L}, \tag{4.59}$$

with $-\frac{N}{2} < \ell \leq \frac{N}{2}$. Since ξ_n is Hermitian,

$$\xi_k = \xi_{-k}^{\dagger}. \tag{4.60}$$

The momentum p_k that is canonically conjugate to ξ_k is not obvious. To find it, we use the Lagrangian formalism. For the Lagrangian

$$\mathcal{L}(\{\xi_k, \dot{\xi}_k\}),$$

the canonically conjugate momentum becomes

$$p_k = \frac{\partial \mathcal{L}}{\partial \dot{\xi}_k}.$$

Using

$$\sum_n e^{i(k+k')na} = N\delta_{k+k',0}, \tag{4.61}$$

then the kinetic energy is T, with

$$\begin{aligned} T &= \frac{M}{2} \sum_n (\dot{\xi}_n)^2 \\ &= \frac{M}{2} \frac{1}{N} \sum_{k,k',n} \dot{\xi}_k \dot{\xi}_{k'} e^{i(k+k')na} \\ &= \frac{M}{2} \sum_k \dot{\xi}_k \dot{\xi}_{-k}. \end{aligned} \tag{4.62}$$

Similarly, the potential energy (PE) becomes

$$\mathrm{PE} = \frac{1}{2}\gamma \sum_n (\xi_{n+1} - \xi_n)^2 = \gamma \sum_k \xi_k \xi_{-k}(1 - \cos ka), \tag{4.63}$$

and the Lagrangian has the form

$$\mathcal{L} = \frac{M}{2} \sum_k \dot{\xi}_k \dot{\xi}_{-k} - \gamma \sum_k \xi_k \xi_{-k}(1 - \cos ka). \tag{4.64}$$

Hence, the conjugate momentum is

$$p_k = \frac{\partial \mathcal{L}}{\partial \dot{\xi}_k} = M\dot{\xi}_{-k}. \tag{4.65}$$

Using Eq. (4.58),

$$\xi_{-k} = \frac{1}{\sqrt{N}} \sum_n \xi_n e^{-i(-k)na},$$

and we have

$$p_k = \frac{1}{\sqrt{N}} \sum_n p_n e^{ikna}, \tag{4.66}$$

thus

$$p_k^\dagger = p_{-k}.$$ (4.67)

Finally, the Hamiltonian (using the results for the vibration frequency obtained in the last section) is

$$H = \frac{1}{2} \sum_k \left[\frac{p_k p_{-k}}{M} + \underbrace{2\gamma \left(1 - \cos ka\right)}_{M\omega_k^2} \xi_k \xi_{-k} \right]$$ (4.68)

$$= \sum_k \left[\frac{1}{2M} p_k^\dagger p_k + \frac{M\omega_k^2}{2} \xi_k^\dagger \xi_k \right],$$ (4.69)

and

$$H = \sum_k H_k,$$

which was our objective.

Using the relations above, we can show that

$$[\xi_k, p_{k'}] = i\hbar \delta_{k,k'}.$$ (4.70)

If we now define

$$a_k^\dagger = \frac{1}{\sqrt{2\hbar M \omega_k}} (M\omega_k \xi_k^\dagger - ip_k)$$ (4.71)

and

$$a_k = \frac{1}{\sqrt{2\hbar M \omega_k}} (M\omega_k \xi_k + ip_k^\dagger),$$ (4.72)

then

$$[a_k, a_{k'}^\dagger] = \delta_{k,k'}$$ (4.73)

and

$$[a_k, a_{k'}] = [a_k^\dagger, a_{k'}^\dagger] = 0.$$ (4.74)

Following our earlier description for the simple harmonic oscillator, we now have

$$H = \sum_k \hbar \omega_k \left(a_k^\dagger a_k + \frac{1}{2} \right)$$

and

$$\omega_k = \sqrt{\frac{2\gamma}{M} (1 - \cos ka)}.$$ (4.75)

We can now make the comparison with the previous description for the simple harmonic oscillator. The operators $a_k^\dagger$ and a_k create and destroy quanta (phonons) in the state k. The number operator $\hat{n}_k = a_k^\dagger a_k$ measures the number of phonons n_k in the state k. The energy

$$E_k = \left(n_k + \frac{1}{2} \right) \hbar \omega_k$$

is viewed as the excitation energy of n_k phonons of energy $\hbar \omega_k$ plus the ground-state energy (zero-point motion $\frac{1}{2} \hbar \omega_k$). The total energy of the vibrating system of cores is the ground-state energy plus the energy of a group of independent phonons excited above the ground state.

For a given k, there is a quantized vibrational or sound wave mode of wavevector k and energy $\hbar \omega_k$. A sound wave with wavevector k is described by n_k phonons excited above the ground state. Therefore, a statement in common use that a phonon is a quantized sound wave is incorrect. It takes n_k phonons to describe a sound wave. The many-body state $|n_{k_1}, n_{k_2}, \ldots\rangle$ for phonons in different k-states can be constructed using creation operators acting on the vacuum $|0\rangle$,

$$|n_{k_1}, n_{k_2}, \ldots\rangle = (a_{k_1}^\dagger)^{n_{k_1}} (a_{k_2}^\dagger)^{n_{k_2}} \ldots |0\rangle . \tag{4.76}$$

The generalization of the one-dimensional case with vibrating cores (in the harmonic approximation) with r atoms per cell in d dimensions leads to the Hamiltonian

$$H = \sum_{\lambda \mathbf{k}} \hbar \omega_{\lambda \mathbf{k}} \left(a_{\lambda \mathbf{k}}^\dagger a_{\lambda \mathbf{k}} + \frac{1}{2} \right), \tag{4.77}$$

where λ is the phonon branch index ($\lambda = 1, 2, \ldots, d \times r$) and $\omega_{\lambda \mathbf{k}}$ are the frequencies from diagonalizing the dynamical matrix.

To generalize, what we have done is to consider the Hamiltonian

$$H = \sum_i H_i(\mathbf{r}_i, \mathbf{p}_i) + \sum_{i,j} V(\mathbf{r}_i, \mathbf{r}_j, \mathbf{p}_i, \mathbf{p}_j), \tag{4.78}$$

which we then transform into the following:

$$H = E_0 + \sum_{\mathbf{k}} E_{\mathbf{k}} c_{\mathbf{k}}^\dagger c_{\mathbf{k}} + \Delta E. \tag{4.79}$$

The first term, E_0, in Eq. (4.79) is the ground-state energy, the second term represents the energy of the elementary excitations, and the third term, ΔE, is the residual energy representing the interactions between the elementary excitations with terms involving more creation and destruction operators. This term can be expressed in terms of lifetimes for individual states $\mathbf{k}$, $\tau_{\mathbf{k}} \sim \hbar / \Delta E_{\mathbf{k}}$. Hence $\tau_{\mathbf{k}}$ measures the decay time of an excitation. For this picture of viewing a system in terms of elementary excitations (quasiparticles and collective excitations) to be useful, we require that

$$\Delta E_{\mathbf{k}} < E_{\mathbf{k}}. \tag{4.80}$$

When Eq. (4.80) is satisfied, then we can view our system as a collection of nearly independent excitations, and the properties of a solid can be viewed by considering the properties of a gas of these elementary excitations responding to a probe. In the example of phonons discussed above, $\Delta E = 0$ because we have made the harmonic approximation. Any anharmonic terms in Eq. (4.7) would lead to interactions between phonons.

4.3 Response functions: heat capacity

In many respects, the response functions corresponding to different probes have similar features. The probe could be temperature or EM radiation, for example. The probes excite elementary excitations which then give rise to the response based on the properties of the system. The heat capacity at constant volume $C_V(T)$ is a response function, giving the responses to a temperature probe by describing the T dependence of the energy U of the system when elementary excitations are created.

$$C_V(T) = \left(\frac{\partial U}{\partial T} \right)_V. \tag{4.81}$$

For a gas of phonons, we compute the thermodynamic average energy U using Bose statistics for phonons (since they satisfy the boson commutator relations given in Eqs. (4.73) and (4.74)) with branch index λ, wavevector $\mathbf{q}$, and energy $\hbar\omega_{\lambda,\mathbf{q}}$:

$$U(T) = \sum_{\lambda,\mathbf{q}} \hbar\omega_{\lambda,\mathbf{q}} \left[\langle n_{\lambda,\mathbf{q}}(T) \rangle + \frac{1}{2} \right], \tag{4.82}$$

where the phonon occupation is given by

$$\langle n_{\lambda,\mathbf{q}}(T) \rangle = \frac{1}{e^{\frac{\hbar\omega_{\lambda,\mathbf{q}}}{k_{\mathrm{B}}T}} - 1}. \tag{4.83}$$

Then

$$C_V(T) = \frac{\partial U}{\partial T} = \sum_{\lambda,\mathbf{q}} \hbar\omega_{\lambda,\mathbf{q}} \frac{e^{\frac{\hbar\omega_{\lambda,\mathbf{q}}}{k_{\mathrm{B}}T}} \left(\frac{\hbar\omega_{\lambda,\mathbf{q}}}{k_{\mathrm{B}}T^2} \right)}{\left(e^{\frac{\hbar\omega_{\lambda,\mathbf{q}}}{k_{\mathrm{B}}T}} - 1 \right)^2}, \tag{4.84}$$

where k_{B} is the Boltzmann's constant. For a sample of volume V,

$$\sum_{\mathbf{q}} \to \frac{V}{(2\pi)^3} \int_{\mathrm{BZ}} d^3\mathbf{q} \tag{4.85}$$

and

$$C_V(T) = \frac{V}{(2\pi)^3} \sum_\lambda \int_{\mathrm{BZ}} \frac{e^{\frac{\hbar\omega_{\lambda,\mathbf{q}}}{k_{\mathrm{B}}T}} k_{\mathrm{B}} \left(\frac{\hbar\omega_{\lambda,\mathbf{q}}}{k_{\mathrm{B}}T}\right)^2}{\left(e^{\frac{\hbar\omega_{\lambda,\mathbf{q}}}{k_{\mathrm{B}}T}} - 1\right)^2} d^3\mathbf{q}. \tag{4.86}$$

In the limit $k_{\mathrm{B}}T \gg \hbar\omega$,

$$\exp\left(\frac{\hbar\omega_{\lambda,\mathbf{q}}}{k_{\mathrm{B}}T}\right) \sim 1 + \frac{\hbar\omega_{\lambda,\mathbf{q}}}{k_{\mathrm{B}}T}. \tag{4.87}$$

If we consider this limit for three dimensions, $\lambda = 1, 2, \ldots, 3r$, and we arrive at the Dulong–Petit law, which is appropriate for high T,

$$C_V(T) = \frac{V}{(2\pi)^3} k_{\mathrm{B}}(3r) \int d^3\mathbf{q} = k_{\mathrm{B}}(3r)N = 3R, \tag{4.88}$$

where R is the gas constant, r is the number of atoms in a cell, and we have used the fact that $\int d^3\mathbf{q}$ gives the volume of the Brillouin zone, which is

$$\int d^3\mathbf{q} = \frac{(2\pi)^3}{V} \times N, \tag{4.89}$$

where N is the number of cells in the crystal. A typical $C_V(T)$ is shown in Fig. 4.10 along with the high T limit of $3R$. The low-temperature T^3 behavior will be discussed in Section 4.4.1, where we discuss the Debye model for phonons.

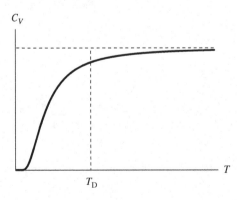

Figure 4.10 Heat capacity of a solid as a function of temperature arising from lattice vibrations with $T_{\mathrm{D}} \approx 300$ K (room temperature).

4.4 Density of states

Often in calculations of response functions such as $C_V(T)$, sums over $\mathbf{q}$ states are converted to integrals over $\mathbf{q}$, as shown in Eqs. (4.85) and (4.86). A further transformation, which is often simplifying and gives more physical insight if the integrand depends only on the frequency explicitly, is to transform the $\mathbf{q}$ integral to an integral over ω or energy. This transformation involves the introduction of the density of states $D(\omega)$. The quantity $D(\omega)d\omega$ gives the number of states with frequencies between ω and $\omega + d\omega$. Hence, for a function $f(\omega(\mathbf{q}))$,

$$\sum_{\mathbf{q}} f(\omega(\mathbf{q})) \to \frac{V}{(2\pi)^3} \int_{\mathrm{BZ}} f(\omega(\mathbf{q})) \, d^3\mathbf{q} \to \int f(\omega)D(\omega)d\omega. \qquad (4.90)$$

For the phonon case, each branch λ can contribute $D_\lambda(\omega)$ to the total density of states $D(\omega)$, so

$$D(\omega) \equiv \sum_{\lambda,\mathbf{q}} \delta(\omega - \omega_{\lambda,\mathbf{q}}) = \sum_\lambda D_\lambda(\omega). \qquad (4.91)$$

In one dimension, the number of q states $dN(q)$ within q and $q+dq$ in the Brillouin zone for a line of length L is

$$dN(q) = 2\frac{L}{2\pi}dq,$$

where a factor of 2 is inserted to limit q to be positive. Hence, the density of q states $W(q)$ becomes

$$W(q) = \frac{dN}{dq} = \frac{L}{\pi}, \qquad (4.92)$$

and the density of states in the frequency domain $D(\omega)$ is

$$D(\omega) = W(q)\frac{dq}{d\omega} = \frac{W(q)}{d\omega/dq}, \qquad (4.93)$$

where $d\omega/dq$ can be identified with the group velocity of the branch of interest.

The generalization to d dimensions is, for each phonon branch λ,

$$D_\lambda(\omega) = \frac{L^d}{(2\pi)^d} \int_s \frac{ds}{|\nabla_\mathbf{q}\omega_\lambda|}, \qquad (4.94)$$

where s is a surface of equal ω in q-space and the integral is over the Brillouin zone.

The one-dimensional chain is an instructive example to consider (Figs. 4.1 and 4.2). Rewriting Eq. (4.22),

$$\omega = \left(\frac{4\gamma}{M}\right)^{\frac{1}{2}} \sin\left(\frac{qa}{2}\right) = \omega_{\max} \sin\left(\frac{qa}{2}\right), \qquad (4.95)$$

hence

$$\frac{dq}{d\omega} = \left(\frac{2}{a}\right) \frac{1}{\left(\omega_{max}^2 - \omega^2\right)^{\frac{1}{2}}}, \tag{4.96}$$

and

$$D(\omega) = \frac{2L}{\pi \, a\omega_{max} \left[1 - \left(\frac{\omega}{\omega_{max}}\right)^2\right]^{\frac{1}{2}}} . \tag{4.97}$$

For $\omega = 0$, we find a constant value for $D(\omega)$,

$$D(0) = \frac{2L}{\pi \, a\omega_{max}}, \tag{4.98}$$

and there is a singularity at $\omega = \omega_{max}$, where the group velocity is zero.

The diatomic one-dimensional chain (Figs. 4.3 and 4.4) is another instructive example. For the acoustic branch (Fig. 4.4), the situation is the same as in Fig. 4.2 for the monatomic lattice, and for $M_2 > M_1$,

$$D_A(\omega) \propto \left[1 - \left(\frac{\omega}{\omega_1}\right)^2\right]^{-\frac{1}{2}}, \tag{4.99}$$

where

$$\omega_1 = \sqrt{\frac{2\gamma}{M_2}},$$

and there is a singularity at ω_1. For the optical branch, the group velocity becomes zero at

$$\omega_2 = \sqrt{\frac{2\gamma}{M_1}}$$

and at the maximum frequency

$$\omega_m = \sqrt{2\gamma \left(\frac{1}{M_1} + \frac{1}{M_2}\right)}.$$

These give rise to singularities in $D(\omega)$, as shown in Fig. 4.11.

4.4.1 The Debye approximation

The continuum approximation in three dimensions assumes a constant sound velocity v_s; that is, $\omega = v_s q$ for $q \leq q_D$ and for each polarization and for all directions (Fig. 4.12). The cutoff q_D is defined by demanding that the volume of $\mathbf{q}$ space considered is the same as

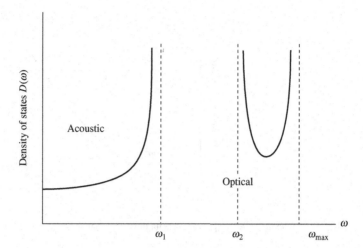

Figure 4.11   Density of phonon states for a diatomic chain.

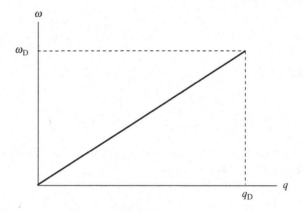

Figure 4.12   The Debye dispersion curve.

the Brillouin zone. This is the Debye[2] approximation. For this case, the density of states is given by

$$D_\lambda(\omega) = \frac{V}{(2\pi)^3} \frac{4\pi q^2}{v_s} = \frac{V}{2\pi^2} \frac{\omega^2}{v_s^3}, \tag{4.100}$$

with $\omega \leq \omega_D$, as shown in Fig. 4.13. In d dimensions, the Debye approximation yields

$$D_\lambda^{\text{Debye}}(\omega) \sim \omega^{d-1}. \tag{4.101}$$

---

[2] P. Debye, "Zur Theorie der spezifischen Waerme," *Ann. Physik.* 39(1912), 789.

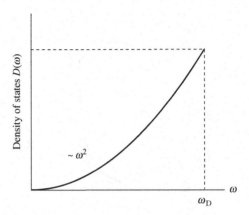

Figure 4.13 Density of states in the Debye approximation for three dimensions.

Since each polarization has N states, if we define ω_D as the frequency of q_D, we have, in three dimensions,

$$N = \int_0^{\omega_D} D_\lambda(\omega)d\omega = \frac{\omega_D^3 V}{6\pi^2 v_s^3} \qquad (4.102)$$

and

$$\omega_D = \left(\frac{N}{V}6\pi^2\right)^{\frac{1}{3}} v_s. \qquad (4.103)$$

Returning to the calculation of $C_V(T)$ for a three-dimensional system with one atom in a unit cell, we have

$$D_\lambda^{\text{Debye}}(\omega) = 3N\left(\frac{\omega^2}{\omega_D^3}\right). \qquad (4.104)$$

Defining

$$x \equiv \frac{\hbar\omega_D}{k_B T} \equiv \frac{T_D}{T}, \qquad (4.105)$$

where $T_D = k\omega_D/k_B$ is called the Debye temperate, Eq. (4.86) gives

$$\begin{aligned}
C_V(T) &= k_B \int_0^{\omega_D} \sum_\lambda \frac{x^2 e^x}{(e^x - 1)^2} D_\lambda^{\text{Debye}}(\omega)d\omega \\
&= \frac{9Nk_B}{\omega_D^3} \int_0^{\omega_D} \frac{x^2 e^x}{(e^x - 1)^2} \omega^2 d\omega \\
&= 9Nk_B \left(\frac{T}{T_D}\right)^3 \int_0^{x_D} \frac{x^4 e^x}{(e^x - 1)^2} dx.
\end{aligned} \qquad (4.106)$$

In the high-temperature limit $T \gg T_D, x \ll 1$,

$$C_V(T) = 9Nk_B \left(\frac{T}{T_D}\right)^3 \int_0^{x_D} \frac{x^4}{x^2} dx = 3Nk_B, \tag{4.107}$$

which is the Dulong–Petit law we obtained previously for $r = 1$. In the low-temperature limit, $T \ll T_D, x \gg 1, T_D/T \to \infty, x_D \to \infty$, and Eq. (4.107) becomes

$$C_V(T) = \left(\frac{12\pi^4}{5}\right) Nk_B \left(\frac{T}{T_D}\right)^3, \tag{4.108}$$

which is the Debye T^3 law, which correctly fits the experimental $C_V(T)$ curve of insulators and is the calculated curve shown in Fig. 4.10. The Debye model works well at low T because in this regime only long-wavelength acoustic modes are excited. The $\omega(\mathbf{q})$ for these modes can be approximated by the continuum model or the Debye model.

4.4.2 The Einstein spectrum

To explain the vanishing heat capacity of solids at low T, Einstein[3] assumed that each mode had the same $\omega = \omega_0$ and viewed a solid as a collection of atoms, with each vibrating at the same frequency

$$\omega_{\lambda\mathbf{q}} = \omega_0. \tag{4.109}$$

The corresponding density of states is

$$D_\lambda^{\text{Einstein}}(\omega) = N\delta(\omega - \omega_0), \tag{4.110}$$

where N is the number of cells. Letting $k_B T_E \equiv \hbar\omega_0$,

$$C_V(T) = 3k_B N \int \frac{x^2 e^x \delta(\omega - \omega_0)}{(e^x - 1)^2} d\omega \tag{4.111}$$

and

$$C_V^{\text{Einstein}}(T) = \frac{3Nk_B \left(\frac{T_E}{T}\right)^2 e^{\frac{T_E}{T}}}{\left(e^{\frac{T_E}{T}} - 1\right)^2}. \tag{4.112}$$

So, at low temperature, a comparison of the models yields

$$C_V^{\text{Einstein}} \sim e^{-\frac{T_E}{T}} \tag{4.113}$$

---

[3] A. Einstein, "Die Plancksche Theorie der Strahlung und die Theorie der spezifischen Wärme," *Ann. Physik.* 22(1907), 180.

and

$$C_V^{\text{Debye}} \sim \left(\frac{T}{T_{\text{D}}}\right)^3. \tag{4.114}$$

As mentioned previously, since acoustic modes are modeled more accurately in the Debye approximation, the heat capacity generally obeys Eq. (4.114); however, Eq. (4.113) does show a vanishing $C_v(T)$, and it has applicability to other situations where the excitations have little wavevector dispersion.

4.5 Critical points and van Hove singularities

In general, densities of states can exhibit sharp structures. This arises because, in the expression for the density of states, e.g. in three dimensions,

$$D_\lambda(\omega) = \frac{V}{(2\pi)^3} \int \frac{ds}{|\nabla_{\mathbf{q}}\omega_\lambda|}, \tag{4.115}$$

there are $\mathbf{q}$-points in the Brillouin zone, where $\nabla_{\mathbf{q}}\omega_\lambda = 0$. These points, where the group velocities are zero, are called critical points,[4] and the resulting structures in the density of states are called van Hove singularities.[5]

As an example, Fig. 4.14 displays the dispersion curve for an optical phonon branch in one dimension, which resembles the optical mode of the diatomic lattice (Fig. 4.4). Because of the zero group velocity at $\omega = \omega_1$ and $\omega = \omega_2$, the density of states exhibits singularities at these frequencies (Fig. 4.11), which correspond to the minimum and maximum frequencies in the optical branch. For three dimensions, $D(\omega)$ exhibits structures that can be characterized in terms of the van Hove singularities.

To explore the influence of critical points on $D(\omega)$ in three dimensions, we examine the topology of the $\omega(\mathbf{q})$ curve near a critical point $\mathbf{q} = \mathbf{q}_{\text{cp}}$, where $\nabla_{\mathbf{q}}\omega_\lambda = 0$. To lowest order, with $i = 1, 2, 3$, we have (assuming the function is analytic and is expanded along principal axes)

$$\omega(\mathbf{q}) = \omega(\mathbf{q}_{\text{cp}}) + \sum_i \alpha_i (\mathbf{q} - \mathbf{q}_{\text{cp}})_i^2. \tag{4.116}$$

If $\alpha_1, \alpha_2, \alpha_3 > 0$, then we have a minimum. If $\alpha_1, \alpha_2, \alpha_3 < 0$, then we have a maximum. If some are positive and some are negative, we have a saddle point. These singularities are classified according to the number of negative α's, as listed in Table 4.1.

There exists a theorem concerning the minimum number of critical points in a branch or band of a dispersion and periodic curve.[6] An analytic function with N variables, which is

---

[4] J. C. Phillips, "Critical points and lattice vibration spectra," *Phys. Rev. B* 104(1956), 1263.
[5] L. Van Hove, "The occurrence of singularities in the elastic frequency distribution of a crystal," *Phys. Rev.* 89(1953), 1189.
[6] L. Van Hove, "The occurrence of singularities in the elastic frequency distribution of a crystal," *Phys. Rev.* 89(1953), 1189.

Table 4.1 Classification of critical points in three dimensions.		
Class	Name	Called
0	M_0	parabolic point (minimum)
1	M_1	saddle point
2	M_2	saddle point
3	M_3	parabolic point (maximum)

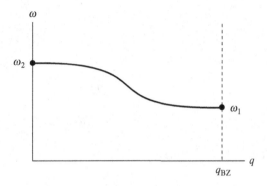

Figure 4.14 There are at least two critical points in one dimension for an analytical periodic function, a maximum at ω_2 and a minimum at ω_1.

periodic in all of them, must have at least C_n^N critical points of type n, where

$$C_n^N = \frac{N!}{n!(N-n)!} \tag{4.117}$$

with $n \le N$.

Therefore, in one dimension, $N = 1$ and $n = 0$ or 1. We have $C_0^1 = 1$ (minimum) and $C_1^1 = 1$ (maximum). These are the critical points which we associated with Fig. 4.14. For two dimensions, $N = 2$ and $n = 0, 1,$ or 2, and we have $C_0^2 = 1$ (minimum), $C_1^2 = 2$ (two saddle points), and $C_2^2 = 1$ (maximum). Hence in two dimensions, we must have at least a maximum, a minimum, and two saddle points. For three dimensions, $C_0^3 = 1$ (one minimum M_0), $C_1^3 = 3$ (three saddle points of type M_1), $C_2^3 = 3$ (three saddle points of type M_2), and $C_3^3 = 1$ (one maximum M_3). Hence, there are at least eight critical points in a three-dimensional spectrum. Most density-of-states functions in three dimensions have more than the minimum number of critical points.

Another theorem associated with the critical points is that $\omega_0 < \omega_1 < \omega_2 < \omega_3$, where ω_0 is the frequency at which the M_0 critical point exists, etc. Different M_1's can be at the same energy, and similarly for the M_2's, so that a three-dimensional $D(\omega)$ can look like it has only four van Hove singularities, as will be discussed later.

The structure in $D(\omega)$ near a critical point can be explored by examining the integral in Eq. (4.115). The topology of the energy surface leads to specific forms for the frequency dependence of $D(w)$ near the critical point, and they will depend on the number of dimensions of space.

In three dimensions, to explore the structure near a critical point, it is convenient to write Eq. (4.116) in cylindrical coordinates, with the assumption that $\alpha_1 = \alpha_2$ with no loss of generality if we are close to the critical point. If we now choose (q, θ, q_z) as our coordinates, where

$$q^2 = q_1^2 + q_2^2 \tag{4.118}$$

and

$$\tan \theta = \frac{q_2}{q_1}, \tag{4.119}$$

$$q_z = q_3, \tag{4.120}$$

then Eq. (4.116) becomes

$$\omega - \omega_c = \alpha_1 q^2 + \alpha_3 q_z^2. \tag{4.121}$$

Near an M_0 singularity, $\omega_c = \omega_0$, $\alpha_1 > 0$, and $\alpha_3 > 0$, and the surface in q-space for ω near ω_0 is an ellipsoid if $\omega > \omega_0$, and there is no solution of Eq. (4.121) if $\omega < \omega_0$. For an M_1 singularity, $\omega_c = \omega_1$, $\alpha_1 > 0$, and $\alpha_3 < 0$, and Eq. (4.121) yields a q-space surface which is a hyperboloid of two sheets if $\omega < \omega_1$. If $\omega > \omega_1$, we have a hyperboloid of one sheet. For an M_2 singularity, $\omega_c = \omega_2$, $\alpha_1 < 0$, and $\alpha_3 > 0$, and, for $\omega < \omega_2$, one has a hyperboloid of one sheet, while for $\omega > \omega_2$, one has a hyperboloid of two sheets. Finally, for an M_3 singularity, $\omega_c = \omega_3$, $\alpha_1 < 0$, and $\alpha_3 < 0$, and the q-space surface is an ellipsoid for $\omega < \omega_3$, and no solution of Eq. (4.121) is possible for $\omega > \omega_3$.

If we choose q_z to be along the principal axis of the structures above, which can all be chosen to be symmetric about the q_z axis, it can be shown that $D(\omega)$ is proportional to the extent of the surface in the q_z direction in the vicinity of the critical point. Hence for an M_0 critical point, using Eq. (4.121) and ignoring the q variable, the q_z range is

$$-\left(\frac{\omega - \omega_0}{\alpha_3}\right)^{\frac{1}{2}} \le q_z \le \left(\frac{\omega - \omega_0}{\alpha_3}\right)^{\frac{1}{2}}. \tag{4.122}$$

The integral in Eq. (4.115) over the q_z variable therefore gives a $D(\omega)$ dependence near M_0 of the form

$$D(\omega) \propto (\omega - \omega_0)^{\frac{1}{2}}. \tag{4.123}$$

A schematic drawing is given in Fig. 4.15.

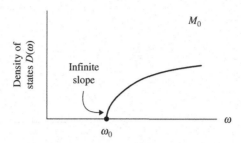

Figure 4.15   Density of states in three dimensions near an M_0 van Hove singularity.

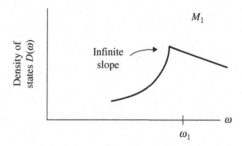

Figure 4.16   Density of states in three dimensions near an M_1 van Hove singularity.

For an M_1 singularity, following the approach above, for $\omega > \omega_1$ ($\alpha_3 < 0$), the surface extends from $q_z = -\infty$ to $q_z = \infty$, except for a gap from

$$q_z = -\left(\frac{\omega_1 - \omega}{\alpha_3}\right)^{\frac{1}{2}} \text{ to } q_z = \left(\frac{\omega_1 - \omega}{\alpha_3}\right)^{\frac{1}{2}} \tag{4.124}$$

(for a hyperboloid of two sheets) and no gap for $\omega > \omega_1$ (a hyperboloid of one sheet). The infinite extent is not relevant since this approximation is appropriate only near the critical point. The outer portions of the surface are finite and smoothly varying. As a result of Eq. (4.124) and this analysis, $D(\omega)$ approaches ω_1 from below with infinite slope and leaves ($\omega > \omega_1$) with finite slope as shown in Fig. 4.16.

Since an M_2 singularity has the opposite ω behavior relative to an M_1 singularity, the $D(\omega)$ approaches ω_2 with finite slope and then drops with infinite slope, as shown in Fig. 4.17. A similar argument can be made when comparing an M_3 maximum with an M_0 minimum. This gives rise to $D(\omega) \propto (\omega_3 - \omega)^{1/2}$ below ω_3 and zero above, as shown in Fig. 4.18.

Figure 4.19 shows the simplest spectrum in three dimensions for a periodic branch. It contains an M_0 at the minimum frequency ω_0, three degenerate M_1 saddle points at ω_1, three degenerate M_2 saddle points at ω_2, and an M_3 maximum at ω_3. Although the eight critical points required for three dimensions are displayed in Fig. 4.19, the degeneracies at ω_1 and ω_2 are not required.

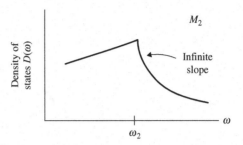

Figure 4.17   Density of states in three dimensions near an M_2 van Hove singularity.

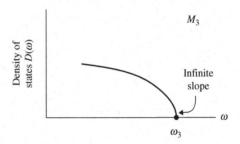

Figure 4.18   Density of states in three dimensions near an M_3 van Hove singularity.

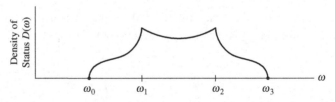

Figure 4.19   The total density of states obtained by including one of each of the critical points. This can be viewed as the simplest $D(\omega)$ for $N = 3$ if the M_1 and M_2 singularities are each taken to be threefold degenerate.

For the Debye spectrum with a minimum in the acoustic branch consistent with

$$\lim_{\mathbf{q}\to 0} \omega(\mathbf{q}) = 0, \tag{4.125}$$

the dispersion curve $\omega(\mathbf{q})$ is not analytic in $\mathbf{q}$. In fact,

$$\omega(\mathbf{q}) = v_s q, \tag{4.126}$$

where the velocity of sound v_s can be different in different directions. Hence the rules above do not apply in this case. Therefore, for the low-frequency acoustic modes, our earlier result $D(\omega) \propto \omega^2$ applies.

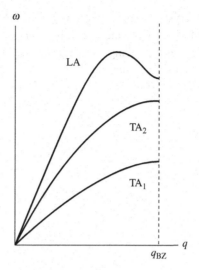

Figure 4.20 The phonon dispersion curve for Al in the (1,1,0) direction (TA_1 and TA_2 are the transverse acoustic modes, and LA is the longitudinal acoustic mode).

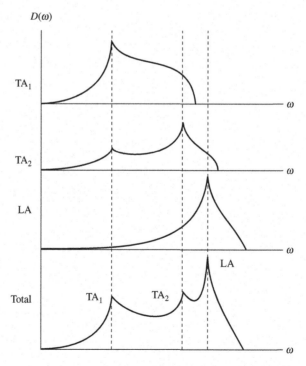

Figure 4.21 Sketch of the density of states for the three different phonon branches of Al and the total density of states.

For a real crystal example, we can consider Al, which is an fcc crystal with one atom per unit cell. Therefore, there are three phonon branches, as shown schematically in Fig. 4.20. The corresponding $D(\omega)$ appears in Fig. 4.21, with van Hove singularities easily recognizable in the spectrum.

I.1. **Crystal structure of MgB$_2$**. MgB$_2$ is a superconductor with a high transition temperature of 39K and has multiple values for the superconducting energy gap. The material has a rather simple structure as given by the model below in Fig. I.1 – the boron atoms (dark balls) form graphitic honeycomb layers (stacked identically on top of each other) and the magnesium atoms (light balls) are located above the center of the B hexagons exactly in between the layers.

Let a be the B–B distance in the layer and c be the distance between the boron layers.

(a) Construct the primitive lattice vectors, the reciprocal lattice, the Wigner–Seitz cell, and the Brillouin zone for MgB$_2$.

(b) What is the volume of the primitive unit cell? What is the volume of the Brillouin zone? Give the locations of the basis atoms in terms of the primitive lattice vectors.

(c) In general, crystal structures have many symmetry operations; those operations can be one of four types: rotation, inversion, reflection, or translation (to find out more about them, read Tinkham's *Group Theory and Quantum Mechanics*). Find, for MgB$_2$, all symmetry operations which leave the crystalline Hamiltonian invariant. (You may neglect spin–orbit interaction and other relativistic effects.)

(d) Draw the irreducible part of the Brillouin zone.

I.2. **The GaN Crystal**. The 2014 Nobel Prize in Physics was given for work on the GaN crystal, specifically for using it in blue light-emitting diodes (LEDs). A model of its crystal structure is given in Fig. I.2, where the Ga atoms are denoted by the light balls and the nitrogen atoms by the dark balls. It has a hexagonal lattice, with $a_1 = a_2 = a$, $a_3 = c$, and the angle between the vectors a_1 and a_2 is 120°.

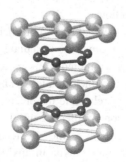

Figure I.1   Crystal structure of MgB$_2$.

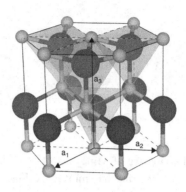

Figure I.2 Crystal structure of GaN.

(a) Write down the explicit expressions of a set of primitive reciprocal lattice vectors b_1, b_2, and b_3 in Cartesian coordinates (in terms of the values a and c). Sketch the first Brillouin zone. What is the volume of the first Brillouin zone in terms of a and c?

(b) How many atoms of each kind are there in the basis of this crystal structure?

(c) In addition to translations, there are many symmetry operations under which the above structure remains invariant. Name any six specific operations among them.

(d) GaN is used in blue LEDs as it is a wide-bandgap semiconductor. Why is it not surprising that this material is an insulator? Explain your answer in terms of the band theory.

I.3. **Born–Oppenheimer approximation for a molecule.** Consider a diatomic molecule with linear dimension a and reduced mass M. Show that the electronic, vibrational, and rotational energy levels can be estimated as terms which are proportional to successively higher orders in the small ratio m/M (m is the electron mass). How is this related to the Born–Oppenheimer approximation discussed in the text?

I.4. **Hartree–Fock approximation.** The Hartree–Fock approximation is a simple yet important model for understanding electron–electron interaction in crystals.

(a) Derive the Hartree–Fock self-consistent field equations from the variational principle using a Slater determinant of single-particle orbitals as a trial many-electron wavefunction.

(b) Consider the total electronic energy of a system (e.g. a molecule) in the Hartree–Fock approximation. Calculate the change in the energy of the system if an electron is promoted from an occupied ith orbital to an unoccupied jth orbital, assuming that the orbitals are unchanged after the excitation. How is this excitation energy related to the eigenvalues of the Hartree–Fock equations? This kind of excitation is called a neutral excitation, which occurs for example in a photo-excitation process, since no electrons are added or removed from the system. The result is known as **Koopmans' theorem**.

I.5. **Born–von Karman boundary condition.** The use of the Born–von Karman boundary condition makes possible a simple and effective description of infinitely extended crystals. With this boundary condition, the number of **k**-states is finite and equivalent in the successively higher Brillouin zone. This leads to different choices of representation for the electron band structure.

(a) Describe and justify the three possible representations of the band structure (that is, the electron energy vs. wavevector) for a crystal: reduced zone scheme, repeated zone scheme, and extended zone scheme.

(b) Show that there are exactly the same number of allowed wavevectors in the first Brillouin zone as there are unit cells in the crystal.

I.6. **Energy bands of elemental solids.** Name three elements in each category which are:

(a) semiconductors (and give their crystal structure);

(b) semimetals (and give their crystal structure);

(c) metals in the fcc structure;

(d) metals in the bcc structure;

(e) metals in the hcp structure.

Consider only elements lighter than Rn. Distinguish metals which are transition metals in (c), (d), and (e).

I.7. **Point group.** Prove the general theorem that only rotations that are multiples of $60°$, $90°$, $120°$, and $180°$ can be symmetry elements of a crystal. (Hint: consider rotation by θ. By writing it as a matrix, show that $1 + 2\cos\theta = $ integer.)

I.8. **Lattice sums.** It is often necessary to do "lattice sums" and sums over "q-space." For lattice vectors R_n, find the results for the following sums:

(a) $\sum\limits_{n} e^{i q \cdot R_n}$,

(b) $\left| \sum\limits_{n} e^{i q \cdot R_n} \right|^2$,

(c) $\sum\limits_{q} e^{i q \cdot R_n}$.

I.9. **Free electron gas.** Treat a metal as a free-electron gas and calculate:

(a) the density of states as a function of energy,

(b) the density of states at the Fermi level in terms of n and E_F,

(c) the electronic heat capacity.

I.10. **Heat capacity from electrons in a semimetal.** Find an expression for the electronic contribution to the heat capacity of a semimetal at very low temperatures for electrons of mass m_e and holes of mass m_h. Let E_e be the energy of the bottom of the electron band and let E_h be the energy of the top of the hole band.

I.11. **Another form of the Kronig–Penney model.** Consider an electron in one dimension in the presence of the potential

$$U(x) = \sum_{m} U_0 \Theta(x - ma)\Theta(ma + b - x), \tag{I.1}$$

where $\Theta(x)$ is a step function ($= 0$ if $x < 0$, $= 1$ if $x > 0$), a is the lattice constant, b is the width of the potential well, and U_0 is the potential height/depth.

(a) Focus on a single unit cell and write down the boundary condition on the Schrödinger equation that leads to Bloch states in this unit cell.

(b) Solve the Schrödinger equation in this cell by taking summation of planewaves and imposing suitable boundary conditions at the positions 0, b, and a. The result is a condition on the Bloch index k.

(c) Take the limit $b \to 0$, $U_0 \to \infty$, and $U_0 b \to W_0 a \frac{\hbar^2 a^{-2}}{m}$. The condition on the Bloch index should become

$$\cos ka = \frac{W_0}{Ka} \sin Ka + \cos Ka, \qquad (I.2)$$

where $K = \sqrt{2m\varepsilon/\hbar^2}$, with ε the energy of the state.

(d) Produce plots of the two lowest-energy bands following from Eq. (I.2) for $a = 1$, $m = 1$, and $W_0 = 1$. Display the bands in the reduced zone scheme and the extended zone scheme.

I.12. **Structure factor.** An energy gap is not necessarily opened up at the Brillouin zone (BZ) edges, due to the vanishing Fourier components of the crystal potential. One example is nearly-free electrons in a hexagonal lattice. The empirical pseudopotential method introduced in the text is closely related to the Fourier components of the crystal field.

(a) Define the term "structure factor" and evaluate it for an hcp lattice. Show that its value leads to a vanishing energy gap for states on the hexagonal face of the Brillouin zone for this structure in the nearly-free electron model (NFEM).

(b) Show explicitly that the first three pseudopotential form factors needed for a band calculation for Si are $V(G^2)$ for $G^2 = 3, 8, 11$ (in units of $(2\pi/a)^2$). What are the values of the G^2's for the next five relevant form factors?

I.13. **Tight-binding model.** Using the tight-binding method, and assuming one spherically symmetric atomic orbital/atom and only nearest-neighbor interactions,

(a) find $E(\mathbf{k})$ for the simple cubic, bcc, and fcc crystal structures,

(b) find $E(\mathbf{k})$ for a two-dimensional (2D) square lattice,

(c) plot $E(\mathbf{k}) = $ constant in the first BZ in part (b).

In (a), (b), and (c), treat the hopping integral as a parameter and the overlap integrals equal to zero for neighboring orbitals.

I.14. **Diatomic chain.** Calculate the band structure of a diatomic chain with the Hamiltonian

$$H = \sum_i \left(\epsilon_1 c_{2i}^\dagger c_{2i} + \epsilon_2 c_{2i+1}^\dagger c_{2i+1} \right) + (-t) \sum_i \left(c_{2i}^\dagger c_{2i+1} + \text{h.c.} \right), \qquad (I.3)$$

where $\epsilon_1 > \epsilon_2$, and $c_i^\dagger$ and c_i are the creation and destruction operators of electronic orbitals at the ith site, respectively. Suppose $t \ll (\epsilon_1 - \epsilon_2)$ and expand $\epsilon(k)$ in $t/(\epsilon_1 - \epsilon_2)$.

I.15. **Tight-binding calculation in 2D.** In high T_c cuprate superconductors, there are layers of Cu–O planes, as shown in Figure I.3. Consider the above 2D square Cu–O crystal with the distance between copper atoms given by a, and the oxygen atoms

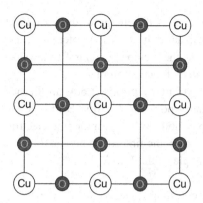

Figure I.3   Cu–O plane in high-temperature cuprate superconductors.

located midway between the copper atoms. Using the tight-binding method and considering only nearest-neighbor interactions (assuming orthogonal orbitals), do the following calculations.

(a) Find $E(k)$, assuming that there is one d-orbital of $(x^2 - y^2)$ symmetry on each copper atom and one p-orbital on each oxygen atom that is pointing along the nearest-neighbor direction. How many electronic bands are there for this model Hamiltonian?

(b) Sketch the energy bands along the [10] and [11] directions.

I.16. **Schrödinger equation of crystal electron.** Find the Schrödinger equation for the periodic part of the Bloch function (i.e. $u_k(r)$).

I.17. **Dirac equation and spin–orbit interaction.** Derive the spin–orbit contribution to the one-electron Hamiltonian starting with the Dirac equation. How does this affect Bloch's theorem? Discuss translational invariance, rotational invariance, inversion symmetry, and time-reversal invariance for this Hamiltonian.

I.18. **Orthogonality of Bloch functions.** Show that two Bloch functions $e^{ik \cdot r} u_k(r)$ and $e^{ik' \cdot r} u_{k'}(r)$ are orthogonal if $k - k'$ is not a reciprocal lattice vector.

I.19. **Parity of Bloch waves.** Consider a crystal potential with inversion symmetry.

(a) Show that for most wavevectors in the first Brillouin zone, the Bloch functions cannot have a definite parity in spite of the crystal inversion symmetry.

(b) There are specific wavevectors for which the Bloch functions can have definite parity. Find the wavefunctions and describe their locations in the first Brillouin zone.

I.20. **f-sum rule.** Examine the problem of deriving the f-sum rule for the case of degenerate bands. Assume $E(k_0)$ is s-fold degenerate. Show that for a crystal with inversion symmetry, the first-order contributions of perturbation theory are zero and one must consider second-order effects (Van Vleck perturbation theory). Discuss the analytic properties of the bands for the $s = 2$ case and define an effective mass tensor (if possible) for this case.

I.21. **Delta function potential.** Solve the Schrödinger equation for a single one-dimensional delta function potential. Under what conditions does a bound state exist? What is the energy of the bound state? Do the same calculation for two one-dimensional delta function potentials separated by a distance a. Describe (without repeating the previous Kronig–Penney calculations) what happens as one goes to the many delta function problems.

I.22. **High symmetry points**. Look up the names (given in Greek letters) of the high symmetry points in the first Brillouin zone for the common lattices given below. Compute and plot $E(k)$ for the free-electron model (FEM) in the following symmetry directions in the first Brillouin zone:

(a) simple cubic, Γ to R;

(b) bcc, Γ to N;

(c) fcc, Γ to X;

(d) hcp, Γ to A.

I.23. **Deep core levels**. Estimate the effect of other atoms on the deep core levels of an atom in a crystal in the following way. Consider a model *simple cubic* crystal with $a = 5\text{Å}$ and one atom per unit cell of atomic number $z = 6$. Calculate roughly the width of the "energy band" formed by the lowest-lying s electrons, and calculate numerically the ratio between this bandwidth and the energy of the atomic s states. Make appropriate approximations in evaluating any integrals that may appear in your answer.

I.24. **Buckyballs**. Buckyballs are molecules with 60 carbon atoms. They crystallize in an fcc lattice. (Assume no band overlap.)

(a) How many filled electron bands are there?

(b) How many partially filled electron bands do you expect?

When this system is doped with potassium, one can form K_3C_{60} in an fcc lattice with the three potassium atoms inserted in the primitive cell.

(c) How many filled electron bands are there for K_3C_{60}?

(d) How many partially filled electron bands do you expect in K_3C_{60}?

I.25. **Density of states**.

(a) Show explicitly that the density of states in three dimensions has the form

$$D(\omega) = \frac{\Omega}{(2\pi)^3} \int \frac{ds}{|\nabla_q \omega|}. \tag{I.4}$$

(b) For a 2D lattice, derive the expression for the behavior of the density of states near critical points giving:

(i) a minimum,

(ii) a maximum,

(iii) a saddle point.

I.26. **Velocity of sound**. The classical formula for the velocity of sound is $v_s = \sqrt{R/\rho}$, where R is the bulk modulus and ρ is the atomic mass density.

(a) Evaluate v_s for a free electron gas using the electron bulk modulus and density.

(b) Are the sound waves the same as ordinary sound waves?

(c) Evaluate this v_s for a gas of He-3 atoms, neutrons, muons, positrons, and protons, assuming in each case a concentration of n fermions/cm^3.

(d) What is k_F for each group of fermions in (c)?

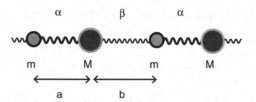

Figure I.4   Spring model of a diatomic chain.

I.27. **Vibrational modes in graphene**. Consider a graphene crystal and let us *ignore* the degree of freedom along the direction that is perpendicular to the carbon atomic plane, i.e. viewing this system as a strictly 2D system.

(a) How many acoustic phonon branches and optical phonon branches are there for this 2D crystal? Why?

(b) Using a Debye model for the phonon density of states, obtain the temperature dependence of its phonon-heat capacity at low temperature.

(c) What is the minimum number of van Hove singularities for each optical phonon branch of such a 2D crystal? Sketch the functional form of an M_0 van Hove singularity in the density of states for an optical branch of this system.

I.28. **Debye model**. Consider a three-dimensional crystal with one atom per unit cell in the Debye approximation.

(a) Find the mean square displacement of an atom at absolute zero temperature.

(b) Derive a formula for the contribution of the zero-point motion to the total energy of the crystal.

(c) Determine an expression for the phonon entropy.

(d) Evaluate (a), (b), and (c) numerically for the case of solid aluminum.

I.29. **Lattice dynamics of a molecular chain**. Consider the following model (see Fig. I.4) for a chain of diatomic molecules, where a = 2b, M and m are masses of the atoms, and α and β are harmonic force constants. In this problem, consider the longitudinal phonons only.

(a) Construct the dynamical matrix and calculate the phonon frequencies, $\omega_\lambda(k)$.

(b) Sketch the phonon dispersion relation and the phonon density of states.

(c) Apply the procedure discussed in the text on second quantization to the system and show that with a suitable choice of a and $a^\dagger$ the Hamiltonian becomes

$$\hat{H} = \sum_{\lambda k}\left(\hat{n}_{\lambda k} + \frac{1}{2}\right)\hbar\omega_{\lambda k}, \tag{I.5}$$

where $\hat{n}_{\lambda k} = a^\dagger_{\lambda k}a_{\lambda k}$.

I.30. **One-dimensional Debye model**. Find the low-temperature heat capacity (within a dimensionless multiplying factor) for a monatomic lattice in one dimension in the Debye approximation. Express your answer in terms of T/T_D, where T is the temperature and T_D is the Debye temperature.

I.31. **Phonons in 2D.** Consider a 2D square lattice with lattice constant a and a basis of three atoms per primitive cell.

   (a) What are the total number of phonon branches for each of the following: longitudinal acoustic, transverse acoustic, longitudinal optic, and transverse optic?

   (b) Assume that the acoustic modes are represented by the Debye model. What is the maximum acoustic phonon frequency in terms of the lattice constant a and the velocity of sound v_s?

   (c) Find the contribution to the low-temperature heat capacity (within a dimensionless coupling constant) from the longitudinal acoustic phonon branch.

   Express your answer in terms of T, T_D, and N, where T is the temperature, T_D is the Debye temperature, and N is the number of cells in the crystal.

I.32. **Zero-point energy correction.** Find a simple expression for the zero-point energy correction to the total energy of a solid in the following two cases.

   (a) Assume a three-dimensional Debye model with one atom per cell where all acoustic modes are degenerate.

   (b) Assume a three-dimensional Einstein model with one atom per cell and all three modes degenerate.

I.33. **Charged harmonic oscillator.** Consider a charged harmonic oscillator in a constant electric field $F = (-e)E$. The Hamiltonian is

$$H = \frac{p^2}{2m} + \frac{\gamma}{2}x^2 + eFx. \tag{I.6}$$

   (a) Show that the Hamiltonian can be written as ($\hbar = 1$)

$$H = \omega\left(A^\dagger A + \frac{1}{2}\right) - \frac{\lambda^2}{\omega}, \tag{I.7}$$

   where $\omega = \sqrt{\gamma/m}$, and give the expressions for λ and A.

   (b) Find $\left[A, A^\dagger\right]$.

   (c) Give an expression for the position operator $X(t)$ and give a physical explanation for its form.

I.34. **Mean position of particles in a one-dimensional potential.** A particle is bound in a one-dimensional potential, $V(x) = \frac{1}{2}m\omega^2x^2 - Cx^3$ (for small x). Describe how the mean position of the particle changes for the different eigenstates when C is small. (Hint: use perturbation theory and show that anharmonic terms will cause expansion.)

PART II

ELECTRON INTERACTIONS, DYNAMICS, AND RESPONSES

5 Electron dynamics in crystals

In the preceding chapters, we discussed the electron states within a single-particle picture in a perfect crystal. This is in some sense equivalent to considering electrons in free space in quantum mechanics; the interaction of the electron with the crystalline potential renormalizes the dispersion relation of the electrons. To understand many phenomena, such as electrical transport in a solid, we would need to describe the behavior of electrons in crystals when some non-periodic perturbation, either from an external applied field or a defect structure, is added to the crystalline periodic field. In this chapter, we introduce in a more rigorous fashion the concepts of group velocity $\mathbf{v}$, effective mass m^*, and the equation of motion for the dynamics of the electron.

Denoting H_0 as the perfect crystal Hamiltonian, the motion of an electron in an additional potential $U(\mathbf{r})$ within a one-electron picture is given by the usual time-dependent Schrödinger equation

$$[H_0 + U(\mathbf{r})]\, \psi(\mathbf{r}, t) = i\hbar \frac{\partial \psi(\mathbf{r}, t)}{\partial t}. \tag{5.1}$$

The static perturbation $U(\mathbf{r})$ may arise from external applied electric or magnetic fields, or come from imperfections in the crystal, such as structural defects and impurities. In many important and common situations, $U(\mathbf{r})$ is slowly varying in space (compared to the atomic separation) but becomes very large as $\mathbf{r} \to \infty$. For example, a uniform applied electric field $\mathcal{E}$ gives rise to $U(\mathbf{r}) = -e\mathcal{E} \cdot \mathbf{r}$, and a uniform magnetic field $\mathbf{B}$ produces a vector potential $\mathbf{A} = \frac{1}{2}\mathbf{B} \times \mathbf{r}$ in the symmetric gauge. Potential perturbations caused by charge impurities, dislocations, surfaces, etc. are also often static and slowly varying compared to the interatomic distance, but they can be very large. These common situations cannot be treated straightforwardly by standard perturbation methods since, particularly for uniform applied fields, the perturbing potential diverges at large distance $\mathbf{r}$ in an extended crystal.

5.1 Effective Hamiltonian and Wannier functions

We develop here a formulation that will allow us to determine the behavior of crystal electrons in an external potential, which is slowly varying in both space and time. The framework is to consider that the problem of an electron in a perfect crystal is solved, and that we transform the dynamics of an electron under the influence of an external potential to an electron in free space with renormalized properties.

The basic assumption or approximation is that the electron moves entirely within the states of a single band. This approach is named the effective Hamiltonian theory or approximation. This is a reasonable approximation for the case where a potential that is nearly

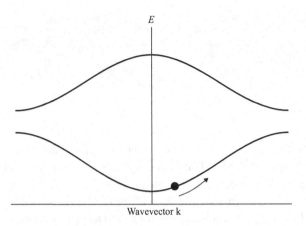

Wavevector k

Figure 5.1 Dynamics of a crystalline electron in a slowly varying field. A slowly varying potential moves an electron in a state with wavevector **k** to nearby states in the same band.

constant in space and smoothly varying in time moves the electron adiabatically between states of nearly the same energy without causing transitions to states in another band, as illustrated in Fig. 5.1.

Electron dynamics is best discussed in terms of localized wavefunctions because we will be thinking in terms of wavepackets. One such set of functions is the so-called Wannier functions.[1]

Let $\phi_{n\mathbf{k}}(\mathbf{r})$ be Bloch functions of wavevector **k** in the band n; the Wannier functions are, in the simplest form, defined by

$$w_n(\mathbf{r} - \mathbf{R}_j) \equiv \frac{1}{\sqrt{N}} \sum_{\mathbf{k}} e^{-i\mathbf{k}\cdot\mathbf{R}_j} \phi_{n\mathbf{k}}(\mathbf{r}), \tag{5.2}$$

where N is the number of unit cells and $\mathbf{R}_j$ is a lattice vector. This expression is for a single non-degenerate band. It can be generalized to a band complex. Moreover, there is considerable freedom in defining Wannier functions. Other possible definitions of Wannier functions include adding a well-chosen phase factor $e^{i\theta_{\mathbf{k}}}$ in each term of the sum on the right-hand side of Eq. (5.2) to maximize its localization in real space.[2]

Wannier functions defined by Eq. (5.2) form a complete set and have the following useful properties:

$$
\begin{align}
\text{(i)} \quad & \phi_{n\mathbf{k}}(\mathbf{r}) = \frac{1}{\sqrt{N}} \sum_j e^{i\mathbf{k}\cdot\mathbf{R}_j} w_n(\mathbf{r} - \mathbf{R}_j), \notag \\
\text{(ii)} \quad & \langle w_{n'}(\mathbf{R}_i) | w_n(\mathbf{R}_j) \rangle = \delta_{n'n}\delta_{ij}, \tag{5.3} \\
\text{(iii)} \quad & w_n(\mathbf{r} - \mathbf{R}_j) \text{ is localized about } \mathbf{R}_j. \notag
\end{align}
$$

[1] G. H. Wannier, "The structure of electronic excitation levels in insulating crystals," *Phys. Rev.* 52(1937), 191.
[2] N. Marzari and D. Vanderbilt, "Maximally localized Wannier functions for composite energy bands," *Phys. Rev. B* 56(1997), 12847.

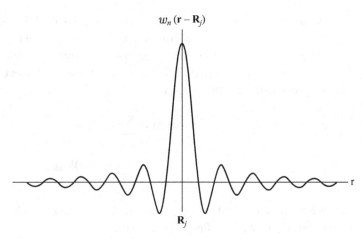

$w_n(\mathbf{r} - \mathbf{R}_j)$

$\mathbf{R}_j$

r

Figure 5.2   Schematic diagram of a Wannier function localized around the lattice site $\mathbf{R}_j$.

The Wannier functions are therefore an orthonormal set of localized functions in real space that span the same functional space as the Bloch states. We have used the Dirac bra–ket notation above. A real-space picture of a Wannier function, given by Eq. (5.2) and computed with planewave Bloch functions, is illustrated in Fig. 5.2.

5.2 Electron dynamics in the effective Hamiltonian approach

We now consider the motion of an electron described by Eq. (5.1) in the effective Hamiltonian approximation. It is most easily discussed in the notation of the state vector $|\psi(t)\rangle$ for the electronic state. Mathematically, $|\psi(t)\rangle$ is just a vector in Hilbert space. It can be expanded in any complete set of "basis" functions that spans the Hilbert space. In particular, the Bloch states $|\phi_{n\mathbf{k}}\rangle$, the position states $|\mathbf{r}\rangle$, or the Wannier functions $|w_n(\mathbf{R})\rangle$ may all be used as separate sets of such basis functions.

For the three sets of basis functions of interest here, completeness of each set of functions gives

$$1 = \sum_{n\mathbf{k}} |\phi_{n\mathbf{k}}\rangle \langle \phi_{n\mathbf{k}}|$$

$$= \sum_{\mathbf{r}} |\mathbf{r}\rangle \langle \mathbf{r}| \tag{5.4}$$

$$= \sum_{n\mathbf{R}_j} |w_n(\mathbf{R}_j)\rangle \langle w_n(\mathbf{R}_j)| .$$

For example, we may express

$$|\psi(t)\rangle = \sum_{\mathbf{r}} |\mathbf{r}\rangle \langle \mathbf{r}| \psi(t)\rangle = \sum_{\mathbf{r}} \psi(\mathbf{r}, t) |\mathbf{r}\rangle \tag{5.5}$$

in expanding the state vector in the basis set of $|\mathbf{r}\rangle$, called the **r**-representation. Here we have denoted the expansion coefficients $\psi(\mathbf{r}, t) \equiv \langle \mathbf{r}|\psi(t)|\mathbf{r}\rangle$. This is just the real-space wavefunction $\psi(\mathbf{r}, t)$ in elementary quantum mechanics as given in Eq. (5.1).

The state vector can equivalently be expanded in the basis set of Bloch states $|\phi_{n\mathbf{k}}\rangle$, and in the Bloch representation we have

$$
\begin{aligned}
|\psi(t)\rangle &= \sum_{n\mathbf{k}} |\phi_{n\mathbf{k}}\rangle \langle \phi_{n\mathbf{k}}|\psi(t)\rangle \\
&\equiv \sum_{n\mathbf{k}} \psi_n(\mathbf{k}, t)|\phi_{n\mathbf{k}}\rangle,
\end{aligned}
\tag{5.6}
$$

where we have denoted the expansion coefficients $\langle \phi_{n\mathbf{k}}|\psi(t)|\mathbf{r}\rangle$ as $\psi_n(\mathbf{k}, t)$. Similarly, in an expansion in Wannier functions, we have

$$
\begin{aligned}
|\psi(t)\rangle &= \sum_{n\mathbf{R}_j} |w_n(\mathbf{R}_j)\rangle \langle w_n(\mathbf{R}_j)|\psi(t)\rangle \\
&\equiv \sum_{n\mathbf{R}_j} \psi_n(\mathbf{R}_j, t)|w_n(\mathbf{R}_j)\rangle,
\end{aligned}
\tag{5.7}
$$

with $\psi_n(\mathbf{R}_j, t) \equiv \langle w_n(\mathbf{R}_j)|\psi(t)|\mathbf{r}\rangle$ denoting the expansion coefficients in the Wannier representation.

We would now like to derive a more simplified equation of motion for the state vector $|\psi(t)\rangle$, starting with

$$
H|\psi\rangle = i\hbar \frac{\partial |\psi\rangle}{\partial t},
\tag{5.8}
$$

where $H = H_0 + U(\mathbf{r})$ is the Hamiltonian operator. Since $U(\mathbf{r})$ is slowly varying in $\mathbf{r}$ and t, we assume that the motion of the electron is well approximated by the Bloch states within a single band. Let us call this the nth band. Then Eq. (5.7) in the effective Hamiltonian approach is taken to be well approximated by

$$
|\psi(t)\rangle = \sum_{j} \psi_n(\mathbf{R}_j, t)|w_n(\mathbf{R}_j)\rangle.
\tag{5.9}
$$

The sum is now restricted to Wannier functions of a single band of index n. Combining Eqs. (5.5) and (5.9), we see that

$$
\psi(\mathbf{r}, t) = \sum_{j} \psi_n(\mathbf{R}_j, t) w_n(\mathbf{r} - \mathbf{R}_j).
\tag{5.10}
$$

The time evolution of the wavefunction $\psi(\mathbf{r}, t)$ is then equivalently given by the time evolution of $\psi_n(\mathbf{R}_j, t)$. In this form, we see that the electron wavepacket (as a solution to Eq. (5.1)) can be viewed as a set of Wannier functions convoluted by an envelope function that depends only on the lattice site index $\mathbf{R}_j$. The electron function $\psi(\mathbf{r}, t)$ may have a lot of internal structure, but as far as the dynamics of the wavepacket are concerned, we only need to know the dynamics of the envelope function (or the wavefunction in the Wannier representation), i.e. the time evolution of $\psi_n(\mathbf{R}_j, t)$ with respect to the crystal lattice positions. We therefore need to develop quantum operators in the Wannier representation.

We start with the usual position $\mathbf{r}_{\text{op}}$ and momentum $\mathbf{p}_{\text{op}}$ operators that have the properties $\mathbf{r}_{\text{op}}|\mathbf{r}\rangle = \mathbf{r}|\mathbf{r}\rangle$, $\mathbf{p}_{\text{op}}|\mathbf{p}\rangle = \mathbf{p}|\mathbf{p}\rangle$, and $[\mathbf{p}_{\text{op}}, \mathbf{r}_{\text{op}}] = -i\hbar$, and define the operators $\mathbf{R}_{\text{op}}$ and $\mathbf{k}_{\text{op}}$ using the relations

$$\mathbf{k}_{\text{op}}|\phi_{n,\mathbf{k}}\rangle \equiv \mathbf{k}|\phi_{n,\mathbf{k}}\rangle$$

and

$$\mathbf{R}_{\text{op}}|w_n(\mathbf{R}_j)\rangle \equiv \mathbf{R}_j|w_n(\mathbf{R}_j)\rangle. \tag{5.11}$$

Then, it is straightforward to show that $\mathbf{R}_{\text{op}}$ and $\mathbf{k}_{\text{op}}$ form a conjugate pair of dynamical variables with

$$[\mathbf{k}_{\text{op}}, \mathbf{R}_{\text{op}}] = -i, \tag{5.12}$$

$$\mathbf{R}_{\text{op}} = i\frac{\partial}{\partial \mathbf{k}}, \tag{5.13}$$

and

$$\mathbf{k}_{\text{op}} = -i\frac{\partial}{\partial \mathbf{R}_j}. \tag{5.14}$$

We now consider the action of a slowly varying potential $U(\mathbf{r})$ on a Wannier function. Since the potential is position dependent, it is well-defined as an operator acting on a function that is dependent on $\mathbf{r}$. We may express the Wannier function in the $\mathbf{r}$-representation and obtain

$$U_{\text{op}}|w(\mathbf{R}_j)\rangle = U_{\text{op}} \sum_{\mathbf{r}} |\mathbf{r}\rangle\langle\mathbf{r}|w(\mathbf{R}_j)\rangle = \sum_{\mathbf{r}} U(\mathbf{r})|\mathbf{r}\rangle\langle\mathbf{r}|w(\mathbf{R}_j)\rangle. \tag{5.15}$$

Since $\langle\mathbf{r}|w(\mathbf{R}_j)\rangle$ is a very localized function around the point $\mathbf{R}_j$, and $U(\mathbf{r})$ is very slowly varying in the scale of the spacing of the $\mathbf{R}$'s, we may write

$$U_{\text{op}}|w(\mathbf{R}_j)\rangle \cong U(\mathbf{R}_j) \sum_{\mathbf{r}} |\mathbf{r}\rangle\langle\mathbf{r}|w(\mathbf{R}_j)\rangle = U(\mathbf{R}_j)|w(\mathbf{R}_j)\rangle. \tag{5.16}$$

Equation (5.16) gives a great simplification to Eq. (5.8). Using the single-band approximation, Eqs. (5.13) and (5.14),

$$H_0|\psi\rangle \cong \sum_{\mathbf{k}} E_n(\mathbf{k})\langle\phi_{n,\mathbf{k}}|\psi\rangle|\phi_{n,\mathbf{k}}\rangle = \sum_{\mathbf{R}_j} \left[E_n\left(-i\frac{\partial}{\partial \mathbf{R}_j}\right) \langle w(\mathbf{R}_j)|\psi\rangle \right]|w(\mathbf{R}_j)\rangle. \tag{5.17}$$

Equation (5.8) can now be rewritten in the Wannier representation, within the slowly varying potential regime, as

$$\sum_{\mathbf{R}} \left[E_n\left(-i\frac{\partial}{\partial \mathbf{R}}\right) + U(\mathbf{R}) \right] \psi(\mathbf{R}, t)|w(\mathbf{R})\rangle = \sum_{\mathbf{R}} \left(i\hbar\frac{\partial}{\partial t}\right) \psi(\mathbf{R}, t)|w(\mathbf{R})\rangle, \tag{5.18}$$

where, for simplicity, we have dropped the index j on the lattice positions. We then have an effective equation for the wavepacket envelope function

$$\left[E_n\left(-i\frac{\partial}{\partial \mathbf{R}}\right) + U(\mathbf{R})\right]\psi(\mathbf{R},t) = i\hbar\frac{\partial}{\partial t}\psi(\mathbf{R},t). \tag{5.19}$$

Effectively, we have the following Hamiltonian for considering the dynamics of the crystal electrons:

$$\mathbf{H}_{\text{eff}} = E_n\left(-i\frac{\partial}{\partial \mathbf{R}}\right) + U(\mathbf{R}) \quad (\mathbf{R}\text{-representation}) \tag{5.20}$$

or

$$\mathbf{H}_{\text{eff}} = E_n(\mathbf{k}) + U\left(i\frac{\partial}{\partial \mathbf{k}}\right) \quad (\text{Bloch representation}). \tag{5.21}$$

Within this formalism, by making use of the slowly varying potential approximation, we have eliminated the electron coordinates $\mathbf{r}$ from the problem, and hence all the details of the electron's interactions with the crystal. The effects of the crystal on the behavior of the electron are incorporated completely in the band structure term in Eq. (5.20). The electron behaves as if it moves only in the potential $U(\mathbf{R})$ but with a renormalized kinetic energy operator in Eq. (5.20) or Eq. (5.21). Making a connection to the usual electron wavefunction, we have

$$\psi(\mathbf{r},t) = \sum_{\mathbf{R}}\psi(\mathbf{R},t)w_n(\mathbf{r}-\mathbf{R}) \tag{5.22}$$

if Eq. (5.20) is used, and

$$\psi(\mathbf{r},t) = \sum_{\mathbf{k}}\psi(\mathbf{k},t)\phi_{n\mathbf{k}}(\mathbf{r}) \tag{5.23}$$

if Eq. (5.21) is used. The two forms are completely equivalent; which one to use is a matter of computational or conceptual convenience. The effective Hamiltonian approach is expected to be a good approximation if the spatial and temporal variations of U are such that (a) $(\nabla U)\cdot\mathbf{d}$ is much less than the bandwidth where d is an interatomic distance, and (b) $\hbar\omega$ is much less than the interband energies where ω is the time variation frequency of U. Under these conditions, a state initially composed of Bloch states within one band will not mix with states of other bands under the influence of the potential U.

The robustness of the approach is illustrated schematically in Fig. 5.3 using a constant applied electric field as an example. The quality $\Delta U(\mathbf{r})$ is the difference between the exact external potential and the piece-wise constant potential taken at the lattice sites where the Wannier functions are located. We see that ΔU is small ($\sim e\mathcal{E}d$) and oscillatory. The largeness (diverging at large distances) of the perturbation is entirely captured in $U(\mathbf{R})$. Moreover, since ΔU is periodic, its effects may in fact be included by redefining the Bloch and Wannier functions as solutions to a new band structure problem, i.e. a periodic crystalline Hamiltonian with ΔU included. Of course, the effects of ΔU in the Wannier functions in Fig. 5.3 are real physical effects that were left out of the theory, e.g. those related to polarization.

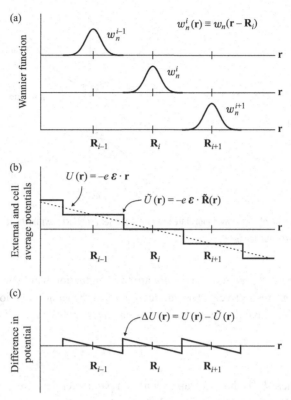

Figure 5.3 Schematic of Wannier functions (a) at different lattice sites $\mathbf{R}_j$ and (b) the potential due to a uniform electric field and (c) the difference between $U(\mathbf{r})$ and $\tilde{U}(\mathbf{r})$. Here $\tilde{\mathbf{R}}(r)$ is a piece-wise constant function that takes the value of the closest value of $\mathbf{R}_j$ at a given position $\mathbf{r}$.

5.3 Shallow impurity states in semiconductors

A good example of an application of the effective Hamiltonian approach is its use in determining the shallow impurity states in semiconductors. Let us consider the simple case of a substitutional impurity (with an extra valence electron from the impurity atom) in a semiconductor that has an isotropic and parabolic conduction band. This would be, for example, the case of Si in GaAs replacing one of the Ga atoms. The extra electron would be attracted to the Si site and in a state that is expected to be comprised of states from the conduction band only (see Fig. 5.4).

We use the single-band approximation and write the impurity wavefunction in terms of the conduction band states:

$$
\begin{aligned}
\psi(\mathbf{r}, t) &= \sum_{\mathbf{k}} \psi(\mathbf{k}, t) \phi_{c\mathbf{k}}(\mathbf{r}) \\
&= \sum_{\mathbf{R}} \psi(\mathbf{R}, t) w_c(\mathbf{r} - \mathbf{R}_j),
\end{aligned}
\tag{5.24}
$$

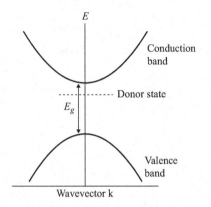

Figure 5.4 Diagram of a shallow donor state in a direct gap semiconductor with a bandgap E_g separating the valence band from the conduction band.

where $\phi_{c\mathbf{k}}(\mathbf{r})$ and $w_c(\mathbf{r} - \mathbf{R})$ are the conduction band Bloch functions and Wannier functions, respectively. The conduction band dispersion relation measured from the top of the valence band, assuming an isotropic and parabolic form, is given by

$$E_c(\mathbf{k}) = E_g + \frac{\hbar^2 k^2}{2m^\star}, \tag{5.25}$$

where E_g is the bandgap and m^* is a parameter describing the band curvature. Application of Eq. (5.20) leads to the following effective Schrödinger equation for the wavefunction in the Wannier or $\mathbf{R}$-representation, assuming a Coulomb interaction (screened by the dielectric constant ϵ of the material) between the additional electron and the positively charged impurity core site,

$$\left[-\frac{\hbar^2}{2m^\star} \nabla^2 + \left(-\frac{e^2}{\epsilon R} \right) \right] \psi(\mathbf{R}) = (E - E_g)\psi(\mathbf{R}). \tag{5.26}$$

This leads to the hydrogenic model for shallow impurity states in the bandgap near the conduction band of semiconductors given in elementary treatments. Table 5.1 compares the calculated results using the effective Hamiltonian (or effective mass) model with experimental impurity levels.

5.4 Motion in external fields

We now proceed to analyze the motion of crystal electrons in an external field using the formalism developed in Section 5.2. The equation of motion for the operators $\mathbf{R}_{\mathrm{op}}$ and $\mathbf{k}_{\mathrm{op}}$ may be obtained using the usual relation from quantum mechanics for operator A,

$$\frac{dA}{dt} = \frac{\partial A}{\partial t} + \frac{1}{i\hbar} [A, H]. \tag{5.27}$$

Table 5.1 Binding energies E_b of the ground (1s) state of shallow donor impurities in some common semiconductors, comparing experimental data with results from the effective Hamiltonian or effective mass theory. The symbol X_Y denotes that the element Y in the host material is substitutionally replaced by an impurity atom X. (After Cardona and Yu, 2010.)[3]

Semiconductor	Theory E_b (meV)	Experiment E_b (meV)
GaAs	5.72	Si_{Ga}: 5.84
		Ge_{Ga}: 5.88
		S_{As}: 5.87
		Se_{As}: 5.79
InSb	0.6	Te_{Sb}: 0.6
CdTe	11.6	In_{Cd}: 14.0
		Al_{Cd}: 14.0
ZnSe	25.7	Al_{Zn}: 26.3
		Ga_{Zn}: 27.9
		F_{Se}: 29.3
		Cl_{Se}: 26.9

Using the effective Hamiltonian in Eqs. (5.20) and (5.21), we obtain for the operator $\mathbf{R}$ the following equation of motion:

$$i\hbar\frac{d\mathbf{R}}{dt} = [\mathbf{R}, H_{\text{eff}}]$$

$$= \left[\mathbf{R}, E_n\left(-i\frac{\partial}{\partial\mathbf{R}}\right) + U(\mathbf{R})\right]$$

$$= \left[\mathbf{R}, E_n\left(-i\frac{\partial}{\partial\mathbf{R}}\right)\right]$$

$$= \left[i\frac{\partial}{\partial\mathbf{k}}, E_n(\mathbf{k})\right]$$

$$= i\frac{\partial}{\partial\mathbf{k}}E_n(\mathbf{k}) - E_n(\mathbf{k})i\frac{\partial}{\partial\mathbf{k}}$$

$$= i\left(\frac{\partial}{\partial\mathbf{k}}E_n(\mathbf{k})\right). \tag{5.28}$$

[3] P. Y. Yu and M. Cardona, *Fundamentals of Semiconductors: Physics and Materials Properties*, 4th ed. (Berlin: Springer, 2010).

Thus, the velocity operator $\mathbf{v}$ is given by

$$\mathbf{v} = \frac{d\mathbf{R}}{dt} = \frac{1}{\hbar}\left(\nabla_\mathbf{k} E_n(\mathbf{k})\right). \tag{5.29}$$

Note that $\mathbf{v}$ does not depend on the form of the perturbation U. It is a general property of the electronic band structure.

For $\mathbf{k}$, again using H_{eff}, we obtain the equation of motion

$$i\hbar\frac{d\mathbf{k}}{dt} = [\mathbf{k}, H_{\text{eff}}]$$

$$= \left[\mathbf{k}, U\left(i\frac{\partial}{\partial\mathbf{k}}\right)\right]$$

$$= \left[-i\frac{\partial}{\partial\mathbf{R}}, U(\mathbf{R})\right]$$

$$= -i\left(\nabla_\mathbf{R} U(\mathbf{R})\right) \tag{5.30}$$

or

$$\hbar\frac{d\mathbf{k}}{dt} = -\nabla_\mathbf{R} U(\mathbf{R}). \tag{5.31}$$

Since $-\nabla U(\mathbf{R})$ is the force $\mathbf{F}$ on the electron, we have

$$\hbar\frac{d\mathbf{k}}{dt} = \mathbf{F}. \tag{5.32}$$

For the special case of a constant force $\mathbf{F}$, Eq. (5.32) can be integrated immediately to

$$\mathbf{k}(t) = \mathbf{k}(t=0) + \mathbf{F}\frac{t}{\hbar}. \tag{5.33}$$

With Eqs. (5.29) and (5.31), we can follow the motion of an electron starting at $t = 0$ in any individual Bloch state with wavevector $\mathbf{k}_0$, or of a wavepacket centered about a given $\mathbf{k}_0$. For example, consider a crystal with the band structure given in Fig. 5.5 in a uniform electric field $\boldsymbol{\mathcal{E}} = \varepsilon\hat{\mathbf{x}}$. The wavevector of a state with initial wavevector at $\mathbf{k}_0 = 0$ at $t = 0$ will change according to Eq. (5.33) to

$$\mathbf{k}(t) = -\frac{|e|\varepsilon t}{\hbar}\hat{\mathbf{x}}. \tag{5.34}$$

The group velocity is given by $\hbar v_x = \frac{\partial E_n}{\partial k_x}$ (see Fig. 5.6), which leads to a time dependence in the position of the particle as

$$x(t) = \int_0^t v_x(t)dt = \int_0^t \frac{dt}{\hbar}\frac{\partial E_n}{\partial k_x} = \frac{1}{\hbar}\int_0^t \frac{dt}{dk_x}\frac{\partial E_n}{\partial k_x}dk_x$$

$$= \frac{1}{(-|e|\varepsilon)}[E_n(k_x(t)) - E_n(k_x(t=0))]. \tag{5.35}$$

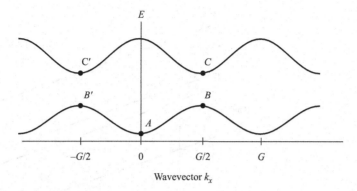

Figure 5.5 Schematic of a band structure in a repeated zone scheme along the k_x direction, where G is the smallest reciprocal lattice vector along this direction.

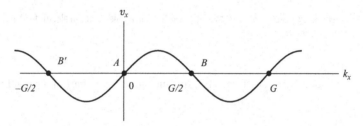

Figure 5.6 Group velocity v_x as a function of k_x for the lower band in Fig. 5.5.

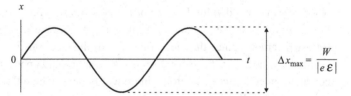

$$\Delta x_{max} = \frac{W}{|e\,\mathcal{E}|}$$

Figure 5.7 Schematic of $x(t)$ of a crystalline electron in a uniform field which shows Bloch oscillation where W is the band width.

Physically, what these equations tell us is that the wavepacket's center of mass wavevector in **k**-space and its center of mass in real space are changing (due to the interaction with the field) as a function of time. When the center of mass wavevector reaches the Brillouin zone boundary, the wavepacket is Bragg-reflected, leading to an oscillatory behavior in **r**(t) as illustrated in Fig. 5.7. As a consequence, in this idealized situation, there is no net current for a perfect crystal. However, for real materials, impurities and zero-point motions of various excitations will introduce scattering that would give rise to a net current even at temperature $T = 0$.

The oscillations in the motion of the electrons given by Eq. (5.35) are called Bloch oscillations. Observation of these oscillations requires that the electron travel many periods

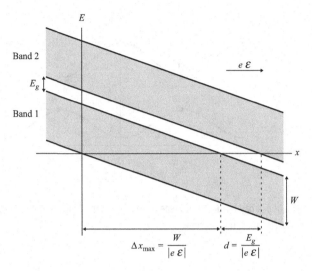

Figure 5.8   Semiclassical picture of Bloch oscillations with $\mathcal{E}$ an applied uniform electric field and W the bandwidth.

without being scattered. The distance traveled in half of a period is given by

$$\Delta x_{\max} = \frac{W}{|e\boldsymbol{\varepsilon}|}, \tag{5.36}$$

where W is the bandwidth. Under normal conditions, with an electric field of the order of V/mm and $W \sim$ few eV, $\Delta x_{\max} \approx$ a few mm, making the Bloch oscillations unobservable for usual samples since the mean free path of the electrons is significantly shorter. Bloch oscillations, however, have been observed in man-made periodic systems, such as semiconductor superlattices and atoms in optical lattices, where the large unit cell dimensions reduce W to small values.[4]

The Bloch oscillation phenomenon may also be understood in a semiclassical picture as illustrated in Fig. 5.8. In a slowly varying potential, the electron in a state at the band extrema effectively runs into a triangular potential barrier with width d that is determined by the bandgap to the next band and by the electric field strength. The electron wave is reflected at the barriers, giving rise to the oscillations. There is, however, a finite probability for tunneling to the next band if $|\mathcal{E}|$ is large enough (i.e. if the barrier is thin enough), leading to phenomena such as Zener tunneling (see Fig. 5.9). Interband transitions are, however, not included in the present effective Hamiltonian formalism, since it is assumed that the wavefunction is always made up of states from a given band or band complexes.

[4] See, for example, M. Raizen, C. Salomon, and Q. Niu, "New light on quantum transport," *Phys. Today* 50(1997), 30.

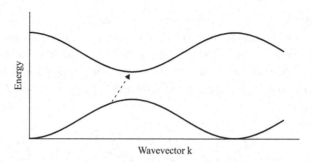

Figure 5.9 Schematic of Zener tunneling with an electron tunneling from the lower band to the upper band under the influence of a strong external field.

5.5 Effective mass tensor

In the example in Figs. 5.5 and 5.6, under the influence of a force $\mathbf{F}$, the wavevector $\mathbf{k}(t)$ of an electron changes as a function of time. The velocity of the electron $\mathbf{v}$ goes to zero and then changes sign as $\mathbf{k}$ crosses the Brillouin zone edge under a constant force. The electron appears as if it has a mass that is a function of $\mathbf{k}$ and becomes negative near the top of the band. This leads us to define the useful concept of the effective mass tensor in electron dynamics (discussed previously in Chapter 3).

As in classical mechanics, mass is defined by $\mathbf{F} = m\mathbf{a}$ or, more generally,

$$\sum_{\beta} \left(\frac{1}{m^*} \right)_{\alpha\beta} F_\beta = a_\alpha. \tag{5.37}$$

Since from Eq. (5.29),

$$\frac{dv_\alpha}{dt} = \frac{1}{\hbar} \frac{d}{dt} \frac{\partial E}{\partial k_\alpha} = \frac{1}{\hbar} \sum_{\beta} \frac{\partial^2 E}{\partial k_\alpha \partial k_\beta} \frac{dk_\beta}{dt}, \tag{5.38}$$

and from Eq. (5.32) $\frac{d\mathbf{k}}{dt} = \frac{\mathbf{F}}{\hbar}$, we have

$$a_\alpha = \sum_{\beta} \frac{1}{\hbar^2} \frac{\partial^2 E}{\partial k_\alpha \partial k_\beta} F_\beta. \tag{5.39}$$

The effective mass tensor is thus given by

$$\left(\frac{1}{m^*} \right)_{\alpha\beta} = \frac{1}{\hbar^2} \frac{\partial^2 E}{\partial k_\alpha \partial k_\beta}. \tag{5.40}$$

The form and properties of the effective mass tensor play an important role in determining the transport properties of materials, especially those of carriers in semiconductors.

In semiconductors, one is mainly dealing with carriers (either electrons or holes) near band extrema, which typically occur near a symmetry point $\mathbf{k}_0$ in the Brillouin zone, e.g.

at $\mathbf{k}_0$ equal to the Γ point for GaAs. There is a way, as discussed in Chapter 3, that allows one to determine the nearby band structure and the effective mass tensor given that we know the energy states at a given point $\mathbf{k}_0$. The method makes use of the knowledge of the electron band structure at a given point of $\mathbf{k}$-space to find the states nearby, and the effective mass tensor, in the so-called $\mathbf{k} \cdot \mathbf{p}$ perturbation theory, resulting in, for the nth band,

$$
\begin{aligned}
\left(\frac{1}{m^*}\right)_{n,\alpha\beta} &= \frac{1}{\hbar^2} \frac{\partial^2 E_n}{\partial k_\alpha \partial k_\beta} = \frac{1}{m}\left[\delta_{\alpha\beta} + \frac{2}{m}\sum_{n'} \frac{p_\alpha^{n'n} p_\beta^{nn'}}{E_n(\mathbf{k}_0) - E_{n'}(\mathbf{k}_0)}\right] \\
&= \frac{1}{m}\left[\delta_{\alpha\beta} + \sum_{n'} f_{nn'}^{\alpha\beta}\right].
\end{aligned}
\tag{5.41}
$$

We have used the notation $p_\alpha^{n'n} = \langle n'\mathbf{k}_0|p_\alpha|n\mathbf{k}_0\rangle$ and defined the quantity f through the second line of Eq. (5.41).

5.6 Equations of motion, Berry phase, and Berry curvature

The equation of motion given in Eq. (5.29) is incomplete when we consider the effects of geometric or Berry phase[5] on the motion of crystal electrons. In the effective Hamiltonian approximation, an electron in a Bloch state $|n\mathbf{k}\rangle$ with wavefunction

$$
\langle \mathbf{r}|n\mathbf{k}\rangle = \psi_{n\mathbf{k}}(\mathbf{r}) = e^{i\mathbf{k}\cdot\mathbf{r}} u_{n\mathbf{k}}(\mathbf{r})
\tag{5.42}
$$

is constrained to move within a single band n. In the presence of a slowly varying static potential $U(\mathbf{r})$, the electron will adiabatically drift in $\mathbf{k}$-space along a path C (see Fig. 5.10). Denoting the electron to initially be in the state $|\psi(t=0)\rangle = |n\mathbf{k}(0)\rangle$, then at some later time t, the state of the electron will evolve into the state $|\psi(t)\rangle$, which within the effective Hamiltonian approximation is $|n\mathbf{k}(t)\rangle$ apart from a phase factor according to the adiabatic theorem of quantum mechanics.[6] (Here $|n\mathbf{k}(t)\rangle$ is given by Eq. (5.42).)

The phase factor that relates $|\psi(t)\rangle$ to $|n\mathbf{k}(t)\rangle$ may be determined by the following analysis. The cell periodic part of a Bloch wavefunction $|u_{n\mathbf{k}}\rangle$ satisfies the equation

$$
\left[\frac{(\mathbf{p} + \hbar\mathbf{k})^2}{2m} + V(\mathbf{r})\right] u_{n\mathbf{k}}(\mathbf{r}) = E_n(\mathbf{k}) u_{n\mathbf{k}}(\mathbf{r}),
\tag{5.43}
$$

or

$$
H(\mathbf{k})|u_{n\mathbf{k}}\rangle = E_n(\mathbf{k})|u_{n\mathbf{k}}\rangle.
\tag{5.44}
$$

[5] M. V. Berry, "Quantal phase factors accompanying adiabatic changes," *Proc. R. Soc. Lond. A* 392(1984), 45.
[6] See, for example, Chapter 10 in D. J. Griffiths, *Introduction to Quantum Mechanics*, 2nd edn. (Upper Saddle River, NJ: Prentice Hall, 2005).

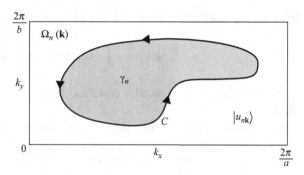

Figure 5.10 Berry curvature in **k**-space. Here $|u_{n\mathbf{k}}\rangle$ is the periodic part of a Bloch state and $\Omega_n(\mathbf{k})$ is the Berry curvature of band n. C denotes a closed loop over which an electron in the nth band picks up a Berry phase of γ_n.

We shall use the solutions $|u_{n\mathbf{k}}\rangle$ in Eq. (5.44) as an instantaneous orthonormal basis to expand $|\psi(t)\rangle$. Equation (5.44) allows, however, an arbitrary **k**-dependent phase factor of $|u_{n\mathbf{k}}\rangle$. To remove this arbitrariness, we make a phase choice, or a gauge, that requires the phase of $|u_{n\mathbf{k}}\rangle$ to be smooth and single-valued along the path C in the parameter or **k**-space. Treating $\mathbf{k}(t)$ as a parameter along the path C, the periodic part of the state of the electron (denoted by $\phi(\mathbf{r}, t) \equiv e^{-i\mathbf{k}\cdot\mathbf{r}} \psi(\mathbf{r}, t)$) evolves according to the time-dependent Schrödinger equation

$$H\big(\mathbf{k}(t)\big) |\phi(t)\rangle = i\hbar \frac{d}{dt} |\phi(t)\rangle. \qquad (5.45)$$

Adiabatically, a system prepared in $|n\mathbf{k}(0)\rangle$ will evolve with $H\big(\mathbf{k}(t)\big)$ and therefore be in the state $|n\mathbf{k}(t)\rangle$ at t.

So, $|\phi(t)\rangle$ can be expressed as

$$|\phi(t)\rangle = \exp\left\{\frac{-i}{\hbar} \int_0^t dt' E_n\big(\mathbf{k}(t')\big)\right\} \exp\big(i\gamma_n(t)\big) |u_{n\mathbf{k}}(t)\rangle. \qquad (5.46)$$

The first exponential factor on the right-hand side of Eq. (5.46) is the familiar dynamical phase factor. The second exponential contains a phase which is in general non-integrable and cannot be written as a function of **k**.

We determine $\gamma_n(t)$ by requiring that $|\phi(t)\rangle$ in Eq. (5.46) satisfy the time-dependent Schrödinger equation. Substituting Eq. (5.46) into Eq. (5.45) leads to the following expression for the time rate of change in $\gamma_n(t)$:

$$\frac{d}{dt}\gamma_n(t) = i\big\langle u_{n\mathbf{k}(t)}\big|\, \nabla_{\mathbf{k}} u_{n\mathbf{k}(t)}\big\rangle \cdot \frac{d\mathbf{k}}{dt}, \qquad (5.47)$$

where

$$\big|\nabla_{\mathbf{k}} u_{n\mathbf{k}(t)}\big\rangle = \sum_i \left[\frac{\big(\big|u_{n(\mathbf{k}+dk\hat{\mathbf{k}}_i)}\big\rangle - |u_{n\mathbf{k}}\rangle\big)}{dk}\right]\hat{\mathbf{k}}_i. \qquad (5.48)$$

Integrating Eq. (5.47) along the path C from some point $\mathbf{k}_0$ with respect to time gives

$$\gamma_n(\mathbf{k}) = i \int_{\mathbf{k}_0}^{\mathbf{k}} \langle u_{n\mathbf{k}'} | \, \boldsymbol{\nabla}_{\mathbf{k}'} u_{n\mathbf{k}'} \rangle \cdot d\mathbf{k}'. \tag{5.49}$$

The normalization of $|u_{n\mathbf{k}}\rangle$ dictates that $\langle u_{n\mathbf{k}} | \, \boldsymbol{\nabla}_{\mathbf{k}} u_{n\mathbf{k}} \rangle = - \langle \boldsymbol{\nabla}_{\mathbf{k}} u_{n\mathbf{k}} | \, u_{n\mathbf{k}} \rangle$, and therefore $\gamma_n(\mathbf{k})$ is real.

Let us define the Berry vector potential $\mathbf{A}_n(\mathbf{k})$ (or the Berry connection) as the quantity

$$\mathbf{A}_n(\mathbf{k}) = i \langle u_{n\mathbf{k}} | \, \boldsymbol{\nabla}_{\mathbf{k}} u_{n\mathbf{k}} \rangle = -\mathrm{Im} \, \langle u_{n\mathbf{k}} | \, \boldsymbol{\nabla}_{\mathbf{k}} u_{n\mathbf{k}} \rangle. \tag{5.50}$$

We note that $\mathbf{A}$ has dimensions of length and is always real, as argued above. The electron's ket vector $|\phi(t)\rangle$, during its adiabatic evolution along the path C in parameter space, acquires an extra phase $\gamma_n(\mathbf{k})$, in addition to the dynamical phase with

$$\gamma_n(\mathbf{k}) = \int_C^{\mathbf{k}} d\mathbf{k}' \cdot \mathbf{A}_n(\mathbf{k}'). \tag{5.51}$$

The form of Eq. (5.51) reveals that we may regard this phase as an Aharonov–Bohm phase caused by an effective "magnetic field"

$$\boldsymbol{\Omega}_n(\mathbf{k}) = \boldsymbol{\nabla}_{\mathbf{k}} \times \mathbf{A}_n(\mathbf{k}), \tag{5.52}$$

which lives in parameter space $\mathbf{k}$. This field $\boldsymbol{\Omega}_n(\mathbf{k})$ is called the Berry curvature.

The Berry vector potential $\mathbf{A}_n(\mathbf{k})$ is gauge dependent. If we make a gauge transformation

$$|u_{n\mathbf{k}}\rangle \to e^{i\mu(\mathbf{k})} |u_{n\mathbf{k}}\rangle, \tag{5.53}$$

where $\mu(\mathbf{k})$ is an arbitrary smooth function of $\mathbf{k}$, then Eq. (5.50) yields

$$\mathbf{A}_n(\mathbf{k}) \to \mathbf{A}_n(\mathbf{k}) - \boldsymbol{\nabla}_{\mathbf{k}} \mu(\mathbf{k}). \tag{5.54}$$

As a consequence, the phase $\gamma_n(\mathbf{k})$ in Eq. (5.51) is gauge dependent, changed by $\mu(\mathbf{k}(0)) - \mu(\mathbf{k}(t))$ in its value after the gauge transformation.

However, for a closed loop C in the parameter space, the extra phase acquired by the state vector $|\phi\rangle$ after it has traversed the loop back to the starting position

$$\gamma_n = \oint_C d\mathbf{k} \cdot \mathbf{A}_n(\mathbf{k}) \tag{5.55}$$

is gauge invariant and is physically observable. The phase γ_n in Eq. (5.55) is known as the Berry phase or geometric phase. Although not gauge invariant, $\gamma_n(\mathbf{k})$ in Eq. (5.51) is also often loosely referred to as the Berry phase.

The Berry curvature $\boldsymbol{\Omega}_n(\mathbf{k})$ defined in Eq. (5.52) is also a gauge-invariant quantity, since $\boldsymbol{\nabla}_{\mathbf{k}} \times (\boldsymbol{\nabla}_{\mathbf{k}} \mu(\mathbf{k})) = 0$ for any smooth function $\mu(\mathbf{k})$. Using Stoke's theorem, we may rewrite the Berry phase for a closed loop in the form

$$\gamma_n = \oint_C d\mathbf{k} \cdot \mathbf{A}_n(\mathbf{k}) = \int_S d\mathbf{s} \cdot \boldsymbol{\Omega}_n(\mathbf{k}), \tag{5.56}$$

where S is any surface enclosed by the closed loop C. Unlike the Berry vector potential, since it is gauge invariant the Berry curvature is observable. The physical origin of the Berry curvature arises from the adiabatic approximation employed, which is equivalent to a projection operation restricting the dynamics of the system to the nth band. The Berry curvature can be viewed as arising from the residual interaction with states in the bands that are not considered in the effective Hamiltonian.

The Berry curvature $\mathbf{\Omega}_n(\mathbf{k})$, an intrinsic local property in $\mathbf{k}$-space of the band structure of a crystal, modifies the motion of the electron. Let us consider the effective Hamiltonian with external potential $U(\mathbf{r})$ in the Bloch representation

$$H = E_n(\mathbf{k}) + U(\mathbf{R}), \tag{5.57}$$

where $\mathbf{R} = i\nabla_{\mathbf{k}}$ is the Wannier coordinate indexing the lattice sites and $E_n(\mathbf{k})$ is the unperturbed band structure. We may perform a gauge transformation to remove the effective Aharonov–Bohm phase or Berry phase in Eq. (5.46), at the expense of adding the Berry vector potential $\mathbf{A}_n(\mathbf{k})$ to $i\nabla_{\mathbf{k}}$ in the argument of U in Eq. (5.57). The transformed Hamiltonian H' is then equal to

$$H' = E_n(\mathbf{k}) + U\big(i\nabla_{\mathbf{k}} + \mathbf{A}_n(\mathbf{k})\big). \tag{5.58}$$

The position operator in this gauge-transformed system is now given by

$$\mathbf{x} = i\nabla_{\mathbf{k}} + \mathbf{A}_n(\mathbf{k}) \tag{5.59}$$

and is no longer simply $\mathbf{R}$. As an operator, the components of $\mathbf{x}$ do not commute with each other. Their commutation relations are given by

$$[x_i, x_j] = i\epsilon^{ijk}\Omega_k^n(\mathbf{k}), \tag{5.60}$$

where ϵ^{ijk} is the Levi-Civita symbol in three dimensions.

We may now derive the equations of motion with the transformed Hamiltonian H' in Eq. (5.58). The equation for $\frac{d\mathbf{k}}{dt}$ may be obtained using

$$\hbar\frac{d\mathbf{k}}{dt} = -i[\mathbf{k}, H'] = -\nabla_{\mathbf{R}}U(\mathbf{R}), \tag{5.61}$$

which is unchanged from Eq. (5.31). However, the equation for the velocity is now given by

$$\hbar\mathbf{v} = -i[\mathbf{x}, H'] = \nabla_{\mathbf{k}}E_n(\mathbf{k}) + \nabla_{\mathbf{x}}U \times \mathbf{\Omega}_n(\mathbf{k}). \tag{5.62}$$

By comparison with Eq. (5.29), there is an extra term involving the Berry curvature. The new term is commonly known as the Luttinger anomalous velocity.[7] For example, for a

---

[7] M. C. Chang and Q. Niu, "Berry phase, hyperorbits, and the Hofstadter spectrum," *Phys. Rev. Lett.* 75(1995), 1348.

crystal in a static electric field $\mathcal{E}$ with $(U(\mathbf{x}) = -e\mathcal{E} \cdot \mathbf{x})$ and a weak magnetic field $\mathbf{B}$, the semiclassical equations of motion become

$$\hbar \frac{d\mathbf{k}}{dt} = e\mathcal{E} + \frac{e}{c}\mathbf{v} \times \mathbf{B} \tag{5.63}$$

and

$$\hbar\mathbf{v} = \nabla_{\mathbf{k}} E_n(\mathbf{k}) - e\mathcal{E} \times \mathbf{\Omega}_n(\mathbf{k}). \tag{5.64}$$

The importance of the contribution of the Berry curvature term to the dynamics of crystal electrons depends on the systems of interest. In many cases, $\mathbf{\Omega}_n(\mathbf{k})$ may be neglected. However, $\mathbf{\Omega}_n(\mathbf{k})$ is nonzero in a wide range of materials, particularly those with broken time-reversal and inversion symmetries, and has given rise to insightful understanding of a number of interesting phenomena, including the anomalous Hall effect, the quantum Hall effect, electric polarization, orbital magnetization, topological insulators, etc.

Let us analyze the general form of the Berry curvature $\mathbf{\Omega}_n(\mathbf{k})$ by examining the velocity expression in Eq. (5.64). The velocity expression should be invariant under time-reversal or spatial inversion operations if the crystal has these symmetries. Under time reversal, $\mathbf{v} \to -\mathbf{v}$, $\mathbf{k} \to -\mathbf{k}$, and $\mathcal{E} \to \mathcal{E}$; under spatial inversion, $\mathbf{v} \to -\mathbf{v}$, $\mathbf{k} \to -\mathbf{k}$, and $\mathcal{E} \to -\mathcal{E}$. Thus, if the crystal has time-reversal symmetry, the invariance form shown in Eq. (5.64) dictates that

$$\mathbf{\Omega}_n(-\mathbf{k}) = -\mathbf{\Omega}_n(\mathbf{k}), \tag{5.65}$$

and, if the crystal has spatial inversion symmetry,

$$\mathbf{\Omega}_n(-\mathbf{k}) = \mathbf{\Omega}_n(\mathbf{k}). \tag{5.66}$$

This leads to the conclusion that, for crystals (such as Al) with both time-reversal and spatial inversion symmetry, the Berry curvature is identically zero throughout the Brillouin zone, with the possible exception of nonzero values at singular points in $\mathbf{k}$-space due to band degeneracy. In such a case, one may use the simpler velocity expression given in Eq. (5.29).

For systems with either broken time-reversal or broken spatial inversion symmetry, the proper equation for the velocity should include the Berry curvature term. In ferromagnetic materials, where time-reversal symmetry is broken, such as fcc Fe under a uniform $\mathcal{E}$-field, the Berry curvature term in Eq. (5.64) gives rise to a current that is perpendicular to $\mathcal{E}$ and constitutes an intrinsic mechanism for the so-called anomalous Hall effect (in which a strong additional Hall resistivity results from the magnetization) in these materials. Presently, this research area is dynamic and evolving.[8]

---

[8] See, for example, the review article by D. Xiao, M. C. Chang, and Q. Niu, "Berry phase effects on electronic properties," *Rev. Mod. Phys.* 82(2010), 1959.

6 Many-electron interactions: the homogeneous interacting electron gas and beyond

Up until now, we have treated the electronic Hamiltonian of solids as an effective Hamiltonian of independent electrons moving in a mean (or average) field due to the ions and other electrons. This is a very useful conceptual approach and can be quantitative for certain physical properties if the appropriate field or self-consistent potential is used, as we shall see in the next chapter on density functional theory.

The proper choice of the mean-field potential $V(\mathbf{r})$ appearing in the one-electron Schrödinger equation

$$\left(-\frac{\hbar^2}{2m}\nabla^2 + V(\mathbf{r})\right)\psi(\mathbf{r}) = \varepsilon\psi(\mathbf{r}) \tag{6.1}$$

is a complex and subtle issue. Underlying this problem is the question of how best to approximate the effects of electron–electron interactions in the exact Hamiltonian H within the Born–Oppenheimer approximation for physical quantities of interest:

$$H = \sum_i \left(\frac{\mathbf{p}_i^2}{2m} + \sum_{\mathbf{R}} V_{\text{ion}}(\mathbf{r}_i - \mathbf{R})\right) + \frac{1}{2}\sum_{i\neq j}\frac{e^2}{|\mathbf{r}_i - \mathbf{r}_j|}, \tag{6.2}$$

where $V_{\text{ion}}(\mathbf{r}_i - \mathbf{R})$ is the ionic potential seen by the ith electron. Very often, an effective one-electron Schrödinger-like equation of the form Eq. (6.1) is obtained by seeking the ground-state energy and wavefunction of an interacting many-electron system using a variational principle approach.

In the Hartree–Fock (HF) treatment, for example, we assume that the ground-state many-electron wavefunction is of the form of a single Slater determinant:

$$\Psi_{\text{HF}}(\mathbf{r}_1\sigma_1, \mathbf{r}_2\sigma_2, ..., \mathbf{r}_N\sigma_N)$$

$$= \frac{1}{\sqrt{N!}}\begin{vmatrix} \phi_1(\mathbf{r}_1\sigma_1) & \phi_1(\mathbf{r}_2\sigma_2) & \phi_1(\mathbf{r}_3\sigma_3) & ... & \phi_1(\mathbf{r}_N\sigma_N) \\ \phi_2(\mathbf{r}_1\sigma_1) & \phi_2(\mathbf{r}_2\sigma_2) & \phi_2(\mathbf{r}_3\sigma_3) & ... & \phi_2(\mathbf{r}_N\sigma_N) \\ \vdots & \vdots & \vdots & & \vdots \\ \phi_N(\mathbf{r}_1\sigma_1) & \phi_N(\mathbf{r}_2\sigma_2) & \phi_N(\mathbf{r}_3\sigma_3) & ... & \phi_N(\mathbf{r}_N\sigma_N) \end{vmatrix}. \tag{6.3}$$

Here N is the number of electrons in the system, and $\mathbf{r}_i$ and σ_i denote the spatial and spin coordinates of the ith electron, respectively. ϕ_λ are single-particle orbitals to be determined in the calculation. Since an electron has spin $\frac{1}{2}$, σ_i may be up or down. Applying the variational principle, i.e. demanding the variation of the ground-state energy with respect to any of the single-particle orbitals ϕ_λ to be zero,

$$\frac{\delta\langle\Psi_{HF}|H|\Psi_{HF}\rangle}{\delta\phi_\lambda} = 0 \tag{6.4}$$

leads to the Hartree–Fock equations

$$-\frac{\hbar^2}{2m}\nabla^2\phi_i(\mathbf{r},\sigma) + V_{\text{ion}}(\mathbf{r})\phi_i(\mathbf{r},\sigma) + V_{\text{H}}(\mathbf{r})\phi_i(\mathbf{r},\sigma)$$
$$+ \sum_{\sigma'}\int V_x(\mathbf{r},\mathbf{r}',\sigma,\sigma')\phi_i(\mathbf{r}',\sigma')d\mathbf{r}' = \varepsilon_i\phi_i(\mathbf{r},\sigma), \tag{6.5}$$

where

$$V_{\text{H}}(\mathbf{r}) = \int d\mathbf{r}'\frac{e^2}{|\mathbf{r}-\mathbf{r}'|}\sum_{\sigma'}\sum_j^{\text{occ}}|\phi_j(\mathbf{r}',\sigma')|^2, \tag{6.6}$$

and

$$V_x(\mathbf{r},\mathbf{r}',\sigma,\sigma') = -\sum_j^{\text{occ}}\frac{e^2}{|\mathbf{r}-\mathbf{r}'|}\phi_j^*(\mathbf{r}',\sigma')\phi_j(\mathbf{r},\sigma). \tag{6.7}$$

Equations (6.5)–(6.7) need to be solved self-consistently (i.e. iteratively). The Hartree–Fock equations differ from the Hartree equations by an additional term V_x on the left-hand side of Eq. (6.5), known as the exchange term. The exchange term arises from the antisymmetric form of the many-fermion wavefunction, which introduces an effective interaction between electrons owing to the Pauli exclusion principle. It is a nonlocal operator of the integral form when acting on a single-particle orbital $\phi_i(\mathbf{r},\sigma)$, i.e.

$$\sum_{\sigma'}\int V_x(\mathbf{r},\mathbf{r}',\sigma,\sigma')\phi_i(\mathbf{r}',\sigma')d\mathbf{r}', \tag{6.8}$$

which, in general, is rather difficult to treat quantitatively.

The eigenvalue ε_i in the Hartree–Fock equation (Eq. (6.5)) is a Lagrange multiplier in the variational procedure, constraining the normalization of the single-particle orbitals. Thus, rigorously speaking, they are not electron excitation energies of the system. However, through the use of Koopmans' theorem, these eigenvalues may be roughly interpreted as electron excitation energies to lowest order in the Coulomb interaction. (We will leave the proof of Koopmans' theorem as a homework exercise for the interested reader.)

The Hartree–Fock method, being a ground-state theory, provides an important first step in treating the effect of electron–electron interactions on the ground-state properties of condensed matter, such as the total energy, and structural and vibrational properties. In particular, the Hartree–Fock equations may be solved exactly for the case of the homogeneous interacting electron gas. In the remainder of this chapter, we shall examine the effects of many-electron interactions on the electronic properties of the electron gas as well as simple metals within the Hartree–Fock approximation and beyond.

6.1 The homogeneous interacting electron gas or jellium model

An important model system that illustrates the role of many-electron interactions in solids without having to deal with the complexity of the atomic structure is that of the uniform interacting electron gas or the jellium model. In this model, we consider a system of interacting electrons in a uniform positive-charged background so that the total system is neutral. The jellium model is then an idealized metal for which we can study the effects of electron interactions on the total energy, electron (or quasiparticle) excitations, and other properties as a function of electron density.

In the zero-order approximation, we neglect electron–electron interactions; the jellium model reduces to that of a non-interacting charged Fermi gas, i.e. the Sommerfeld model. In the non-interacting limit, the Hamiltonian can be written as

$$H_0 = \sum_i \frac{\mathbf{p}_i^2}{2m} + V^{\mathrm{PB}} = \sum_i H_0^i, \tag{6.9}$$

where V^{PB} is the potential due to the positive background. For an N electron paramagnetic system, we may take the ground-state wavefunction to be in the form

$$\Psi_0 = \prod_{i=1}^N \phi_i(\mathbf{r}_i, \sigma_i), \tag{6.10}$$

with

$$\phi_i(\mathbf{r}, \sigma) = \phi_{\mathbf{k}\lambda}^0(\mathbf{r}) = \frac{1}{\sqrt{\Omega}} e^{i\mathbf{k}\cdot\mathbf{r}} \chi_\lambda, \tag{6.11}$$

where Ω is the volume of the sample and χ_λ is a spin function with $\lambda = \uparrow$ or $\downarrow$. The value of $\mathbf{k}$ is within the Fermi sphere, i.e. $k \leq k_{\mathrm{F}}$, with k_{F} the Fermi wavevector defined through

$$N = 2 \sum_{k \leq k_{\mathrm{F}}} 1, \tag{6.12}$$

where the factor 2 is due to spin, and the orbital energy is independent of λ given by

$$\varepsilon_0(\mathbf{k}) = \frac{\hbar^2 k^2}{2m}.$$

The total energy of the system is

$$E_0 = 2 \sum_{k \leq k_{\mathrm{F}}} \varepsilon_0(\mathbf{k}). \tag{6.13}$$

The jellium model has only one variable physical parameter, the density of the electrons $n = \frac{N}{\Omega}$ or $k_{\mathrm{F}} = (3\pi^2 n)^{1/3}$ in the paramagnetic phase. In the study of the interacting electron gas, it is often useful to express physical quantities in terms of another related parameter r_s (the typical interelectron distance in units of the Bohr radius a_0):

$$r_s = \left(\frac{3}{4\pi}\right)^{\frac{1}{3}} \left(\frac{1}{n}\right)^{\frac{1}{3}} a_0^{-1}. \tag{6.14}$$

The dimensionless parameter r_s is known as the electron gas parameter and may be viewed as the effective radius of the volume occupied by the electron in units of the Bohr radius. In the literature, the density dependence of physical quantitites is traditionally expressed in terms of r_s. For example, the Fermi wavevector $k_F = 1.92/r_s$ (a_0^{-1}), the Fermi energy $E_F^0 = \hbar^2 k_F^2/2m = 3.68/r_s^2$ (Ry), and the plasma frequency $\hbar\omega_p = 3.46/r_s^{3/2}$ (Ry). Here Ry denotes the energy unit of one Rydberg. In the Sommerfeld model, the average kinetic energy per electron, $\langle E_{KE}\rangle$, is given by

$$\langle E_{KE}\rangle = \frac{2}{N}\sum_{k<k_F} \varepsilon_k^0 = \frac{3}{5}E_F^0 = \frac{2.21}{r_s^2} \text{ (Ry)}. \tag{6.15}$$

The average potential energy per electron $\langle E_{PE}\rangle$ in the interacting system can be estimated as

$$\langle E_{PE}\rangle \approx \frac{e^2}{r_s a_0} = \frac{2}{r_s} \text{ (Ry)}. \tag{6.16}$$

Thus, $r_s \sim \langle E_{PE}\rangle/\langle E_{KE}\rangle$ is a measure of the ratio of the average potential energy to the kinetic energy in the electron gas. A system with $r_s < 1$ corresponds to one with high density and dominant kinetic energy; a system with $r_s > 1$ corresponds to one with low density and dominant potential energy.

Since r_s is a measure of the ratio of the Coulomb energy to the kinetic energy of the electron gas, we may use it as a small parameter (if r_s is small) to do perturbation theory to determine the effects of Coulomb interaction on the properties of the electron gas. However, real materials typically have values of r_s between 1 and 5 so that $\langle E_{KE}\rangle \sim \langle E_{PE}\rangle$, making the system strongly interacting, and making it difficult to apply traditional perturbation theory. Figure 6.1 illustrates the range of r_s. The importance of treating many-electron effects going beyond Hartree–Fock will generally depend on r_s and the dimensionality of the system under consideration. For example, in three dimensions and for $r_s > 100$, electron correlation is so strong that the electrons crystallize into a crystalline arrangement forming an electronic lattice (the Wigner crystal).[1] As we will see below, one can determine the exact correlation energy in the limit of extremely large r_s.

[1] E. Wigner, "On the interaction of electrons in metals," *Phys. Rev.* 46(1934), 1002.

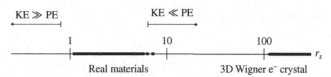

Figure 6.1 Schematic of the range of the electron gas parameter r_s on log scale showing the ranges where the kinetic energy (KE) dominates over the potential energy (PE) and vice versa. Typical metals in three dimensions have r_s between 1 and 6. At extremely high r_s, the electrons form a Wigner lattice.

6.2 Hartree–Fock treatment of the interacting electron gas

To gain some insight into electron interaction effects, we solve the uniform interacting electron gas problem analytically within the Hartree–Fock approach. We shall assume that the single-particle orbitals in Eq. (6.3) are planewaves times a spin function, i.e. the orbitals are specified by the quantum numbers $\mathbf{k}$ and λ with

$$\phi_i^{\text{HF}} = \phi_{\mathbf{k}\lambda} = \frac{1}{\sqrt{\Omega}} e^{i\mathbf{k}\cdot\mathbf{r}} \chi_\lambda, \tag{6.17}$$

where χ_λ is the spin function $\begin{pmatrix} 1 \\ 0 \end{pmatrix}$ for $\lambda = \uparrow$ and $\begin{pmatrix} 0 \\ 1 \end{pmatrix}$ for $\lambda = \downarrow$, and show that $\phi_{\mathbf{k}\lambda}$ are the solutions of the self-consistent Hartree–Fock equations (6.5)–(6.7). We note that planewave solutions result in a uniform electron charge density. This results in $V_{\text{ion}}(\mathbf{r}) + V_{\text{H}} = 0$, since the electron charge density cancels the positively charged background density. Only the exchange term V_x given by Eq. (6.7) survives among the potential energy terms in Eq. (6.5).

Equation (6.5) reduces to

$$-\frac{\hbar^2}{2m}\nabla^2\phi_i(\mathbf{r},\sigma) + \sum_{\sigma'}\int V_x(\mathbf{r},\mathbf{r}',\sigma,\sigma')\phi_i(\mathbf{r}',\sigma')d\mathbf{r}' = \varepsilon_i\phi_i(\mathbf{r},\sigma). \tag{6.18}$$

Substituting in ϕ_i from Eq. (6.17), we have

$$\frac{\hbar^2 k^2}{2m}\phi_{\mathbf{k}\lambda}(\mathbf{r}) - \frac{1}{\Omega^{3/2}}\sum_{k'<k_F}\sum_{\lambda'}\int d\mathbf{r}' \frac{e^2}{|\mathbf{r}-\mathbf{r}'|}e^{-i\mathbf{k}'\cdot\mathbf{r}'}e^{i\mathbf{k}'\cdot\mathbf{r}}e^{i\mathbf{k}\cdot\mathbf{r}}\delta_{\lambda'\lambda}\chi_\lambda = \varepsilon_\lambda(\mathbf{k})\phi_{\mathbf{k}\lambda}(\mathbf{r}).$$

$$\tag{6.19}$$

The second term on the left-hand side, the exchange term, when summed over λ' may be rearranged (assuming equal occupation of $\mathbf{k}\uparrow$ and $\mathbf{k}\downarrow$ orbitals, i.e. a paramagnetic electron gas), giving

$$\text{exchange term} = -\frac{1}{\Omega^{3/2}} \sum_{k' < k_{\mathrm{F}}} \int d\mathbf{r}' \frac{e^2}{|\mathbf{r} - \mathbf{r}'|} e^{-i\mathbf{k}' \cdot (\mathbf{r}' - \mathbf{r})} e^{-i\mathbf{k} \cdot (\mathbf{r} - \mathbf{r}')} e^{i\mathbf{k} \cdot \mathbf{r}} \chi_\lambda$$

$$= -\frac{1}{\Omega} \sum_{k' < k_{\mathrm{F}}} \int d\mathbf{r}' \frac{e^2}{|\mathbf{r} - \mathbf{r}'|} e^{-i(\mathbf{k} - \mathbf{k}') \cdot (\mathbf{r} - \mathbf{r}')} \phi_{\mathbf{k}\lambda}(\mathbf{r})$$

$$= \left(-\frac{1}{\Omega} \sum_{k' < k_{\mathrm{F}}} \frac{4\pi e^2}{|\mathbf{k} - \mathbf{k}'|^2} \right) \phi_{\mathbf{k}\lambda}(\mathbf{r}). \tag{6.20}$$

Equation (6.19) becomes

$$\left\{ \frac{\hbar^2 k^2}{2m} - \int_{k' < k_{\mathrm{F}}} \frac{d\mathbf{k}'}{(2\pi)^3} \frac{4\pi e^2}{|\mathbf{k} - \mathbf{k}'|^2} \right\} \phi_{\mathbf{k}\lambda}(\mathbf{r}) = \varepsilon_\lambda(\mathbf{k}) \phi_{\mathbf{k}\lambda}(\mathbf{r}). \tag{6.21}$$

We have thus shown that the planewaves $\phi_{\mathbf{k}\lambda}$ are solutions to the Hartree–Fock equation in the case of the uniform electron gas,

$$H_{\mathrm{HF}} \phi_{\mathbf{k}\lambda}(\mathbf{r}) = \varepsilon_\lambda(\mathbf{k}) \phi_{\mathbf{k}\lambda}(\mathbf{r}), \tag{6.22}$$

with eigenvalues (we shall drop the subscript λ from now on since we are considering a paramagnetic system)

$$\varepsilon_{\mathrm{HF}}(\mathbf{k}) = \frac{\hbar^2 k^2}{2m} - \int_{k' < k_{\mathrm{F}}} \frac{d\mathbf{k}'}{(2\pi)^3} \frac{4\pi e^2}{|\mathbf{k} - \mathbf{k}'|^2}, \tag{6.23}$$

where the integral can be evaluated to give

$$\varepsilon_{\mathrm{HF}}(\mathbf{k}) = \frac{\hbar^2 k^2}{2m} - \frac{2e^2}{\pi} k_{\mathrm{F}} F\left(\frac{k}{k_{\mathrm{F}}} \right). \tag{6.24}$$

Here $F(x)$ is the function

$$F(x) = \frac{1}{2} + \frac{(1 - x^2)}{4x} \ln\left| \frac{1 + x}{1 - x} \right|. \tag{6.25}$$

Figure 6.2 illustrates the function $F(x)$ and the Hartree–Fock energy eigenvalues for a paramagnetic electron gas. If one invokes Koopmans' theorem, then the eigenvalues are a rough estimate of the electron (or quasiparticle) excitation energy of the system. The eigenvalues describe the energy for exciting a single-particle-like excitation within the Hartree–Fock approximation, including only the exchange interaction. The single-particle energy can be written as a sum of two terms,

$$\varepsilon_{\mathrm{HF}}(\mathbf{k}) = \varepsilon_0(\mathbf{k}) + \Sigma_x^{\mathrm{HF}}(\mathbf{k}) \tag{6.26}$$

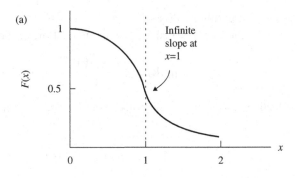

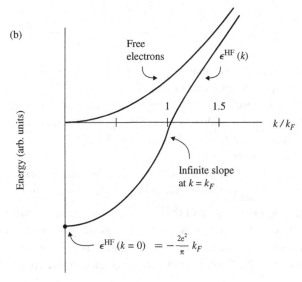

Figure 6.2 (a) The function $F(x)$ given in Eq (6.25) with infinite slope at $x = 1$. (b) Electron energy dispersion relation of a free electron gas and an interacting electron gas in the Hartree–Fock (HF) approximation in the paramagnetic phase.

with

$$\Sigma_x^{\mathrm{HF}}(\mathbf{k}) = -\frac{2e^2}{\pi} k_{\mathrm{F}} F\left(\frac{k}{k_{\mathrm{F}}}\right).\tag{6.27}$$

The quantity Σ_x given in Eq. (6.27) is called the self energy of the excited particle. Owing to this exchange self energy, the single-particle states are lowered in energy and the occupied bandwidth $W_{\mathrm{HF}} = \varepsilon_{\mathrm{HF}}(k_{\mathrm{F}}) - \varepsilon_{\mathrm{HF}}(0)$ increases from the independent-particle value of $\hbar^2 k_{\mathrm{F}}^2/2m$ to $\hbar^2 k_{\mathrm{F}}^2/2m + e^2 k_{\mathrm{F}}/\pi$.

From the functional form of $F(x)$, we note that $F(1) = 1/2$ and its derivative is

$$\frac{d}{dx}F(x) = \frac{1}{4x}\left(\frac{1+x^2}{x}\ln\left|\frac{1+x}{1-x}\right| - 2\right),\tag{6.28}$$

which has a logarithmic divergence as $x \to 1$ or $k \to k_F$. This divergence leads to behaviors for the electrons that are peculiar to the Hartree–Fock results. The band velocity

$$\mathbf{v}_{HF}(\mathbf{k}) = \frac{1}{\hbar} \frac{\partial \varepsilon_{HF}(\mathbf{k})}{\partial \mathbf{k}} \to \infty \text{ as } k \to k_F. \tag{6.29}$$

The effective mass in the vicinity of k_F can be calculated using the relation

$$\frac{m}{m_{HF}^*} = \frac{\partial \varepsilon_{HF}}{\partial \varepsilon_0(\mathbf{k})},$$

which gives

$$\frac{m}{m_{HF}^*} = 1 + \frac{m}{\hbar^2 k} \frac{\partial}{\partial k} \Sigma_x^{HF}(\mathbf{k}). \tag{6.30}$$

Thus, the effective mass $m_{HF}^* \to 0$ as $k \to k_F$. Both behaviors are unphysical. Therefore single-particle excitation calculations based on the Hartree–Fock approximation typically give poor results for metallic systems. Some other consequences of the singular behavior of $F(x)$ at $x = 1$ include a vanishing of the electron density of states $D(\varepsilon)$ at $\varepsilon = \varepsilon_F$ and an electronic specific heat that behaves like $C_{e\ell}(T) \sim \frac{T}{\ln|T|}$ instead of a linear T dependence at low temperature T.

The above unphysical behavior can be traced back to the divergence of the Fourier transform $\frac{4\pi e^2}{\Omega k^2}$ of the Coulomb interaction $\frac{e^2}{r}$, as k goes to zero. The divergence reflects the long-range nature of the Coulomb interaction. Physically, in a metallic system, $\frac{e^2}{r}$ will be screened by the other electrons, especially at large values of r. A screened Coulomb potential would eliminate the divergence as $k \to 0$. For example, within Thomas–Fermi screening, the Coulomb potential between two electrons is given by (see Chapter 8)

$$V(r) = \frac{e}{r} e^{-K_s r}, \tag{6.31}$$

which would yield a non-divergent $V(k) = \frac{4\pi e^2}{\Omega(k^2 + K_s^2)}$, where K_s is the Thomas–Fermi screening wavevector. In general, a treatment going beyond the Hartree–Fock approximation is required to obtain accurate excited-state electronic properties.

The above discussion shows that a Slater determinant of planewave single-particle orbitals indeed solves the Hartree–Fock equations. However, this solution need not be the lowest energy or the ground-state solution. Overhauser showed that, at low densities (i.e. high r_s), there are solutions in the form of charge or spin density waves that are lower in energy than the uniform density planewave solution.

6.3 Ground-state energy: Hartree–Fock and beyond

The Hartree–Fock approximation, being a variational approach for the ground-state wavefunction, is expected to be more accurate for the ground-state energy as opposed to individual single-particle excitation energies. The ground-state energy per particle for an

N-particle system is given by

$$E_{GS} = \frac{1}{N}\langle \Psi|H|\Psi \rangle, \tag{6.32}$$

where H is the full Hamiltonian given by Eq. (6.2) and Ψ is the many-body ground-state wavefunction. For the interacting electron gas treated in the Hartree–Fock approximation, the ground-state energy per particle defined by Eq. (6.32) is

$$E_{GS} = E_{KE} + E_{X}, \tag{6.33}$$

where E_{KE} is the kinetic energy per particle, which for our planewave solutions is the same as the non-interacting kinetic energy, and E_{X} is the exchange energy per particle with (for the paramagnetic case)

$$E_{X} = \frac{1}{2}\left(\frac{2}{N}\right)\sum_{k<k_F} \Sigma_x^{HF}(\mathbf{k}). \tag{6.34}$$

Since the Hartree–Fock wavefunction is a Slater determinant of planewaves, the kinetic energy is the same as the Sommerfeld model (i.e. the non-interacting case) given by Eq. (6.15) for an electron gas of density given by r_s. The exchange energy may be evaluated using Eqs. (6.27) and (6.34):

$$\begin{aligned}
E_{X} &= \frac{1}{N}\sum_{k<k_F}\Sigma_x^{HF}(k) = \frac{1}{N}\frac{\Omega}{(2\pi)^3}\int d\mathbf{k}\left[-\frac{2e^2}{\pi}k_F F\left(\frac{k}{k_F}\right)\right] \\
&= -\frac{1}{n}\frac{e^2 k_F^4}{\pi^3}\int_0^1 x^2 F(x)dx = -\frac{3}{4}\frac{e^2 k_F}{\pi}.
\end{aligned} \tag{6.35}$$

The ground-state energy per particle, as expressed in terms of the electron gas parameter r_s, is equal to

$$E_{GS}^{HF} = \frac{2.21}{r_s^2} - \frac{0.916}{r_s} \text{ (Ry)}. \tag{6.36}$$

For $r_s = 2$, which is typical of metals like Al, we see that the kinetic energy is nearly the same as the exchange energy, showing that the system is not weakly interacting in this sense.

The Hartree–Fock method yields two terms for the expression of the electron excitation energies and ground-state energy per particle, given by Eq. (6.26) and Eq. (6.36), respectively. We expect that a more accurate theoretical treatment of the many-body wavefunction, going beyond Hartree–Fock, would result in more terms in both expressions. In the limit of small r_s (i.e. high density), the kinetic energy dominates, and we may expect the ground-state energy per particle (and other quantities) to be expandable in a series in powers of r_s. It is reasonable to expect E_{GS} to be of the form

$$E_{GS} = \frac{2.21}{r_s^2} - \frac{0.916}{r_s} + A + B\ln r_s + Cr_s + \dots \text{ (Ry)} \tag{6.37}$$

in the regime of $r_s < 1$.

Using the Rayleigh–Schrödinger perturbation theory for the uniform interacting electron gas, Gell-Mann and Brueckner showed in 1957[2] that indeed E_{GS} may be expanded in the above form with A $= -0.096$ and B $= 0.0622$. The portion of the energy going beyond Hartree–Fock is traditionally called the correlation energy. Thus, the expression for E_{GS} is usually divided into

$$E_{GS} = E_{GS}^{HF} + E_C = E_{KE} + E_X + E_C. \tag{6.38}$$

For the interacting electron gas at small r_s,

$$E_C = -0.096 + 0.0622 \ln(r_s) + \cdots, \tag{6.39}$$

in units of Rydberg.

Figure 6.3 depicts the calculated correlation energy of the interacting electron gas as a function of r_s, using the results of many-body perturbation theory given by Eq. (6.39). Since the result is derived from perturbation theory, it is accurate only in the limit of $r_s < 1$. Indeed, a better wavefunction (i.e. a better description of electron–electron interaction effects) can only lower the ground-state energy from that of Hartree–Fock. The correlation energy should always be negative. The positive values of E_C in Eq. (6.39) and Fig. 6.3 at large r_s are a sign of the breakdown of perturbation theory. For real metals with $2 < r_s < 6$, the expression for E_C in Eq. (6.39) is inaccurate.

For intermediate values of r_s, there are several formulas for E_C available in the literature. One of the simplest and yet highly accurate expressions for E_C is the Wigner interpolation formula. Wigner was able to arrive at a formula that is good for the range of r_s of real materials by using simple and sound physical reasoning. He considered the interacting electron gas at very low density, i.e. $r_s \gg 1$. In this limit, the potential energy dominates over the kinetic energy and the electrons condense into an electron crystal in three dimensions at $r_s \geq 100$. (Such Wigner crystals have been observed in dilute electron systems, such as electrons on a liquid He surface.) To calculate the ground-state energy of the electron crystal, let us divide the crystal into Wigner–Seitz cells. For a close-packed structure, the Wigner–Seitz cell may be replaced by a sphere of volume

$$V_{WS} = \frac{4\pi}{3} r_s^3. \tag{6.40}$$

The energy per electron is the energy of the electron in one of the Wigner–Seitz spheres, which consists of a kinetic energy term E_{KE} and a potential energy term E_{PE}. E_{KE} comes from the zero-point motion of the electron and scales like $r_s^{-3/2}$, which may be neglected in the $r_s \to \infty$ limit. The potential energy (which by definition is the exchange plus correlation energy) comes from the Coulomb interaction of the charges involved and has two contributions:

$$E_{PE} = E_{e-p} + E_{p-p}. \tag{6.41}$$

---

[2] M. Gell-Mann and K. A. Brueckner, "Correlation energy of an electron gas at high density," *Phys. Rev.* 106(1957), 364.

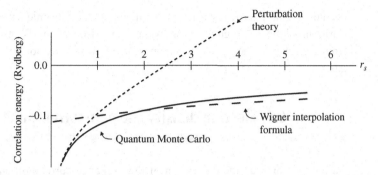

Figure 6.3 Correlation energy of the interacting electron gas as a function of r_s from different calculations.

Here E_{e-p} is the interaction energy between the electron at the center of the Wigner–Seitz sphere and the positive neutralizing background density $n_p = 3/(4\pi r_s^3)$, and it is given by

$$E_{e-p} = \int_{WS} d\mathbf{r} \left(-\frac{e^2}{r} \right) \left(\frac{3}{4\pi r_s^3} \right) = -\frac{3}{r_s} \text{ (Ry)}. \tag{6.42}$$

The second term on the right-hand side of Eq. (6.41), E_{p-p}, is the self energy of the positive neutralizing background charge, given by

$$E_{p-p} = \frac{6}{5 r_s} \text{ (Ry)}. \tag{6.43}$$

Combining Eq. (6.42) and Eq. (6.43) yields

$$E_{PE} = -\frac{1.8}{r_s} \text{ (Ry)}. \tag{6.44}$$

Using the fact that $E_{PE} = E_X + E_C$ and the Hartree–Fock result of $E_X = -0.916/r_s$, we then deduce that in the $r_s \to \infty$ limit, the correlation energy is

$$E_C(r_s \gg 1) = -\frac{0.88}{r_s} \text{ (Ry)}. \tag{6.45}$$

Using an estimate of E_C ($r_s = 1$) $= -0.10$ Ry from experiment, Wigner arrived at a simple interpolation formula of[3]

$$E_C(r_s) = -\frac{0.88}{r_s + 7.8} \text{ (Ry)}. \tag{6.46}$$

The Wigner interpolation formula has been proved to be accurate and useful in many applications, and although many other correlation energy formulas have been derived over the years, the Wigner interpolation formula remains among the best. Very accurate values of E_C for the electron gas over a wide range of r_s of interest ($r_s \approx 0–100$) have now

---

[3] There was an error in the original 1934 Wigner paper. The correct formula is given here.

been obtained numerically using modern computation techniques such as quantum Monte Carlo simulations.[4] Figure 6.3 shows a comparison of the Wigner formula with results from perturbation theory and quantum Monte Carlo simulations of the interacting electron gas.

6.4 Electron density and pair-correlation functions

Before we continue the discussion of the exchange-correlation energy E_{xc} and use it in applications, let us understand more of its physical origin. For this purpose, we introduce the concept of a pair-correlation function in a quantum many-body system.

We consider an N-body fermion wavefunction $\Psi(x_1, x_2, x_3, \ldots, x_N)$, where $x_i \equiv (\mathbf{r}_i, \sigma_i)$, and is normalized such that

$$\int |\Psi(x_1, x_2, \ldots, x_N)|^2 d^N x = 1. \tag{6.47}$$

The electron density operators are denoted as

$$\hat{\rho}(\mathbf{r}) = \sum_i \delta(\mathbf{r} - \mathbf{r}_i)\delta_{\sigma_i,\uparrow} + \sum_i \delta(\mathbf{r} - \mathbf{r}_i)\delta_{\sigma_i,\downarrow} = \hat{\rho}_\uparrow(\mathbf{r}) + \hat{\rho}_\downarrow(\mathbf{r}). \tag{6.48}$$

The electron density is given by

$$\rho(\mathbf{r}) = \langle \Psi | \hat{\rho}(\mathbf{r}) | \Psi \rangle = \sum_i \sum_\sigma \int \Psi^*(x_1, \ldots, x_N)\delta(\mathbf{r} - \mathbf{r}_i)\Psi(x_1, \ldots, x_N) d^N x$$

$$= N \sum_\sigma \int |\Psi(\mathbf{r}\sigma, x_2, x_3, \ldots, x_N)|^2 dx_2 \ldots dx_N. \tag{6.49}$$

The factor of N results from all the particles being identical. We have $\int \rho(\mathbf{r})d\mathbf{r} = N$, as required.

The pair-correlation function $g(\mathbf{r}, \mathbf{r}')$ is defined such that $g(\mathbf{r}, \mathbf{r}')\rho(\mathbf{r}')$ is the density of electrons at $\mathbf{r}'$, given that there is one electron at $\mathbf{r}$. Hence the combination $\rho(\mathbf{r})g(\mathbf{r}, \mathbf{r}')\rho(\mathbf{r}')$ is the distribution of particle pairs in the system. That is,

$$\rho(\mathbf{r})g(\mathbf{r}, \mathbf{r}')\rho(\mathbf{r}') = \left\langle \Psi | \sum_{i<j} \delta(\mathbf{r} - \mathbf{r}_i)\delta(\mathbf{r}' - \mathbf{r}_j) | \Psi \right\rangle$$

$$= \sum_{i<j} \int |\Psi(x_1, x_2, \ldots, x_N)|^2 \delta(\mathbf{r} - \mathbf{r}_i)\delta(\mathbf{r}' - \mathbf{r}_j) dx_1 \ldots dx_N$$

$$= \frac{N(N-1)}{2} \left\langle \int |\Psi(\mathbf{r}\sigma, \mathbf{r}'\sigma', x_3, \ldots, x_N)|^2 dx_3 \ldots dx_N \right\rangle_{\text{spin}}. \tag{6.50}$$

[4] D. M. Ceperley and B. J. Alder, "Ground state of the electron gas by a stochastic method," *Phys. Rev. Lett.* 45(1980), 566.

Here $\langle\rangle_{\text{spin}}$ denotes integration over the spin degree of freedom of particles 1 and 2. We may denote

$$\rho_2(\mathbf{r},\mathbf{r}') = \left\langle \int |\Psi(\mathbf{r}\sigma,\mathbf{r}'\sigma',x_3,\ldots,x_N)|^2 dx_3\ldots dx_N \right\rangle_{\text{spin}}, \qquad (6.51)$$

which is the two-particle density matrix. Thus, $g(\mathbf{r},\mathbf{r}')$ is succinctly expressed as

$$g(\mathbf{r},\mathbf{r}') = \frac{N(N-1)}{2}\frac{\rho_2(\mathbf{r},\mathbf{r}')}{\rho(\mathbf{r})\rho(\mathbf{r}')}. \qquad (6.52)$$

A number of general properties of the pair-correlation function can be seen from the above expressions. For example,

$$\int g(\mathbf{r},\mathbf{r}')\rho(\mathbf{r})\rho(\mathbf{r}')d\mathbf{r}d\mathbf{r}' = \frac{N(N-1)}{2} \qquad (6.53)$$

and

$$g(\mathbf{r},\mathbf{r}') = g(\mathbf{r}',\mathbf{r}). \qquad (6.54)$$

The sum rule (Eq. (6.53)) arises because the integrand is the distribution of particle pairs; integrating it over all volume gives the total number of pairs in the system, which is $N(N-1)/2$. The reciprocity relation (Eq. (6.54)) comes from the symmetry of the first line in Eq. (6.50).

Physically, the density of electrons at $\mathbf{r}'$ would be uncorrelated to the occupation of an electron at $\mathbf{r}$ in a material if $\mathbf{r}'$ were very far away from $\mathbf{r}$. Thus we have the relation

$$g(\mathbf{r},\mathbf{r}' \to \infty) = 1. \qquad (6.55)$$

Also, we may decompose the pair-correlation function into parallel and antiparallel spin components,

$$g(\mathbf{r},\mathbf{r}') = g_{\uparrow\uparrow}(\mathbf{r},\mathbf{r}') + g_{\uparrow\downarrow}(\mathbf{r},\mathbf{r}'), \qquad (6.56)$$

by separating the integration over spin coordinates in Eq. (6.50) into $\sigma = \sigma'$ and $\sigma \neq \sigma'$ partial sums, i.e.

$$g_{\uparrow\uparrow}(\mathbf{r},\mathbf{r}') = \frac{N(N-1)}{2\rho(\mathbf{r})\rho(\mathbf{r}')}\sum_{\sigma,\sigma'}\int |\Psi(\mathbf{r}\sigma,\mathbf{r}'\sigma',x_3,\ldots,x_N)|^2\delta_{\sigma\sigma'}dx_3\ldots dx_N \qquad (6.57)$$

and

$$g_{\uparrow\downarrow}(\mathbf{r},\mathbf{r}') = \frac{N(N-1)}{2\rho(\mathbf{r})\rho(\mathbf{r}')}\sum_{\sigma,\sigma'}\int |\Psi(\mathbf{r}\sigma,\mathbf{r}'\sigma',x_3,\ldots,x_N)|^2(1-\delta_{\sigma\sigma'})dx_3\ldots dx_N. \qquad (6.58)$$

For electrons (or any fermions), we have the relation

$$g_{\uparrow\uparrow}(\mathbf{r}_1,\mathbf{r}_2) \to 0 \text{ as } \mathbf{r}_1 \to \mathbf{r}_2, \qquad (6.59)$$

which is a consequence of the Pauli exclusion principle since the wavefunction in Eq. (6.57) vanishes as $\mathbf{r} \to \mathbf{r}'$ for $\sigma = \sigma'$. However, the value of $g_{\uparrow\downarrow}$ at $\mathbf{r} = \mathbf{r}'$ will depend on the details of electron correlations in a specific system. Another useful sum rule is

$$\int d\mathbf{r}' \left[g(\mathbf{r}, \mathbf{r}') - 1 \right] \rho(\mathbf{r}') = -1. \tag{6.60}$$

This may be proven by noting that

$$\int d\mathbf{r}' g(\mathbf{r}, \mathbf{r}') \rho(\mathbf{r}') = N - 1 \tag{6.61}$$

(since $g(\mathbf{r}, \mathbf{r}')\rho(\mathbf{r}')$ by definition is the density at $\mathbf{r}'$ given an electron at $\mathbf{r}$), and that $\int \rho(\mathbf{r}')d\mathbf{r}' = N$. Equation (6.60) is then the statement that $(N-1) - N = -1$.

6.5 $g(\mathbf{r}, \mathbf{r}')$ of the interacting electron gas

For a homogeneous system like an electron gas, the pair-correlation function $g(\mathbf{r}, \mathbf{r}')$ depends only on the distance between $\mathbf{r}$ and $\mathbf{r}'$ and not on the positions separately:

$$g(\mathbf{r}, \mathbf{r}') = g(|\mathbf{r} - \mathbf{r}'|) = g_{\uparrow\uparrow}(|\mathbf{r} - \mathbf{r}'|) + g_{\uparrow\downarrow}(|\mathbf{r} - \mathbf{r}'|). \tag{6.62}$$

We now evaluate these correlation functions for the interacting electron gas within the Hartree–Fock approximation to gain some insight into their behavior. For simplicity, we work with a paramagnetic electron gas, i.e. the number of spin-up electrons is equal to the number of spin-down electrons. (We note that it can be shown that at sufficiently low density, the electron gas goes into a magnetic phase.) The electron density given by Eq. (6.49) for a homogeneous system is then a constant $\rho(\mathbf{r}) = \rho_0 = \frac{N}{\Omega}$ and the pair-correlation function is

$$\begin{aligned} g(|\mathbf{r} - \mathbf{r}'|) &= \frac{N(N-1)}{2\rho_0^2} \rho_2(|\mathbf{r} - \mathbf{r}'|) \\ &= \frac{N(N-1)}{2\rho_0^2} \left\langle \int \left| \Psi(\mathbf{r}\sigma, \mathbf{r}'\sigma', x_3, \ldots, x_N) \right|^2 dx_3 \ldots dx_N \right\rangle_{\text{spin}}. \end{aligned} \tag{6.63}$$

Use of one-electron orthonormal orbitals in the Hartree–Fock Slater determinant yields

$$g(|\mathbf{r} - \mathbf{r}'|) = \frac{N(N-1)}{2\rho_0^2} \left[\frac{1}{N(N-1)} \right] \left\langle \sum_{i,j}^{\text{occ}} \left| \begin{matrix} \phi_i(\mathbf{r}\sigma) & \phi_i(\mathbf{r}'\sigma') \\ \phi_j(\mathbf{r}\sigma) & \phi_j(\mathbf{r}'\sigma') \end{matrix} \right|^2 \right\rangle_{\text{spin}}. \tag{6.64}$$

The sum over (i, j) in Eq. (6.64) is carried out over all occupied orbitals.

For the homogeneous electron gas,

$$\phi_i(\mathbf{r}\sigma) = \frac{e^{i\mathbf{k}\cdot\mathbf{r}}}{\sqrt{\Omega}} \chi_{s_i}(\sigma), \tag{6.65}$$

where χ_{s_i} is the spin function with $s_i = \uparrow$ or $\downarrow$, and

$$\langle \chi_{s_i} | \chi_{s_j} \rangle = \delta_{s_i s_j}. \tag{6.66}$$

(We note here that s_i is the quantum number denoting the spin state and σ is the coordinate in spin space.) Thus,

$$g(|\mathbf{r} - \mathbf{r}'|) = \frac{1}{2\rho_0^2} \sum_{\mathbf{k}_i s_i, \mathbf{k}_j s_j} \left\langle \left| \begin{matrix} \phi_{\mathbf{k}_i s_i}(\mathbf{r}\sigma) & \phi_{\mathbf{k}_i s_i}(\mathbf{r}'\sigma') \\ \phi_{\mathbf{k}_j s_j}(\mathbf{r}\sigma) & \phi_{\mathbf{k}_j s_j}(\mathbf{r}'\sigma') \end{matrix} \right|^2 \right\rangle_{\text{spin}}. \tag{6.67}$$

We make use of the result

$$\left\langle \left| \phi_{\mathbf{k}_i s_i}(\mathbf{r}\sigma) \phi_{\mathbf{k}_j s_j}(\mathbf{r}'\sigma') - \phi_{\mathbf{k}_i s_i}(\mathbf{r}'\sigma') \phi_{\mathbf{k}_j s_j}(\mathbf{r}\sigma) \right|^2 \right\rangle_{\text{spin}} = \frac{2 - 2e^{i(\mathbf{k}_i - \mathbf{k}_j) \cdot (\mathbf{r} - \mathbf{r}')} \delta_{s_i s_j}}{\Omega} \tag{6.68}$$

and obtain within Hartree–Fock (using $N_\uparrow = N_\downarrow = \frac{N}{2}$ and noting that there are four combinations for (s_i, s_j): $(\uparrow, \uparrow)$, $(\downarrow, \downarrow)$, $(\uparrow, \downarrow)$, and $(\downarrow, \uparrow)$)

$$g_{\uparrow\downarrow}^{\text{HF}}(\mathbf{r}, \mathbf{r}') = \frac{1}{N^2} \sum_{\mathbf{k}_i, \mathbf{k}_j}^{\text{occ}} 2 = \frac{1}{N^2} 2 \left(\frac{N}{2}\right)^2 = \frac{1}{2}, \tag{6.69}$$

which is independent of $|\mathbf{r} - \mathbf{r}'|$, and

$$g_{\uparrow\uparrow}^{\text{HF}}(\mathbf{r}, \mathbf{r}') = \frac{1}{N^2} \sum_{\mathbf{k}_i, \mathbf{k}_j}^{\text{occ}} \left[2 - 2e^{i(\mathbf{k}_i - \mathbf{k}_j) \cdot (\mathbf{r} - \mathbf{r}')} \right] = \frac{2}{N^2} \left[\left(\frac{N}{2}\right)^2 - \left(\sum_{\mathbf{k}} e^{i\mathbf{k} \cdot (\mathbf{r} - \mathbf{r}')}\right)^2 \right]$$

$$= \frac{1}{2} \left[1 - f(\mathbf{r} - \mathbf{r}')^2 \right], \tag{6.70}$$

where

$$f(\mathbf{r}) = \frac{2}{N} \sum_{\mathbf{k}}^{\text{occ}} e^{i\mathbf{k} \cdot \mathbf{r}} = \frac{3}{(rk_{\text{F}})^3} \left[\sin(rk_{\text{F}}) - (rk_{\text{F}}) \cos(rk_{\text{F}}) \right] = \frac{3}{rk_{\text{F}}} j_1(rk_{\text{F}}), \tag{6.71}$$

and $j_1(rk_{\text{F}})$ is the first Bessel function. We see explicitly that $g_{\uparrow\uparrow}^{\text{HF}}$ depends only on $|\mathbf{r} - \mathbf{r}'|$. Since $f(r)$ in Eq. (6.71) goes to 1 as $r \to 0$, we have

$$\lim_{\mathbf{r} \to 0} g_{\uparrow\uparrow}^{\text{HF}}(|\mathbf{r}|) \to 0. \tag{6.72}$$

For $r \to \infty$, we have $f \to 0$ and obtain

$$\lim_{\mathbf{r} \to \infty} g_{\uparrow\uparrow}^{\text{HF}}(|\mathbf{r}|) \to \frac{1}{2}. \tag{6.73}$$

Combining Eqs. (6.69) and (6.70) yields

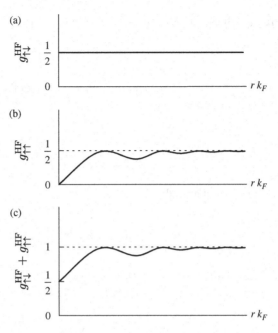

Pair-correlation function of a homogeneous electron gas in the paramagnetic phase in the Hartree–Fock approximation: (a) $g^{HF}_{\uparrow\downarrow}(r)$, (b) $g^{HF}_{\uparrow\uparrow}(r)$, (c) $g^{HF}(r) = g^{HF}_{\uparrow\downarrow}(r) + g^{HF}_{\uparrow\uparrow}(\mathbf{r})$.

$$g^{HF}(r) = g^{HF}_{\uparrow\uparrow}(r) + g^{HF}_{\uparrow\downarrow}(r) = 1 - \frac{1}{2}\left[\frac{3}{rk_F}j_1(rk_F)\right]^2, \tag{6.74}$$

with

$$g^{HF}(r \to 0) = \frac{1}{2}$$

and

$$g^{HF}(r \to \infty) = 1,$$

for a paramagnetic electron gas. The $|\mathbf{r} - \mathbf{r}'|$ dependence of $g^{HF}_{\uparrow\uparrow}$, $g^{HF}_{\uparrow\downarrow}$, and g^{HF} is illustrated in Fig. 6.4. We see that they have all the general properties expected of the pair-correlation functions of a many-body fermionic system. The extent over which $g^{HF}_{\uparrow\uparrow}$ deviates from $\frac{1}{2}$ is dictated by $\frac{1}{k_F}$. On the other hand, $g^{HF}_{\uparrow\downarrow}$ is a constant of value $\frac{1}{2}$ because the Hartree–Fock Slater determinant ground-state wavefunction only has exchange correlation between electrons of like spin; there is no correlation between electrons of unlike spin. Thus, for the paramagnetic phase, the chance of finding a spin-down electron at any value of $\mathbf{r}'$ is independent of whether there is a spin-up electron at $\mathbf{r}$ and is equal to $\frac{1}{2}$ within this approximation.

In a more accurate description of the many-body wavefunction than that of Hartree–Fock, there is correlation between unlike spins, due to the Coulomb repulsion between the electrons. These correlations reduce $g_{\uparrow\downarrow}(r)$ to less than $\frac{1}{2}$ as $r \to 0$. And $g_{\uparrow\uparrow}(r)$ would change from that of $g_{\uparrow\uparrow}^{HF}(r)$. This must be the case since the sum rule (Eq. (6.60)) is required. The difference between the exact g and g^{HF} leads to a change in the ground-state energy and gives rise to the correlation energy.

6.6 The exchange-correlation hole

From the definition of the pair-correlation function in Section 6.4, the electron density at position $\mathbf{r}'$ as seen by an electron at position $\mathbf{r}$ is given by

$$\rho_{\mathbf{r}}(\mathbf{r}') = g\left(\mathbf{r}, \mathbf{r}'\right) \rho\left(\mathbf{r}'\right). \tag{6.75}$$

The difference between the electron density seen by an electron at $\mathbf{r}$, $\rho_{\mathbf{r}}(\mathbf{r}')$, and that of the electron density $\rho(\mathbf{r}')$ measured, for example, in an X-ray scattering experiment, is given by

$$\Delta\rho_{\mathbf{r}}(\mathbf{r}') = \left[\rho_{\mathbf{r}}(\mathbf{r}') - \rho(\mathbf{r}')\right] = \left[g\left(\mathbf{r}, \mathbf{r}'\right) - 1\right]\rho(\mathbf{r}'). \tag{6.76}$$

Integrating over volume and making use of the sum rule (Eq. (6.60)) yields

$$\int \Delta\rho_{\mathbf{r}}(\mathbf{r}')d\mathbf{r}' = \int \left[g\left(\mathbf{r}, \mathbf{r}'\right) - 1\right]\rho(\mathbf{r}')d\mathbf{r}' = -1. \tag{6.77}$$

Equation (6.77) tells us that there is a depletion of electron density near the position of an electron, and that the integrated missing density is exactly equal to one electron. This is physically expected, since the Pauli exclusion principle, as well as Coulomb repulsion, reduces the density of electrons near a specific site already occupied by an electron. Conservation of particle number requires that the missing density be equal to one, since there is already one electron at site $\mathbf{r}$. This depletion of (or hole in) electron density given by Eq. (6.76) is called the exchange-correlation hole.

The exchange-correlation hole moves with the electron and, in general, changes size and shape as the electron travels through a system of inhomogeneous electron density. However, Eq. (6.77) always holds; that is, the integrated depletion is that of a missing electron. Thus, the electron density $\rho_{\mathbf{r}}(\mathbf{r}')$ seen by a given electron at $\mathbf{r}$ is always different from that of $\rho(\mathbf{r}')$. For example, in an interacting homogeneous electron gas, although $\rho(\mathbf{r}')$ is a constant, $\rho_{\mathbf{r}}(\mathbf{r}')$ is highly inhomogeneous. The structure of the inhomogeneity is dictated by the size and shape of the pair-correlation function. Figure 6.5 illustrates schematically the difference between the two densities, $\rho(\mathbf{r}')$ and $\rho_{\mathbf{r}}(\mathbf{r}')$.

The evaluation of the pair-correlation function and the exchange-correlation hole for a real material is non-trivial since it requires knowing the many-electron ground-state wavefunction. In particular, a reasonable description of $g_{\uparrow\downarrow}(\mathbf{r}, \mathbf{r}')$ means incorporation of

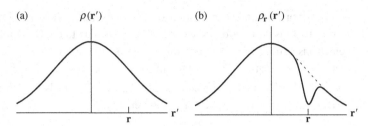

Figure 6.5 Schematic of $\rho(\mathbf{r}')$ and $\rho_{\mathbf{r}}(\mathbf{r}')$. (a) charge density $\rho(\mathbf{r}')$ of a system as measured by an external probe. (b) charge density $\rho_{\mathbf{r}}(\mathbf{r}')$ of the same system as seen by an electron at $\mathbf{r}$.

correlation effects going beyond Hartree–Fock. One approach is to use variational quantum Monte Carlo simulations to determine the ground-state wavefunction. Figure 6.6 shows some results from such calculations for diamond. Different panels depict $g_{\uparrow\uparrow}(\mathbf{r},\mathbf{r}')$ and $g_{\uparrow\downarrow}(\mathbf{r},\mathbf{r}')$ as functions of $\mathbf{r}'$ on the (110) plane, with $\mathbf{r}$ fixed either on the bond center or the interstitial region of the diamond crystal. The pictures show a suppression from the value of $\frac{1}{2}$ for $g_{\uparrow\uparrow}$, as $\mathbf{r} \to \mathbf{r}'$, reaching zero at $\mathbf{r} = \mathbf{r}'$, showing a deep exchange hole. On the other hand, the suppression of $g_{\uparrow\downarrow}$ is not nearly as pronounced because diamond is only a moderately correlated electron system. Both $g_{\uparrow\uparrow}(\mathbf{r},\mathbf{r}')$ and $g_{\uparrow\downarrow}(\mathbf{r},\mathbf{r}')$ show anisotropy and strong variations as a function of $\mathbf{r}$, owing to the electron density inhomogeneity of the material.

6.7 The exchange-correlation energy

The exchange-correlation energy of an electron is the amount of lowering of the energy of the electron, as the many-body wavefunction describes more correlations of the motions of the electron as we improve on the Hartree approximation (that is, the energy gain due to both exchange and correlation effects). This would be the lowering of the potential energy if the change in the kinetic energy could be neglected. In that case, we may calculate this energy difference by evaluating the potential energy using a charge density around an electron at $\mathbf{r}$, with and without the exchange-correlation hole contribution. Without the exchange-correlation hole, the electron density is $\rho(\mathbf{r}')$ and the potential energy due to electron–electron interaction for an electron at $\mathbf{r}$ is

$$\varepsilon_0(\mathbf{r}) = e^2 \int \frac{\rho(\mathbf{r}')\, d\mathbf{r}'}{|\mathbf{r} - \mathbf{r}'|}. \tag{6.78}$$

With the exchange-correlation hole, the potential energy for the same electron at $\mathbf{r}$ is

$$\varepsilon(\mathbf{r}) = e^2 \int \frac{\rho_{\mathbf{r}}(\mathbf{r}')d\mathbf{r}'}{|\mathbf{r} - \mathbf{r}'|} = e^2 \int \frac{g(\mathbf{r},\mathbf{r}')\rho(\mathbf{r}')d\mathbf{r}'}{|\mathbf{r} - \mathbf{r}'|}. \tag{6.79}$$

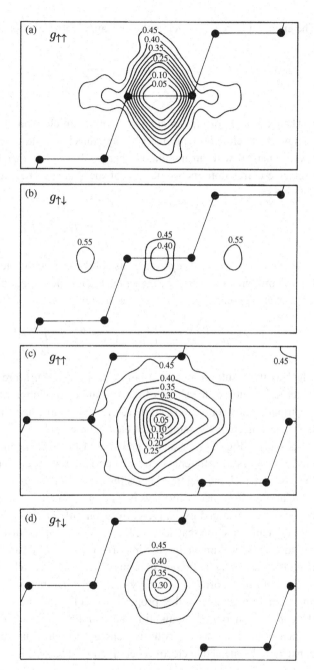

Figure 6.6 The pair-correlation function in diamond for (a) parallel spin and (b) opposite spin with an electron at the bond center, and for (c) parallel spin and (d) opposite spin with an electron at the tetrahedral interstitial site. The atoms and bonds are schematically represented by the lines and dots in the [111] direction. (After Fahy, Wang and Louie, 1990.)[5]

[5] S. Fahy, X.W. Wang, and S.G. Louie, "Pair-correlation function and single-particle occupation numbers in diamond and silicon," *Phys. Rev. Lett.* 65(1990), 1478.

The energy difference, which is the exchange-correlation energy for the electron at $\mathbf{r}$, is

$$\varepsilon_{xc}(\mathbf{r}) = \varepsilon(\mathbf{r}) - \varepsilon_0(\mathbf{r}) = e^2 \int \frac{\left[\rho_{\mathbf{r}}(\mathbf{r}') - \rho(\mathbf{r}')\right]}{|\mathbf{r} - \mathbf{r}'|} d\mathbf{r}' = e^2 \int \frac{\left[g(\mathbf{r},\mathbf{r}') - 1\right]\rho(\mathbf{r}')}{|\mathbf{r} - \mathbf{r}'|} d\mathbf{r}'.$$

(6.80)

The exchange-correlation energy $\varepsilon_{xc}(\mathbf{r})$ of an electron at $\mathbf{r}$ is then just the interaction energy of the electron with its exchange-correlation hole, which can be considered positively charged with an integrated charge of $-e$. The quantity $\varepsilon_{xc}(\mathbf{r})$ is also called the exchange-correlation energy density of the system. The exchange-correlation energy of the total system is

$$E_{xc} = \int \varepsilon_{xc}(\mathbf{r})\rho(\mathbf{r})d\mathbf{r}.$$

(6.81)

We may apply this concept to the case of the interacting electron gas within the Hartree–Fock approximation. Since the system is homogeneous, $\varepsilon_{xc}(\mathbf{r})$ is the same everywhere and is the exchange-correlation energy per particle. Using Eqs. (6.80) and (6.74),

$$\varepsilon_{xc}^{HF} = e^2 \rho_0 \int \frac{d\mathbf{r}}{r} \left[g^{HF}(\mathbf{r}) - 1\right] = -\frac{6e^2 k_F}{\pi} \int_0^\infty \frac{j_1^2(x)}{x} dx = -\frac{3}{4} \frac{e^2 k_F}{\pi},$$

(6.82)

which is the result we obtained before in Eq. (6.35). More accurate theory will yield a better description of the shape of the exchange-correlation hole as well as a change in the electron kinetic energy, and will therefore give a more accurate value for $\varepsilon_{xc}(\mathbf{r})$. However, the basic concept is the same as developed here.

As seen in Fig. 6.6, the shape and size of the exchange-correlation hole change as an electron moves to different positions in a material. A particularly interesting situation is the case in which the electron leaves the crystal and stays near the surface. The exchange-correlation hole by necessity stays behind in the crystal. The Coulomb interaction between the positively charged exchange-correlation hole and the electron that leaves the crystal results in an attraction that scales like the inverse of the distance from the surface, giving rise to the classical image force. Similarly, the attractive van der Waals interaction between neutral objects may be framed in terms of exchange-correlation effects.

The exchange-correlation energy plays an important role in determining conceptually and quantitatively many other properties of solids. We now give a simple example of how the cohesive energy of the alkali metals can be estimated. The cohesive energy per atom (which is the same as the cohesive energy per electron for the alkali metal case if we consider only one active electron per alkali metal atom) is defined by

$$E_{coh} = E_{atom} - E_{GS},$$

(6.83)

where E_{atom}, using the pseudopotential concept (discussed in Chapter 7), is the binding energy of the outer active electrons in the atom for the isolated case, and E_{GS} is the energy per atom in the solid state. The cohesive energy E_{coh} is positive for a stable solid.

Table 6.1 Calculated cohesive energies of the alkali metals within the interacting electron gas model for electron–electron interaction using different approximations – Hartree, Hartree–Fock (HF), and Hartree–Fock plus Wigner correlations (HF + Wigner). All energies are in kcal/mol.

	$E_{coh}^{Hartree}$	E_{coh}^{HF}	$E_{coh}^{HF+Wigner}$	expt.[6]
Li	unbound	17.0	41.4	36.5
Na	unbound	6.8	30.3	26.0
K	unbound	4.3	25.7	22.6
Rb	unbound	3.4	24.4	18.9
Cs	unbound	2.9	23.3	18.8

For the alkali metal elements, E_{atom} is just the negative of the ionization potential and for the solid

$$E_{GS} = E_{KE} + E_{e\text{-ion}} + E_{e\text{-e}}. \tag{6.84}$$

We may use the interacting homogeneous electron gas as a model for evaluating Eq. (6.84). The kinetic energy is

$$E_{KE} = \frac{2.21}{r_s^2} \left(\frac{m_e}{m_b} \right), \tag{6.85}$$

where m_b is the conduction band effective mass of the alkali metal, and $E_{e\text{-ion}}$ is the bottom of the conduction band. Both m_b and $E_{e\text{-ion}}$ may be obtained by a band structure calculation. The electron–electron interaction term $E_{e\text{-e}}$ may be approximated by using the results for the homogeneous electron gas, since the charge density of the conduction electrons of a simple metal is expected to be nearly uniform.

$$E_{e\text{-e}} = E_{Hartree} + E_x + E_{cor} = \frac{1.2}{r_s} + \left(-\frac{0.916}{r_s^2} \right) + \left(-\frac{0.88}{7.8 + r_s} \right) \text{(Ry)}. \tag{6.86}$$

For the correlation energy in the above expression, we have used the Wigner interpolation formula. If a crystal is formed, E_{GS} is a minimum at the equilibrium lattice constant r_s^{eq}, and hence E_{coh} may be determined by setting

$$\left. \frac{\partial E_{GS}(r_s)}{\partial r_s} \right|_{r_s = r_s^{eq}} = 0. \tag{6.87}$$

In the case of very light elements, there is a contribution $E_{zero\text{-point}}$ to the cohesive energy due to the zero-point motion of the atoms confined to the crystal, which can be evaluated

---

[6] Experimental values from H. Brooks, "Cohesive energy of alkali metals," *Phys. Rev.* 91(1953), 1027.

via knowledge of the phonon spectrum. Table 6.1 shows the results for the calculated E_{coh} of alkali metals with this simple model using Hartree, Hartree–Fock, and Hartree–Fock + Wigner correlation models. We see that the alkali metals are unbounded within the Hartree-only approximation. Both exchange and correlation effects are needed to give a reasonable description of the cohesive properties of this simplest class of metals.

Density functional theory (DFT)

In this chapter, we turn to first-principles calculations of the properties and behavior of real materials. Tremendous progress has been made in the past decades in employing *ab initio* or first-principles theories and numerical computations to explain and predict the properties, and even the existence, of condensed matter systems. These first-principles studies have been particularly successful in investigating materials and reduced-dimensional systems with a moderate amount of electron correlation.

As discussed in previous chapters, the structure and properties of condensed matter are basically dictated by the outer valence electrons of its constituent atoms. The mutual interactions of these electrons and their interactions with the ions determine the electronic structure of the system, which in turn determines many of the properties of the material. Understanding material properties from first principles involves solving an interacting quantum many-body problem with the Hamiltonian (see Chapter 2)

$$H_{\text{tot}} = \sum_j \frac{\mathbf{P}_j^2}{2M_j} + \sum_i \frac{\mathbf{p}_i^2}{2m} + \sum_{j<j'} \frac{Z_j Z_{j'} e^2}{|\mathbf{R}_j - \mathbf{R}_{j'}|} + \sum_{i<i'} \frac{e^2}{|\mathbf{r}_i - \mathbf{r}_{i'}|} + \sum_{i,j} \frac{-Z_j e^2}{|\mathbf{R}_j - \mathbf{r}_i|}. \quad (7.1)$$

For simplicity, we have omitted spin and relativistic effects in our discussion here. Generally, exact solutions to this problem are impractical, and often undesirable, since the many-electron wavefunctions would be so complicated that it would be difficult to obtain a physical understanding from the exact solutions. In this chapter, we discuss one particular formalism, density functional theory, for solving this many-electron problem in an *ab initio* fashion for properties related to the electronic ground state of the system. Nowadays, density functional theory is arguably the most popular approach for the first-principles study of the ground-state properties of molecules and solids, with applications to many disciplines ranging from physics to chemistry to the biological and engineering sciences. This formalism provides a means to treat electron–electron interactions and transforms the many-electron problem to one of a self-consistent-field one-particle problem for ground-state properties, which is exact in principle.

For an interacting many-electron system, it is useful to distinguish the ground-state properties from the electron excited-state or spectroscopic properties. At low temperature, properties of a crystal such as the structure, cohesive energies, structural and vibrational properties, and structural phase stability are ground-state properties because they are determined collectively by all the electrons in the ground state. Many of these properties can be obtained by knowing the ground-state total energy of the system as a function of the atomic coordinates. However, electronic excitation properties such as those measured in photoemission, transport, and tunneling experiments involve creating an excited particle (electron or hole) above the ground state. It should be thought of as an $N+1$ particle problem that requires a different theoretical treatment. Understanding the optical properties from first principles is a still different problem because it is an $N+2$ particle problem. In this case, we need to include the electron–hole interaction, which can be very important in many systems, particularly in lower-dimensional structures. (See Chapters 9 and 16.) The excited-state response or spectroscopic properties are best treated within the concept and formalism of the interacting-particle Green's functions, which will be discussed in Chapter 13.

7.1 The ground state and density functional formalism

As a quantum many-body problem, one of the major challenges in determining material properties is the proper treatment of the electron–electron interaction. For ground-state properties, it is shown that this many-body problem can be exactly reduced to one of solving a self-consistent-field one-particle problem using the so-called density functional theory (DFT). Unlike other theories, such as the Hartree or Hartree–Fock approximations, DFT as formulated by Hohenberg, Kohn, and Sham is in principle exact in calculating the total energy, the electron charge density distribution, and other related ground-state properties (such as structural and vibrational properties) of an interacting many-electron system.

Although there had been previous calculations of material properties based on electron densities, in 1964, Hohenberg and Kohn[1] demonstrated that, for an interacting many-electron system in an external static potential $v(\mathbf{r})$, the ground-state energy E_v can be expressed as a functional of the charge density $\rho(\mathbf{r})$, i.e. $E_v[\rho]$, rather than the usual formulation that $E_v[\Psi(\mathbf{r}_1, \mathbf{r}_2, \ldots, \mathbf{r}_N)]$, where Ψ is the many-body wavefunction with N electrons. The external potential $v(\mathbf{r})$ is normally taken to be the ionic potential. The theory has also been extended to the case of time-dependent external potentials. We shall restrict our discussion here to the time-independent case. For an interacting many-electron system in a static potential $v(\mathbf{r})$, the following three fundamental statements are shown.

I. The ground-state energy can be written as $E_v[\rho] = \int v(\mathbf{r})\rho(\mathbf{r})d\mathbf{r} + F[\rho]$, where $\rho(\mathbf{r})$ is the electron density and $F[\rho]$ is a universal functional of the density, independent of $v(\mathbf{r})$.

II. $E_v[\rho]$ is a minimum for the correct physical density, with $\rho(\mathbf{r})$ satisfying the constraint $N = \int \rho(\mathbf{r})d\mathbf{r}$.

III. $\rho(\mathbf{r})$ and hence $E_v[\rho]$ can in principle be exactly obtained from the solution of an associated one-electron problem with an effective potential $v_{\mathrm{eff}}(\mathbf{r})$.

We demonstrate statement I by considering the simple case of a paramagnetic system with a non-degenerate ground state. This is achieved using the technique of proof by contradiction. Proofs of the more general cases may be found in the literature. Consider a

[1] P. Hohenberg and W. Kohn, "Inhomogeneous electron gas," *Phys. Rev.* 136(1964), B864.

collection of N electrons enclosed in a large box moving under the influence of an external potential $v(\mathbf{r})$ and their mutual Coulomb repulsion. The Hamiltonian for the electrons is

$$H = \sum_i \frac{\mathbf{p}_i^2}{2m} + \sum_i v(\mathbf{r}_i) + \frac{1}{2} \sum_{i \neq j} \frac{e^2}{|\mathbf{r}_i - \mathbf{r}_j|} \tag{7.2}$$
$$\equiv T + V + U,$$

where we have defined T, V, and U as the operators for the total kinetic energy, the interaction energy with the external potential, and the electron–electron interaction energy of the system, respectively. In the ground state $|\Psi\rangle$, the electron density is

$$\rho(\mathbf{r}) = \langle \Psi | \hat{\rho}(\mathbf{r}) | \Psi \rangle, \tag{7.3}$$

with

$$\hat{\rho}(\mathbf{r}) = \sum_i \delta(\mathbf{r} - \mathbf{r}_i), \tag{7.4}$$

where i runs over all electrons.

Since the ground state is non-degenerate, $\rho(\mathbf{r})$ is a unique functional of $v(\mathbf{r})$. For statement I to hold, we need the converse to be true. That is, $v(\mathbf{r})$ is a unique functional of $\rho(\mathbf{r})$, apart from a trivial additive constant. Let us assume that the contrary holds. That is, there exists another external potential $v'(\mathbf{r})$, such that its ground state $|\Psi'\rangle$ gives the same density, i.e. $\rho'(\mathbf{r}) = \rho(\mathbf{r})$. By assumption, $v'(\mathbf{r}) - v(\mathbf{r})$ is not a constant, thus $|\Psi'\rangle$ is not equal to $|\Psi\rangle$, because each satisfies a different Schrödinger equation:

$$H | \Psi \rangle = E | \Psi \rangle \tag{7.5}$$

and

$$H' | \Psi' \rangle = E' | \Psi' \rangle. \tag{7.6}$$

By the minimal property of the ground state for the primed system,

$$E' = \langle \Psi' | H' | \Psi' \rangle < \langle \Psi | H' | \Psi \rangle, \tag{7.7}$$

but

$$\langle \Psi | H' | \Psi \rangle = \langle \Psi | H + V' - V | \Psi \rangle = E + \int [v'(\mathbf{r}) - v(\mathbf{r})] \rho(\mathbf{r}) d\mathbf{r}, \tag{7.8}$$

leading to the inequality

$$E' < E + \int [v'(\mathbf{r}) - v(\mathbf{r})] \rho(\mathbf{r}) d\mathbf{r}. \tag{7.9}$$

Similarly, for the unprimed system, we obtain

$$E < E' + \int [v(\mathbf{r}) - v'(\mathbf{r})] \rho'(\mathbf{r}) d\mathbf{r}. \tag{7.10}$$

Adding Eqs. (7.9) and (7.10), and making use of the assumption that $\rho'(\mathbf{r}) = \rho(\mathbf{r})$, we have the contradictory result that

$$E + E' < E + E'. \tag{7.11}$$

This contradiction thus proves that $v(\mathbf{r})$ is a unique functional of $\rho(\mathbf{r})$; that is, given a physical electron density, there is a unique mapping to an external potential. Now we may define a universal functional that is valid for any number of particles and any external potential,

$$F[\rho] = \langle \Psi[\rho] | T + U | \Psi[\rho] \rangle. \tag{7.12}$$

This is because knowing ρ implies knowing v, which in turn means knowing the exact ground state $\Psi[\rho]$. The ground-state energy of the system is then a functional of the electron density and is given by

$$E_v[\rho] = \int v(\mathbf{r}) \rho(\mathbf{r}) d\mathbf{r} + F[\rho]. \tag{7.13}$$

Statement II may also be proven using Eq. (7.13) and a variational principle of quantum mechanics: the energy of the system is a minimum at the ground-state wavefunction $|\Psi\rangle$ relative to an arbitrary variation of $|\Psi\rangle$, keeping the number of particles constant. If $F[\rho]$ were a known and sufficiently simple functional of $\rho(\mathbf{r})$, the problem of determining the ground-state energy and density would be quite straightforward. However, $F[\rho]$ is not explicitly known.

7.2 The Kohn–Sham equations

Statement III of the Kohn–Sham formulation[2] of DFT states that there exists a set of one-body equations that exactly determine $\rho(\mathbf{r})$. In fact, as shown below, these equations are formally similar to other self-consistent field equations such as the Hartree equations, but they yield the exact density in principle. These equations are derived using the following theoretical construction developed by Kohn and Sham. In the expression for the ground-state energy (Eq. (7.13)), $F[\rho]$ contains both the electron kinetic energy and the electron–electron interaction energy, which in general are very difficult to evaluate and for which there are no known explicit expressions in terms of the density $\rho(\mathbf{r})$. However, we do know the contributions to $F[\rho]$ in some limiting cases. If exchange and correlation effects are neglected, the interaction energy is just the classical Hartree term

$$E_{\text{Hartree}} = \frac{e^2}{2} \int \frac{\rho(\mathbf{r}) \rho(\mathbf{r}')}{|\mathbf{r} - \mathbf{r}'|} d\mathbf{r} d\mathbf{r}'. \tag{7.14}$$

[2] W. Kohn and L. J. Sham, "Self-consistent equations including exchange and correlation effects," *Phys. Rev.* 140(1965), A1133.

And, for a noninteracting system of the same density $\rho(\mathbf{r})$, the kinetic energy of the system $T_s[\rho]$ may be evaluated from the single-particle orbitals $\varphi_i(\mathbf{r})$ with

$$T_s[\rho] = -\frac{\hbar^2}{2m} \sum_i^{\text{occ}} \int \varphi_i^*(\mathbf{r}) \nabla^2 \varphi_i(\mathbf{r})\, d\mathbf{r} \qquad (7.15)$$

and

$$\rho(\mathbf{r}) = \sum_i^{\text{occ}} \varphi_i^*(\mathbf{r}) \varphi_i(\mathbf{r}). \qquad (7.16)$$

For many systems, one expects $F[\rho]$ to be close to the sum of Eqs. (7.14) and (7.15). This reasoning prompted Kohn and Sham to express Eq. (7.13) in the form

$$E_v[\rho] = \int v(\mathbf{r})\rho(\mathbf{r})d\mathbf{r} + T_s[\rho] + \frac{1}{2}e^2 \int \frac{\rho(\mathbf{r})\rho(\mathbf{r}')}{|\mathbf{r}-\mathbf{r}'|} d\mathbf{r} d\mathbf{r}' + E_{\text{xc}}[\rho]. \qquad (7.17)$$

All of our ignorance of $F[\rho]$ for the interacting system (going beyond $T_s[\rho]$ and E_{Hartree}) is now contained in the new term $E_{\text{xc}}[\rho]$. The energy $E_{\text{xc}}[\rho]$ is called the exchange-correlation energy functional within DFT. It is important to note that $E_{\text{xc}}[\rho]$ contains the electron–electron interaction energy beyond the Hartree approximation and also contains part of the kinetic energy of the interacting electrons, since $T_s[\rho]$ is, by definition, the kinetic energy of a system of non-interacting electrons of the same density.

To arrive at a useful theoretical framework, one can further show that the electron charge density and total energy may be obtained by solving an effective system of non-interacting electrons, making the DFT approach practical. In the Kohn–Sham theoretical construct, two systems are considered – the origin material system of interest with interacting electrons, and an associated, fictitious system with the same number of electrons but non-interacting, moving in some effective one-body potential $v_{\text{eff}}(\mathbf{r})$ (see Fig. 7.1). The electron density of the non-interacting system is used in evaluating the energy functional (Eq. (7.17)) of the real system. Making use of the variational principle (statement II above) that the physical ground-state density minimizes the total energy functional, we may vary the non-interacting system until the functional (Eq.(7.17)) is minimized and arrive at the physical charge density and ground-state energy of the real interacting electron system. This variation gives rise to a set of Euler–Lagrange equations that governs the single-particle orbitals and energies of the non-interacting system.

For the non-interacting system with potential $v'(\mathbf{r})$ and N electrons, ρ' is determined by solving

$$\left[-\frac{\hbar^2}{2m}\nabla^2 + v'(\mathbf{r}) \right] \varphi_i'(\mathbf{r}) = \epsilon_i' \varphi_i'(\mathbf{r}), \qquad (7.18)$$

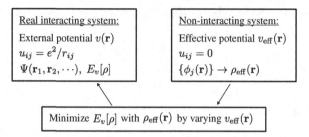

Figure 7.1 Kohn–Sham formulation of DFT. The charge density of a noninteracting system is used to minimize the energy functional to arrive at a set of one-electron equations to determine the charge density and total energy of the interacting system.

and taking $\rho'(\mathbf{r}) = \sum_{i=1}^{N} |\varphi_i'(\mathbf{r})|^2$ (with the states ordered in increasing eigenvalues). The non-interacting kinetic energy of charge density ρ' is then given by

$$T_s[\rho'] = \sum_i^{\text{occ}} \epsilon_i' - \int v'(\mathbf{r})\rho'(\mathbf{r})dr. \tag{7.19}$$

For the interacting system, we may write the total electron energy (Eq. (7.17)) when we evaluate with the trial density $\rho'(\mathbf{r})$ as

$$E_v[\rho'] = \left[\sum_i^{\text{occ}} \epsilon_i' - \int v'(\mathbf{r})\rho'(\mathbf{r})dr \right] + \int v(\mathbf{r})\rho'(\mathbf{r})dr + \frac{1}{2}e^2 \int \frac{\rho'(\mathbf{r})\rho'(\mathbf{r}')}{|\mathbf{r} - \mathbf{r}'|}drdr' + E_{\text{xc}}[\rho']. \tag{7.20}$$

The two terms in large square brackets on the right-hand side of Eq. (7.20) taken together is again $T_s[\rho']$ and is given in terms of single-particle orbital properties. To obtain the physical density and total energy, we minimize $E_v[\rho']$ with respect to ρ', i.e. for $\rho' \to \rho' + \delta\rho'$, we determine

$$E_v \to E_v + \delta E_v. \tag{7.21}$$

A change in $\rho'(\mathbf{r})$, which is equivalent to a change in $v'(\mathbf{r})$, will induce a change in all terms in Eq. (7.20), leading to

$$\delta E_v[\rho'] = \sum_i^{\text{occ}} \delta\epsilon_i' + \int \delta\rho'(\mathbf{r}) \left\{ -v'(\mathbf{r}) - \int \frac{\delta v'(\mathbf{r}')}{\delta\rho'(\mathbf{r})}\rho'(\mathbf{r}')dr' + v(\mathbf{r}) \right.$$

$$\left. + e^2 \int \frac{\rho'(\mathbf{r}')}{|\mathbf{r} - \mathbf{r}'|}dr' + \frac{\delta E_{\text{xc}}[\rho']}{\delta\rho'(\mathbf{r})} \right\} d\mathbf{r}. \tag{7.22}$$

For the density $\rho'(\mathbf{r})$ (or equivalently $v'(\mathbf{r})$) that minimizes $E_v[\rho]$, we need to have Eq. (7.22) equal to 0. Since, to first order,

$$\sum_i \delta\epsilon_i = \sum_i \langle \varphi_i'|\delta v'|\varphi_i' \rangle = \int \delta v'(\mathbf{r}')\rho'(\mathbf{r}')dr' = \int \delta\rho'(\mathbf{r}) \int \frac{\delta v'(\mathbf{r}')}{\delta\rho'(\mathbf{r})}\rho'(\mathbf{r}')drdr', \tag{7.23}$$

setting Eq. (7.22) equal to zero leads to the condition that v' at the extremum takes on the expression

$$v'(\mathbf{r}) = v(\mathbf{r}) + e^2 \int \frac{\rho(\mathbf{r}')d\mathbf{r}'}{|\mathbf{r} - \mathbf{r}'|} + \frac{\delta E_{\text{xc}}[\rho]}{\delta\rho(\mathbf{r})} \equiv v_{\text{eff}}(\mathbf{r}). \tag{7.24}$$

This theoretical construct thus gives rise to a set of Euler–Lagrange equations that governs the single-particle orbitals and energies of the associated non-interacting system:

$$\left\{ \frac{p^2}{2m} + v(\mathbf{r}) + v_{\text{H}}(\mathbf{r}) + v_{\text{xc}}(\mathbf{r}) \right\} \varphi_i(\mathbf{r}) = \epsilon_i\varphi_i(\mathbf{r}) \tag{7.25}$$

with

$$\rho(\mathbf{r}) = \sum_i^{\text{occ}} |\varphi_i(\mathbf{r})|^2, \tag{7.26}$$

$$v_{\text{H}}(\mathbf{r}) = e^2 \int \frac{\rho(\mathbf{r}')}{|\mathbf{r} - \mathbf{r}'|} d\mathbf{r}', \tag{7.27}$$

and

$$v_{\text{xc}}(\mathbf{r}) = \frac{\delta E_{\text{xc}}}{\delta\rho(\mathbf{r})}. \tag{7.28}$$

This set of equations for the fictitious non-interacting system is known as the Kohn–Sham equations. Besides the external potential $v(\mathbf{r})$, the effective potential v_{eff} for the non-interacting system has two other terms: a Hartree term $v_{\text{H}}(\mathbf{r})$ and an exchange-correlation term $v_{\text{xc}}(\mathbf{r})$, which is given by the functional derivative of E_{xc} with respect to the electron charge density. In principle, if E_{xc} were known, a self-consistent solution to the Kohn–Sham equations would give the exact electron density and ground-state energy of the interacting system as a function of the atomic coordinates, as well as a host of other properties that are related to the ground state.

The DFT formalism, in particular in the Kohn–Sham framework, is a tremendous conceptual and technical simplification since the $3N$-dimensional many-electron wavefunction is eliminated from the problem. However, the existence of the universal functional was only deduced by a proof by contradiction, and the exact E_{xc} remains unknown, although a number of useful approximations, such as the local density approximation (LDA) or the generalized gradient approximation (GGA), have been developed over the years.

Since E_{xc} is unknown, approximations must be made. A common approach is to write that

$$E_{\text{xc}} = \int \rho(\mathbf{r})\varepsilon_{\text{xc}}(\mathbf{r})d\mathbf{r}, \tag{7.29}$$

where $\varepsilon_{\text{xc}}(\mathbf{r})$, an exchange-correlation energy density, is assumed to be a function of the local density $\rho(\mathbf{r})$ in the LDA, or a function of $\rho(\mathbf{r})$ and $\nabla\rho(\mathbf{r})$ in the GGA. These two levels of approximations are exact in the uniform density limit and may employ data from homogeneous electron gas calculations. They have allowed for very accurate *ab initio* computation of the properties of many materials.

In the LDA, the exchange-correlation energy functional is taken as

$$E_{xc}(\rho) \approx \int \rho(\mathbf{r})\varepsilon_{xc}(\rho(\mathbf{r}))d\mathbf{r}, \qquad (7.30)$$

where ε_{xc} is the exchange-correlation energy density of a uniform electron gas with density of $\rho(\mathbf{r})$. The two approximations employed in the LDA are: (1) local density dependence, and (2) use of data from a uniform interacting electron gas. The exchange-correlation potential is then given by

$$v_{xc}(\rho(\mathbf{r})) = \frac{d}{d\rho}(\rho\varepsilon_{xc}(\rho))|_{\rho(\mathbf{r})}. \qquad (7.31)$$

Neglecting correlations and considering only the exchange energy of a uniform electron gas, this potential is given by (see Chapter 6)

$$v_x(\rho(\mathbf{r})) = -\frac{e^2}{\pi}(3\pi^2)^{1/3}\rho^{1/3}(\mathbf{r}). \qquad (7.32)$$

In general, when correlations effects are included, v_{xc} in the LDA takes on the form of

$$v_{xc}(\rho) = -\frac{3}{2}\frac{e^2}{\pi}(3\pi^2)^{1/3}\rho^{1/3}\alpha(\rho), \qquad (7.33)$$

where $\alpha(\rho)$ is a function of ρ that depends on the approximation used in the calculation of the correlation energy of the electron gas.

We would like to emphasize that, within this framework, the individual Kohn–Sham eigenvalues ε_i and eigenfunctions $\varphi_i(\mathbf{r})$ of Eq. (7.25) do not correspond physically to the excitation energies and amplitudes of the electrons of the interacting system. Only the electron charge density and total energy are rigorously meaningful. In particular, the Kohn–Sham eigenvalues are just Lagrange multipliers in the Kohn–Sham variational construct, and, in general, they are not equal to the electron excitation (or quasiparticle) energies of a system even if the exact exchange-correlation functional E_{xc} is known. We can see this from the following simple example. Consider the case of the interacting homogeneous electron gas. In this case, because the density is homogeneous, both the Hartree potential and the exchange-correlation potential in the Kohn–Sham equation, Eq. (7.25), are independent of the coordinates of the electrons. Equation (7.25) reduces to

$$\left\{\frac{p^2}{2m} + \text{constant}\right\}\varphi_i(\mathbf{r}) = \varepsilon_i\varphi_i(\mathbf{r}). \qquad (7.34)$$

The solution to Eq. (7.34) is always that of a free electron dispersion for the Kohn–Sham eigenvalues. That is, if we were to interpret the Kohn–Sham eigenvalues as electron excitation energies, then the band structure and therefore the effective mass m^* and occupied bandwidth would be those of the free electrons, independent of the interaction strength. Also, the lifetime of a quasiparticle created in a particular momentum state would be infinite. This is clearly incorrect. It was this misuse of the Kohn–Sham eigenvalues for the

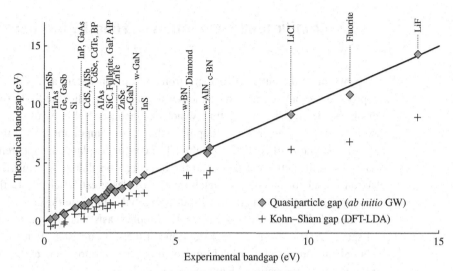

Figure 7.2 Comparison of calculated bandgaps from DFT-LDA and from the *ab initio* GW approach with experimentally measured bandgaps for common semiconductors and insulators. (After S. G. Louie, 1997.)[5]

band structure that led to the famous bandgap problem in semiconductors and insulators. That is, the Kohn–Sham eigenvalues give bandgaps that are significantly different from the measured values. Given this caveat, the Kohn–Sham band structure often provides a good starting point for the understanding of the electronic structure of materials. Moreover, it can be shown that the highest occupied Kohn–Sham orbital energy is in fact the ionization energy of a physical system.

The appropriate quasiparticle excitation energies are needed for determining the bandgaps, which require the inclusion of the self energy of the electron, i.e. many-electron effects as discussed in Chapter 6. The Kohn–Sham bandgap problem for materials was resolved by the development of an *ab initio* approach for quasiparticle excitations in solids by Hybertsen and Louie[3] based on the so-called GW approximation to the electron self energy,[4] in which the self energy is taken to first order in the screened Coulomb interaction W in a Feynman diagram series expansion, as opposed to the bare Coulomb interaction between electrons. Figure 7.2 compares the theoretical calculated bandgaps from DFT and from the *ab initio* GW approach with experimentally measured bandgaps for common semiconductors and insulators. In materials, the screened Coulomb interaction in general is significantly weaker than the bare Coulomb interaction, leading to the accuracy and versatility of the GW approach.

[3] M. S. Hybertsen and S. G. Louie, "First-principles theory of quasiparticles: calculation of bandgaps in semi-conductors and insulators," *Phys. Rev. Lett.* 55(1985), 1418; ibid., "Electron correlation in semiconductors and insulators: bandgaps and quasiparticle energies," *Phys. Rev. B* 34(1986), 5390.

[4] L. Hedin, "New method for calculating the one-particle Green's function with application to the electron-gas problem," *Phys. Rev.* 139(1965), A796.

[5] S. G. Louie, "First-principles theory of electron excitation energies in solids, surfaces, and defects," in *Topics in Computational Materials Science*, ed. C. Y. Fong (Singapore: World Scientific, 1997) p. 96.

7.3 *Ab initio* pseudopotentials and density functional theory

The behavior of the outer valence electrons of the constituent atoms determines most of the properties of a material. Thus, another important ingredient in first-principles calculations is the electron–ion interaction, which can be described in terms of pseudopotentials. As discussed in Chapter 3, pseudopotentials play an important role in the theory of the electronic structure of condensed matter. The concept explains the apparent weak interaction between the active electrons and the ion cores in solids (e.g. it justifies a nearly-free electron model for the simple metals) and provides great efficiency in the computation of the properties of real materials. Pseudopotentials eliminate the core electrons from the problem and allow the use of significantly simpler basis sets (e.g. planewaves or uniform real-space grids) in the numerical solutions of the self-consistent-field equations (e.g. the Kohn–Sham equations). For DFT studies, it is necessary to generate the pseudopotentials within the density functional Kohn–Sham framework.

Although the use of pseudopotentials dates back to the 1930s, they are best understood in terms of the Phillips–Kleinman cancellation theorem.[6] Phillips and Kleinman argued that the wavefunctions of the valence and higher-energy electronic states are expected to be smooth away from the atomic sites and oscillatory with atomic character in the core regions (since valence states are orthogonal to the core states, that is, $\langle \phi_c | \phi_v \rangle = 0$, where c and v denote core and valence states, respectively). The one-electron wavefunctions (e.g. the Kohn–Sham orbitals in Eq. (7.25)) may then be written in the form

$$|\varphi_i(\mathbf{r})\rangle = |\phi_i(\mathbf{r})\rangle + \sum_c |f_c\rangle \langle f_c | \phi_i(\mathbf{r})\rangle, \tag{7.35}$$

where $|\phi(\mathbf{r})\rangle$ is a smooth pseudowavefunction and $|f_c\rangle$ are normalized Bloch sums of core states with energy ε_c. Solving the one-particle Schrödinger or Kohn–Sham equation

$$\left\{ \frac{p^2}{2m} + V \right\} |\varphi_i(\mathbf{r})\rangle = \varepsilon_i |\varphi_i(\mathbf{r})\rangle, \tag{7.36}$$

where V is a one-particle potential, is equivalent to solving (by grouping terms) the following equation for the pseudowavefunction $\phi_i(\mathbf{r})$:

$$\left\{ \frac{p^2}{2m} + V + V_R(\varepsilon_i) \right\} |\phi_i(\mathbf{r})\rangle = \varepsilon_i |\phi_i(\mathbf{r})\rangle. \tag{7.37}$$

The additional term V_R, given by

$$V_R(\varepsilon) = \sum_c (\varepsilon - \varepsilon_c) |f_c(\mathbf{r})\rangle \langle f_c(\mathbf{r}')|, \tag{7.38}$$

---

[6] J. C. Phillips and L. Kleinman, "New method for calculating wavefunctions in crystals and molecules," *Phys. Rev.* 116(1959), 287.

is a nonlocal, energy-dependent operator. Since the core states are, in general, significantly lower in energy than those of the valence and higher-energy states of interest, V_R is effectively a repulsive potential with negligible energy dependence that cancels a large part of the original strong electron–ion interaction V. The resulting net potential seen by the electrons is then a weak pseudopotential

$$V_p = V + V_R. \tag{7.39}$$

Within this framework, the solutions to Eq. (7.37) yield the same eigenvalues as the original Eq. (7.36) and wavefunctions that are similar outside the core region. The above arguments provide the conceptual basis for pseudopotentials and also show that pseudopotentials are not unique, since a different choice for the second term on the right-hand side of Eq. (7.35) would lead to the same eigenvalues but different pseudopotentials. This ambiguity has led to a number of different construction methods to optimize the accuracy and computational efficiency of pseudopotentials for different classes of materials.

In general, construction of *ab initio* pseudopotentials involves first solving the all-electron problem for a given atom in a particular configuration, where Eq. (7.36) is solved within a specific approximation (e.g. DFT in the LDA). The resulting valence electron wavefunctions $\varphi_i(\mathbf{r})$ may be used to form a set of pseudowavefunctions $\phi_i(\mathbf{r})$ by joining $\varphi_i(\mathbf{r})$ to a properly chosen smooth, nodeless function for $\mathbf{r}$ less than a certain cutoff radius from the nucleus, as illustrated in Fig. 7.3 for the case of the $3s$ and $3p$ states of sodium. By inverting the Schrödinger equation (for a given $\phi_i(\mathbf{r})$ and ε_i), we obtain an equation for $\phi_i(\mathbf{r})$:

$$\left\{ \frac{p^2}{2m} + V_p^i \right\} |\phi_i(\mathbf{r})\rangle = \varepsilon_i |\phi_i(\mathbf{r})\rangle. \tag{7.40}$$

The corresponding pseudopotential $V_p^i(\mathbf{r})$ is constructed for the ith (e.g. s, p, d) state of this particular element. This procedure is usually carried out for an atom in a configuration appropriate in the condensed state. The pseudopotential V_p^i contains the effects of screening of the valence electrons in the atomic configuration from which it is constructed. An *ab initio* ionic pseudopotential describing the intrinsic interaction of an electron with an atom stripped of the outer valence electrons may next be obtained by unscreening the atomic pseudopotential, i.e. subtracting off the screening Hartree and exchange-correlation potential resulting from the pseudocharge density of the valence electrons. Various constraints (such as norm conservation of the orbitals) and other procedures have been developed for the construction of pseudopotentials. The resulting *ab initio* ionic pseudopotentials are in general very accurate and transferable and have been used successfully in first-principles calculations (DFT and beyond) for different properties and in different environments.

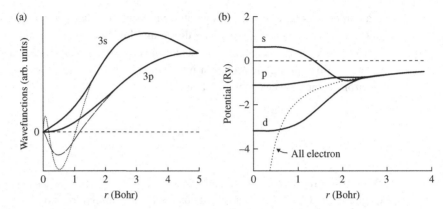

Figure 7.3 (a) Construction of pseudowavefunctions (solid curves) from all-electron atomic wavefunctions (dashed curves) of sodium. (b) *Ab initio* ionic pseudopotentials of sodium. (Figure courtesy of J. R. Chelikowsky.)

7.4 Some applications of DFT to electronic, structural, vibrational, and related ground-state properties

The structure and other ground-state properties of condensed matter have been studied within the density functional formalism employing different approximations for $E_{xc}[\rho]$ and different computational techniques both with and without the use of pseudopotentials. The combination of DFT and the *ab initio* pseudopotential method is, however, a particularly powerful and versatile approach for calculating the ground-state properties of materials. We now give some selected examples of such studies.

Figure 7.4 shows results from the work of Yin and Cohen, who first showed that structural energies may be accurately determined using the *ab initio* pseudopotential density functional approach. By computing the total energy of the system

$$E_{total} = E_{el} + E_{ion\text{-}ion} \tag{7.41}$$

for different structures (where E_{el} is the DFT electronic ground-state energy from Eq. (7.17) and $E_{ion\text{-}ion}$ is the classical electrostatic interaction energy among the ions), one can determine the stable structure and various structural parameters, such as the cohesive energies, lattice constants, and bulk and elastic moduli, with high accuracy. Cohesive energies within a few percent of the experimental values are now obtainable with available exchange-correlation functionals, but further treatment of many-electron effects is required to achieve chemical accuracy. Relative energies are even more accurate, typically yielding lattice constants within one percent and bulk moduli within a few percent of experimental values.

Knowledge of the equation of state for the various structures further allows for the investigation of structural phase transitions under pressure by evaluating the Gibbs free energy

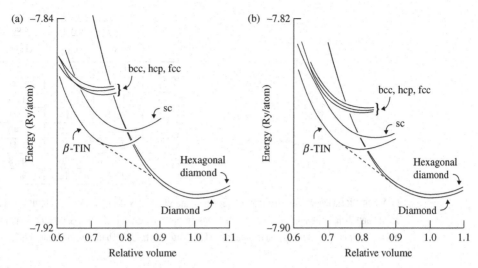

Figure 7.4   Total energy per atom of Si (a) and Ge (b) in different crystal structures. Relative volume is defined as the calculated volume compared to the lowest energy volume in the diamond structure. The negative of the common tangent (dashed line) constructed between the diamond and β-tin structure curves gives the critical transition pressure between the two structures. (After Yin and Cohen, 1982.)[7]

G of the system

$$G = E_{\text{tot}} + PV - TS. \tag{7.42}$$

In particular, at low temperature, the critical transition pressure between the two adjacent structures in Fig. 7.4 is given by the negative of the common tangent of the two equations of state. This is a consequence of the fact that, at pressure equal to the critical transition pressure P_c, the Gibbs free energy of the two phases is equal, and for $T \approx 0$ (the entropy term TS may thus be neglected),

$$E_{\text{tot}}^{(1)} + P_c V^{(1)} = E_{\text{tot}}^{(2)} + P_c V^{(2)}, \tag{7.43}$$

leading to

$$P_c = \frac{-\partial E_{\text{tot}}}{\partial V} = -\left[\frac{E_{\text{tot}}^{(2)} - E_{\text{tot}}^{(1)}}{V^{(2)} - V^{(1)}} \right]. \tag{7.44}$$

Such analysis has allowed the calculation of not only the critical pressure, but also the volume discontinuity at the transition, and has provided an *ab initio* understanding and prediction of the high-pressure phases of matter.

The structural parameters of complex materials, surfaces, defects, clusters, molecules, and nanostructured systems can be determined by minimizing the total energy or the forces

---

[7] M. T. Yin and M. L. Cohen, "Theory of static structural properties, crystal stability, and phase transformations: application to Si and Ge," *Phys. Rev. B* 26(1982), 5668.

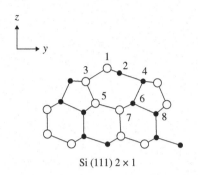

	y-coordinate (Å)		z-coordinate (Å)	
	Theory	Exp.	Theory	Exp.
1	0.00	0.00	0.00	0.00
2	1.11	1.12	−0.47	−0.38
3	−1.99	−2.00	−1.26	−1.26
4	3.30	3.26	−1.21	−1.19
5	−1.14	−1.10	−3.48	−3.46
6	2.37	2.40	−3.41	−3.39
7	1.16	1.20	−4.08	−4.06
8	−2.14	−2.12	−4.36	−4.26

Figure 7.5 (a) Schematic side view of the geometric structure of the Si(111) 2 × 1 surface. Surface atoms 1 and 2 form π-bonded chains. The solid and open circles denote atoms in different (110) planes. (b) Atomic coordinates from energy minimization as compared to data from LEED. (After Northrup, Hybertsen, and Louie, 1991.)[8]

on the atoms:

$$\mathbf{F}_i = -\frac{\partial E_{\text{total}}}{\partial \mathbf{R}_i}. \tag{7.45}$$

For nonperiodic systems, a common approach is to carry out the electronic structure calculations by employing a supercell scheme[9] in which the surface, defect, or molecule is repeated periodically to mimic the isolated structure. As an illustration, Fig. 7.5 compares the calculated geometric structure of the Si(111) 2 × 1 surface with experimental coordinates from low-energy electron diffraction (LEED) measurements. This surface is quite interesting both in terms of its structure and its electronic and optical properties. At low temperature, the atoms on the Si(111) surface rearrange (doubling the surface unit cell) from the ideally terminated geometry to form chains of π-bonded atoms with the position of the atoms on the chains undergoing significant buckling. Figure 7.5 shows excellent agreement between the calculated and experimental geometry. Similarly, by employing DFT geometry determination, understanding of the structural and bonding properties of much larger and more complex systems may be achieved.

The ability to calculate the ground-state total energy or forces on the atoms allows for the first-principles study of the lattice dynamics of a system within the Born–Oppenheimer approximation (i.e. the electrons remain in the ground state as the ions move). The phonon frequencies and eigenvectors are typically obtained in two schemes: (i) the frozen phonon method or (ii) the linear response method. Conceptually, in the frozen phonon approach (see Fig. 7.6), one considers a distortion of the form $\mathbf{u}_i(\mathbf{k}) = \mathbf{u}_0 \cos(\mathbf{k} \cdot \mathbf{R}_i + \delta_{\mathbf{k}})$ frozen into the ideal structure of a crystal, and computes the change in the total energy of the system $\Delta E(\mathbf{u}_0)$ as a function of the distortion amplitude u_0. Upon an expansion of the change in

[8] J. E. Northrup, M. S. Hybertsen, and S. G. Louie, "Many-body calculation of the surface state energies for Si(111)2 × 1," *Phys. Rev. Lett.* 66(1991), 500.

[9] M. L. Cohen, M. Schlüter, J. R. Chelikowsky, and S. G. Louie, "Self-consistent pseudopotential method for localized configurations: Molecules," *Phys. Rev.* B12 (1975), 5575.

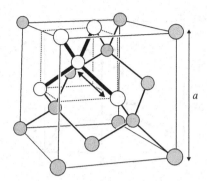

Schematic picture of a frozen phonon calculation, corresponding to the distortion of a zone-center optical phonon mode in the diamond structure.

energy in a Taylor series,

$$\delta E(\mathbf{u}_0) = K_2 u_0^2 + K_3 u_0^3 + K_4 u_0^4 + \cdots, \tag{7.46}$$

the harmonic contribution to the phonon frequency is given by the coefficient K_2, and the higher-order coefficients determine the anharmonic contributions. The advantage of the frozen phonon approach is that the anharmonic contributions may be studied in a straightforward manner. However, the size of the supercell in the total energy calculation is dictated by the wavelength of the frozen distortion. In practice, it is only possible to study a certain discrete number of phonons with wavevectors $\mathbf{k}$ that are at symmetry points or along high-symmetry directions in the Brillouin zone that do not lead to a very large supercell.

Density functional perturbation theory is an alternate approach to phonons. In such a scheme, the atomic force constants

$$\mathbf{C}_{ij} = \frac{\partial^2 E_{\text{tot}}}{\partial \mathbf{R}_i \partial \mathbf{R}_j} \tag{7.47}$$

are evaluated using linear response theory, and the dynamical matrix is then diagonalized to obtain the phonon dispersion. For example, the electronic contribution to the atomic force constant may be shown to be equal to

$$\mathbf{C}_{ij}^{e\ell} = \int \left[\frac{\delta \rho(\mathbf{r})}{\delta \mathbf{R}_i} \frac{\partial V_{\text{ion}}}{\partial \mathbf{R}_j} + \rho(\mathbf{r}) \frac{\partial^2 V_{\text{ion}}}{\partial \mathbf{R}_i \partial \mathbf{R}_j} \right] d\mathbf{r}. \tag{7.48}$$

This approach allows the computation of phonon properties throughout the Brillouin zone, but only within the harmonic approximation. Figure 7.7 shows the linear response results for the phonon dispersion relation and phonon density of states for silicon and germanium. Both the frozen phonon and linear response approaches have given very accurate phonon results for a variety of materials, ranging from metals to semiconductors to complex oxides. For some systems, anharmonic terms can modify the phonon structure significantly.

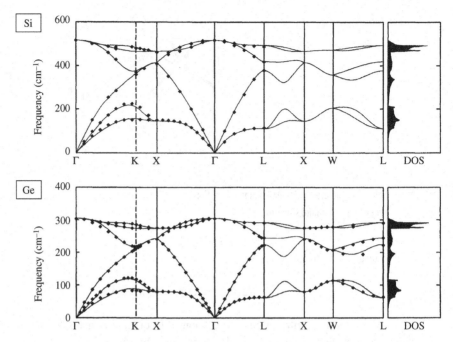

Figure 7.7 Calculated phonon dispersion relations of Si and Ge (solid curves) as compared to experimental data. (After Giannozzi *et al.*, 1991.)[10]

Knowing the forces on the atoms from first principles, e.g. through Eq. (7.45), we can solve the equation of motions for the atoms

$$M_\ell \ddot{\mathbf{R}}_\ell = -\frac{\partial E_{\text{tot}}}{\partial \mathbf{R}_\ell}. \tag{7.49}$$

Such *ab initio* molecular dynamics simulations yield the dynamical properties of a system of interest within the approximation that the ions are classical particles, and, by keeping the ions at finite temperature with the electrons on the Born–Oppenheimer surface, finite temperature effects may be calculated. These kinds of DFT-based molecular dynamics simulations have found important applications in studies of amorphous and liquid systems, phase stability and melting of solids, dynamics of defects and impurities, structure and properties of clusters and nanostructures, chemical reactions on surfaces, etc.

An example of another important physical quantity that is obtainable from DFT is the electron–phonon coupling matrix element, $M_{\mathbf{q}\alpha}(\mathbf{k} \rightarrow \mathbf{k}')$, which describes the strength of the scattering of an electron from state $\mathbf{k}$ to state $\mathbf{k}'$ via the emission or absorption of a phonon with wavevector $\mathbf{q}$ and branch index α. This interaction will be discussed in detail in Chapter 10, and it is central to phenomena such as electrical resistivity, thermal resistivity, and phonon-mediated superconductivity. The electron–phonon matrix element

[10] P. Giannozzi, S. de Gironcoli, P. Pavone, and S. Baroni, "Ab initio calculation of phonon dispersions in semiconductors," *Phys. Rev. B* 43(1991), 7231.

Table 7.1 Calculated electron–phonon coupling constant λ, electrical resistivity ρ (in $\mu\Omega$ cm), and thermal resistivity w (in K cm/W) from DFT at 273 K as compared to experiment (after Savrasov, Savrasov, and Andersen, 1994).[11]

	Al	Nb	Mo
λ^{cal}	0.44	1.26	0.42
λ^{exp}	0.42	1.33	0.44
ρ^{cal}	2.35	13.67	4.31
ρ^{exp}	2.42	13.30	4.88
w^{cal}	0.42	2.17	0.73
w^{exp}	0.42	1.93	0.72

Table 7.2 Physical quantities as derivatives of total energy E with respect to different perturbations – electric field $\mathcal{E}$, magnetic field $\mathbf{B}$, atomic position $\mathbf{R}$, and strain ϵ_{ik}. The symbols i, j, k are Cartesian coordinates and α, β are atomic indices.

Polarizability:	$\alpha_{ij} \sim -\dfrac{\partial^2 E}{\partial \mathcal{E}_i \partial \mathcal{E}_j}$
Hyperpolarizability:	$\beta_{ijk} \sim -\dfrac{\partial^3 E}{\partial \mathcal{E}_i \partial \mathcal{E}_j \partial \mathcal{E}_k}$
Born charge:	$Z^*_{\alpha ij} \sim \dfrac{\partial^2 E}{\partial \mathcal{E}_i \partial R_{\alpha j}}$
Piezoelectric tensor:	$\gamma^*_{ijk} \sim \dfrac{\partial^2 E}{\partial \mathcal{E}_i \partial \epsilon_{jk}}$
Raman tensor:	$\mathcal{R}_{ijk\alpha} \sim \dfrac{\partial^3 E}{\partial \mathcal{E}_i \partial \mathcal{E}_j \partial R_{\alpha k}}$
Elastic constant:	$C_{ijkl} \sim \dfrac{\partial^2 E}{\partial \epsilon_{ij} \partial \epsilon_{kl}}$
Dynamical matrix:	$M_{\alpha i \beta j} \sim \dfrac{\partial^2 E}{\partial R_{\alpha i} \partial R_{\beta j}}$
Magnetic susceptibility:	$\chi^B_{ij} \sim -\dfrac{\partial^2 E}{\partial B_i \partial B_j}$

can be expressed in the form (see Chapter 10)

$$M_{\mathbf{q}\alpha}(\mathbf{k} \to \mathbf{k}') = \langle \varphi_{\mathbf{k}'} | H_I | \varphi_{\mathbf{k}} \rangle , \qquad (7.50)$$

with H_I proportional to the change in the potential seen by an electron in the presence of a phonon. Within DFT, H_I may be computed using either a frozen phonon or a linear response approach. Table 7.1 compares the DFT results for the electron–phonon coupling constant (λ), the electrical resistivity (ρ), and the thermal resistivity (w) with experimental results for the metals Al, Nb, and Mo.

[11] S. Y. Savrasov, D. Y. Savrasov, and O. K. Andersen, "Linear-response calculations of electron–phonon interactions," *Phys. Rev. Lett.* 72(1994), 372.

A number of other phenomena or response functions of materials can be formulated as changes in total energy E of the system in response to the combination of different perturbations. These quantities therefore can be and have been obtained within standard DFT calculations. Table 7.2 provides a partial list of such quantities and their expressions as derivatives of the total energy E.

8 The dielectric function for solids

The dielectric response function is a central concept in the understanding of many phenomena in condensed matter systems. It is essential for understanding the response of a material to electromagnetic probes. Similar concepts are used in the screening of perturbations from charged impurities and defects and of the various many-body (e.g. electron–electron) interactions in a solid. It is thus ubiquitous in the discussion of the electronic, transport, optical, and magnetic properties of solids.

In this chapter, we shall focus on the longitudinal dielectric function (i.e. the density-density response function) of an interacting many-electron system. The extension to the transverse dielectric function (i.e. the current–current response function) will be discussed in Chapter 9 on the optical properties of solids. We will illustrate the importance of the dielectric function with several applications.

8.1 Linear response theory

Suppose an external potential $\delta V_{\text{ext}}(\mathbf{r}', t')$ is applied to a system of electrons. If the perturbation is weak, which is the case under many circumstances, one may assume that the total change in potential $\delta V(\mathbf{r}, t)$ in the material is linearly related to the external potential by

$$\delta V(\mathbf{r}, t) = \int \epsilon^{-1}(\mathbf{r}, \mathbf{r}', t - t') \delta V_{\text{ext}}(\mathbf{r}', t') d\mathbf{r}' dt', \tag{8.1}$$

where ϵ^{-1} is called the inverse dielectric function, or equivalently,

$$\delta V_{\text{ext}}(\mathbf{r}, t) = \int \epsilon(\mathbf{r}, \mathbf{r}', t - t') \delta V(\mathbf{r}', t') d\mathbf{r}' dt'. \tag{8.2}$$

The dielectric function $\epsilon(\mathbf{r}, \mathbf{r}', t - t')$ is in general a two-point function of the spatial coordinates because of the inhomogeneity of the charge distribution in a real material. The $t - t'$ dependence of the dielectric function is a consequence of time translation symmetry.

It is often useful to consider ϵ in momentum and frequency space. A Fourier transform of Eq. (8.2) with respect to time yields

$$\delta V_{\text{ext}}(\mathbf{r}, \omega) = \int \epsilon(\mathbf{r}, \mathbf{r}', \omega) \delta V(\mathbf{r}', \omega) d\mathbf{r}'. \tag{8.3}$$

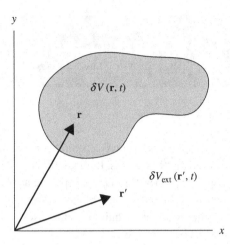

The total potential $\delta V(\mathbf{r},t)$ in a material (a cross section in the x–y plane is denoted by the shaded area) is different from that of the bare applied potential $\delta V_{\text{ext}}(\mathbf{r},t)$ due to the effects of screening.

Furthermore, if one uses the convention for the Fourier transform of a two-point function as

$$f(\mathbf{r}, \mathbf{r}') = \frac{1}{\Omega} \sum_{\mathbf{q}\mathbf{q}'} e^{i\mathbf{q}\cdot\mathbf{r}} f(\mathbf{q}, \mathbf{q}') e^{-i\mathbf{q}'\cdot\mathbf{r}'} \tag{8.4}$$

then Eq. (8.3) becomes

$$\delta V_{\text{ext}}(\mathbf{q}, \omega) = \sum_{\mathbf{q}'} \epsilon(\mathbf{q}, \mathbf{q}', \omega) \delta V(\mathbf{q}', \omega) d\mathbf{q}'. \tag{8.5}$$

For a crystal with translation vectors $\mathbf{R}$, the discrete spatial translational symmetry of the system dictates that

$$\epsilon(\mathbf{r}, \mathbf{r}', \omega) = \epsilon(\mathbf{r} + \mathbf{R}, \mathbf{r}' + \mathbf{R}, \omega). \tag{8.6}$$

Equation (8.6) implies that $\epsilon(\mathbf{q}, \mathbf{q}', \omega)$ is equal to zero unless $\mathbf{q} - \mathbf{q}' = \mathbf{G}$, where $\mathbf{G}$ is a reciprocal lattice vector of the crystal.

The linear response relationship between the total and external potential for a crystal in momentum–frequency space is then given by

$$\delta V_{\text{ext}}(\mathbf{q} + \mathbf{G}, \omega) = \sum_{\mathbf{G}'} \epsilon(\mathbf{q} + \mathbf{G}, \mathbf{q} + \mathbf{G}', \omega) \delta V(\mathbf{q} + \mathbf{G}', \omega), \tag{8.7}$$

with $\mathbf{q}$ restricted to the first Brillouin zone of the crystal. It is traditional to express the dielectric function for crystals in the form of a matrix in the reciprocal lattice vectors, i.e.

$$\delta V_{\text{ext}}(\mathbf{q} + \mathbf{G}, \omega) = \sum_{\mathbf{G}'} \epsilon_{\mathbf{G}\mathbf{G}'}(\mathbf{q}, \omega) \delta V(\mathbf{q} + \mathbf{G}', \omega). \tag{8.8}$$

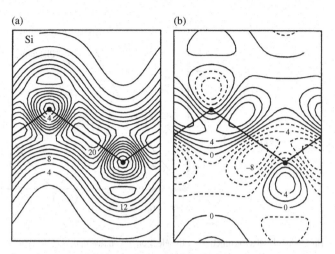

(a)   (b)

Figure 8.2 Local-field effects of a uniform external field on a silicon crystal. (a) Valence electron charge density $\rho(\mathbf{r})$ contour plot of Si on the [110] plane. (b) Change in valence electron density $\delta\rho(\mathbf{r})$ in the presence of a uniform electric field (along the vertical direction) on the same plane as in (a). (After Hybertsen and Louie, 1987.)[1]

This equation is valid for any periodic system within linear response theory. The fact that ϵ is a matrix with off-diagonal elements in $\mathbf{G}$ and $\mathbf{G}'$ reflects the charge inhomogeneity of the crystal. A consequence is that a long-wavelength perturbation, i.e. an external potential of the form $\delta V_{\text{ext}}(\mathbf{q})$ with $\mathbf{q}$ inside the first Brillouin zone, can give rise to short-wavelength fields in a crystal because

$$\delta V(\mathbf{q} + \mathbf{G}, \omega) = \sum_{\mathbf{G}'} \epsilon^{-1}_{\mathbf{G}\mathbf{G}'}(\mathbf{q}, \omega) \delta V_{\text{ext}}(\mathbf{q} + \mathbf{G}', \omega). \qquad (8.9)$$

The effects of the off-diagonal elements of the dielectric matrix $\epsilon_{\mathbf{G}\mathbf{G}'}(\mathbf{q}, \omega)$ are termed local-field effects. Figure 8.2 illustrates the response of a silicon crystal to a uniform external electric field, showing that a long-wavelength perturbation leads to a short-wavelength response.

The dielectric matrix defined through Eq. (8.7) describes the general response of a system to an external scalar potential perturbation. However, in the long wavelength limit, it may also be used to describe the macroscopic dielectric response to electromagnetic radiation for a cubic system or polycrystalline sample. For optical frequencies, as we shall see in Chapter 9, one can relate the long wavelength ($q \to 0$) $\epsilon(\omega)$ to other optical constants, such as the complex index of refraction $N(\omega)$. For low frequencies, e.g. in ac transport experiments, $\epsilon(\omega)$ may be related to the conductivity $\sigma(\omega)$. Thus, by knowing the dielectric function, one can obtain the optical, transport, and a number of other properties of

[1] M. S. Hybertsen and S. G. Louie, "Ab initio static dielectric matrices from the density functional approach I: Formulation and application to semiconductors and insulators," *Phys. Rev. B* 35(1987), 5585; ibid., "Ab initio static dielectric matrices from the density functional approach II: Calculation of the screening response in diamond, Si, Ge and LiCl," *Phys. Rev. B* 35(1987), 5602.

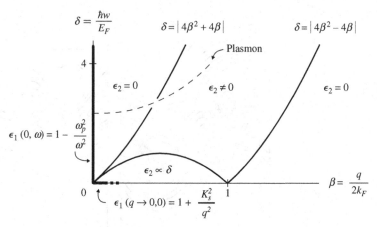

Figure 8.3 The Lindhard dielectric function for the electron gas on the (q, ω) or (β, δ) plane, showing the behavior of ϵ_1 along the two axes and regions where ϵ_2 is nonzero. The dashed line shows the plasmon dispersion.

materials. In general, ϵ is a complex quantity $\epsilon = \epsilon_1 + i\epsilon_2$, with the real part $\epsilon_1(\omega)$ and the imaginary part $\epsilon_2(\omega)$ satisfying the Kramers–Kronig relations.

For a homogeneous system, which is invariant under infinitesimal spatial translations, the dielectric function takes the form $\epsilon(\mathbf{r} - \mathbf{r}', t - t')$. It is straightforward to show that ϵ is diagonal in $\mathbf{q}$-space, i.e. it is of the form $\epsilon(\mathbf{q}, \omega)$ with $\mathbf{q}$ extending over all values.

The diagonal form of the dielectric matrix is often used for metals while retaining the dependence of the function on the direction of the wavevector $\mathbf{q}$. This approximation is also sometimes used for semiconductors and insulators, even though the electron charge density distribution in these materials is far from uniform. The off-diagonal elements of ϵ are, in fact, very important for many phenomena. They are often neglected just because of the computational efforts needed for their evaluation.

For truly homogeneous, isotropic systems such as the electron gas, ϵ is only a function of the magnitude of the wavevector $\mathbf{q}$ and of the frequency ω. Even in that case, $\epsilon(q, \omega)$ is a highly complicated function on the q–ω plane, as illustrated in Fig. 8.3. As will be discussed in Section 8.4, for the three-dimensional electron gas in the long-wavelength limit, it simplifies to

$$\epsilon(q \to 0, \omega) = 1 - \frac{\omega_p^2}{\omega^2}, \tag{8.10}$$

with $\omega_p^2 = \frac{4\pi n e^2}{m}$, where ω_p is the classical plasma frequency, n is the electron density, and m is the electron mass, and in the static limit with $q \ll k_F$,

$$\epsilon(q, \omega = 0) = 1 + \frac{K_s^2}{q^2}, \tag{8.11}$$

with $K_s^2 = 4\pi e^2 D(\varepsilon_F)$, where $D(\varepsilon_F)$ is the density of states at the Fermi energy ε_F. Equation (8.11) is called the Thomas–Fermi screening function in elementary treatments.

8.2 Self-consistent field framework

The polarizability $\chi(\mathbf{r}, \mathbf{r}', t - t')$ of an interacting electron system relates a small external perturbation, $\delta V_{\text{ext}}(\mathbf{r}, t)$, to the resulting change in the electron density

$$\delta \rho_{\text{ind}}(\mathbf{r}, t) = \int \chi(\mathbf{r}, \mathbf{r}', t - t') \delta V_{\text{ext}}(\mathbf{r}', t') d\mathbf{r}' dt', \tag{8.12}$$

which, for simplicity, we write as

$$\delta \rho = \chi \delta V_{\text{ext}}. \tag{8.13}$$

In a self-consistent field approach to the electronic structure of solids, one can similarly define an independent-particle polarizability χ_0 as the response to the total perturbing potential

$$\delta \rho = \chi_0 \delta V, \tag{8.14}$$

where δV is the change in the self-consistent field potential seen by the electrons. For example, within the density functional formalism in Chapter 7,

$$\delta V = \delta V_{\text{ext}} + \delta V_{\text{H}} + \delta V_{\text{xc}} = \delta V_{\text{ext}} + V_c \delta \rho + K_{\text{xc}} \delta \rho, \tag{8.15}$$

where V_c is the bare Coulomb interaction, V_{xc} is the exchange-correlation potential within the density functional Kohn–Sham formulation, and K_{xc} is the functional derivative of V_{xc} with respect to the density. Using Eqs. (8.12), (8.14), and (8.15), the full polarizability χ is related to the independent-particle polarizability by

$$\chi = (1 - \chi_0 V_c - \chi_0 K_{\text{xc}})^{-1} \chi_0. \tag{8.16}$$

We now can make the connection to the dielectric response function, which is defined by $\delta V = \epsilon^{-1} \delta V_{\text{ext}}$. However, we must pay attention to the induced portion of the screening potential, which depends upon the nature of the external probe or the processes to which the dielectric function is applied. In the above analysis, we have implicitly assumed that the source of the external potential is distinguishable from the electrons (e.g. due to an impurity). If the probe of the screening potential is a test charge, then it is affected only by the electrostatic term (i.e. the Hartree term) in Eq. (8.15), and the resulting dielectric matrix is given by

$$\epsilon_{\text{tc}}^{-1} = 1 + V_c \chi. \tag{8.17}$$

This is called the test-particle–test-particle response function. If the probe is an electron, then the whole screening potential in Eq. (8.15) is effective and the dielectric response function is

$$\epsilon_{\text{el}}^{-1} = 1 + (V_c + K_{\text{xc}}) \chi. \tag{8.18}$$

The usual random phase approximation (RPA) response function is obtained by neglecting the exchange-correlation contribution in Eq. (8.18) and in χ given in Eq. (8.16). The result is

$$\epsilon_{\text{RPA}}^{-1} = 1 + V_c(1 - \chi_0 V_c)^{-1}\chi_0, \tag{8.19}$$

which is equivalent to the more common form of $\epsilon_{\text{RPA}} = 1 - V_c\chi_0$.

8.3 The RPA dielectric function within DFT

In this section, we consider the RPA dielectric function of a solid in some detail within the framework of a self-consistent field approach, such as density functional theory (DFT). Within the Kohn–Sham framework of DFT, the ground state of the associated independent-particle system from which the density is derived is given by the Slater determinant constructed from the Kohn–Sham orbitals

$$\Psi_0 \left(\{\mathbf{r}_\ell\} \right) = \frac{1}{\sqrt{N!}} \left| \phi_1 \quad \phi_2 \quad \cdots \quad \phi_N \right|, \tag{8.20}$$

where $\phi_\ell \equiv (\phi_\ell(\mathbf{r}_1) \quad \phi_\ell(\mathbf{r}_2) \quad \cdots \quad \phi_\ell(\mathbf{r}_N))^T$, with T denoting the transpose operation of taking a row to a column vector, and the density is

$$\rho(\mathbf{r}) = \langle \Psi_0 | \hat{\rho}(\mathbf{r}) | \Psi_0 \rangle = \sum_{l=1}^{N} \phi_\ell^*(\mathbf{r})\phi_\ell(\mathbf{r}), \tag{8.21}$$

with $\hat{\rho}(\mathbf{r}) = \sum_i \delta(\mathbf{r} - \mathbf{r}_i)$.

An excited state with energy E_{ij} corresponds to promoting an electron from an occupied orbital ϕ_i to an unoccupied orbital ϕ_j, which we denote as

$$\Psi_{ij} \left(\{\mathbf{r}_\ell\} \right) = \frac{1}{\sqrt{N!}} \left| \phi_1 \quad \cdots \quad \phi_{i-1} \quad \phi_j \quad \phi_{i+1} \quad \cdots \quad \phi_N \right|, \tag{8.22}$$

with energy $\epsilon_j - \epsilon_i$ relative to the ground state (i.e. $E_{ij} - E_0 = \epsilon_j - \epsilon_i$).

We consider a small potential perturbation with amplitude δV, which is turned on adiabatically. We write the corresponsing perturbation Hamiltonian as

$$\hat{\mathcal{H}}_I = e^{-i\omega t}e^{\alpha t} \sum_i \delta V(\mathbf{r}_i) = \int \delta V(\mathbf{r}, t)\hat{\rho}(\mathbf{r}) \, d\mathbf{r}, \tag{8.23}$$

where $\alpha = 0^+$.

Within first-order time-dependent perturbation theory, the perturbed ground state is given by

$$|\Psi_0'\rangle = |\Psi_0\rangle + \sum_{ij} \frac{|\Psi_{ij}\rangle \langle \Psi_{ij}| \, \hat{\mathcal{H}}_I \, |\Psi_0\rangle}{E_0 - E_{ij} + \hbar\omega + i\hbar\alpha}, \tag{8.24}$$

and $\langle \Psi'_0(t) |$ can be obtained by taking the complex conjugate of $| \Psi'_0 \rangle$ and switching $\omega \rightarrow -\omega$, since $\hat{\mathcal{H}}_I$ is not Hermitian.

The change in charge distribution to first order in δV is

$$
\delta \rho(\mathbf{r}) = \langle \Psi'_0 | \hat{\rho}(\mathbf{r}) | \Psi'_0 \rangle - \langle \Psi_0 | \hat{\rho}(\mathbf{r}) | \Psi_0 \rangle
$$

$$
= \sum_{ij} \left[\frac{\langle \Psi_0 | \hat{\rho}(\mathbf{r}) | \Psi_{ij} \rangle \langle \Psi_{ij} | \hat{\mathcal{H}}_I | \Psi_0 \rangle}{E_0 - E_{ij} + \hbar\omega + i\hbar\alpha} + \frac{\langle \Psi_0 | \hat{\mathcal{H}}_I | \Psi_{ij} \rangle \langle \Psi_{ij} | \hat{\rho}(\mathbf{r}) | \Psi_0 \rangle}{E_0 - E_{ij} - \hbar\omega - i\hbar\alpha} \right]. \quad (8.25)
$$

Now, since

$$
\langle \Psi_0 | \hat{\rho} | \Psi_{ij} \rangle = \phi_i^*(\mathbf{r}) \phi_j(\mathbf{r}), \quad (8.26)
$$

we have

$$
\delta \rho(\mathbf{r}) = \sum_{ij} \left[\frac{\phi_i^*(\mathbf{r}) \phi_j(\mathbf{r}) \int \phi_j^*(\mathbf{r}') \phi_i(\mathbf{r}') \delta V(\mathbf{r}', t) \, d\mathbf{r}'}{E_0 - E_{ij} + \hbar\omega + i\hbar\alpha} \right.
$$

$$
\left. + \frac{\phi_i(\mathbf{r}) \phi_j^*(\mathbf{r}) \int \phi_j(\mathbf{r}') \phi_i^*(\mathbf{r}') \delta V(\mathbf{r}', t) \, d\mathbf{r}'}{E_0 - E_{ij} - \hbar\omega - i\hbar\alpha} \right]. \quad (8.27)
$$

Using $E_{ij} - E_0 = \epsilon_j - \epsilon_i$ and performing an ensemble averaging at finite temperature, we have

$$
\delta \rho(\mathbf{r}) = \sum_{ij} f_i (1 - f_j) \left[\frac{\phi_i^*(\mathbf{r}) \phi_j(\mathbf{r}) \int \phi_j^*(\mathbf{r}') \phi_i(\mathbf{r}') \delta V(\mathbf{r}', t) d\mathbf{r}'}{\epsilon_i - \epsilon_j + \hbar\omega + i\hbar\alpha} \right.
$$

$$
\left. + \frac{\phi_i(\mathbf{r}) \phi_j^*(\mathbf{r}) \int \phi_j(\mathbf{r}') \phi_i^*(\mathbf{r}') \delta V(\mathbf{r}', t) d\mathbf{r}'}{\epsilon_i - \epsilon_j - \hbar\omega - i\hbar\alpha} \right], \quad (8.28)
$$

where f is the Fermi–Dirac distribution function.

Using the definition of the independent-particle polarizability $\chi_0(\mathbf{r}, \mathbf{r}')$ in Eq. (8.14), and switching the indices $i \leftrightarrow j$ in the second term of Eq. (8.28), we arrive at

$$
\chi_0(\mathbf{r}, \mathbf{r}', \omega) = \sum_{ij} (f_i - f_j) \frac{\phi_i^*(\mathbf{r}) \phi_j(\mathbf{r}) \phi_j^*(\mathbf{r}') \phi_i(\mathbf{r}')}{\epsilon_i - \epsilon_j + \hbar\omega + i\hbar\alpha}. \quad (8.29)
$$

Transforming to reciprocal space, the resulting expression for χ_0 is

$$
\chi^0_{\mathbf{G}\mathbf{G}'}(\mathbf{q}, \omega) = \frac{1}{\Omega} \sum_{n,n',\mathbf{k}\sigma} (f_{n,\mathbf{k}} - f_{n',\mathbf{k}+\mathbf{q}})
$$

$$
\times \frac{\langle n, \mathbf{k} | e^{-i(\mathbf{q}+\mathbf{G})\cdot\mathbf{r}} | n', \mathbf{k} + \mathbf{q} \rangle \langle n', \mathbf{k} + \mathbf{q} | e^{i(\mathbf{q}+\mathbf{G}')\cdot\mathbf{r}} | n, \mathbf{k} \rangle}{\epsilon_{n,\mathbf{k}} - \epsilon_{n',\mathbf{k}+\mathbf{q}} + \hbar\omega + i\hbar\alpha}. \quad (8.30)
$$

Here, the one-particle states are labeled by Bloch wavevector $\mathbf{k}$ and band index n. Conservation of crystal momentum has also been used in the matrix elements. Application of Eq. (8.30) together with Eq. (8.16) and either Eq. (8.17) or Eq. (8.18) yields the desired dielectric response functions.

8.4 The homogeneous electron gas

In the case of an homogeneous electron gas, the one-particle states are planewaves and the matrix elements in Eq. (8.30) are equal to unity, resulting in the following RPA dielectric function from Eq. (8.19):

$$\epsilon(q,\omega) = 1 - \frac{4\pi e^2}{\Omega q^2} \sum_{k\sigma} \frac{f(\mathbf{k}) - f(\mathbf{k}+\mathbf{q})}{\varepsilon(\mathbf{k}) - \varepsilon(\mathbf{k}+\mathbf{q}) + \hbar\omega + i\hbar\alpha}, \tag{8.31}$$

with $\alpha \to 0^+$. The various limiting forms of Eq. (8.31) are very instructive and have useful applications in simple metallic systems. We shall consider the dielectric response function at $T = 0$, since in most circumstances $k_B T \ll E_F$ for a metal.

Long-wavelength static screening. We consider $\omega = 0$ and $q \ll k_F$ of the electron gas. This corresponds to the Thomas–Fermi limit. Let $\mu = \frac{\mathbf{q} \cdot \mathbf{k}}{|\mathbf{q}||\mathbf{k}|}$, and define

$$\epsilon(\mathbf{q},\omega) = \epsilon_1(\mathbf{q},\omega) + i\epsilon_2(\mathbf{q},\omega), \tag{8.32}$$

where ϵ_1 and ϵ_2 are the real and imaginary parts of the dielectric function. Since $q \ll k_F$ to lowest order in q, we may write

$$f(\mathbf{k}) - f(\mathbf{k}+\mathbf{q}) = -q\mu \frac{\partial f}{\partial k} = q\mu \delta(k - k_F) \tag{8.33}$$

and

$$\varepsilon(\mathbf{k}+\mathbf{q}) - \varepsilon(\mathbf{k}) = q\mu \frac{\partial \varepsilon}{\partial k}. \tag{8.34}$$

In this limit,

$$\epsilon_1(\mathbf{q},\omega = 0) = 1 + \frac{4\pi e^2}{\Omega q^2} \sum_{k\sigma} \frac{q\mu \delta(\mathbf{k} - \mathbf{k}_F)}{q\mu \frac{\partial \varepsilon}{\partial k}} = 1 + \frac{4\pi e^2 D(E_F)}{q^2} = 1 + \frac{K_s^2}{q^2}, \tag{8.35}$$

where $D(E_F)$ is the density of states per unit volume at the Fermi energy and K_s is called the Thomas–Fermi wavevector, with

$$K_s = \sqrt{4\pi e^2 D(E_F)}. \tag{8.36}$$

Long-wavelength finite frequency. For $\omega \neq 0$ and $q \ll k_F$, that is, $\varepsilon(\mathbf{q}) = \frac{\hbar^2 q^2}{2m} \ll \hbar\omega$ and $qv_F = \frac{\hbar q k_F}{m} \ll \hbar\omega$, we need to approach Eq. 8.31 in a different way. By substituting arguments $\mathbf{k}$ by $\mathbf{k} - \mathbf{q}$, and only taking the real part of ϵ, we can get

$$\epsilon_1(\mathbf{q},\omega) = 1 - \frac{8\pi e^2}{\Omega q^2} \sum_{k} f(\mathbf{k}) \left[\frac{1}{\varepsilon(\mathbf{k}) - \varepsilon(\mathbf{k}+\mathbf{q}) + \hbar\omega} - \frac{1}{\varepsilon(\mathbf{k}-\mathbf{q}) - \varepsilon(\mathbf{k}) + \hbar\omega} \right]. \tag{8.37}$$

Define $\Lambda(\mathbf{k}, \mathbf{q}) = \varepsilon(\mathbf{k} + \mathbf{q}) - \varepsilon(\mathbf{k}) = \frac{\hbar^2 \mathbf{k} \cdot \mathbf{q}}{m} + \frac{\hbar^2 q^2}{2m}$, and we can expand Eq. 8.37 in series of $\Lambda/\hbar\omega$,

$$
\begin{aligned}
\epsilon_1(\mathbf{q}, \omega) &= 1 - \frac{8\pi e^2}{\Omega q^2} \sum_{\mathbf{k}} f(\mathbf{k}) \left[\frac{1}{\hbar\omega - \Lambda(\mathbf{k}, \mathbf{q})} - \frac{1}{\hbar\omega + \Lambda(\mathbf{k}, \mathbf{q})} \right] \\
&= 1 - \frac{16\pi e^2}{\Omega q^2} \sum_{\mathbf{k}} f(\mathbf{k}) \frac{1}{\hbar\omega} \left[\frac{\Lambda}{\hbar\omega} + \left(\frac{\Lambda}{\hbar\omega} \right)^3 + \cdots \right] \\
&= 1 - \frac{\omega_p^2}{\omega^2} \left\{ 1 + \frac{1}{\hbar^2 \omega^2} \left[\frac{3}{5}(q v_{\mathrm{F}})^2 + \varepsilon^2(\mathbf{q}) \right] + \mathcal{O}\left(\frac{1}{\omega^4} \right) \right\},
\end{aligned}
\tag{8.38}
$$

where we have transformed the $\mathbf{k}$ summation into an integral.

As $q \to 0, \epsilon_1(q = 0, \omega) = 1 - \frac{\omega_p^2}{\omega^2}$, as in classical long-wavelength plasma screening. Since

$$
\delta V(\mathbf{q}, \omega) = \frac{\delta V_{\text{ext}}(\mathbf{q}, \omega)}{\epsilon(\mathbf{q}, \omega)},
\tag{8.39}
$$

we see that an infinitesimal perturbation at a specific $(\mathbf{q}, \omega)$ can lead to a large response if $\epsilon(\mathbf{q}, \omega)$ vanishes. This is the condition for normal modes in a system. Note that $\delta V(\mathbf{q}, \omega)$ can also be expressed in terms of an electron charge density response $\delta\rho(\mathbf{q}, \omega)$. The condition $\epsilon(\mathbf{q}, \omega) = 0$ thus determines the plasmon dispersion relation of the electron system. For the electron gas, using Eq. (8.39) for long-wavelength oscillations, we have the following relation for the plasmons within the RPA:

$$
\omega^2(\mathbf{q}) = \omega_p^2 \left(1 + \frac{3}{5} \frac{q^2 v_{\mathrm{F}}^2}{\hbar^2 \omega_p^2} \right).
\tag{8.40}
$$

To obtain the imaginary part of $\epsilon(\mathbf{q}, \omega)$ in evaluating Eq. (8.31), we make use of the mathematical identity $\lim_{\nu \to 0^+} \frac{1}{x + i\nu} = P\left(\frac{1}{x}\right) - i\pi \delta(x)$. That is,

$$
\epsilon_2(\mathbf{q}, \omega) = \frac{4\pi^2 e^2}{\Omega q^2} \sum_{\mathbf{k}} (f(\mathbf{k}) - f(\mathbf{k} + \mathbf{q})) \, \delta(\varepsilon(\mathbf{k} + \mathbf{q}) - \varepsilon(\mathbf{k}) - \hbar\omega).
\tag{8.41}
$$

We see that ϵ_2 is zero unless $\varepsilon(\mathbf{k} + \mathbf{q}) - \varepsilon(\mathbf{k}) = \hbar\omega$ for some $\mathbf{k}$ in the Fermi sphere. This is the condition for energy conservation. Thus, in general, ϵ_1 gives information on virtual processes, and ϵ_2 gives information on real processes in a system.

For a free electron gas, $\varepsilon(\mathbf{k})$ is given by the free electron dispersion relation, and the energy conservation condition becomes

$$
\frac{\hbar^2}{2m} \left(q^2 + 2\mathbf{k} \cdot \mathbf{q} \right) = \hbar\omega.
\tag{8.42}
$$

This means that, for a given ω, there are minimum and maximum values of q that define the range over which $\epsilon_2(\mathbf{q}, \omega)$ is nonzero. This region of nonzero $\epsilon_2(\mathbf{q}, \omega)$ in the $(\mathbf{q}, \omega)$ plane is

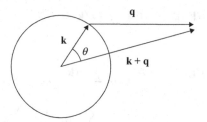

Figure 8.4 Schematic of the Fermi sphere of an electron gas, showing scattering of an occupied state $\mathbf{k}$ to an empty state $\mathbf{k} + \mathbf{q}$.

called the continuum Landau damping region. (See Fig. 8.3.) In this region, electron–hole excitations, satisfying both conservation of energy and momentum with ω and $\mathbf{q}$, exist in the system.

The Lindhard dielectric function. For a free electron gas at $T = 0$, Eq. (8.31) may be evaluated analytically to obtain the so-called Lindhard dielectric function in a closed form.[2] Defining the dimensionless variables $\beta = \frac{q}{2k_F}$ and $\delta = \frac{\hbar\omega}{E_F}$, the resulting expressions for the real and imaginary parts of the dielectric function are, respectively,

$$\epsilon_1(\beta, \delta) = 1 + \frac{K_s^2}{q^2} \frac{1}{8\beta} \left\{ 4\beta + \left[1 - \left(\beta - \frac{\delta}{4\beta} \right)^2 \right] \ln \left| \frac{1 + (\beta - \delta/4\beta)}{1 - (\beta - \delta/4\beta)} \right| \right.$$
$$\left. + \left[1 - \left(\beta + \frac{\delta}{4\beta} \right)^2 \right] \ln \left| \frac{1 + (\beta + \delta/4\beta)}{1 - (\beta + \delta/4\beta)} \right| \right\} \tag{8.43}$$

and

$$\epsilon_2(\beta, \delta) = \frac{\pi K_s^2}{q^2} \frac{1}{8\beta} \begin{cases} \delta & \text{if } \beta < 1, \ \delta \leq |4\beta^2 - 4\beta|, \\ 1 - (\beta - \delta/4\beta)^2 & \text{if } |4\beta^2 - 4\beta| < \delta < |4\beta^2 + 4\beta|, \\ 0 & \text{if } \delta > |4\beta^2 + 4\beta|, \\ 0 & \text{if } \delta < |4\beta^2 - 4\beta| \text{ and } \beta > 1. \end{cases} \tag{8.44}$$

The boundary for $\epsilon_2 \neq 0$ arises from the conservation of energy and momentum as discussed above. In terms of β and δ, the conservation condition is

$$4\beta^2 + 4\beta \frac{k}{k_F} \mu = \delta. \tag{8.45}$$

At a given $\delta = \hbar\omega/E_F$, the minimum value of β (or q) is determined when $k = k_F$ and $\mu = \cos(\theta = 0) = 1$, as seen in Fig. 8.3. This gives the boundary $4\beta^2 + 4\beta = \delta$ on the (δ, β) plane. The maximum value of β (or q) is determined when $k = k_F$ and $\mu = \cos(\theta = \pi) = -1$ or $|4\beta^2 - 4\beta| = \delta$.

[2] J. Lindhard, "On the properties of a gas of charged particles," *K. Dan. Vidensk. Selsk. Mat. Fys. Medd.* 28(1954), 8.

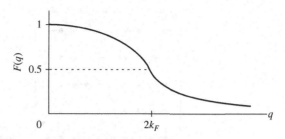

Figure 8.5 Behavior of the function $F(q)$ given in Eq. (8.48).

Denoting $\gamma = \delta/4\beta$, ϵ_1 can be rewritten in a form that is useful for various expansions:

$$\epsilon_1(\beta,\delta) = 1 + \frac{K_s^2}{q^2}\frac{1}{8\beta}\left\{4\beta + (1 - \beta^2 - \gamma^2)\ln\left|\frac{(1+\beta)^2 - \gamma^2}{(1-\beta)^2 - \gamma^2}\right|\right.$$

$$\left. + 2\beta\gamma\ln\left|\frac{(1-\gamma)^2 - \beta^2}{(1+\gamma)^2 - \beta^2}\right|\right\}. \tag{8.46}$$

The Lindhard dielectric function is very useful as a starting point in analyzing the various properties of metallic systems. In the static ($\omega = 0$) limit, it takes the form from Eq. (8.46):

$$\epsilon_1(q, \omega = 0) = 1 + \frac{K_s^2}{q^2}F(\beta), \tag{8.47}$$

with

$$F(\beta) = \frac{1}{2} + \frac{1-\beta^2}{4\beta}\ln\left|\frac{1+\beta}{1-\beta}\right|. \tag{8.48}$$

Since $\beta = \frac{q}{2k_F}$, $F(q \to 0) = 1$ and has a value of $1/2$ and a slope of minus infinity at $q = 2k_F$, as illustrated in Fig. 8.5.

Because $F(q)$ is always less than one and decays rapidly to zero at $q > 2k_F$, Eq. (8.47) shows that the Thomas–Fermi screening function given in Eq. (8.35) in general gives rise to overscreening at large q (or short wavelength) in the analysis of $\delta V(q) = \frac{\delta V_{\text{ext}}(q)}{\epsilon(q)}$. The effective screening length in an electron gas $\lambda_q = \frac{1}{K_s\sqrt{F(q)}}$ increases as q increases. This is because, at length scales smaller than the Fermi wavelength ($q > 2k_F$), the electrons cannot rearrange to screen the perturbation.

8.5 Some simple applications

Sound velocity in metals. We may use the Lindhard dielectric function to deduce the velocity of sound in metals and understand why it is of remarkably constant value for the various metals. Let us consider N ions per unit volume. The ions are of mass M and charge

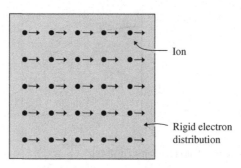

Figure 8.6 Schematic of an array of ions in a crystal undergoing long-wavelength oscillations in the presence of a rigid uniform distribution of electrons.

$-Ze$ at the lattice points, as illustrated in Fig. 8.6. Since the overall system is neutral, then there are also $n = ZN$ electrons per unit volume. First let us keep the electrons rigid (i.e. no screening) as the ions move. The long-wavelength characteristic frequency of vibration for the ions is given by the bare ionic plasma frequency

$$\Omega_p^2 = \frac{4\pi N(Ze)^2}{M} = \left(\frac{Zm}{M}\right)\frac{4\pi (ZN)e^2}{m} = \left(\frac{Zm}{M}\right)\omega_p^2. \tag{8.49}$$

Here m is the electron mass and ω_p is the electron plasma frequency $\omega_p = \sqrt{\frac{4\pi ne^2}{m}}$. The bare ionic plasma frequency Ω_p is independent of q and is significantly lower than ω_p because of the heavier atomic mass M as compared to the electron mass m. The fast dynamics of the electrons, however, will screen the interaction, replacing e^2 by $\frac{e^2}{\epsilon(q,\omega)}$. Since $\frac{m}{M} \ll 1$, we have $\Omega_p \ll \omega_p$. Hence the screening may be included by replacing e^2 by $\frac{e^2}{\epsilon(q,\omega=0)}$ for any lattice motions. In particular, for sound waves in metal, we have $q \to 0$, and the renormalized (or dressed) phonon dispersion is given by

$$\omega^2(q) = \frac{\Omega_p^2}{\epsilon(q)} = \frac{\Omega_p^2}{1 + K_s^2 F(q/2k_F)/q^2}. \tag{8.50}$$

In the above expression, we have used Eq. (8.47). At long wavelength (i.e. $q \to 0$), Eq. (8.50) leads to a sound wave dispersion relation of

$$\omega(q) = \frac{\Omega_p}{K_s}q = v_s q. \tag{8.51}$$

Thus, the sound velocity v_s is given by

$$v_s = \sqrt{\frac{Zm}{M}}\frac{\omega_p}{K_s} = \sqrt{\frac{1}{3}\frac{Zm}{M}}\, v_F. \tag{8.52}$$

To arrive at the final expression for v_s, we have made use of the relation $K_s = \sqrt{4\pi e^2 D(\epsilon_F)}$ and the free electron gas density of states and Fermi velocity v_F. Since $\frac{m}{3M}$ is typically of the order of 10^{-6}, Eq. (8.52) demonstrates why the sound velocity of typical metals is

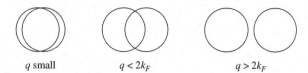

Figure 8.7 Schematic of two spheres in **k**-space with radius k_F separated by a distance of q. The points at the intersection of the two spheres' surfaces yield a large contribution to $\epsilon(q, \omega = 0)$.

around 10^5 cm/s and is about 10^3 times smaller than the Fermi velocity of the electrons. For example, for the alkali metal potassium, Eq. (8.52) gives a calculated sound velocity $v_s^{\mathrm{cal}} = 1.8 \times 10^5$ cm/s as compared to a measured velocity of $v_s^{\mathrm{expt}} = 2.2 \times 10^5$ cm/s in the [100] direction.

However, the simple analysis given above cannot be extended directly to semiconductors or insulators. In an insulating material, because of a bandgap separating the occupied from the unoccupied bands, it is straightforward to show that Eq. (8.31) yields a static dielectric function that is a constant as q approaches zero, i.e. $\epsilon^{\mathrm{semic}}(q \to 0, \omega = 0) = \epsilon_0$. (See the discussion below.) This behavior leads to $\omega^2(q) = \Omega_p^2/\epsilon_0$ in the above simple analysis, which is a constant independent of q. This incorrect behavior for the sound waves stems from the neglect of the off-diagonal elements of the dielectric matrix, which are more important for insulating materials. Inclusion of the full matrix will recover the correct behavior for the sound waves, i.e. $\omega(q \to 0) = 0$, as this must be the case owing to the translation symmetry of space.

The Kohn effect. Near $q = 2k_F$, the static Lindhard dielectric function has a peculiar behavior, which leads to some important consequences that are observable in metals. This behavior arises from the fact that the function $F(q)$ has a logarithmic singularity at $q = 2k_F$. The first derivative of $F(q)$ is, in fact, equal to minus infinity at $2k_F$. The physical origin of this singularity can be seen from the diagrams in Fig. 8.7. For $q < 2k_F$, it is possible to draw a vector **q** such that both ends lie in the Fermi surface. Thus, the energy denominator for the dielectric function can be small. But for $q > 2k_F$, it is not possible to take an electron from a filled state **k** to an empty state **k** + **q** with approximate conservation of energy. This means that the energy denominator in Eq. (8.31) at $\omega = 0$ is always large and the contribution of all processes to $\epsilon(q)$ is small.

A sudden drop in $F(\mathbf{q})$ results in a sudden reduction in screening $\epsilon(q)$ at $q = 2k_F$. Equation (8.50) implies that there will be a sudden increase in $\omega(q)$ for the phonons at $q = 2k_F$. This phenomenon, known as the Kohn effect,[3] is observed in certain metals. The lower the dielectric screening value, the stiffer the response of the electrons to the ion motions, and the higher the lattice vibration frequency.

Induced charge near a charge impurity. The carriers in a metal will rearrange to screen out the Coulomb potential generated by the introduction of a charge impurity. The screening charge density distribution may be calculated from the dielectric response

---

[3] W. Kohn, "Image of the Fermi surface in the vibration spectrum of a metal," *Phys. Rev. Lett.* **2**(1959), 393.

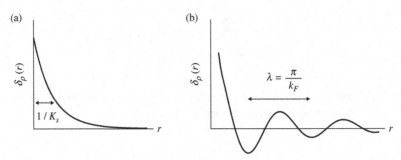

Figure 8.8 Schematic of the induced electron density around a positive impurity point charge at the origin evaluated with (a) the Thomas–Fermi approximation and (b) the Lindhard dielectric function.

function since (again in matrix form)

$$\delta V_{\text{ind}} = \delta V - \delta V_{\text{ext}} = \left(\frac{1}{\epsilon} - 1\right)\delta V_{\text{ext}}, \tag{8.53}$$

where δV_{ext} is the potential due to the impurity, and the induced charge density within the RPA is related to the induced potential via Poisson's equation,

$$\nabla^2 \delta V_{\text{ind}}(\mathbf{r}) = -4\pi e \delta\rho(\mathbf{r}). \tag{8.54}$$

Figure 8.8 illustrates the induced charge distribution near a positive point charge at the origin with density $\delta\rho_{\text{ext}} = \delta(\mathbf{r})$, using either the Thomas–Fermi or the Lindhard dielectric function. Within the simplified Thomas–Fermi model, the screening charge $\delta\rho(\mathbf{r})$ has an exponential form with a decay length given by $1/K_s$. The Lindhard dielectric function, however, yields a result that oscillates at large distance ($r > \frac{1}{k_F}$) in the form

$$\delta\rho(\mathbf{r}) \sim \frac{1}{r^3}\cos(2k_F r). \tag{8.55}$$

This oscillatory behavior, called Friedel oscillations, arises from the singularity in $\epsilon(q)$ at $q = 2k_F$, which has its physical origin at the sharp cutoff of occupied electron states at $k = k_F$. The wavelength of the oscillations at large distance is given by $\lambda = \pi/k_F$. The electron density profile near the surface of a simple metal also often shows similar oscillations for the same physical reason (see Fig. 8.9). A number of other phenomena in metal physics have their origins in the same screening effects. A good example is the distance dependence of the interaction between two spins in a metal. The interaction oscillates in sign with a period of $\lambda = \pi/k_F$. These kinds of interactions are termed RKKY interactions after Ruderman, Kittel, Kasuya, and Yosida.

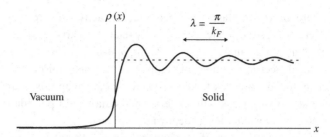

Figure 8.9 Schematic of electron density profile $\rho(x)$ of a simple metal near the surface.

8.6 Some other properties of the dielectric function

Dispersion relations and sum rules. There are many relations that the dielectric response function of a material satisfies which are useful. We will mention a few here. The first set is the Kramers–Kronig relations. For each component $(\mathbf{G}, \mathbf{G}')$ of the dielectric matrix, one has

$$\epsilon_1(\mathbf{q}, \omega) - 1 = \frac{1}{\pi} P \int_{-\infty}^{\infty} \epsilon_2(\mathbf{q}, \omega') \frac{d\omega'}{\omega' - \omega} \tag{8.56}$$

and

$$\epsilon_2(\mathbf{q}, \omega) = \frac{1}{\pi} P \int_{-\infty}^{\infty} [1 - \epsilon_1(\mathbf{q}, \omega')] \frac{dw'}{\omega' - \omega}. \tag{8.57}$$

Similarly, the components of the inverse of the dielectric matrix satisfy

$$\operatorname{Re}\epsilon^{-1}(\mathbf{q}, \omega) - 1 = \frac{1}{\pi} P \int_{-\infty}^{\infty} \operatorname{Im}[\epsilon^{-1}(\mathbf{q}, \omega')] \frac{d\omega'}{\omega' - \omega} \tag{8.58}$$

and

$$\operatorname{Im}\epsilon^{-1}(\mathbf{q}, \omega) = \frac{1}{\pi} P \int [1 - \operatorname{Re}\epsilon^{-1}(\mathbf{q}, \omega')] \frac{d\omega'}{\omega' - \omega}. \tag{8.59}$$

These relations may be derived with the assumptions that the system satisfies causality, linear response, and conservation of energy. They are valid for any systems in general as long as the above assumptions hold.

The dielectric matrix also satisfies a number of sum rules. Two common ones for the diagonal elements are

$$\int_0^{\infty} d\omega \, \omega \operatorname{Im} \epsilon^{-1}(\mathbf{q}, \omega) = -\frac{\pi}{2} \omega_p^2 \tag{8.60}$$

and

$$\int_0^{\infty} d\omega \, \omega \operatorname{Im} \epsilon(\mathbf{q}, \omega) = \frac{\pi}{2} \omega_p^2, \tag{8.61}$$

where ω_p is the plasma frequency of the active electrons in the range of frequency of interest. These sum rules are basically related to particle conservation. The sum rules can be of great help in analyzing and relating optical and charge particle energy loss experiments to the dielectric functions of solids, and they play a useful role in calculations, as well as enabling one to check the consistency of a given approximation. For example, evaluating the above sum rules with experimental data would provide a measure of the number of active electrons in the frequency range probed.

The generalization of the sum rule in Eq. (8.60) to the off-diagonal elements of the dielectric matrix yields

$$\int_0^\infty d\omega \, \omega \, \mathrm{Im} \, \epsilon_{\mathbf{GG'}}^{-1}(\mathbf{q}, \omega) = -\frac{\pi}{2}\omega_p^2 \frac{(\mathbf{q}+\mathbf{G}) \cdot (\mathbf{q}+\mathbf{G'})}{|\mathbf{q}+\mathbf{G}|^2} \frac{\rho(\mathbf{G}-\mathbf{G'})}{\rho(0)}, \qquad (8.62)$$

where $\rho(\mathbf{G})$ is the $\mathbf{G}$th Fourier component of the electron charge density $\rho(\mathbf{r})$. These sum rules may be derived from considering the double commutation $\left[[H, \hat{\rho}_{\mathbf{q}+\mathbf{G}}], \hat{\rho}_{\mathbf{q}+\mathbf{G'}}\right]$, where $\hat{\rho}_{\mathbf{q}+\mathbf{G}}$ is the $\mathbf{q}+\mathbf{G}$ component of the Fourier transform of the density operator $\hat{\rho}(\mathbf{r})$.

Many-body dielectric function. In Sections 8.2 and 8.3, we discussed the dielectric function within the framework of self-consistent field theories. It is also possible to derive a very similar expression for the many-body dielectric function in terms of matrix elements between exact eigenstates of the many-body system. Consider the Fourier transform of the electron density operator $\hat{\rho}(\mathbf{r})$ of a homogeneous system,

$$\hat{\rho}(\mathbf{q}) = \int e^{-i\mathbf{q}\cdot\mathbf{r}} \hat{\rho}(\mathbf{r}) \, d\mathbf{r}. \qquad (8.63)$$

This is often called the particle density fluctuation operator. Using perturbation theory along the lines given in Section 8.3, we arrive at (for $T = 0$)

$$\epsilon^{-1}(\mathbf{q}, \omega) = 1 - \frac{4\pi e^2}{\Omega q^2} \sum_n |\langle n| \, \hat{\rho}(\mathbf{q}) \, |0\rangle|^2 \left[\frac{1}{\hbar\omega + \hbar\omega_{n0} + i\hbar\alpha} + \frac{1}{-\hbar\omega + \hbar\omega_{n0} - i\hbar\alpha} \right], \qquad (8.64)$$

where $|n\rangle$ are the exact eigenstates of the system with $|0\rangle$ the ground state, $\alpha = 0^+$, and $\hbar\omega_{n0} = E_n - E_0$.

Ground-state electron interaction energy. If we know the dielectric function, we may obtain the ground-state interaction energy. Recall that

$$\lim_{\alpha \to 0^+} \frac{1}{x \pm i\alpha} = P\frac{1}{x} \mp i\pi \delta(x). \qquad (8.65)$$

The above mathematical identity gives (from Eq. (8.64)) the following for the imaginary part of ϵ^{-1}:

$$\mathrm{Im} \, \epsilon^{-1}(\mathbf{q}, \omega) = \frac{4\pi e^2}{\Omega q^2} \sum_n |\langle n| \, \hat{\rho}(\mathbf{q}) \, |0\rangle|^2 \left[\delta(\omega + \omega_{n0}) - \delta(\omega - \omega_{n0})\right]. \qquad (8.66)$$

Integrating Eq. (8.66) over positive ω yields

$$\int_0^\infty d\omega \, \mathrm{Im} \, \epsilon^{-1}(\mathbf{q}, \omega) = -\frac{4\pi^2 e^2}{\Omega q^2} \sum_n \left| \langle n | \hat{\rho}(\mathbf{q}) | 0 \rangle \right|^2 = -\frac{4\pi^2 e^2}{\Omega q^2} \langle 0 | \hat{\rho}^\dagger(\mathbf{q}) \hat{\rho}(\mathbf{q}) | 0 \rangle .$$

$$(8.67)$$

Now the Coulomb interaction energy of the electron is

$$E_{\mathrm{int}} = \left\langle 0 \left| \frac{1}{2} \sum_{i \neq j} \frac{e^2}{|\mathbf{r}_i - \mathbf{r}_j|} \right| 0 \right\rangle \tag{8.68}$$

$$= \left\langle 0 \left| \sum_q' \frac{2\pi e^2}{\Omega q^2} \left[\hat{\rho}^\dagger(\mathbf{q}) \, \hat{\rho}(\mathbf{q}) - n_0 \right] \right| 0 \right\rangle, \tag{8.69}$$

where n_0 is the average electron density. Comparing Eq. (8.67) with Eq. (8.69), we have

$$E_{\mathrm{int}} = -\sum_q' \left\{ \frac{1}{2\pi} \int_0^\infty d\omega \, \mathrm{Im} \, \epsilon^{-1}(\mathbf{q}, \omega) + \frac{2\pi n_0 e^2}{q^2} \right\}. \tag{8.70}$$

Energy loss of a fast-moving charge particle. The dielectric function may be used to understand and calculate the energy loss spectrum of a fast charged particle through an electron system. Let us assume that the particle with charge $z|e|$ is distinguishable from the electrons in our system, and we have the following initial and final scattered states:

$$\begin{aligned} \text{Initial:} \quad & \left| e^{i\mathbf{K} \cdot \mathbf{R}} \right\rangle | 0 \rangle \\ \text{Final:} \quad & \left| e^{i(\mathbf{K} \pm \mathbf{k}) \cdot \mathbf{R}} \right\rangle | n \rangle , \end{aligned} \tag{8.71}$$

where $\left| e^{i\mathbf{K} \cdot \mathbf{R}} \right\rangle$ is the wavefunction of the high-energy fast particle (which can be taken as a planewave) and $| n \rangle$ is the nth excited state of the electron system. The interaction term between the fast particle and the system may be written as

$$H_{\mathrm{I}} = \int \hat{\rho}(\mathbf{r}) \, V_c(\mathbf{r} - \mathbf{R}) \, d\mathbf{r} = -Ze^2 \int \frac{1}{|\mathbf{R} - \mathbf{r}|} \sum_q \hat{\rho}(\mathbf{q}) \, e^{i\mathbf{q} \cdot \mathbf{r}} \, d\mathbf{r} \tag{8.72}$$

and can be simplified to

$$H_{\mathrm{I}} = \sum_k \left(-\frac{4\pi Ze^2}{k^2} \right) e^{-i\mathbf{k} \cdot \mathbf{R}} \hat{\rho}^\dagger(\mathbf{k}). \tag{8.73}$$

Using Fermi's golden rule, the rate of transition from the initial to all possible final states, having an energy transfer $\hbar\omega$ and a momentum transfer $\hbar\mathbf{k}$, is

$$w(\mathbf{k}, \omega) = \frac{2\pi}{\hbar} \sum_n \left| \langle n | H_{\mathrm{I}} | 0 \rangle \right|^2 \delta(E_n - E_0 - \hbar\omega). \tag{8.74}$$

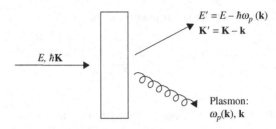

Figure 8.10 Schematic of energy loss of a fast charge particle by emitting a plasmon.

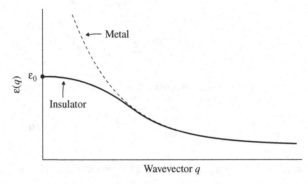

Figure 8.11 Schematic figure of the difference in the static dielectric function $\epsilon(q)$ between a metal and an insulator.

Combining Eq. (8.73) and Eq. (8.66), we may express w in terms of Im $\epsilon^{-1}(\mathbf{k}, \omega)$. The resulting loss power spectrum $P(\mathbf{k}, \omega)$ is then given by

$$P(\mathbf{k}, \omega) \sim -\frac{1}{k^2} \text{ Im} \left[\frac{1}{\epsilon(\mathbf{k}, \omega)} \right] = \frac{1}{k^2} \frac{\epsilon_2(\mathbf{k}, \omega)}{\epsilon_1^2(\mathbf{k}, \omega) + \epsilon_2^2(\mathbf{k}, \omega)} . \tag{8.75}$$

Thus peaks in the loss spectrum occur whenever $\epsilon_1(\mathbf{k}, \omega) \approx \epsilon_2(\mathbf{k}, \omega) \approx 0$. This is the condition for having a plasmon excitation in the system. The physical interpretation is that the fast charged particle loses energy to the system by creating one or more plasmons. (See Fig. 8.10.)

Dielectric constant ϵ_0 of an insulator. Thus far, we have mostly discussed the dielectric screening behavior of a metallic system. There are major differences in the nature of screening between a metal and an insulator, particularly in the long-wavelength static limit. This is schematically illustrated in Fig. 8.11. In a metal, $\epsilon(\mathbf{q} \to 0)$ diverges as $1/q^2$ in the small q limit, but in an insulator, $\epsilon(\mathbf{q} \to 0)$ becomes a constant ϵ_0. This difference arises from the fact that, in an insulator, there is an energy gap between the filled and empty electronic states.

An important consequence of having $\epsilon(\mathbf{q}) \to \epsilon_0$ as $q \to 0$ is that, at large distance r from a charged center in an insulator, the Coulomb potential is given by $V(r) \sim 1/\epsilon_0 r$. Thus there is incomplete screening in the insulators, leading to phenomena such as localized impurity states (e.g. donor or acceptor states in semiconductors) and bound excitons.

The dielectric constant $\epsilon_0 = \epsilon(\mathbf{q} \to 0)$ of an insulator satisfies approximately the following simple relation:

$$\epsilon_0 = 1 + \left(\frac{\hbar\omega_p}{\langle E_g \rangle}\right)^2, \tag{8.76}$$

where ω_p is the plasma frequency associated with the valence band electrons and $\langle E_g \rangle$ is an average bandgap. This relation may be understood from the following analysis in an extreme tight-binding limit. In this model, the insulator has M occupied flat bands, and the unoccupied bands are all separated from the filled bands by energy $\langle E_g \rangle$.

The RPA dielectric function $\epsilon(\mathbf{q}, \omega = 0)$ is given by (from Eqs. (8.19) and (8.30))

$$\epsilon(\mathbf{q} \to 0) = 1 - \frac{4\pi e^2}{\Omega q^2} \sum_{\mathbf{k} n n' \sigma} \frac{|\langle n'\mathbf{k} + \mathbf{q}|\, e^{i\mathbf{q}\cdot\mathbf{r}}\, |n\mathbf{k}\rangle|^2}{\varepsilon(n\mathbf{k}) - \varepsilon(n'\mathbf{k} + \mathbf{q})}, \tag{8.77}$$

where n denotes the occupied bands, n' denotes the empty bands, and $\mathbf{k}$ is within the first Brillouin zone. Using $\varepsilon(n\mathbf{k}) - \varepsilon(n'\mathbf{k} + \mathbf{q}) = -\langle E_g \rangle$, where the number of occupied bands is M and the number of $\mathbf{k}$ vectors is the number of unit cells N, we have

$$\epsilon(\mathbf{q} \to 0) = 1 + \frac{4\pi e^2}{\Omega q^2} \frac{NM}{\langle E_g \rangle} 2 \sum_{n'} |\langle n'\mathbf{k} + \mathbf{q}|\, e^{i\mathbf{q}\cdot\mathbf{r}}\, |n\mathbf{k}\rangle|^2. \tag{8.78}$$

Next we use a well-known theorem of atomic physics, the Thomas–Reiche–Kuhn sum rule:

$$\sum_{n'} (E_n - E_{n'}) \left| \langle n'|\, e^{i\mathbf{q}\cdot\mathbf{r}}\, |n\rangle \right|^2 = \frac{\hbar^2 q^2}{m}. \tag{8.79}$$

Combining Eq. (8.78) and Eq. (8.79), we have the relation for the flat-band insulator

$$\epsilon(\mathbf{q} \to 0) = 1 + 4\pi \left(\frac{2NM}{\Omega}\right) \frac{e^2}{m} \frac{\hbar^2}{\langle E_g \rangle^2}$$
$$= 1 + \left(\frac{\hbar\omega_p}{\langle E_g \rangle}\right)^2, \tag{8.80}$$

since $\frac{2NM}{\Omega} = n$ is the density of valence band electrons.

II.1. **Wannier functions**.

  (a) Estimate in one-dimension the degree of localization of a Wannier function using planewaves as an approximation for the band states.

  (b) Show that the matrix elements of the crystal Hamiltonian H_0 between Wannier functions coming from different bands must vanish.

  (c) Show that the matrix element $\langle R|H_0|R'\rangle$ depends only upon $(R - R')$, where R is the site index for the Wannier function.

II.2. **Bandstructure and dynamics of electrons**.

  (a) For a one-dimensional crystal with lattice constant a, consider a single band with dispersion

$$E(k) = E_1 + (E_2 - E_1)\sin^2\frac{ka}{2}, \tag{II.1}$$

    with E_1 and $E_2 > 0$.

   (i) Give an expression for the electron group velocity.

   (ii) Give an expression for the electron effective mass.

   (iii) Consider the case in which an electric field E is applied and there is a single electron in the band (assume no scattering). Using the equation of motion

$$\hbar\frac{dk}{dt} = eE \tag{II.2}$$

    and the information above, find the time it takes for the electron to execute one complete oscillation and the range of the distance covered in this oscillation.

  (b) Consider an energy band of a two-dimensional square lattice given by

$$E(k_x, k_y) = -A(\cos k_x a + \cos k_y a), \tag{II.3}$$

    where A is a positive constant, and a is the lattice constant. Suppose that an electron occupying a state near the bottom of this band is subjected to a slowly varying static potential $V(x, y)$. Write down, within the effective Hamiltonian formalism, the differential equation that the wavefunction of the electron obeys within the Wannier representation. Repeat for the corresponding equation in the Bloch representation.

II.3. **Electrons in an electric field**. Consider a one-dimensional energy band given by $E(k) = E_0\cos ka$. The band is empty except for a single electron in a state which is a superposition of Bloch states of wavevectors very close to $k = \pi/2a$ (halfway to the Brillouin zone edge). The spread is small compared to the size of the Brillouin zone, but large enough so that the real-space wavepacket is well localized compared to the macroscopic crystal.

(a) What is the velocity of the position space wavepacket?

(b) At time $t = 0$, an electric field of magnitude $\mathcal{E}$ is applied in the $+\hat{x}$ direction. Neglecting all dissipative effects (such as lattice vibration and imperfections), find $\bar{k}(t)$, the average wavevector, as a function of time.

(c) Find the velocity and position of the real-space wavepacket as a function of time. Assume the packet is localized at $x = 0$ initially. Are there any qualitative differences between this motion and that of a classical free electron in a uniform field?

II.4. **Velocity and effective mass of crystal electrons**. A one-dimensional tight-binding band has the form (a is the lattice constant)

$$E(k) = E_0 \cos ka. \tag{II.4}$$

Find the velocity and effective mass for a particle in this band for wavevectors:

(a) $k = 0 + \delta k$,

(b) $k = (\pi/a) - \delta k$,

(c) $k = (\pi/2a) - \delta k$,

(d) $k = (\pi/2a) + \delta k$,

where $\delta k \cdot a \ll 1$. In each case, also give the limiting results for $\delta k = 0$.

II.5. **Electron Dynamics**.

(a) Derive the expression for the energy band of an fcc crystal with a conventional cubic unit cell with edge length a. Use the tight-binding model assuming one s-orbital per atom and a nearest-neighbor interaction of hopping strength τ.

(b) Use the above energy band to answer the following questions. Let an electron at rest $(k = 0)$ at $t = 0$ feel a uniform electric field E which is constant in time.

  (i) Find its trajectory in real space; that is, derive the expression for $x(t), y(t)$, and $z(t)$.

  (ii) Sketch the trajectory in the **[111]** direction.

  (iii) Estimate δR, the amplitude of the Bloch oscillation, for a realistic laboratory electric field and a bandwidth of 5 eV.

II.6. **Berry (or geometric) phase in the spin dynamics of an electron in a magnetic field**. Consider an electron located and pinned at the origin in real space, subjected to a magnetic field $B(t)$, which is of constant magnitude but changing direction very slowly. The magnetic field sweeps out a closed loop on the surface of a sphere of radius $|B(t)|$ in a period T.

(a) Show that, at any given time t, the eigenstate representing spin-up along $B(t)$ has the form

$$\chi_\uparrow = \begin{pmatrix} \cos(\theta/2) \\ e^{i\phi} \sin(\theta/2) \end{pmatrix}, \tag{II.5}$$

where θ and ϕ are the spherical coordinates of the vector $B(t)$.

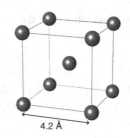

Figure II.1   Bcc lattice structure.

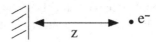

Figure II.2   An electron outside a metal surface.

(b) Evaluate the Berry connection $A(\theta,\phi)$ and the Berry curvature $\boldsymbol{\Omega}(\theta,\phi)$ for the system with the electron in the spin-up state.

(c) Show that the Berry phase acquired in one period by the electron wavefunction is

$$\gamma = -\frac{1}{2}\Delta, \qquad (II.6)$$

where Δ is the solid angle subtended at the origin by the closed loop formed by $\boldsymbol{B}(t)$.

II.7. **Hartree–Fock approximation for ferromagnetic electron gas**. Consider the ground state of an interacting ferromagnetic electron gas. The particles are in planewave states, but all spins point in the same direction.

(a) Find the expression (in terms of r_s) for the ground-state energy of the system in the Hartree–Fock approximation.

(b) Compare the result in (a) with the paramagnetic state (equal amount of up and down spin electrons) discussed in the text in the same approximation. Are there values of r_s for which the ferromagnetic state is lower in energy?

II.8. **Exchange-correlation hole**.

(a) Sodium atoms crystallize into a bcc metal with nearly homogeneous electron density and a lattice constant of $a = 4.2\text{Å}$ (a is the length of the edge of the bcc cube as shown in Fig. II.1). Thus we may consider metallic properties in the jellium model.

(i) What is the value of r_s for metallic sodium?

(ii) Within the Hartree–Fock approximation, what is the *approximate* size of the exchange hole associated with each electron?

(b) Suppose an electron is removed from a metal at a distance z away from the surface, as shown in Fig. II.2.

(i) Find the expression for the z-dependence of the exchange-correlation energy of the electron outside of the metal in the limit of large z. You may assume that the exchange-correlation hole is very small in extent compared to the value of z and that it is confined to the surface.

  (ii) How would you expect the answer in (i) to change if a more realistic exchange-correlation hole function were used? Why?

II.9. **Wigner electron crystal**. In the discussion in the text of the Wigner electron crystal, it was shown that the potential energy per electron is given by $-1.8/r_s$ (in Ry). We have, however, neglected the kinetic energy term.

  (a) Use the Wigner sphere construct to show that the electron feels a harmonic potential (in Ry),

$$V(r) = -\frac{3}{r} + \frac{r^2}{r_s^3}, \tag{II.7}$$

   where r is the displacement from the equilibrium lattice site.

  (b) Find the kinetic energy per particle from the zero-point motion of the electrons by assuming that each electron moves independently of the others.

  (c) What is the equilibrium volume per electron within this model? What is the bulk modulus?

II.10. **Kramers–Kronig relations**.

  (a) Derive and discuss the Kramers–Kronig relations

$$\epsilon_1(\omega) = \frac{1}{\pi} P \int_{-\infty}^{\infty} \frac{\omega' \epsilon_2(\omega')}{\omega'^2 - \omega^2} d\omega' + \text{const.}, \tag{II.8}$$

$$\epsilon_2(\omega) = -\frac{\omega}{\pi} P \int_{-\infty}^{\infty} \frac{\epsilon_1(\omega')}{\omega'^2 - \omega^2} d\omega'. \tag{II.9}$$

  (b) Show that the Kramers–Kronig relations are satisfied by $\epsilon_1(\omega)$ and $\epsilon_2(\omega)$ for the dielectric function

$$\epsilon(\omega) = \lim_{\gamma \to 0^+} \left(1 - \frac{\omega_p^2}{\omega^2 + i\omega\gamma} \right). \tag{II.10}$$

II.11. **Sum rules of dielectric function**. For a free electron gas (no impurities) in the $q \to 0$ limit, $\epsilon = 1 - \omega_p^2/\omega^2$.

  (a) Find $\epsilon_2(\omega)$ and compute $\int_0^{\infty} \omega \epsilon_2(\omega) d\omega$. (You may use the Lindhard dielectric function or other physical reasoning.)

  (b) Find $\text{Im}\left(\frac{1}{\epsilon(\omega)}\right)$ and compute $\int_0^{\infty} \omega \text{Im}\left(\frac{1}{\epsilon(\omega)}\right) d\omega$. (Note: Although the integrals in (a) and (b) are done for the electron gas case, the results are those of two well-known sum rules in many-electron physics which are valid for any dielectric function.)

II.12. **Induced charge in a metal**. Using the Lindhard dielectric function, perform a calculation to show that the induced charge density arising from an impurity in a metal is proportional to $r^{-3} \cos(2k_F r)$ at large r.

II.13. **Zeros and poles of dielectric function**. Show (as rigorously as you can) that the zeros of the dielectric function give the longitudinal normal modes, while the poles give the transverse modes of the system.

II.14. **Polar insulators**.
  (a) Derive $\epsilon(\omega) = \epsilon(\infty)\frac{\omega^2 - \omega_l^2}{\omega^2 - \omega_t^2}$, where ω_l and ω_t are the longitudinal and transverse optical phonon frequencies, respectively.
  (b) Calculate and plot the frequency-dependent reflectivity using the dielectric function in (a).
  (c) What does the reflectivity look like in the polariton region?
  (d) Why is $\omega_l > \omega_t$?

II.15. **Density functional theory**. Show that, in the Kohn–Sham formulation of the density functional theory, the eigenvalues ε_i's which appear in the Kohn–Sham equations satisfy

$$\varepsilon_i = \frac{\partial E}{\partial n_i}, \qquad (\text{II.11})$$

where E is the total electronic energy and n_i is the occupation number for the ith single-particle orbital. Discuss the possible significance of this relation.

II.16. **Kohn effect**. Expand on the discussion in the text of the Kohn effect from the point of view of screening; that is, discuss the influence of the behavior of the Lindhard dielectric function at $q = 2k_F$ on the phonon spectrum of a metal.

PART III

OPTICAL AND TRANSPORT
PHENOMENA

Electronic transitions and optical properties of solids

9.1 Response functions

Measuring the optical properties of matter has had an enormous influence on physics and other areas of science. It has made perhaps the largest impact of all experimental fields. Because of the optical studies of atoms, great theoretical advances such as those of Einstein, Bohr, and Pauli were made possible. This field was one of the early experimental areas of investigation in condensed matter physics, and probing matter with photons remains one of the principal methods for learning about the nature of matter.

By changing the wavelength or frequency ω and the intensity of the probing light, different physical properties can be evaluated. The response of the system can be measured in different ways. For example, for solids (see Fig. 9.1), the incident light having an intensity $I(\omega)$ can excite elementary excitations, and the response of the solid can be evaluated by measuring the reflection $R(\omega)$, absorption $A(\omega)$, and transmission $T(\omega)$ of the light. These are response functions, and others can be defined. By dealing with the complex dielectric function

$$\epsilon(\omega) = \epsilon_1(\omega) + i\epsilon_2(\omega) \tag{9.1}$$

it is possible to compute the various linear optical response functions, assuming a linear response to the probe and assuming that the wavevector $\mathbf{q}$ of the light is small for the physical process of interest.

Previously, we examined the longitudinal Lindhard dielectric function and the self-consistent field dielectric response function within density functional theory. To explain the responses to electromagnetic (EM) waves, we need the transverse ϵ. For cubic crystals or polycrystalline samples, as $\mathbf{q} \to 0$, $\epsilon_{\text{transverse}} \to \epsilon_{\text{longitudinal}}$. For EM waves, a useful relation is $q \sim E$ (in eV) $\times 5 \times 10^4$ cm^{-1}. Therefore, for infrared, visible, and UV photons, $q_{\text{photon}} \ll q_{\text{Brillouin zone}}$, which is typically 10^8 cm^{-1}, so we may assume that $q \approx 0$. These lower-energy photons are useful for exploring the band structure energies of a material. For higher-energy photons, such as X-rays, $q_{\text{photon}} \approx q_{\text{Brillouin zone}}$. The higher-energy photons are useful for determining lattice structures and electron charge density distributions of materials. Hence, the use of synchrotron radiation and X-rays for crystallography and structural biology is common. The very low energy part of the EM spectrum such as the far infrared is used to probe other excitations such as phonons.

The standard expressions for the current density and displacement field, $\mathbf{j}$ and $\mathbf{D}$, in materials involve the conductivity and the transverse dielectric functions, σ and ϵ, respectively.

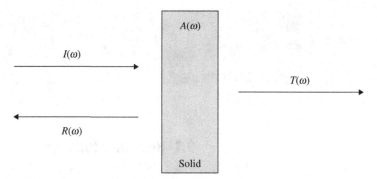

Figure 9.1 Experimental setup for measuring the optical properties of a material.

They can be expressed within linear response theory as

$$\mathbf{j}(\mathbf{r}, t) = \int \sigma(\mathbf{r}, \mathbf{r}', t - t')\mathbf{E}(\mathbf{r}', t')dt'\,d\mathbf{r}' \tag{9.2}$$

and

$$\mathbf{D}(\mathbf{r}, t) = \int \epsilon(\mathbf{r}, \mathbf{r}', t - t')\mathbf{E}(\mathbf{r}', t')dt'\,d\mathbf{r}', \tag{9.3}$$

where $\mathbf{j}$, $\mathbf{D}$, and $\mathbf{E}$ describe the current density and EM fields, and $(\mathbf{r}, t)$ are the space and time coordinates. Since the wavelength of the applied EM field of optical light is much greater than the size of an atom, and $\epsilon(\mathbf{r}, \mathbf{r}')$ is nearly diagonal compared to the wavelength (we are assuming locality), we may assume that ϵ is proportional to $\delta(\mathbf{r}, \mathbf{r}')$ and is independent of $\mathbf{r}$ for a homogeneous system by taking an appropriate spatial average. Then the equation for $\mathbf{D}$ becomes

$$\mathbf{D}(\mathbf{r}, t) = \int \epsilon(t - t')\mathbf{E}(\mathbf{r}, t')dt', \tag{9.4}$$

and we have a similar expression for $\mathbf{j}(\mathbf{r}, t)$. For steady-state conditions, Eq. (9.4) and the one for $\mathbf{j}(\mathbf{r}, t)$ can be converted to the frequency domain

$$\mathbf{j}(\mathbf{r}, \omega) = \sigma(\omega)\mathbf{E}(\mathbf{r}, \omega) \tag{9.5}$$

and

$$\mathbf{D}(\mathbf{r}, \omega) = \epsilon(\omega)\mathbf{E}(\mathbf{r}, \omega). \tag{9.6}$$

In general, σ and ϵ are tensors, and thus the vectors $\mathbf{j}$ and $\mathbf{E}$ do not necessarily point in the direction of the applied field $\mathbf{D}$. However, for an isotropic sample (e.g. one with cubic symmetry, or a polycrystalline sample), σ and ϵ are diagonal and can be considered to be scalars in the present discussion:

$$\sigma_{ij}(\omega) = \sigma(\omega)\delta_{ij} \tag{9.7}$$

and

$$\epsilon_{ij}(\omega) = \epsilon(\omega)\delta_{ij}. \tag{9.8}$$

Following standard conventions for defining $\sigma(\omega)$ and $\epsilon(\omega)$, we consider a non-magnetic material with $\mathbf{B} = \mathbf{H}$ and $\mathbf{M} = 0$. The corresponding wave equations are commonly used:

$$(\text{convention 1}) \quad \nabla^2\mathbf{E} = \frac{-\omega^2}{c^2}\left[\epsilon^0(\omega) + \frac{4\pi i}{\omega}\sigma^0(\omega)\right]\mathbf{E},$$

$$(\text{convention 2}) \quad \nabla^2\mathbf{E} = \frac{-\omega^2}{c^2}\epsilon(\omega)\mathbf{E},$$

$$(\text{convention 3}) \quad \nabla^2\mathbf{E} = \frac{-\omega^2}{c^2}\left[1 + \frac{4\pi i}{\omega}\sigma(\omega)\right]\mathbf{E}. \tag{9.9}$$

Here, convention 1 associates $\mathbf{D} = \epsilon^0\mathbf{E}$ with only the bound electrons and $\mathbf{j} = \sigma^0\mathbf{E}$ with only the free electrons. Convention 2 associates $\mathbf{D} = \epsilon\mathbf{E}$ with all the electrons and uses $\mathbf{j} = \frac{1}{c}\frac{\partial\mathbf{D}}{\partial t}$ for the current density. Convention 3 associates $\mathbf{j} = \sigma\mathbf{E}$ with all the electrons and sets $\epsilon = 1$. In the literature, convention 2 is often used in semiconductor physics, whereas convention 3 is often used in metal physics.

Using the wave equations in Eq. (9.9) and assuming the following form for the $\mathbf{E}$-field of the EM wave,

$$\mathbf{E} = \frac{\mathbf{E}_0}{2}[e^{i(\mathbf{q}\cdot\mathbf{r}-\omega t)} + e^{-i(\mathbf{q}\cdot\mathbf{r}-\omega t)}], \tag{9.10}$$

we have (from convention 1)

$$\frac{\omega}{q} = \frac{c}{\left[\epsilon^0(\omega) + \frac{4\pi i\sigma^0(\omega)}{\omega}\right]^{\frac{1}{2}}}, \tag{9.11}$$

which gives the dispersion relation for the EM wave in a solid. From convention 2, we have a different-looking form of the dispersion relation,

$$\frac{\omega}{q} = \frac{c}{\sqrt{\epsilon}} = \frac{c}{N}, \tag{9.12}$$

with N being the complex refractive index, which is given by

$$N = n + ik = \sqrt{\epsilon}, \tag{9.13}$$

where n is the usual index of refraction, k is the extinction coefficient, which measures the damping of the wave in the media, and ϵ is the complex dielectric function given in Eq. (9.1) within convention 2. By relating the first two conventions, we have $\epsilon_1 = \epsilon_0$ and $\epsilon_2 = \frac{4\pi\sigma_0}{\omega}$. On the other hand, relating this to convention 3, we have

$$\epsilon = 1 + \frac{4\pi i\sigma}{\omega}. \tag{9.14}$$

Nonzero damping arises when $k \neq 0$. Since

$$N^2 = (n + ik)^2 = n^2 - k^2 + 2nki, \tag{9.15}$$

we have

$$\epsilon_1 = n^2 - k^2$$

and

$$\epsilon_2 = 2nk. \tag{9.16}$$

Two useful optical constants, assuming normal incidence and assuming that the sample width d satisfies $d \gg \lambda$, are the reflection R, given by

$$R = \frac{(n - 1)^2 + k^2}{(n + 1) + k^2}, \tag{9.17}$$

and the absorption coefficient η (which describes the fractional EM energy absorbed per unit distance and unit time), given by

$$\eta = \frac{2\omega k}{c}. \tag{9.18}$$

For example, for the case of insulators or pure semiconductors at zero temperature with no free carriers, the expressions for the various quantities for a material with a bandgap E_g and for $\omega < E_g$ are

$$\sigma_1 = 0, \tag{9.19}$$

$$\epsilon_1 = n^2 - k^2, \tag{9.20}$$

and

$$\epsilon_2 = 2nk = \frac{4\pi\sigma_1}{\omega}. \tag{9.21}$$

Therefore, we have

$$\epsilon_2 = \sigma_1 = 0 \implies k = 0, \tag{9.22}$$

and

$$n = \sqrt{\epsilon_1}. \tag{9.23}$$

This indicates that the material is transparent, but it does not preclude some reflection, which at normal incidence is given by

$$R = \frac{(n - 1)^2}{(n + 1)^2}, \tag{9.24}$$

which is large if $\epsilon_1 > 1$. For large ω ($\omega > E_g$), ϵ_2 and hence σ_1 are finite and we have absorption.

9.2 The Drude model for metals

For simple metals, we can apply the theory of Paul Drude, which was developed around 1900.[1] This model is that of a free electron gas with "dirt" used to scatter electrons, resulting in a scattering time of τ. To analyze this system, we use the relaxation time approximation. From this model we can derive the dielectric function by starting with the equation of motion for the electron including the relaxation time τ. Hence,

$$m\ddot{\mathbf{x}} = -e\mathbf{E} - \frac{m\dot{\mathbf{x}}}{\tau},$$

or (by taking the time Fourier transform)

$$-m\omega^2 \mathbf{x}(\omega) = -im\omega\gamma\,\mathbf{x}(\omega) - e\mathbf{E}(\omega), \tag{9.25}$$

where $\gamma \equiv \frac{1}{\tau}$. Solving for $\mathbf{x}$ gives us

$$m\mathbf{x}(\omega) = \frac{-e\mathbf{E}(\omega)}{-\omega^2 + i\omega\gamma}. \tag{9.26}$$

The polarization per unit volume $\mathbf{P}(\omega)$ is then given by

$$\mathbf{P}(\omega) = -ne\mathbf{x}(\omega) = \alpha\mathbf{E}(\omega). \tag{9.27}$$

We may obtain $\epsilon(\omega)$ through the relation

$$\epsilon = 1 + 4\pi\alpha. \tag{9.28}$$

This allows us to identify the real and imaginary parts of ϵ for this model as

$$\epsilon_1(\omega) = 1 - \frac{\omega_p^2 \tau^2}{1 + \omega^2 \tau^2} \tag{9.29}$$

and

$$\epsilon_2(\omega) = \frac{\omega_p^2 \tau}{\omega(1 + \omega^2 \tau^2)}, \tag{9.30}$$

where ω_p is the plasma frequency $\sqrt{\frac{4\pi ne^2}{m}}$. These functions are shown in Fig. 9.2. This calculation also yields

$$\sigma_1(\omega) = \frac{\omega\epsilon_2(\omega)}{4\pi} = \frac{\omega_p^2 \tau}{4\pi(1 + \omega^2 \tau^2)}. \tag{9.31}$$

---

[1] P. Drude, "Zur Elektronentheorie der metalle," *Ann. Physik* 306(1900), 566.

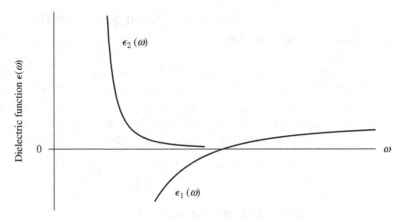

Figure 9.2   Schematic figure for $\epsilon_1(\omega)$ and $\epsilon_2(\omega)$ within the Drude theory. At high ω, $\epsilon_1(\omega) \rightarrow 1$.

To get all the desired optical properties, we need only use the relations for the optical constants n and k. Thus, with two material parameters, γ (typical value $\sim 10^{13}/\text{sec}$) and ω_p (typical value $\sim 10^{15}/\text{sec}$), we can calculate the optical properties of simple metals.

We now explore the reflectivity of a metal in three characteristic frequency regions. For region I, which is for very low ω, we consider $\omega \ll \omega_p$ and $\omega\tau \ll 1$, corresponding to many scatterings in one EM wave period. In this regime, for the typical values given above, the equations become

$$\epsilon_1(\omega) \approx -\omega_p^2\tau^2 \sim -10^4 \tag{9.32}$$

and

$$\epsilon_2(\omega) \approx \frac{\omega_p^2\tau}{\omega} = \frac{\omega_p^2\tau^2}{\omega\tau} \gg |\epsilon_1(\omega)|. \tag{9.33}$$

This dominance of $\epsilon_2(\omega)$ allows us to calculate $N(\omega)$ as follows

$$N(\omega) = \frac{\omega_p\tau}{\sqrt{\omega\tau}}\left(\frac{1+i}{\sqrt{2}}\right). \tag{9.34}$$

This condition sets the real and imaginary parts of $N(\omega)$ equal and, since $N = n + ik$, then $|n| = |k| \gg 1$. Thus, we can calculate the reflection or reflectivity as

$$R = \frac{(n-1)^2 + k^2}{(n+1)^2 + k^2} \approx 1 - \frac{2}{n} = 1 - \frac{2\sqrt{2\omega\tau}}{\omega_p\tau} \approx 1, \tag{9.35}$$

indicating a large reflectivity in this regime. The reflectivity can be written in terms of the conductivity, using Eq. (9.14), as

$$R(\omega) = 1 - \sqrt{\frac{2\omega}{\pi\sigma_1(0)}}, \tag{9.36}$$

which is the reason why highly conducting metals ($\sigma \gg 1$) reflect well at low ω. Equation (9.36) is often written as

$$1 - R(\omega) \sim \sqrt{\omega}, \tag{9.37}$$

and is known as the Hagen–Rubens relation,[2] and this low-frequency region is referred to as the Hagen–Rubens region.

For region II, $\omega\tau \gg 1$ and $\omega < \omega_p$, corresponding to the electron oscillating in the EM field many times before being scattered. The corresponding equations are

$$\epsilon_1(\omega) = 1 - \frac{\omega_p^2 \tau^2}{1 + \omega^2 \tau^2} \approx 1 - \frac{\omega_p^2}{\omega^2} \tag{9.38}$$

and

$$\epsilon_2(\omega) \approx \frac{\omega_p^2}{\omega^2(\omega\tau)} \ll |\epsilon_1(\omega)|. \tag{9.39}$$

This gives a reflectivity of

$$R(\omega) \approx 1 - \frac{2}{\omega_p \tau} \approx 1, \tag{9.40}$$

and explains why simple metals reflect in this frequency range as well. The reflectance is independent of ω, which explains the shininess and colorlessness of many metals.

For region III, which comprises very high ω, $\omega > \omega_p$, the corresponding equations are

$$\epsilon_1(\omega) = 1 - \frac{\omega_p^2 \tau^2}{1 + \omega^2 \tau^2} > 0, \tag{9.41}$$

with

$$k(\omega) \to 0 \tag{9.42}$$

and

$$n(\omega) = \sqrt{\epsilon_1} \to 1, \tag{9.43}$$

which implies that the metal is transparent. Figure 9.3 summarizes these results with a plot of $1 - R(\omega)$ vs. ω. The abrupt increase in the transmission (since $k(\omega) = 0$ in this high ω range) near ω_p, as shown in Fig. 9.3, is an excellent experimental measure of the plasma frequency.

For the Hagen–Rubens region, we can also estimate other relevant properties. The absorption coefficient $\eta = \frac{2\omega k}{c}$ becomes very large because

$$k(\omega) = \frac{\omega_p \tau}{\sqrt{2\omega\tau}}. \tag{9.44}$$

---

[2] The relation was discovered by Ernst B. Hagen and Heinrich Rubens in 1903.

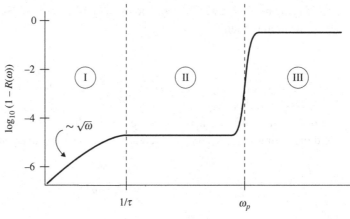

Figure 9.3   Frequency dependence of $\log_{10}(1 - R(\omega))$ within the Drude theory.

Also, the EM wave skin depth

$$\delta = \frac{c}{\omega k} = \frac{c}{\omega} \frac{\sqrt{2\omega\tau}}{\omega_p \tau} \tag{9.45}$$

can be related to the dc conductivity $\sigma_1(0) = \frac{ne^2\tau}{m}$, and we have

$$\delta = \frac{c}{\sqrt{2\pi\sigma_1(0)\omega}} \sim \omega^{-\frac{1}{2}}. \tag{9.46}$$

Hence, the simple Drude model, which is based on intraband transitions arising from scattering centers, gives the essential features of the optical properties of simple metals.

9.3 The transverse dielectric function

As discussed in Section 9.1, when examining the physics of optical properties, the appropriate dielectric function to use for determining response functions for transverse EM probes is the "transverse dielectric function." In our previous discussions, in Chapter 8 for example, we considered a longitudinal field with $\mathbf{q} \parallel \mathbf{E}$, because we assumed that

$$\nabla^2\phi = -4\pi\,\delta\rho \implies \nabla \cdot \mathbf{E} = -4\pi\,\delta\rho. \tag{9.47}$$

As $q \to 0$, this approximation is acceptable for transverse EM waves in a cubic or homogeneous system. However, more generally for an optical response, we need to consider a current–current response function derivation of the self-consistent field dielectric function.

For an EM field $\mathbf{E}(\omega)$, the current density is given by

$$\mathbf{j}_i = \sum_j \sigma_{ij} \mathbf{E}_j, \tag{9.48}$$

where the conductivity tensor can be shown from perturbation theory:

$$\sigma_{ij} = -i \frac{e^2 \hbar^2}{m^2 \omega} \sum_{\substack{n\mathbf{k} \\ n'\mathbf{k}'}} \frac{f_0(E_{n\mathbf{k}}) - f_0(E_{n'\mathbf{k}'})}{E_{n'\mathbf{k}'} - E_{n\mathbf{k}} - \hbar\omega - i\eta} \langle n'\mathbf{k}' | \nabla_i | n\mathbf{k} \rangle \langle n\mathbf{k} | \nabla_j | n'\mathbf{k}' \rangle, \tag{9.49}$$

where $\eta = 0^+$. Using the relation (with $\eta = 0^+$)

$$\frac{1}{x + i\eta} = P\left(\frac{1}{x}\right) - i\pi \delta(x), \tag{9.50}$$

we obtain

$$\mathrm{Re}\,\sigma_{ij} = \frac{\pi e^2 \hbar^2}{m^2 \omega} \sum_{\substack{n\mathbf{k} \\ n'\mathbf{k}'}} \left[f_0(E_{n\mathbf{k}}) - f_0(E_{n'\mathbf{k}'}) \right] \langle n'\mathbf{k}' | \nabla_i | n\mathbf{k} \rangle \langle n\mathbf{k} | \nabla_j | n'\mathbf{k}' \rangle \delta(E_{n'\mathbf{k}'} - E_{n\mathbf{k}} - \hbar\omega). \tag{9.51}$$

This is the Kubo–Greenwood expression[3] for the real part of the conductivity, which can be derived using different methods. Note that this formula is very general for quantum systems. For example, $n\mathbf{k}$ need not be band states. Using Eq. (9.14), the diagonal part of Eq. (9.51) yields ($\hat{e}$ = unit vector $\parallel \vec{E}$)

$$
\begin{aligned}
\epsilon_2(\omega) &= \frac{4\pi}{\omega} \mathrm{Re}\, \hat{e} \cdot \overleftrightarrow{\sigma} \cdot \hat{e} \\
&= \frac{4\pi^2 e^2 \hbar^2}{m^2 \omega^2} \sum_{\substack{n\mathbf{k} \\ n'\mathbf{k}'}} \left[f_0(E_{n\mathbf{k}}) - f_0(E_{n'\mathbf{k}'}) \right] |\langle n'\mathbf{k}' | \nabla \cdot \hat{e} | n\mathbf{k} \rangle|^2 \delta(E_{n'\mathbf{k}'} - E_{n\mathbf{k}} - \hbar\omega).
\end{aligned}
\tag{9.52}
$$

Note that $\epsilon_2(\omega)$ now depends on $\hat{e}$. For a cubic system, we may write

$$|\langle n'\mathbf{k}' | \hat{e} \cdot \nabla | n\mathbf{k} \rangle|^2 = \tfrac{1}{3} |\langle n'\mathbf{k}' | \nabla | n\mathbf{k} \rangle|^2, \tag{9.53}$$

which is independent of $\hat{e}$.

Fermi's golden-rule derivation of $\epsilon_2(\omega)$. We may arrive at the expression in Eq. (9.52) using perturbation theory. In a perturbing EM field,

$$H = \frac{1}{2} \sum_i \frac{1}{m} \left(\mathbf{p}_i + \frac{|e|\mathbf{A}(\mathbf{r}_i)}{c} \right)^2 + \sum_i e\phi(\mathbf{r}_i) + \sum_i V(\mathbf{r}_i) = H_0 + H_1, \tag{9.54}$$

---

[3] R. Kubo, "Statistical-mechanical theory of irreversible processes I: General theory and simple applications to magnetic and conduction problems," *J. Phys. Soc. Japan* 12(1957); D. A. Greenwood, "The Boltzmann equation in the theory of electrical conduction in metals," *Proc. Phys. Soc.* 71(1958), 585.

with H_0 the unperturbed Hamiltonian, and

$$H_1 = \frac{1}{2c}\frac{|e|}{m}\sum_i (\mathbf{p}_i \cdot \mathbf{A}(\mathbf{r}_i) + \mathbf{A}(\mathbf{r}_i) \cdot \mathbf{p}_i) + \frac{1}{2c^2}\sum_i \frac{e^2}{m}A^2(\mathbf{r}_i) + \sum_i e\phi(\mathbf{r}_i). \qquad (9.55)$$

If we choose the Coulomb gauge

$$\mathbf{\nabla} \cdot \mathbf{A} = 0 \qquad (9.56)$$

and

$$\phi = 0, \qquad (9.57)$$

then we can write for the ith electron (in the limit of small $\mathbf{A}$)

$$\tilde{H}_1^i = -i\frac{|e|\hbar}{2mc}\mathbf{A}(\mathbf{r}_i) \cdot \mathbf{\nabla}_i - i\frac{|e|\hbar}{2mc}\mathbf{\nabla}_i \cdot \mathbf{A}(\mathbf{r}_i) + \frac{e^2}{2mc^2}A^2(\mathbf{r}_i)$$

$$\approx -i\frac{|e|\hbar}{mc}\mathbf{A}(\mathbf{r}_i) \cdot \mathbf{\nabla}_i = \frac{|e|}{mc}\mathbf{A}(\mathbf{r}_i) \cdot \mathbf{p}_i. \qquad (9.58)$$

For an EM wave of the form

$$\mathbf{A} = \frac{\mathbf{A}_0}{2}[e^{i(\mathbf{q}\cdot\mathbf{r}-\omega t)} - e^{-i(\mathbf{q}\cdot\mathbf{r}-\omega t)}], \qquad (9.59)$$

we have

$$\mathbf{E} = -\frac{1}{c}\frac{\partial \mathbf{A}}{\partial t} \qquad (9.60)$$

or

$$\mathbf{A} = \frac{c}{i\omega}\mathbf{E}.$$

Therefore, $\tilde{H}_1^i = H_1^i + H_1^{i*}$, with

$$H_1^i = \frac{-|e|\hbar}{m\omega}e^{i(\mathbf{q}\cdot\mathbf{r}-\omega t)}\frac{\mathbf{E}_0}{2} \cdot \mathbf{\nabla}. \qquad (9.61)$$

We can now use Fermi's golden rule to derive $\epsilon_2(\omega)$ for insulators. For absorption for insulators at $T = 0$, the absorption coefficient η is defined through

$$\eta \equiv \frac{\text{energy absorbed/time--volume}}{\text{incident energy/area--time}} = \frac{-\frac{dI}{dt}}{I\left(\frac{c}{n}\right)}, \qquad (9.62)$$

where the incident energy is

$$I \sim e^{-2x/\delta}, \qquad (9.63)$$

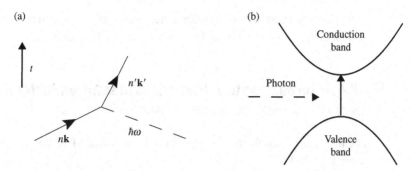

Figure 9.4   Direct optical absorption process in an insulator. (a) Feynman diagram. (b) Band diagram.

where δ is the distance over which **E** is reduced by $\frac{1}{e}$. Using

$$-\frac{dI}{dt} = \frac{-dI}{dx}\frac{dx}{dt} = \frac{2I}{\delta}\left(\frac{c}{n}\right),$$ (9.64)

we obtain, using Eqs. (9.63), (9.16), and (9.18),

$$\eta = \frac{-\frac{dI}{dt}}{I\left(\frac{c}{n}\right)} = \frac{2}{\delta} = \frac{2k\omega}{c} = \frac{\omega}{nc}\epsilon_2.$$ (9.65)

If we define $W_{\text{tot}}(\omega) \equiv$ number of photons absorbed/volume–time, then the energy absorbed/time–volume $= W_{\text{tot}}(\omega)\hbar\omega$. Hence, for an EM wave of the form in Eq. (9.59), which has energy density $u = \frac{1}{8\pi}(nE_0)^2$,

$$\frac{\text{energy incident}}{\text{area–time}} = uv = \frac{1}{8\pi}(nE_0)^2\left(\frac{c}{n}\right)$$

$$= \frac{ncE_0^2}{8\pi},$$ (9.66)

and

$$\eta = \frac{8\pi\,\hbar\omega}{ncE_0^2}W_{\text{tot}}(\omega)$$ (9.67)

This gives

$$\epsilon_2(\omega) = \frac{8\pi\,\hbar}{E_0^2}W_{\text{tot}}(\omega).$$ (9.68)

We may now obtain $\epsilon_2(\omega)$ from Eq. (9.68) by considering the transition rate from $(n\mathbf{k}) \to (n'\mathbf{k}')$,

$$w(n\mathbf{k} \to n'\mathbf{k}') = \frac{2\pi}{\hbar}|\langle n'\mathbf{k}'|H_1|n\mathbf{k}\rangle|^2\delta(E_{n'\mathbf{k}'} - E_{n\mathbf{k}} - \hbar\omega)(f_{n\mathbf{k}} - f_{n'\mathbf{k}'})$$

$$= \frac{\pi\,\hbar e^2 E_0^2}{2m^2\omega^2}\left|\hat{\mathbf{e}}\cdot\mathbf{M}_{nn'}(\mathbf{k},\mathbf{k}')\right|^2\delta(E_{n'\mathbf{k}'} - E_{n\mathbf{k}} - \hbar\omega)(f_{n\mathbf{k}} - f_{n'\mathbf{k}'}),$$ (9.69)

with **M** an interband transition matrix element from H_1. (See Section 9.4.) W_{tot} is the sum of all possible ($n\mathbf{k} \to n'\mathbf{k}'$), and we obtain Eq. (9.52) from Eq. (9.68).

9.4 Interband optical transitions in semiconductors and insulators

For insulating crystals, the interband matrix element **M** in Eq. (9.69) may be evaluated using Eq. (9.61),

$$H_1 = \frac{1}{2}\frac{e\hbar}{m\omega}E_0 e^{i(\mathbf{q}\cdot\mathbf{r}-\omega t)}\hat{\mathbf{e}}\cdot\nabla, \tag{9.70}$$

and we have

$$\mathbf{M}_{nn'}(\mathbf{k},\mathbf{k}') \equiv \langle n'\mathbf{k}'|e^{i\mathbf{q}\cdot\mathbf{r}}\nabla|n\mathbf{k}\rangle = \int u^*_{n'\mathbf{k}'}e^{-i\mathbf{k}'\cdot\mathbf{r}}e^{i\mathbf{q}\cdot\mathbf{r}}\nabla(u_{n\mathbf{k}}e^{i\mathbf{k}\cdot\mathbf{r}})d^3r$$

$$= \int u^*_{n'\mathbf{k}'}(\mathbf{r})e^{-i(\mathbf{k}'-\mathbf{q}-\mathbf{k})\cdot\mathbf{r}}(\nabla + i\mathbf{k})u_{n\mathbf{k}}(\mathbf{r})d^3r. \tag{9.71}$$

Since u is a periodic function, Eq. (9.71) gives the result that **M** is zero unless

$$\mathbf{k}' - \mathbf{q} - \mathbf{k} = \mathbf{G}. \tag{9.72}$$

This is the origin of the selection rule of conservation of crystal momentum. For $\mathbf{q} \sim 0$, $\mathbf{k}' \sim \mathbf{k}$, we have

$$\epsilon_2(\omega) = \frac{4\pi^2 e^2 \hbar^2}{m^2 \omega^2}\sum_{vc}\sum_{\mathbf{k}}|\hat{\mathbf{e}}\cdot M_{vc}(\mathbf{k})|^2\delta(E_c - E_v - \hbar\omega). \tag{9.73}$$

Here v denotes the occupied valence bands and c the empty conduction bands. The real part of the dielectric function $\epsilon_1(\omega)$ can be obtained using the Kramers–Kronig relation.

We now introduce the joint density of states $J(E)$ in analogy with the usual density of states $g(E)$ that is given for a paramagnetic system by (Ω is the sample volume)

$$g(E) = \frac{1}{\Omega}\sum_{n\mathbf{k}}\delta(E - E_{n\mathbf{k}}) = \frac{2}{(2\pi)^3}\int d^3k\,\delta(E - E_{n\mathbf{k}}). \tag{9.74}$$

The integral for $\epsilon_2(\omega)$ is over equal-energy-difference surfaces in **k**-space defined by $E_{c\mathbf{k}} - E_{v\mathbf{k}} - \hbar\omega = 0$. In many cases, $\hat{\mathbf{e}}\cdot\mathbf{M}_{vc}$ can be assumed to be a slowly varying function of **k** (except for the case $M = 0$ at some **k**-point, due to symmetry), and we can define the joint density of states as

$$J_{cv}(\omega) = \sum_{vc\mathbf{k}}\delta[E_c(\mathbf{k}) - E_v(\mathbf{k}) - \hbar\omega] = \frac{2}{(2\pi)^3}\int_s ds\frac{1}{\nabla_{\mathbf{k}}(E_{c\mathbf{k}} - E_{v\mathbf{k}})}, \tag{9.75}$$

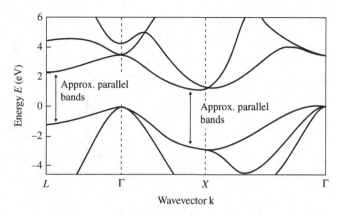

Figure 9.5 An example of parallel bands leading to critical points and van Hove singularities in the joint density of states.

where $s = $ surface of constant $\hbar\omega = E_{c\mathbf{k}} - E_{v\mathbf{k}}$, and within this simplification

$$\epsilon_2(\omega) \sim \frac{1}{\omega^2} \sum_{vc} |\hat{\mathbf{e}} \cdot M_{vc}|^2 J_{cv}(\omega), \tag{9.76}$$

where M_{vc} is a constant for a given (v, c) pair of bands. The joint density of states gives a measure of the density of full and empty states of equal energy difference. This quantity plays an important role in determining the interband contribution to ϵ_2 and the optical properties of solids for certain classes of materials for which $M_{vc} \approx$ constant.

Similarly to the electron or phonon analysis of $g(E)$, we define critical points in $\mathbf{k}$-space by the condition

$$\nabla_{\mathbf{k}}(E_{c\mathbf{k}} - E_{v\mathbf{k}}) = 0, \tag{9.77}$$

which associates a critical point with regions in $\mathbf{k}$-space where the electronic energy bands are parallel, as shown in Fig. 9.5.

Hence, near a critical point energy $\hbar\omega(\mathbf{k}_0)$, we can expand the energy difference in principal axes to obtain in three dimensions

$$\hbar\omega(\mathbf{k}) = \hbar\omega_0(\mathbf{k}_0) + \sum_{\ell=1}^{3} \alpha_\ell (\mathbf{k} - \mathbf{k}_0)_\ell^2. \tag{9.78}$$

Depending on the signs of the α_ℓ, we obtain van Hove singularities of different types (as in the case of $g(E)$), M_0, M_1, M_2, and M_3 in $J_{cv}(E)$ for a three-dimensional crystal.

In many applications involving the optical properties of semiconductors, the behavior of the absorption edge plays a very important role. The optical absorption edge of direct gap materials, where the valence band maximum and conduction band minimum lie at the same $\mathbf{k}$, is quite different from those of the indirect gap materials. Direct transitions constitute the lowest-energy transitions of the absorption spectrum of direct gap materials and determine the shape of the absorption edge, as in the case of GaAs.

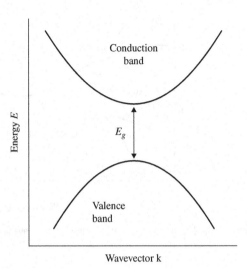

Figure 9.6 A direct bandgap material with parabolic bands.

For this case, near the band edges for parabolic bands (see Fig. 9.6),

$$E_{c\mathbf{k}} - E_{v\mathbf{k}} = E_g + \frac{\hbar^2 k^2}{2m_c} - \frac{\hbar^2 k^2}{2m_v} = E_g + \frac{\hbar^2 k^2}{2\mu}, \tag{9.79}$$

where μ is the reduced mass of the electron ($m_e = m_c$) and hole ($m_h = -m_v$). This yields

$$J_{cv}(\omega) = \frac{1}{(2\pi)^3}\left(\frac{2\mu}{\hbar^2}\right)^{3/2}(\hbar\omega - E_g)^{1/2} \tag{9.80}$$

and, near the edge,

$$\epsilon_2(\omega) \sim \frac{1}{\omega^2}(\hbar\omega - E_g)^{\frac{1}{2}} \tag{9.81}$$

if one assumes that M is a constant.

Dipole-allowed vs. dipole-forbidden transitions. We now consider dipole-allowed vs. dipole-forbidden transitions. In general, $\epsilon_2(\omega)$ (Eq. (9.73)) will depend on the matrix element $|\hat{\epsilon} \cdot \mathbf{M}|^2$, with (where the u functions are normalized in a unit cell)

$$\begin{aligned}
M_{vc}(\mathbf{k}) &= \langle c, \mathbf{k} + \mathbf{q}| \, e^{i\mathbf{q}\cdot\mathbf{r}}\boldsymbol{\nabla} \, |v, \mathbf{k}\rangle \\
&= \int u^*_{c,\mathbf{k}+\mathbf{q}}(\boldsymbol{\nabla} + i\mathbf{k})u_{v,\mathbf{k}}(\mathbf{r})d^3\mathbf{r} \\
&= \int u^*_{c,\mathbf{k}+\mathbf{q}}\boldsymbol{\nabla} u_{v,\mathbf{k}}d^3\mathbf{r} + i\mathbf{k}\int u^*_{c,\mathbf{k}+\mathbf{q}}u_{v\mathbf{k}}d^3\mathbf{r} \\
&\equiv \mathbf{M}^d_{vc}(\mathbf{k}) + i\mathbf{k}M^f_{vc}(\mathbf{k}).
\end{aligned} \tag{9.82}$$

Usually, the second term of the sum can be neglected as compared with the first term because of the orthogonality of Bloch functions with the same $\mathbf{k}$, since $\mathbf{q} \approx 0$. $\mathbf{M}^d_{vc}$ gives

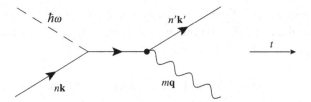

Figure 9.7   Phonon-assisted transition caused by a photon of frequency ω through the absorption or emission of a phonon having wavevector **q**.

the standard dipole selection rule: $M_{vc}^d = 0$ between states of the same parity. For dipole-forbidden transitions, M_{vc}^f in the second term will determine the transition intensity. Hence, for dipole-allowed transitions (e.g. GaAs, in which the top of the valence band is p-like and the bottom of the conduction band is s-like),

$$\epsilon_2 \sim \frac{1}{\omega^2} |M_{vc}^d|^2 \sqrt{\hbar\omega - E_g}, \tag{9.83}$$

and, for dipole-forbidden transitions (e.g. NiO and the cuprates), owing to the extra **k** factor in the second term in Eq. (9.82),

$$\epsilon_2 \sim \frac{1}{\omega^2} |M_{vc}^f|^2 (\hbar\omega - E_g)^{3/2}. \tag{9.84}$$

We note also that if $M_{cv}^d(\mathbf{k})$, although zero at $\mathbf{k}_0$, has a linear k-dependent term, it will contribute to give an effective total M^f in Eq. (9.84).

Indirect optical transitions. Up to this point, we have only considered direct transitions. The wavevector or crystal momentum selection rule from the interaction $H_1 \sim \mathbf{A} \cdot \mathbf{\nabla}$ is

$$\mathbf{k}_f = \mathbf{k}_i + \mathbf{q}_{\text{photon}} + \mathbf{G}. \tag{9.85}$$

However, if we have an indirect bandgap material, the absorption edge corresponding to the indirect gap energy range arises from transitions involving impurities (which remove the above selection rule) in the system or through the emission or absorption of phonons (or other elementary excitations). (See Fig. 9.7.) Therefore, we cannot consider only the term $\mathbf{A} \cdot \mathbf{\nabla}$.

For phonon-assisted transitions, the **k**-selection rule is

$$\mathbf{k}_f = \mathbf{k} + \mathbf{q}_{\text{photon}} \pm \mathbf{q}_{\text{phonon}} + \mathbf{G}. \tag{9.86}$$

The perturbation Hamiltonian becomes

$$H_1 = H_{\text{electron–photon}} + H_{\text{electron–phonon}} = H_{\text{rad}} + H_{\text{lat}}. \tag{9.87}$$

Since H_1 involves terms separately with a photon and a phonon, we need to evaluate it to second order to obtain an indirect optical transition. The transition amplitude involves summing up all virtual processes that lead to a final state of an electron in the conduction band and an electron missing (hole created) in the valence band. For the system in its ground state with N_p photons and n_q phonons, the initial state is

$$|i\rangle = |\text{electrons in their ground state, } N_p \text{ photons, and } n_q \text{ phonons}\rangle, \tag{9.88}$$

and the final state is

$$|f\rangle = |\text{excited electron–hole pair, } N_p - 1 \text{ photons, and } n_q \pm 1 \text{ phonons}\rangle. \tag{9.89}$$

Second-order perturbation theory gives the transition matrix element square for use in Fermi's golden rule as

$$\left| \sum_m \frac{\langle f| H_{\text{lat}} |m\rangle \langle m| H_{\text{rad}} |i\rangle}{E_m - E_i} + (H_{\text{rad}} \leftrightarrow H_{\text{lat}}) \text{ term} \right|^2, \tag{9.90}$$

where $|m\rangle$ is an intermediate state, and, as discussed in Chapter 10,

$$H_{\text{lat}} \sim \left(a_{-\mathbf{q}}^\dagger + a_{\mathbf{q}}\right)c_{\mathbf{k+q}}^\dagger c_{\mathbf{k}}, \tag{9.91}$$

with

$$\langle n_\mathbf{q} - 1| H_{\text{lat}} |n_\mathbf{q}\rangle \sim \sqrt{n_\mathbf{q}} \quad \text{(absorption)} \tag{9.92}$$

and

$$\langle n_\mathbf{q} + 1| H_{\text{lat}} |n_\mathbf{q}\rangle \sim \sqrt{n_\mathbf{q} + 1} \quad \text{(emission)}, \tag{9.93}$$

where the thermal averaged value $\langle n_\mathbf{q}\rangle$ is

$$\langle n_\mathbf{q}\rangle = \frac{1}{e^{\beta\hbar\omega(\mathbf{q})} - 1}, \tag{9.94}$$

and $\beta = \frac{1}{k_B T}$. Owing to the difference in the energy denominator, typically only one of the two terms in Eq. (9.90) dominates. Since for absorption,

$$\langle n_\mathbf{q}\rangle = \frac{1}{e^{\beta\hbar\omega(\mathbf{q})} - 1}, \tag{9.95}$$

and for emission,

$$\langle n_\mathbf{q}\rangle + 1 = \frac{1}{1 - e^{-\beta\hbar\omega(\mathbf{q})}}, \tag{9.96}$$

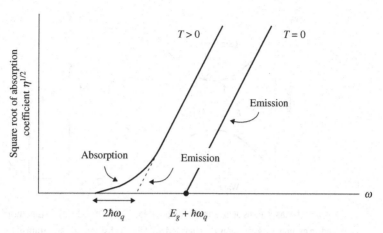

Figure 9.8 Phonon absorption and emission processes in optical absorption spectra. For $T > 0$, the emission onset is at lower energy because, in general, E_g decreases with T, and both emission and absorption processes can be appreciable.

we have, for a simple three-band model with a single intermediate band β and all bands parabolic,

$$\varepsilon_2(\omega) \sim \frac{1}{\omega^2} \int d^3\mathbf{k}_1 d^3\mathbf{k}_2 \left| \frac{\langle c\mathbf{k}_2 | H_{\text{lat}} |\beta\mathbf{k}_1\rangle \langle \beta\mathbf{k}_1 | H_{\text{rad}} |v\mathbf{k}_1\rangle}{E_\beta(\mathbf{k}_1) - E_v(\mathbf{k}_1) - \hbar\omega} \right|^2 \delta(E_c(\mathbf{k}_2)$$
$$- E_v(\mathbf{k}_1) - \hbar\omega \pm \hbar\omega_\mathbf{q})$$
$$\sim \frac{1}{\omega^2} \left(\hbar\omega - E_g - \hbar\omega_\mathbf{q}\right)^2 \left(\langle n_\mathbf{q}\rangle + 1\right) \quad \text{(for emission of phonon),} \qquad (9.97)$$

or

$$\sim \frac{1}{\omega^2} \left(\hbar\omega - E_g + \hbar\omega_\mathbf{q}\right)^2 \left(\langle n_\mathbf{q}\rangle\right) \quad \text{(for absorption of phonon).} \qquad (9.98)$$

As seen in the above expressions and in Fig. 9.8, there are characteristic frequency and temperature dependences for phonon-assisted indirect transitions that are distinct from direct transitions.

9.5 Electron–hole interaction and exciton effects

Up to this point, we have not considered the Coulomb interaction between the excited electron and hole involved in an optical transition. If the electron and hole bind, this is analogous to the case of positronium, and the resulting excitation is called an exciton. Even without bound states, there are "exciton effects" owing to the correlations between the two quasiparticles created. Often excitons are broadly classified as Wannier excitons, when the

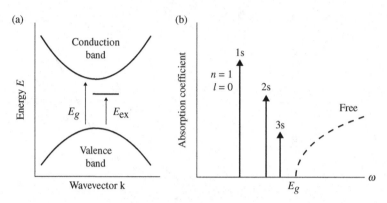

Figure 9.9 Schematic of energy levels of excitons in a two-parabolic band model: (a) Quasiparticle band structure with bandgap E_g and an exciton state at energy E_{ex} indicated. (b) Sketch of absorption spectrum of a dipole-allowed direct bandgap semiconductor with the lowest three excitonic transitions shown.

electron and hole are weakly bound, i.e. the electron and hole spatial correlation is large compared to the lattice constant and Frenkel excitons, when they are strongly bound.

Wannier excitons can often be described approximately by an effective mass equation, which at the simplest level is given by

$$\left(\frac{\mathbf{p}^2}{2\mu} - \frac{e^2}{\epsilon r}\right)F(\mathbf{r}) = EF(\mathbf{r}), \tag{9.99}$$

where

$$E = E_{\mathrm{ex}} - E_g \tag{9.100}$$

is the exciton binding energy and $F(\mathbf{r})$ is its wavefunction with $\mathbf{r}$ the electron–hole relative coordinates. Here E_{ex} is the energy of the excitonic state and E_g is the bandgap (see Fig. 9.9), with

$$\frac{1}{\mu} = \frac{1}{m_h} + \frac{1}{m_e}$$

and

$$\mathbf{r} = \mathbf{r}_e - \mathbf{r}_h. \tag{9.101}$$

More rigorously, an exciton state is a coherent superposition of many free electron–hole excitation configurations given in the form

$$|\psi_{\mathrm{ex}}\rangle = \sum_i \alpha_i |\phi_i\rangle, \tag{9.102}$$

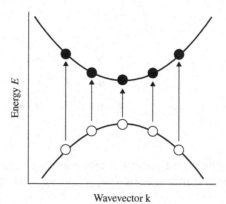

Energy E

Wavevector k

Figure 9.10 Schematic picture of an exciton state composed of many independent-particle transitions or electron–hole pairs represented by the arrows.

where

$$|\phi_i\rangle = c^\dagger_{c,\mathbf{k+q}} c_{v,\mathbf{k}} |G\rangle, \qquad (9.103)$$

and $c^\dagger$ and c are the one-particle (or, more rigorously, quasiparticle) creation and annihilation operators, and $|G\rangle$ represents the electronic ground state. This is depicted in Fig. 9.10.

To determine the exciton energies and wavefunctions, we solve the many-body Schrödinger equation, $H|\psi_{\text{ex}}\rangle = E_{\text{ex}}|\psi_{\text{ex}}\rangle$, where H is the many-electron Hamiltonian. To solve this problem, we need to find $\langle\phi_i|H|\phi_j\rangle$. For example, the ground-state Slater determinant is described by

$$\Psi_G(1,\ldots,N) = \mathcal{A}\{a_1(1)a_2(2)\ldots a_N(N)\}, \qquad (9.104)$$

where $\mathcal{A}$ is the antisymmetry operator, a_i are single-particle orbitals, and 1 denotes a composite index $\mathbf{x}_1 = (\mathbf{r}_1, \sigma_1)$. In the occupation number formalism, this can be written as (where the orbitals are positioned in order of increasing energy)

$$|G\rangle = |1111\ldots 0000\rangle. \qquad (9.105)$$

It is convenient to separate the Hamiltonian into two parts ($H = H_1 + H_2$), the one-electron Hamiltonian H_1 and the electron–electron interaction term H_2, where

$$H_1 = \sum_{i=1}^{N} h_1(\mathbf{r}_i) = \sum_{i=1}^{N} \left[\frac{\mathbf{p}_i^2}{2m} + V(\mathbf{r}_i) \right] \qquad (9.106)$$

and

$$H_2 = \frac{1}{2} \sum_{i,j} h_2(i,j) = \frac{1}{2} \sum_{i\neq j} \frac{e^2}{r_{ij}}. \qquad (9.107)$$

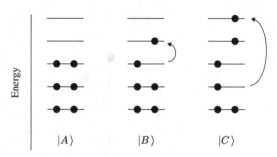

Figure 9.11 Illustration of the different types of Slater determinants required to create all the matrix elements needed to solve the exciton problem. The dots indicate occupied orbitals and arrows indicate a transition creating a free electron–hole pair.

To solve this problem, which we will do in the next section, we need to calculate the matrix elements between three types of Slater determinants, as shown in Fig. 9.11. These matrix elements can be worked out if $|A\rangle$ is a given single Slater determinant, as

$$\langle A| H_1 |A\rangle = \sum_i \langle a_i| h_1 |a_i\rangle$$

and

$$\langle A| H_2 |A\rangle = \sum_{ij} \frac{1}{2}\Big[\langle a_i a_j| h_2 |a_i a_j\rangle - \langle a_i a_j| h_2 |a_j a_i\rangle\Big], \tag{9.108}$$

where

$$\langle a_i a_j| h_2 |a_i a_j\rangle = \int a_i^*(\mathbf{x}_1) a_j^*(\mathbf{x}_2) h_2(\mathbf{r}_2, \mathbf{r}_1) a_i(\mathbf{x}_1) a_j(\mathbf{x}_2) d\mathbf{x}_1 d\mathbf{x}_2. \tag{9.109}$$

For $|B\rangle = c_{b_k}^\dagger c_{a_k} |A\rangle$, as depicted in Fig. 9.11,

$$\langle A| H_1 |B\rangle = \langle a_k| h_1 |b_k\rangle \tag{9.110}$$

and

$$\langle A| H_2 |B\rangle = \sum_j \Big[\langle a_k a_j| h_2 |b_k a_j\rangle - \langle a_k a_j| h_2 |a_j b_k\rangle\Big]. \tag{9.111}$$

For $|C\rangle = c_{c_k}^\dagger c_{a_k} c_{c_\ell}^\dagger c_{a_\ell} |A\rangle$, as depicted in Fig. 9.11,

$$\langle A| H_1 |C\rangle = 0 \tag{9.112}$$

and

$$\langle A| H_2 |C\rangle = \langle a_k a_\ell| h_2 |c_k c_\ell\rangle - \langle a_k a_\ell| h_2 |c_\ell c_k\rangle. \tag{9.113}$$

9.5.1 Weak binding limit in the two-band model

To illustrate the physical properties of excitons, we examine the weak binding limit in a two-band model. We also note that the basis states or determinants defined above can have finite center of mass momentum (see below). This is shown schematically in Fig. 9.12.

For an insulating material with weak spin–orbit coupling (i.e. the spin is not coupled to the orbital motion), we may construct the ground state as

$$|G\rangle = \mathcal{A}\left\{\phi_{v,\mathbf{k}_1}(\mathbf{r}_1)\alpha(1)\phi_{v,\mathbf{k}_1}(\mathbf{r}_2)\beta(2)\ldots\phi_{v,\mathbf{k}_{\frac{N}{2}}}(\mathbf{r}_N)\beta(N)\right\}, \tag{9.114}$$

where v is the band index, and $\alpha = \uparrow, \beta = \downarrow$ represent spin-up and -down spinor functions. Hence, the free electron–hole excited configurations take the form

$$\left|\phi_{c\mathbf{k}_e s_e, v\mathbf{k}_h s_h}\right\rangle = c^\dagger_{c\mathbf{k}_e s_e} c_{v\mathbf{k}_h s_h}|G\rangle, \tag{9.115}$$

where, at this point, we are not limiting ourselves to optical excitations, so $\mathbf{k}_e$ and $\mathbf{k}_h$ do not need to be the same and the spin can flip. The wavevector $\mathbf{k}_{ex} = \mathbf{k}_e - \mathbf{k}_h$ characterizes the center of mass momentum of the free electron–hole (e–h) pair excitation. The exciton state can now be written as

$$|\psi_{ex}\rangle = \sum_i A_i |\phi_i\rangle, \tag{9.116}$$

where $|\phi_i\rangle$ is given by Eq. (9.115).

We now need to solve for the coefficients A_i in Eq. (9.116) by solving the many-electron Schrödinger equation in order to obtain the exciton states by taking matrix elements of the Hamiltonian, $H = H_1 + H_2$, given in Eqs. (9.106) and (9.107). This Hamiltonian is spin independent (since we neglect spin–orbit interaction, which can be important for heavier atoms). Therefore, we should have a well-defined spin configuration for the excited states. Combining two spin $\frac{1}{2}$ particles (an electron and a hole) yields an $S = 1$ triplet state or

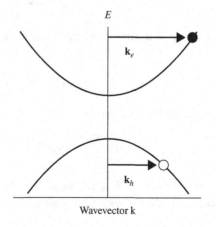

Figure 9.12   Schematic picture showing a nonzero center of mass momentum for a free electron–hole pair configuration.

an $S = 0$ singlet state. We note here that the spin state of a hole is opposite to that of the electron state being emptied.

For $S = 1$, we may have schematically (with the second spin being that of the hole)

$$S_z = 1: |\uparrow\rangle\,|\uparrow\rangle,$$

$$S_z = 0: \frac{1}{\sqrt{2}}\left(|\uparrow\rangle\,|\downarrow\rangle - |\downarrow\rangle\,|\uparrow\rangle\right),$$

$$S_z = -1: |\downarrow\rangle\,|\downarrow\rangle,$$

and for $S = 0$,

$$S_z = 0: \frac{1}{\sqrt{2}}\left(|\uparrow\rangle\,|\downarrow\rangle + |\downarrow\rangle\,|\uparrow\rangle\right). \tag{9.117}$$

Letting M denote a specific spin state, we have within a two-band model (i.e. neglecting the sum over v and c)

$$\left|\psi_{\mathrm{ex}}^M\right\rangle = \sum_{\mathbf{k}_e, \mathbf{k}_h} A_{\mathbf{k}_e, \mathbf{k}_h} \left|\phi_{\mathbf{k}_e, \mathbf{k}_h}^M\right\rangle, \tag{9.118}$$

where ϕ^M is a composite Slater determinant of the form given by combining Eqs. (9.115) and (9.117). The Hamiltonian also has crystalline translational symmetry, and, therefore, the exciton has a well-defined center of mass wavevector $\mathbf{k}_{\mathrm{ex}}$. So $\mathbf{k}_{\mathrm{ex}}$ is a good quantum number and each of the terms in Eq. (9.118) must have $\mathbf{k}_{\mathrm{ex}} = \mathbf{k}_e - \mathbf{k}_h$. This implies that $\mathbf{k}_e - \mathbf{k}_h = \mathbf{k}_e' - \mathbf{k}_h'$ in Eq. (9.118). In other words, the exciton wavefunction is made up of transitions with the same change in $\mathbf{k}$. Thus, we can write the exciton wavefunction as

$$\left|\psi_{\mathrm{ex}}^M\right\rangle = \sum_{\mathbf{k}} A(\mathbf{k}) \left|\phi_{\mathbf{k}+\frac{1}{2}\mathbf{k}_{\mathrm{ex}}, \mathbf{k}-\frac{1}{2}\mathbf{k}_{\mathrm{ex}}}^M\right\rangle. \tag{9.119}$$

We define the following "envelope function" in real space:

$$F(\mathbf{r}) = \sum_k A(\mathbf{k})e^{i\mathbf{k}\cdot\mathbf{r}}. \tag{9.120}$$

We may now derive a differential equation for $F(\mathbf{r})$ using the many-body Schrödinger equation

$$H\psi_{\mathrm{ex}} = E_{\mathrm{ex}}\psi_{\mathrm{ex}}, \tag{9.121}$$

which becomes the matrix equation

$$\sum_{\mathbf{k}'} H_{\mathbf{k},\mathbf{k}'} A(\mathbf{k}') = E_{\mathrm{ex}} A(\mathbf{k}). \tag{9.122}$$

At this point, we need to describe the matrix elements of H in Eq. (9.122). The diagonal elements can be written as (using Koopmans' theorem)

$$\left\langle\phi_{\mathbf{k}_h, \mathbf{k}_e}\left|H\right|\phi_{\mathbf{k}_e, \mathbf{k}_h}\right\rangle = E_c(\mathbf{k}_e) - E_v(\mathbf{k}_h) + \langle G|H|G\rangle, \tag{9.123}$$

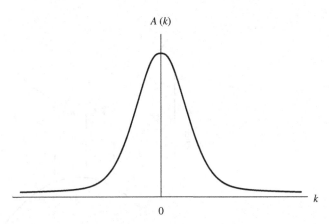

The coefficient $A(\mathbf{k})$ for the exciton state in Eq. (9.119) is schematically plotted in **k**-space. In the weak binding approximation, the distribution is very narrow.

where E_c and E_v are the conduction and valence band state energies, respectively, and, without any loss in generality, we set the last term (which is the ground-state energy) to zero. Using the list of matrix elements above for the basis functions (free electron–hole pair excitations) $|A\rangle$, $|B\rangle$, and $|C\rangle$, the off-diagonal matrix elements become

$$\left\langle \phi^M_{\mathbf{k}_e,\mathbf{k}_h} \middle| H \middle| \phi^M_{\mathbf{k}'_e,\mathbf{k}'_h} \right\rangle = \left\langle \phi_{c\mathbf{k}_e}\phi_{v\mathbf{k}'_h} \middle| \frac{-e^2}{r_{12}} \middle| \phi_{c\mathbf{k}'_e}\phi_{v\mathbf{k}_h} \right\rangle - 2\delta_M \left\langle \phi_{c\mathbf{k}_e}\phi_{v\mathbf{k}'_h} \middle| \frac{-e^2}{r_{12}} \middle| \phi_{c\mathbf{k}_h}\phi_{v\mathbf{k}'_e} \right\rangle,$$

$$(9.124)$$

where $\delta_M = 1$ if $S = 0$, and $\delta_M = 0$ if $S = 1$.

The first term is attractive and is called the direct term because it resembles a direct interaction between two charges of opposite sign if $\mathbf{k}_e = \mathbf{k}'_e$ and $\mathbf{k}_h = \mathbf{k}'_h$. Similarly, the second term is called the exchange term because it looks like a Hartree–Fock exchange interaction. Note that the exchange term is repulsive and is nonzero only for the spin singlet case.

As we discussed above, a definite center of mass momentum implies that $\mathbf{k}_e - \mathbf{k}_h = \mathbf{k}'_e - \mathbf{k}'_h$ for all transitions that make up the exciton state. If we make the weak binding approximation (that is, the wavefunction of the bound electron–hole pair in real space is large), which amounts to a narrow distribution in **k**-space, then $\mathbf{k}_e - \mathbf{k}'_e \approx 0$. This is shown in Figs. 9.13 and 9.14.

The direct term can be written as

$$\left\langle \phi_{c\mathbf{k}_e}\phi_{v\mathbf{k}'_h} \middle| \frac{-e^2}{r_{12}} \middle| \phi_{c\mathbf{k}'_e}\phi_{v\mathbf{k}_h} \right\rangle = \left\langle \phi_{c\mathbf{k}_e}T\phi_{v\mathbf{k}_h} \middle| \frac{-e^2}{r_{12}} \middle| \phi_{c\mathbf{k}'_e}T\phi_{v\mathbf{k}'_h} \right\rangle, \qquad (9.125)$$

where T is the time-reversal operator. This interaction is shown diagrammatically in Fig. 9.15.

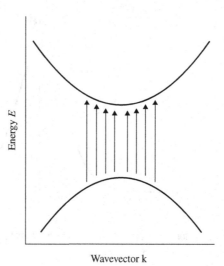

Wavevector k

Figure 9.14 All the transitions that make up the exciton state in the weak binding limit are close together in **k**-space.

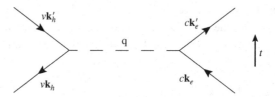

Figure 9.15 Diagram of the direct interaction in Eq. (9.124).

If we explicitly write out the direct term, we have

$$\langle\rangle_{\text{dir}} = \iint d\mathbf{r}_1 d\mathbf{r}_2 \left(\frac{-e^2}{r_{12}}\right) e^{-i(\mathbf{k}_e-\mathbf{k}'_e)\cdot(\mathbf{r}_1-\mathbf{r}_2)} u^*_{c\mathbf{k}_e}(\mathbf{r}_1) u_{c\mathbf{k}'_e}(\mathbf{r}_1) u^*_{v\mathbf{k}'_h}(\mathbf{r}_2) u_{v\mathbf{k}_h}(\mathbf{r}_2). \quad (9.126)$$

The term involving the periodic part of the Bloch function u's (we shall call it $g(\mathbf{r}_1, \mathbf{r}_2)$) can be seen in the weak binding approximation as $g(\mathbf{r}_1, \mathbf{r}_2) \approx \rho_{c\mathbf{k}_e}(\mathbf{r}_1) \rho_{v\mathbf{k}_h}(\mathbf{r}_2)$ because all the **k**-points are near each other in **k**-space. The function g is periodic and can be expanded in planewaves of reciprocal lattice vectors

$$g(\mathbf{r}_1, \mathbf{r}_2) = \sum_{m,n} a_{m,n} e^{i\mathbf{G}_m\cdot\mathbf{r}_1} e^{-i\mathbf{G}_n\cdot\mathbf{r}_2}. \quad (9.127)$$

If the density is approximately constant in a unit cell, then $a_{m,n} = 1$ for $m = n = 0$, and $a_{m,n} \approx 0$ otherwise. So, the direct term can now be written as

$$\langle\rangle_{\text{dir}} \approx \iint d\mathbf{r}_1 d\mathbf{r}_2 \left(\frac{-e^2}{r_{12}}\right) e^{-i(\mathbf{k}_e-\mathbf{k}'_e)\cdot(\mathbf{r}_1-\mathbf{r}_2)}. \quad (9.128)$$

Similarly, for the exchange term in Eq. (9.124), we may argue that, to a good approximation because u_{ck} is orthogonal to u_{vk},

$$\langle \, \rangle_{\text{ex}} = J_{\text{ex}}(\mathbf{k}_{\text{ex}})\delta_M, \tag{9.129}$$

which is independent of $\mathbf{k}_e - \mathbf{k}'_e$ (which leads to a $\delta(\mathbf{r})$ in real space). Collecting all the terms, we get the final equation for the exciton eigenvalues and eigenfunctions (i.e. Eq. (9.122)) in the form

$$\left[E_c\left(\mathbf{k} + \frac{1}{2}\mathbf{k}_{\text{ex}}\right) - E_v\left(\mathbf{k} - \frac{1}{2}\mathbf{k}_{\text{ex}}\right) - E_{\text{ex}} \right] A(\mathbf{k})$$
$$+ \sum_{\mathbf{k}'} \left[-\int d\mathbf{r}' \frac{e^2}{r'} e^{-i(\mathbf{k}-\mathbf{k}')\cdot\mathbf{r}'} + J_{\text{ex}}(\mathbf{k}_{\text{ex}})\delta_M \right] A(\mathbf{k}') = 0. \tag{9.130}$$

We now introduce some new notation to simplify the evaluation and interpretation of our results. We begin by defining

$$E_{cv}(\mathbf{k}, \mathbf{k}_{\text{ex}}) = E_c\left(\mathbf{k} + \frac{1}{2}\mathbf{k}_{\text{ex}}\right) - E_v\left(\mathbf{k} - \frac{1}{2}\mathbf{k}_{\text{ex}}\right). \tag{9.131}$$

The envelope wavefunction F was introduced previously in Eq. (9.120). Since our goal is to derive an equation for E_{ex} and F, we use these definitions and Eq. (9.130) to obtain

$$\left[E_{cv}(\mathbf{k}, \mathbf{k}_{\text{ex}}) \right] A(\mathbf{k}) + \left[-\int d\mathbf{r}' \frac{e^2}{r'} F(\mathbf{r}') e^{-i\mathbf{k}\cdot\mathbf{r}'} + J_{\text{ex}}(\mathbf{k}_{\text{ex}})\delta_M F(0) \right] = E_{\text{ex}} A(\mathbf{k}). \tag{9.132}$$

Since mathematically for any function with variable $-i\nabla$,

$$f(-i\nabla)F(\mathbf{r}) = \sum_{\mathbf{k}} f(\mathbf{k})A(\mathbf{k})e^{i\mathbf{k}\cdot\mathbf{r}} \tag{9.133}$$

and

$$\sum_{\mathbf{k}} \int d\mathbf{r}' \frac{e^2}{r'} F(\mathbf{r}') e^{-i\mathbf{k}\cdot\mathbf{r}'} e^{i\mathbf{k}\cdot\mathbf{r}} = \frac{e^2}{r} F(\mathbf{r}), \tag{9.134}$$

if we multiply Eq. (9.132) by $e^{-i\mathbf{k}\cdot\mathbf{r}}$ and then sum over $\mathbf{k}$, we obtain

$$\left[E_{cv}(-i\nabla, \mathbf{k}_{\text{ex}}) - \frac{e^2}{r} + J_{\text{ex}}(\mathbf{k}_{\text{ex}})\delta_M \delta(\mathbf{r}) \right] F(\mathbf{r}) = E_{\text{ex}} F(\mathbf{r}). \tag{9.135}$$

This equation is the effective mass equation describing excitons. Two assumptions were used to derive Eq. (9.135): a trial wavefunction represented by a linear combination of Slater determinants describing free electron–hole pair excitations, and the weak binding approximation.

It is useful at this point to make a few comments regarding the effective mass equation (Eq. (9.135)). The excitation energy E_{ex} is the energy needed to create an exciton. At this

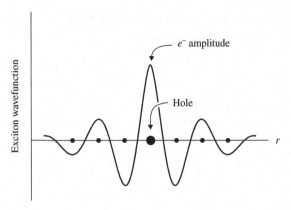

Figure 9.16 Schematic drawing of the exciton wavefunction $\psi_{\text{ex}}(\mathbf{r}_e,\mathbf{r}_h)$ in real space, with $\mathbf{r}$ the electron–hole relative coordinates assuming a fixed position for the hole.

point, we can go beyond our rather simple treatment since the electron–hole interaction will be screened by the other electrons by changing $-\frac{e^2}{r} \to -\frac{e^2}{\epsilon r}$, where ϵ is the dielectric function of the material. In most cases, the static ϵ can be used since the relevant binding energy $E_b = E_g - E_{\text{ex}}$ is small compared to the bandgap energy. Only the relative coordinates between the electron and the hole remain within Eq. (9.135). We can expand the interband transition energy term in the form

$$E_{cv}(-i\nabla, \mathbf{k}_{\text{ex}}) = \sum_{\substack{\alpha,\beta \\ n,m}} B_{\alpha\beta}^{nm} \nabla_\alpha^n \nabla_\beta^m, \tag{9.136}$$

with the coefficients $B_{\alpha\beta}^{nm}$ related to the band effective masses. Also, coming back to the definition of $A(\mathbf{k})$ in Eq. (9.119), one can show that within our approximations the full exciton state is given by

$$|\psi_{\text{ex}}\rangle = \sum_{\mathbf{R}_1} \sum_{\mathbf{R}_2} F(\mathbf{R}_1 - \mathbf{R}_2) w_1^\dagger w_2 |G\rangle, \tag{9.137}$$

where now the w_i's are the Wannier function annihilation operators. The electron–hole amplitude or wavefunction in real space is $\psi_{\text{ex}}(\mathbf{r}_e, \mathbf{r}_h) = \sum_{\mathbf{R}_1, \mathbf{R}_2} F(\mathbf{R}_1 - \mathbf{R}_2) w_c(\mathbf{r}_e - \mathbf{R}_1) w_v(\mathbf{r}_h - \mathbf{R}_h)$, with w_c and w_v the Wannier functions. The exciton wavefunction in real space $\psi_{\text{ex}}(\mathbf{r}_e, \mathbf{r}_h = 0)$ is illustrated schematically in Fig. 9.16.

9.5.2 Excitonic effects in the isotropic two-band model

As depicted in Fig. 9.17, we may assume for illustration purposes that both bands in an isotropic two-band model can be approximated as parabolas, $E_{cv}(-i\nabla) = E_g - \frac{\hbar^2}{2\mu}\nabla^2$, where $\frac{1}{\mu} = \frac{1}{m_e} + \frac{1}{m_h}$, and that $k_{\text{ex}} = 0$, which is the case for the optical transitions. Putting

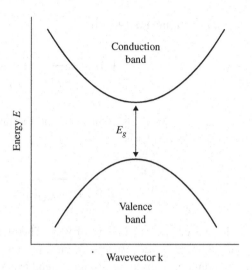

Figure 9.17 The isotropic two-band model.

all of these into the effective mass equation gives

$$\left[\frac{-\hbar^2}{2\mu}\nabla^2 - \frac{e^2}{\epsilon r} + J_{cv}(0)\delta_M\delta(\mathbf{r})\right]F(\mathbf{r}) = (E_{\text{ex}} - E_g)F(\mathbf{r}), \tag{9.138}$$

where $(E_{\text{ex}} - E_g) = E$ is the negative of the exciton binding energy. In the weak binding approximation, $\epsilon =$ constant and J is typically negligible because of the large spatial extent of the exciton envelope function F. Equation (9.138) resembles the Schrödinger equation for a hydrogen atom. Therefore, in three-dimensional systems, we expect bound states ($E < 0$) of the form

$$E = -\frac{R}{n^2}, \tag{9.139}$$

where $R = \frac{\mu e^4}{2\hbar^2\epsilon^2}$. This energy is typically < 0.01 Ry for most bulk semiconductors. The wavefunction F becomes a hydrogen-like wavefunction $F_{n,\ell,m}(\mathbf{r})$. For the unbound states $E > 0$, the wavefunctions are again hydrogen-like, $F_{s,l,m}(\mathbf{r}) = R_{s,l}(\mathbf{r})Y_{l,m}$, where $s = \sqrt{\frac{2\mu E}{\hbar^2}}$. It should be noted that for $E \gg R$, $F \to e^{i\mathbf{k}_e\cdot\mathbf{r}}$ and $A(\mathbf{k}) \to \delta(\mathbf{k} - \mathbf{k}_e)$.

These results for most bulk semiconductors are consistent with the weak binding approximation. For example, the $1s$ exciton wavefunction has the form

$$F_{100} = \frac{1}{\sqrt{\pi a_{\text{ex}}^3}}e^{-\frac{r}{a_{\text{ex}}}}, \tag{9.140}$$

with a spread determined by $a_{\text{ex}} = a_0\epsilon\frac{m_e}{\mu}$, where a_0 is the Bohr radius. Now, in general, $\frac{m_e}{\mu} > 1$ and $\epsilon \approx 10 - 20$ for semiconductors. So, $a_{\text{ex}} \approx 100a_0$. Thus, for $A(\mathbf{k})$, we have (by Fourier-transforming Eq. (9.140))

$$A(\mathbf{k}) = A_0 \frac{1}{\left(1 + (ka_{\mathrm{ex}})^2\right)^2}, \qquad (9.141)$$

where $A_0 = 2\pi a_{\mathrm{ex}}^2/\sqrt{\pi a_{\mathrm{ex}}^3}$ is a constant, and the ratio of A at the Brillouin zone edge compared to the zone center is

$$\frac{A(k_{\mathrm{BZ}})}{A(0)} = \frac{1}{\left(1 + (k_{\mathrm{BZ}}a_{\mathrm{ex}})^2\right)^2} \approx 10^{-4} - 10^{-6}, \qquad (9.142)$$

where we have used $k_{\mathrm{BZ}} \approx 1$ in units of Bohr. Therefore, the weak binding approximation is consistent. For this model, J_{ex} affects only s-states since it is a delta function at $\mathbf{r} = 0$, and it is important only for more strongly bound excitons.

Returning to the optical response function, we have from Eq. (9.69) the expression $\epsilon_2(\omega) = \frac{8\pi\hbar}{E_0^2} W_{\mathrm{tot}}$ for the imaginary part of ϵ, where W_{tot} is the number of photons absorbed per volume per unit time. In the independent electron picture without considering excitonic effects, this becomes (rewriting Eq. (9.73) and using the subscript F to denote free or non-interacting particles)

$$\epsilon_{2F}(\omega) = \frac{4\pi^2 e^2 \hbar^2}{m^2 \omega^2} \int_{\mathrm{BZ}} \frac{2d^3k}{(2\pi)^3} \left|\hat{\mathbf{e}} \cdot \mathbf{M}_{cv}(\mathbf{k})\right|^2 \delta\big[E_c(\mathbf{k}) - E_v(\mathbf{k}) - \hbar\omega\big], \qquad (9.143)$$

where $\hat{\mathbf{e}}$ is the polarization vector of the photon and M is the interband transition matrix element. With excitonic effects included, the probability of a transition from the ground state to an exciton state (from Fermi's golden rule) is

$$P_{|G\rangle \to |\psi_{\mathrm{ex}}\rangle} = \frac{2\pi}{\hbar} \left| \langle \psi_G | H_I | \psi_{\mathrm{ex}} \rangle \right|^2 \delta(E_{\mathrm{ex}} - \hbar\omega), \qquad (9.144)$$

where $H_I = \sum_i \frac{|e|}{mc} \mathbf{A}(\mathbf{r}_i) \cdot \mathbf{p}_i$. Using the rules introduced previously for the matrix elements (in particular, $\langle A| \sum_i h_i |B\rangle = \langle a_k| h_1 |b_k\rangle$), we find

$$P_{|G\rangle \to |\psi_{\mathrm{ex}}\rangle} = \frac{\pi}{2\hbar} \left(\frac{e\mathcal{A}_0}{mc}\right)^2 \delta_{\mathbf{k}_{\mathrm{ex}}} \delta_M \hbar^2 |\sum_{\mathbf{k}} A(\mathbf{k})\hat{\mathbf{e}} \cdot \mathbf{M}_{cv}(\mathbf{k})|^2 \delta(E_{\mathrm{ex}} - \hbar\omega), \qquad (9.145)$$

where $\mathcal{A}_0$ is the amplitude of the vector potential of the EM wave, $A(\mathbf{k})$ is the exciton wavefunction in $\mathbf{k}$-space, and

$$\mathbf{M}_{cv}(\mathbf{k}) = \int u_c^*(\mathbf{k} + \mathbf{q}, \mathbf{r})(\nabla + i\mathbf{k})u_v(\mathbf{k}, \mathbf{r})d^3r \qquad (9.146)$$

is the interband matrix element, as before.

For dipole-allowed transitions, the first term on the right-hand side of Eq. (9.146) for the matrix element (i.e. the $\mathbf{V}$ term in the integrand) is assumed to be constant independent of $\mathbf{k}$, and one neglects the $i\mathbf{k}$ term for $\mathbf{M}$. Since $\sum_{\mathbf{k}} A(\mathbf{k}) = F(\mathbf{r} = 0)$, Eq. (9.145) becomes

$$P_{|G\rangle|\psi_{\text{ex}}\rangle} \propto \delta_{k_{\text{ex}}}\delta_M |\hat{\mathbf{e}} \cdot \mathbf{M}_{cv}(0)|^2 |F(0)|^2 \delta(E_{\text{ex}} - \hbar\omega). \tag{9.147}$$

Since $F(0) \neq 0$ only for s-states for the hydrogenic functions where $F_{n00}(0) = \frac{1}{\sqrt{\pi a_{\text{ex}}^3 n^3}}$, this gives transition intensity to the excitons with principal quantum number n as $I_n \propto \frac{1}{n^3}$. Thus, we expect a series of lines at $E = E_g - \frac{R}{n^2}$ for $n = 1, 2, \dots$, with the intensity decreasing as $\frac{1}{n^3}$ (Fig. 9.9(b)). At high values of n, we can express the absorption intensity in terms of the density of states

$$D(E) = 2\left(\frac{dn}{dE}\right) = \frac{n^3}{R}, \tag{9.148}$$

yielding the result that the intensity/energy $= I_n(E)D(E)$ goes to a constant for transitions near E_g. This is shown as the shoulder in Fig. 9.18 near E_g.

For the unbound excitons ($E > 0$), we have for dipole-allowed bands

$$\epsilon_2(\omega) = \epsilon_{2F}(\omega)|F_{s00}(0)|^2 = \epsilon_{2F}(\omega)\frac{\pi \chi e^{\pi \chi}}{\sinh(\pi \chi)}, \tag{9.149}$$

where $\chi = \sqrt{\frac{R}{\hbar\omega - E_g}}$. There are two interesting limiting cases: (1) $\hbar\omega - E_g \gg R$, $\epsilon_2(\omega) \to \epsilon_{2F}(\omega)$, and (2) $\hbar\omega - E_g \ll R$, $\epsilon_2(\omega) \to 2\pi \chi \epsilon_{2F}(\omega) \approx \frac{\sqrt{\hbar\omega - E_g}}{\sqrt{\hbar\omega - E_g}} =$ constant, which is the same constant that approaches from the $E < 0$ side. This behavior is sketched in Fig. 9.18.

Up to this point, we have considered the case of dipole-allowed transitions where optical transitions lead to optically active (in linear response) excitons of hydrogen atom-like states with $\ell = 0$ (s-states). But in the case of dipole-forbidden transitions, we shall see that the optically active excitons must have $\ell = 1$, and they form hydrogen atom-like p-states. We

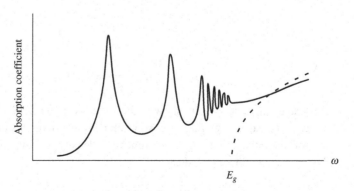

Figure 9.18 Plot of a typical optical spectrum with excitonic effects included.

begin with Eq. (9.146) and rewrite M in the form

$$\mathbf{M} = \mathbf{M}^{\text{allowed}} + i\mathbf{k}M^{\text{forbidden}}. \tag{9.150}$$

For the dipole-forbidden case, $\mathbf{M}^{\text{allowed}} = 0$. We use only the second term in Eq. (9.150) for Eq. (9.145), and since

$$\sum_{\mathbf{k}} A(\mathbf{k}) \, \mathbf{k} = \frac{\partial}{\partial \mathbf{r}} F(\mathbf{r})|_{\mathbf{r}=0}, \tag{9.151}$$

the probability of creating an exciton state in the dipole-forbidden case is then

$$P_{|G\rangle \to |\psi_{\text{ex}}\rangle} = \frac{\pi}{2\hbar} \left(\frac{e\mathcal{A}_0}{mc} \right) \hbar^2 \delta_{\mathbf{k}_{\text{ex}}} \delta_M \left| M^{\text{forbidden}} \right|^2 \left| \hat{\mathbf{e}} \cdot \frac{\partial}{\partial \mathbf{r}} F(\mathbf{r}) \right|_{\mathbf{r}=0}^2 \delta(E_{\text{ex}} - \hbar\omega). \tag{9.152}$$

Since there is only a nonzero transition probability for

$$\frac{\partial}{\partial \mathbf{r}} F(\mathbf{r}) \bigg|_{\mathbf{r}=0} \neq 0, \tag{9.153}$$

which is the case for hydrogen atom-like p-states with

$$\frac{\partial}{\partial \mathbf{r}} F_{n1m}(\mathbf{r}) \bigg|_{\mathbf{r}=0} = \left(\frac{n^2 - 1}{\pi a_{\text{ex}}^5 n^5} \right)^{\frac{1}{2}}. \tag{9.154}$$

This results in having peaks in the absorption spectra at energies

$$E = E_g - \frac{R}{n^2} \tag{9.155}$$

for $n = 2, 3, 4, 5, \ldots, \infty$. Here, R is again the effective Rydberg for the exciton. The intensity of each absorption line is proportional to the transition probability

$$I_n \sim \frac{n^2 - 1}{n^5} \tag{9.156}$$

and

$$I_n \sim \frac{1}{n^3} \text{ as } n \to \infty. \tag{9.157}$$

The absorption spectra are shown schematically in Fig. 9.19. Note that as the energy approaches the gap energy E_g, the absorption per unit energy becomes constant. This occurs because the intensity of each state times the density of states is a constant (as discussed above for the dipole-allowed case):

$$\lim_{n \to \infty} I_n D(E_n) = \text{constant}. \tag{9.158}$$

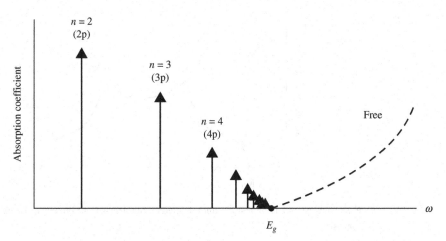

Figure 9.19   Absorption spectra for dipole-forbidden transitions of exciton states.

For the case of unbound states, ϵ_2 is given by (again using hydrogenic wavefunctions)

$$\epsilon_2(\omega) = \epsilon_{2F}(\omega)\frac{\pi\,\chi(1+\chi^2)e^{\pi\chi}}{\sinh\pi\chi}, \qquad (9.159)$$

where ϵ_{2F} is the independent electron ϵ_2 and $\chi = \sqrt{\frac{R}{(\hbar\omega - E_g)}}$.

There are again two cases of interest. Case I is $\hbar\omega - E_g \gg R$ (high energy), $\epsilon_2(\omega) \sim \epsilon_{2F}(\omega)$. Case II is $\hbar\omega - E_g \ll R$ (near the continuum absorption edge), $\epsilon_2(\omega) \sim \epsilon_{2F}(\omega)\chi^3$ = constant. The form of the absorption for energies approaching E_g from above is also a constant. An example of a system with dipole-forbidden transitions is Cu_2O. In Cu_2O, the absorption peaks fit very well into a hydrogen-like series[4] $\nu_n = 17\,250 - \frac{786}{n^2}$ cm^{-1} with $n = 2, 3, \ldots$

So far, we have only been considering an onset of absorption with an M_0 critical point or singularity in the joint density of states. However, excitonic effects at van Hove singularities of kinds M_1, M_2, and M_3 can be significant. For the independent electron model, a model dipole-allowed absorption spectrum in three dimensions is shown in Fig. 9.20. The independent electron case is modified in several ways because of excitonic effects. The effect of excitons near the van Hove singularity M_3 is to reduce the oscillator strength (Fig. 9.21)

$$\epsilon_2(\omega) = \epsilon_{2F}(\omega)\frac{\pi x' e^{\pi x'}}{\sinh\pi x'}, \qquad (9.160)$$

where $x' = \sqrt{\frac{R}{E_3 - \hbar\omega}}$, and there are no discrete lines above E_3 like there are below E_0. For the other two cases, near M_1 excitonic effects enhance $\epsilon_2(\omega)$ and near M_2 excitonic effects weaken $\epsilon_2(\omega)$. In GaAs this leads to an apparent shift in the position of an absorption peak by about 0.5 eV in the higher-frequency region, even though no bound exciton is formed.

---

[4] P. W. Baumeister, "Optical absorption of cuprous oxide," *Phys. Rev.* 121(1961), 359.

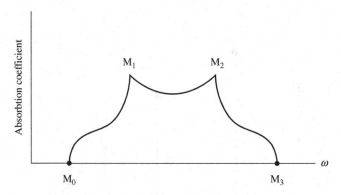

Figure 9.20 Model of dipole-allowed absorption spectrum at different critical points for a three-dimensional independent electron system.

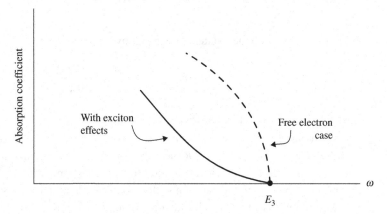

Figure 9.21 Absorption at M_3 van Hove singularity in the joint density of states is decreased due to excitonic effects.

Moving beyond the strict hydrogenic model, we note that the exchange interaction term, which is given by

$$J = J(\mathbf{k}_{ex})\delta(\mathbf{r})\delta_M, \qquad (9.161)$$

is repulsive and short-ranged. Since photons cannot flip spin, we expect only singlet states to be excited optically. However, if spin–orbit interactions are allowed and strong, the singlet and triplet states will mix and extra weak transitions will be activated. In first-order perturbation theory, the energy splitting between singlet and triplet states is given for s-like states by (see Fig. 9.22)

$$\Delta E \sim \langle \psi_{ex} | J | \psi_{ex} \rangle \sim |F(0)|^2 \sim \frac{1}{n^3}. \qquad (9.162)$$

Hence, the exchange energy ΔE for Wannier excitons is small except for $n = 1$ and for strongly bound excitons.

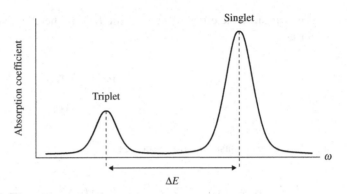

Figure 9.22 Single-triplet splitting of excitons.

For many materials, the electron bands are anisotropic. In these cases, the effective masses and dielectric constants are tensors instead of scalars. For many materials, they are described in terms of two constants representing an in-plane and an out-of-plane component,

$$\mu \to (\mu_{\parallel}, \mu_{\perp}) \tag{9.163}$$

and

$$\epsilon \to (\epsilon_{\parallel}, \epsilon_{\perp}). \tag{9.164}$$

For systems of this type, the anisotropic properties are characterized by the parameter γ,

$$\gamma = \frac{\epsilon_{\parallel}\mu_{\parallel}}{\epsilon_{\perp}\mu_{\perp}}, \tag{9.165}$$

with

$$0 \le \gamma \le 1. \tag{9.166}$$

This problem is difficult to solve analytically in general and is usually done numerically. For the specific case of $\gamma = 0$ (the case where $\mu_{\perp} \to \infty$, which corresponds to the two-dimensional case, e.g. the two-dimensional electron gas, quantum wells, etc.), the excitation energy for bound excitons (with dipole-allowed bands) is

$$E_{\text{ex}} = E_g - \frac{R}{\left(n + \frac{1}{2}\right)^2}, \tag{9.167}$$

for $n = 0, 1, 2, \ldots$ For $n = 0$, the excitation energy becomes

$$E_{\text{ex}} = E_g - 4R. \tag{9.168}$$

This is an example of how, in reduced dimensions, Coulomb interaction and correlation effects become more important.

If we consider the effects of a magnetic field in the effective mass approximation, we may use

$$\mathbf{p}_e \rightarrow \mathbf{p}_e - \frac{e}{c}\mathbf{A}(\mathbf{r}_e), \tag{9.169}$$

$$\mathbf{p}_h \rightarrow \mathbf{p}_h + \frac{e}{c}\mathbf{A}(\mathbf{r}_h), \tag{9.170}$$

and the effective mass equation becomes

$$\left[\frac{1}{2m_e}\left(\mathbf{p}_e - \frac{e}{c}\mathbf{A}(\mathbf{r}_e)\right)^2 + \frac{1}{2m_h}\left(\mathbf{p}_h + \frac{e}{c}\mathbf{A}(\mathbf{r}_h)\right)^2 - \frac{e^2}{\epsilon|\mathbf{r}_e - \mathbf{r}_h|}\right]F(\mathbf{r}_e, \mathbf{r}_h)$$
$$= \left(E_{\text{ex}} - E_g\right)F(\mathbf{r}_e, \mathbf{r}_h). \tag{9.171}$$

For a nearly uniform magnetic field, the canonical transformation $F(\mathbf{r}_e, \mathbf{r}_h) = e^{i\left(\mathbf{K}+\frac{e}{c}\mathbf{A}(\mathbf{r})\right)\cdot\mathbf{R}}\phi(\mathbf{r})$ may be used, and Eq. (9.171) becomes

$$\left[\frac{p^2}{2\mu} + \frac{e}{c}\left(\frac{1}{m_h} - \frac{1}{m_c}\right)\mathbf{A}(\mathbf{r})\cdot\mathbf{p} + \frac{e^2}{2\mu c^2}\mathbf{A}^2(\mathbf{r}) - \frac{e^2}{\epsilon r} - \frac{2e\hbar}{Mc}\mathbf{K}\cdot\mathbf{A}(\mathbf{r})\right]\phi(\mathbf{r})$$
$$= \left[E_{\text{ex}} - E_g - \frac{\hbar^2 K^2}{2M}\right]\phi(\mathbf{r}), \tag{9.172}$$

where $\mathbf{r} = \mathbf{r}_e - \mathbf{r}_h$, $\mathbf{R} = \frac{m_e\mathbf{r}_e + m_h\mathbf{r}_h}{m_e + m_h}$, $M = m_e + m_h$, $\mu = \left(\frac{1}{m_c} + \frac{1}{m_h}\right)^{-1}$, and $\mathbf{K}$ = momentum of the center of mass motion. Rearranging terms, Eq. (9.172) becomes

$$\left(H_0 + H_z + H_d + H_s\right)\phi(\mathbf{r}) = \left(E_{\text{ex}} - E_g - \frac{\hbar^2 K^2}{2M}\right)\phi(\mathbf{r}), \tag{9.173}$$

where $H_0 = \frac{p^2}{2\mu} - \frac{e^2}{\epsilon r}$ is the unperturbed Hamiltonian. For the unperturbed Hamiltonian, we get

$$E_{\text{ex}}^0 = E_g + \frac{\hbar^2 K^2}{2M} - \frac{\mu e^4}{2n^2 \hbar^2 \epsilon^2}, \tag{9.174}$$

and the other terms correspond to (using the gauge $\mathbf{A} = -\frac{1}{2}\mathbf{r} \times \mathbf{B}$)

(1) Zeeman splitting: $H_z = \frac{e}{c}\left(\frac{1}{m_h} - \frac{1}{m_e}\right)\mathbf{A}(\mathbf{r})\cdot\mathbf{p} = \frac{eB}{2c}\left(\frac{1}{m_h} - \frac{1}{m_e}\right)L_z$, $\quad$ (9.175)

(2) diamagnetic term: $H_d = \frac{e^2}{2\mu c^2}A^2(\mathbf{r}) = \frac{e^2 B^2}{8\mu c^2}(x^2 + y^2)$, $\quad$ (9.176)

and

$$\text{(3) magneto-Stark effect: } H_s = -\frac{2e}{Mc}\mathbf{K}\cdot\mathbf{A}(\mathbf{r}) = -\frac{2e}{Mc}\mathbf{B}\cdot(\mathbf{r}\times\mathbf{K})$$

$$= -\frac{2e}{Mc}\mathbf{r}\cdot(\mathbf{K}\times\mathbf{B}) \tag{9.177}$$

In the H_s term, $\mathbf{K}\times\mathbf{B}$ acts as an effective electric field and gives rise to a Stark effect.

The study of excitons and shallow impurity states in a magnetic field in semiconductors is equivalent to studying atomic physics in extremely high magnetic fields. The relevant parameter to measure the effect of an external $\mathbf{B}$-field on a system with binding energy R^* and an effective cyclotron frequency ω_c^* is

$$\delta = \frac{\hbar\omega_c^*}{2R^*} = \frac{\hbar\frac{eB}{\mu^*c}}{2\frac{\mu^*e^4}{2\hbar\epsilon^2}} = \left(\frac{\hbar\omega_c}{2R_y}\right)\left(\frac{\epsilon}{\frac{\mu^*}{m}}\right)^2. \tag{9.178}$$

For GaAs, $\delta = 1$ at about a magnetic field of 7 Tesla. For an atomic system, for $\delta = 1$, we need

$$B = B_{\text{GaAs}}\left(\frac{\frac{\mu^*}{m}}{\epsilon}\right)^{-2} \approx 7\left(\frac{10}{0.07}\right)^2 T = 10^5 T. \tag{9.179}$$

Magnetic field studies of exciton and impurity states in semiconductors may thus have relevance in astrophysics because there are atoms and molecules in such large fields in outer space.

Electron–phonon interactions

The interplay between electrons and lattice vibrations is one of the most interesting and important physical processes in the study of solids. The static lattice with atoms assumed frozen at their equilibrium positions yields the crystal potential, which governs many electronic properties such as band structure effects. With vibrations of the lattice comes a host of new effects resulting from the interactions of electrons and phonons. These include the scattering of electrons by phonons, which gives rise to a temperature-dependent resistivity (Fig. 10.1). Other first-order processes include the non-radiative recombination of electron–hole pairs and their generation, which is responsible for ultrasonic attenuation (Fig. 10.2).

Second-order electron–phonon processes lead to extraordinary effects. The emission and subsequent reabsorption of virtual phonons is the fundamental process for the creation of polarons in insulators and the mass enhancement of electrons in all types of solids (Fig. 10.3(a)). This process is also responsible for part of the temperature dependence of energy bandgaps. Another second-order process is the electron–electron interaction via the exchange of a virtual phonon, which can lead to bipolarons, Cooper pairs, and superconductivity (Fig. 10.4). The emission of virtual electron–hole pairs (Fig. 10.3(b)) by phonons and the subsequent destruction of the electron–hole pair renormalize the phonon energy. This can lead to softening (lowering of energy) of a phonon mode, which in turn can cause a structural phase transition.

10.1 The rigid-ion model

The rigid-ion model is one of the simplest ways of formulating and understanding electron–phonon interactions. The idea is that each ion (or atom) is assumed to contribute rigidly to the total potential seen by the electrons in the solid. As an ion is moved away from its equilibrium position, the potential is no longer periodic, and this will scatter an electron from one Bloch state to another. Let $V(\mathbf{r})$ be the static crystal potential which changes because of vibrations, i.e.

$$V(\mathbf{r}) \to V(\mathbf{r}, t) = \sum_{\ell, a} V_a\left(\mathbf{r} - \mathbf{R}_\ell^a(t)\right), \tag{10.1}$$

where

$$\mathbf{R}_\ell^a(t) = \mathbf{R}_\ell + \boldsymbol{\tau}_a + \delta\mathbf{R}_\ell^a(t) = \mathbf{R}_\ell^a + \delta\mathbf{R}_\ell^a(t), \tag{10.2}$$

(a) Emission   (b) Absorption

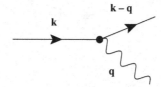
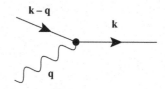

Figure 10.1   Emission and absorption of a phonon by an electron.

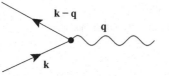
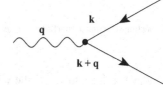

Figure 10.2   Emission of a phonon in non-radiative recombination of an electron–hole pair (left) and generation (right) of an electron–hole pair.

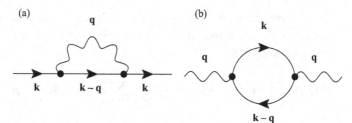

Figure 10.3   Emission and absorption of a virtual phonon by an electron (a) and emission and reabsorption of a virtual electron–hole pair by a phonon (b).

where ℓ is the cell index, a is the atomic index, and $\boldsymbol{\tau}_a$ is the vector to atoms in the basis. We assume the displacement is much less than the atomic spacing d, i.e.

$$\xi \equiv \delta\mathbf{R}_\ell^a \ll d. \tag{10.3}$$

For typical vibrations, it can be shown that

$$\frac{\xi}{d} \sim \left(\frac{m}{M}\right)^{1/4} \tag{10.4}$$

and the excitation or phonon energy E_{ph} scales as

$$\frac{E_{\text{ph}}}{E_{\text{el}}} \sim \left(\frac{m}{M}\right)^{1/2} \lesssim 10^{-2}, \tag{10.5}$$

where E_{el} is a typical electron energy and $\frac{m}{M}$ is the ratio of the electron and atomic masses. Because of these scaling arguments, we can assume that the large-scale band structure is

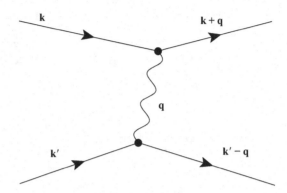

Figure 10.4   Electron–electron interaction by the exchange of a virtual phonon.

not modified. However, the vibrations of the ions do affect the dynamics of electrons. For example, in the case of a metal, electron–phonon interactions can cause a "wrinkle" in the band structure within a few meV of the Fermi energy. This represents a modification of the effective mass of the electrons by the phonons but only at energy scales comparable to phonon energies.

The derivation of the electron–phonon interaction matrix element proceeds as follows. Assuming a small displacement, at a specific time t, we can use the following Taylor expansion approach:

$$V(\mathbf{r}, t) = \sum_{\ell, a} V_a\left(\mathbf{r} - \mathbf{R}_\ell^a - \delta\mathbf{R}_\ell^a(t)\right) \approx \sum_{\ell, a} \left\{ V_a\left(\mathbf{r} - \mathbf{R}_\ell^a\right) + \delta\mathbf{R}_\ell^a \cdot \nabla_{\mathbf{r}} V_a\left(\mathbf{r} - \mathbf{R}_\ell^a\right) \right\}. \quad (10.6)$$

Now the total Hamiltonian is $H_{\text{tot}} = H_{\text{el}} + H_{\text{ph}} + H_{\text{el-ph}}$, and we can write the electron–phonon contribution as

$$H_{\text{el-ph}} = \sum_{\ell, a} \delta\mathbf{R}_\ell^a \cdot \nabla_{\mathbf{r}} V_a\left(\mathbf{r} - \mathbf{R}_\ell^a\right). \quad (10.7)$$

By expanding the displacement in phonon coordinates (see Chapter 4), we have

$$\delta\mathbf{R}_\ell^a = \sum_{\mathbf{q}\alpha} A_{\mathbf{q}\alpha}^a \hat{\boldsymbol{\epsilon}}_{\mathbf{q}\alpha}^a e^{i\mathbf{q}\cdot\mathbf{R}_\ell^a} \left(a_{\mathbf{q}\alpha} + a_{-\mathbf{q}\alpha}^\dagger\right), \quad (10.8)$$

where

$$A_{\mathbf{q}\alpha}^a = \sqrt{\frac{\hbar}{2M_a N \omega_{\mathbf{q}\alpha}}}, \quad (10.9)$$

where $\mathbf{q}$ is the wavevector, a labels the atomic species, α is the phonon branch index, and $\hat{\boldsymbol{\epsilon}}$ is the polarization vector. The phonon destruction and creation operators are represented by $a_{\mathbf{q}\alpha}$ and $a_{\mathbf{q}\alpha}^\dagger$. They are not related to the atomic species label using the same letter a.

In the language of second quantization,

$$H_{\text{el}} = \sum_{n\mathbf{k}} \varepsilon_{n\mathbf{k}} c_{n\mathbf{k}}^{\dagger} c_{n\mathbf{k}}, \tag{10.10}$$

$$H_{\text{ph}} = \sum_{\mathbf{q}\alpha} \left(a_{\mathbf{q}\alpha}^{\dagger} a_{\mathbf{q}\alpha} + \frac{1}{2} \right) \hbar\omega_{\mathbf{q}\alpha}, \tag{10.11}$$

and

$$
\begin{aligned}
H_{\text{el-ph}} &= \sum_{n,\mathbf{k},n',\mathbf{k}'} \langle n'\mathbf{k}' | H_{\text{el-ph}} | n\mathbf{k} \rangle c_{n'\mathbf{k}'}^{\dagger} c_{n\mathbf{k}} \\
&= \sum_{n,\mathbf{k},n',\mathbf{k}',\mathbf{q},\alpha} M_{n\mathbf{k}\to n'\mathbf{k}'}^{a\mathbf{q}} c_{n'\mathbf{k}'}^{\dagger} c_{n\mathbf{k}} (a_{\mathbf{q}\alpha} + a_{-\mathbf{q}\alpha}^{\dagger}).
\end{aligned} \tag{10.12}
$$

Equation (10.12) defines the electron–phonon matrix element M.

Suppressing the band indices n and n' by using an extended zone scheme so that $\mathbf{k}$ and $\mathbf{k}'$ are not limited to the first Brillouin zone, Eq. (10.7) becomes

$$H_{\text{el-ph}} = \sum_{\ell,a} \delta\mathbf{R}_{\ell}^{a} \cdot \nabla_{\mathbf{r}} V_a(\mathbf{r} - \mathbf{R}_{\ell}^{a}) = \sum_{a} \sum_{\mathbf{q}\alpha} A_{\mathbf{q}\alpha}^{a} \hat{\epsilon}_{\mathbf{q}\alpha}^{a} \cdot \left[\sum_{\ell} e^{i\mathbf{q}\cdot\mathbf{R}_{\ell}^{a}} \nabla_{\mathbf{r}} V_a(\mathbf{r} - \mathbf{R}_{\ell}^{a}) \right] (a_{\mathbf{q}\alpha} + a_{-\mathbf{q}\alpha}^{\dagger}). \tag{10.13}$$

We now need $\langle \mathbf{k}' | H_{\text{el-ph}} | \mathbf{k} \rangle$ to evaluate the quantities in Eq. (10.12). Using the atomic form factor of the potential,

$$V_a(\mathbf{q}) = \frac{1}{\Omega_a} \int e^{-i\mathbf{q}\cdot\mathbf{r}} V_a(\mathbf{r}) d\mathbf{r} = \frac{wN}{\Omega} \int e^{-i\mathbf{q}\cdot\mathbf{r}} V_a(\mathbf{r}) d\mathbf{r}, \tag{10.14}$$

where Ω_a is the average volume per atom, w is the number of atoms per cell, Ω is the crystal volume, and N is the number of cells in the crystal. To evaluate the matrix element we use the relations

$$\nabla_{\mathbf{r}} V_a(\mathbf{r}) = \frac{i}{wN} \sum_{\mathbf{q}'} \mathbf{q}' e^{i\mathbf{q}'\cdot\mathbf{r}} V_a(\mathbf{q}') \tag{10.15}$$

and

$$\nabla_{\mathbf{r}} V_a(\mathbf{r} - \mathbf{R}_{\ell}^{a}) = \frac{i}{wN} \sum_{\mathbf{q}'} \mathbf{q}' e^{i\mathbf{q}'\cdot\mathbf{r}} V_a(\mathbf{q}') e^{-i\mathbf{q}'\cdot\mathbf{R}_{\ell}^{a}}. \tag{10.16}$$

So, Eq. (10.13) becomes

$$H_{\text{el-ph}} = \sum_{\ell,a,\alpha,\mathbf{q},\mathbf{q}'} \frac{i}{wN} A_{\mathbf{q}\alpha}^{a} \hat{\epsilon}_{\mathbf{q}\alpha}^{a} \cdot \mathbf{q}' e^{i\mathbf{q}'\cdot\mathbf{r}} e^{i(\mathbf{q}-\mathbf{q}')\cdot\mathbf{R}_{\ell}^{a}} V_a(\mathbf{q}') (a_{\mathbf{q}\alpha} + a_{-\mathbf{q}\alpha}^{\dagger}). \tag{10.17}$$

Since

$$\sum_{\ell} e^{i(\mathbf{q}-\mathbf{q}')\cdot\mathbf{R}_{\ell}^{a}} = N\delta(\mathbf{q} - \mathbf{q}' + \mathbf{G}), \tag{10.18}$$

we obtain

$$\langle \mathbf{k}'|H_{\text{el-ph}}|\mathbf{k}\rangle = \sum_{\mathbf{G}} \left[\frac{1}{w} e^{-i\mathbf{G}\cdot\boldsymbol{\tau}_a} V_a(\mathbf{q}+\mathbf{G}) \right] \hat{\epsilon}^a_{\mathbf{q}\alpha} \cdot i(\mathbf{q}+\mathbf{G}) \langle \mathbf{k}'|e^{i(\mathbf{q}+\mathbf{G})\cdot\mathbf{r}}|\mathbf{k}\rangle. \quad (10.19)$$

Hence, the electron–phonon part of the Hamiltonian in the second quantization form is given by

$$H_{\text{el-ph}} = \sum_{\mathbf{k},\mathbf{k}',\mathbf{q},\alpha} M_{\mathbf{q}\alpha}(\mathbf{k}-\mathbf{k}')c^\dagger_{\mathbf{k}'}c_{\mathbf{k}}(a_{\mathbf{q}\alpha} + a^\dagger_{-\mathbf{q}\alpha}), \quad (10.20)$$

and the electron–phonon matrix element is given by

$$M_{\mathbf{q}\alpha}(\mathbf{k}-\mathbf{k}') = \frac{i}{w} \sum_{a,\mathbf{G}} \sqrt{\frac{\hbar}{2M_a N\omega_{\mathbf{q}\alpha}}} \hat{\epsilon}_{\mathbf{q}\alpha} \cdot (\mathbf{q}+\mathbf{G}) e^{i\mathbf{G}\cdot\boldsymbol{\tau}_a} V_a(\mathbf{q}+\mathbf{G})\langle \mathbf{k}'|e^{i(\mathbf{q}+\mathbf{G})\cdot\mathbf{r}}|\mathbf{k}\rangle. \quad (10.21)$$

10.2 Electron–phonon matrix elements for metals, insulators, and semiconductors

Let us consider an example appropriate for metals, which is to assume that $|\mathbf{k}'\rangle$ and $|\mathbf{k}\rangle$ are planewaves, as one expects in a pseudopotential approach. Hence,

$$\langle \mathbf{k}'|e^{i(\mathbf{q}+\mathbf{G})\cdot\mathbf{r}}|\mathbf{k}\rangle = \delta_{\mathbf{k}',\mathbf{k}+\mathbf{q}+\mathbf{G}} \quad (10.22)$$

and

$$M_{\mathbf{q}\alpha}(\mathbf{k} \to \mathbf{k}') = \frac{i}{w} \sum_{a,\mathbf{G}} \sqrt{\frac{\hbar}{2M_a N\omega_{\mathbf{q}\alpha}}} \hat{\epsilon}_{\mathbf{q}\alpha} \cdot (\mathbf{q}+\mathbf{G}) e^{i\mathbf{G}\cdot\boldsymbol{\tau}_a} V_a(\mathbf{q}+\mathbf{G})\delta_{\mathbf{k}',\mathbf{k}+\mathbf{q}+\mathbf{G}}. \quad (10.23)$$

Therefore, if we know the pseudopotential form factor, we can, in principle, get $M_{\mathbf{q}\alpha}$.

It is useful to distinguish the following two different processes. The first is the so-called "normal" process or N-process, where $\mathbf{k}'$ is not far from $\mathbf{k}$. This implies that

$$\mathbf{k}' = \mathbf{k} + \mathbf{q}, \quad (10.24)$$

with $\mathbf{G} = 0$, and assuming $w = 1$:

$$M_{\mathbf{q}\alpha}(\mathbf{k} \to \mathbf{k} + \mathbf{q}) = iA_{\mathbf{q}\alpha}\hat{\epsilon}_{\mathbf{q}\alpha} \cdot \mathbf{q}V(\mathbf{q}). \quad (10.25)$$

For these processes, there is no coupling if $\hat{\epsilon} \perp \mathbf{q}$. Hence, transverse phonons are not effective in scattering electrons.

The second process with $\mathbf{G} \neq 0$ is called an Umklapp process or U-process (Fig. 10.5). The relative values of the matrix elements for these two processes are

$$\frac{M_N}{M_U} = \frac{\hat{\epsilon}_\alpha \cdot \mathbf{q}V(\mathbf{q})}{\hat{\epsilon}_\alpha \cdot (\mathbf{q}+\mathbf{G})V(\mathbf{q}+\mathbf{G})}. \quad (10.26)$$

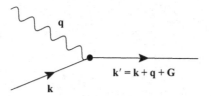

Figure 10.5   Scattering of an electron in an Umklapp process. If $\mathbf{G} = 0$, this is a normal process.

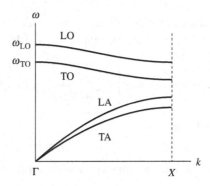

Figure 10.6   Phonon dispersion relation for a polar diatomic crystal (schematic).

Usually $V(\mathbf{q} + \mathbf{G}) < V(\mathbf{q})$ for a smooth pseudopotential, but Umklapp processes allow coupling to transverse phonon modes, and there are many of these processes (i.e. many $\mathbf{G}$ vectors) that contribute to the properties of metals.

Another example is an ionic insulator or polar semiconductor where the energies of the longitudinal optical (LO) and transverse optical (TO) modes at long wavelengths are non-degenerate. This splitting of the phonon energies (Fig. 10.6) may be shown to be given by the ratio of the low- and high-frequency $q = 0$ dielectric constants

$$\frac{\omega_{\mathrm{LO}}^2}{\omega_{\mathrm{TO}}^2} = \frac{\epsilon_0}{\epsilon_\infty}. \tag{10.27}$$

For charged ions and two atoms per unit cell, small q terms dominate because of the long-range Coulomb interaction ($\sim \frac{1}{q^2}$ in three dimensions). Here we use

$$V_1(q) = -V_2(q) = \frac{4\pi Z e^2}{\epsilon_\infty q^2 \Omega_a} \tag{10.28}$$

at small q, where Z is the charge of one of the ions and ϵ_∞ is the electronic dielectric constant of the polar material.

The polarization vectors of the two atoms for the longitudinal mode have the form

$$\hat{\epsilon}_{\mathrm{LO}}^{(1)} = \hat{q} = -\hat{\epsilon}_{\mathrm{LO}}^{(2)}, \tag{10.29}$$

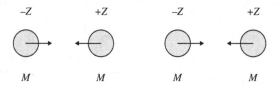

Figure 10.7 Motion of ions in a diatomic ionic crystal (having charge Z and mass M) for long-wavelength LO phonon modes.

since the displacements are in opposite directions for nearest-neighbor atoms. (See Fig. 10.7.) Using Eq. (10.21), with $w = 2$ and $M_1 = M_2 = M$,

$$M_{\mathrm{LO}}(\mathbf{k} \to \mathbf{k} + \mathbf{q}) = \frac{i}{2} \left[\sqrt{\frac{\hbar}{2MN\omega_{\mathrm{LO}}}} \hat{q} \cdot \mathbf{q} V_1(\mathbf{q}) - \sqrt{\frac{\hbar}{2MN\omega_{\mathrm{LO}}}} \hat{q} \cdot \mathbf{q} V_2(\mathbf{q}) \right]. \quad (10.30)$$

Using the relation

$$\omega_{\mathrm{LO}}^2 = \omega_{\mathrm{TO}}^2 + \Omega_p^2, \quad (10.31)$$

where Ω_p is the ionic plasma frequency and Eq. (10.27), we obtain

$$M_{\mathrm{LO}}(\mathbf{k} \to \mathbf{k} + \mathbf{q}) = i \left[\frac{2\pi e^2 \hbar \omega_{\mathrm{LO}}}{q^2 \Omega} \left(\frac{1}{\epsilon_\infty} - \frac{1}{\epsilon_0} \right) \right]^{1/2}. \quad (10.32)$$

Here, ϵ_0 is the dielectric constant including both the electronic and phonon contributions. This is the electron–phonon matrix element Fröhlich obtained by using an empirical approach, and this coupling is often referred to as the Fröhlich interaction or Fröhlich coupling.[1]

Our third example is the long-wavelength coupling of electrons to lattice vibrations in covalent semiconductors. This is called deformation potential coupling. This electron–phonon coupling is appropriate for excited electrons (or holes) in the conduction (valence) bands of semiconductors and insulators in the $\mathbf{q} \approx 0$ limit (Fig. 10.8) to describe how carriers are scattered to nearby states.

Consider the scattering of carriers near a band extremum in the $\mathbf{q} \to 0$ limit. Within the framework of an effective Hamiltonian or effective mass theory (see Chapter 5), we express the conduction electron energy as

$$E(\mathbf{k}) = [E(\mathbf{k}) - E_C] + E_C(\mathbf{r}), \quad (10.33)$$

where the bracketed term on the right-hand side is the band energy relative to the energy of the conduction band minimum E_C, and the last term is the change in the conduction band minimum energy for a deformed crystal at location $\mathbf{r}$.

A periodic change of the lattice constant arising from a compressional wave will cause a periodic change in $E_C(\mathbf{r})$. The perturbation (from the effective mass or effective Hamiltonian point of view) will scatter the Bloch states, and this is associated with the coupling

[1] H. Fröhlich, "Electrons in lattice fields," *Adv. Phys.* 3(1954), 325.

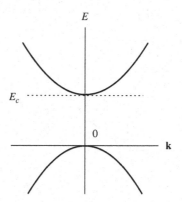

Figure 10.8 Energy band structure for carriers near band extrema.

of the electrons to the lattice or phonons

$$H_{\text{el-ph}} = E_C(\mathbf{r}) - E_C = \delta E_C(\mathbf{r}). \tag{10.34}$$

This is then the perturbation potential that scatters an electron from one state to another. We may write $\delta E_C(\mathbf{r})$ in terms of the dilation $\Delta(\mathbf{r})$ of the crystal volume Ω:

$$\Delta(\mathbf{r}) \equiv \frac{\delta\Omega(\mathbf{r})}{\Omega}. \tag{10.35}$$

Hence,

$$\delta E_C(\mathbf{r}) = \frac{\partial E_C}{\partial\Omega}\delta\Omega = \Omega\frac{\partial E_C}{\partial\Omega}\Delta(\mathbf{r}) = C_1\Delta(\mathbf{r}), \tag{10.36}$$

where C_1 is defined as the deformation potential with units of energy.[2] For example, consider a unit volume around an electron at $\mathbf{r}$. Suppose the unit volume is strained (Fig. 10.9), then the original volume

$$\Omega = \hat{x} \cdot (\hat{y} \times \hat{z}) \tag{10.37}$$

would change to

$$\Omega' = \hat{x}' \cdot (\hat{y}' \times \hat{z}') \tag{10.38}$$

and

$$\Delta = \frac{\Omega' - \Omega}{\Omega}. \tag{10.39}$$

We now need to express the dilation of the crystal near the position of an electron $\mathbf{r}$ in terms of the distortion of the crystal lattice sites. That is,

$$\Delta(\mathbf{r}) = f(\{\delta\mathbf{R}_\ell(\mathbf{r})\}, \mathbf{r}), \tag{10.40}$$

[2] J. Bardeen and W. Shockley, "Deformation potentials and mobilities in non-polar crystals," *Phys. Rev.* 80(1950), 72.

(a)   (b)

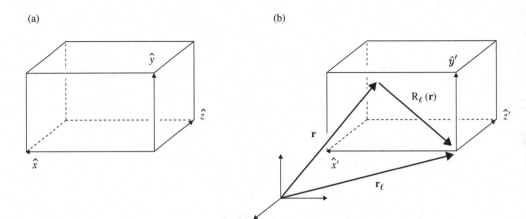

(a) Unstrained unit volume. (b) Unit volume under strain.

where the nearby lattice sites $\mathbf{r}_\ell$ (see Fig. 10.9(b)) when measured from the electron are
denoted as

$$\mathbf{R}_\ell(\mathbf{r}) = \mathbf{r}_\ell - \mathbf{r}. \tag{10.41}$$

Upon a uniform strain, the position of $\mathbf{R}_\ell(\mathbf{r})$ will change:

$$\mathbf{R}_\ell(\mathbf{r}) \rightarrow \mathbf{R}'_\ell(\mathbf{r}) = \mathbf{R}_\ell(\mathbf{r}) + \delta\mathbf{R}_\ell(\mathbf{r}). \tag{10.42}$$

Under uniform or slowly varying dilation, the atoms will move away from their unstrained
positions. The change in the lattice sites for the atoms near the electron will scale linearly
with the distance from the electron and

$$\delta\mathbf{R}_\ell(\mathbf{r}) \rightarrow 0 \text{ as } \mathbf{r}_\ell(\mathbf{r}) \rightarrow \mathbf{r}. \tag{10.43}$$

We may write

$$\mathbf{R}'_\ell(\mathbf{r}) = \mathbf{R}_\ell(\mathbf{r}) + \sum_\mu R_\mu \frac{\partial \delta\mathbf{R}(\mathbf{r})}{\partial R_\mu}\bigg|_\mathbf{r}. \tag{10.44}$$

Therefore, we can express the dilation as

$$\Delta(\mathbf{r}) = \frac{\Omega' - \Omega}{\Omega} = -\nabla_\mathbf{r} \cdot \delta\mathbf{R}(\mathbf{r}). \tag{10.45}$$

The electron–phonon coupling comes from the dependence of $\vec{\nabla}_\mathbf{r} \cdot \delta\mathbf{R}(r)$ on the position
of the electrons. Therefore, the perturbing Hamiltonian, using Eqs. (10.34) and (10.36),
becomes

$$H_{\text{el-ph}} = -C_1 \nabla_\mathbf{r} \cdot \delta\mathbf{R}(\mathbf{r}). \tag{10.46}$$

We now express $\delta\mathbf{R}(\mathbf{r})$ in phonon coordinates:

$$\delta\mathbf{R}(\mathbf{r}) = \sum_{\ell\mathbf{q}} \sqrt{\frac{\hbar}{2\rho\Omega\omega_\ell(\mathbf{q})}} \hat{q} e^{i\mathbf{q}\cdot\mathbf{r}} (a_{\mathbf{q}\ell} + a^\dagger_{-\mathbf{q}\ell}), \tag{10.47}$$

where the index ℓ counts longitudinal modes only and ρ is the mass density. From Eq. (10.45), we have

$$\Delta(\mathbf{r}) = -i\sum_{\ell\mathbf{q}}\sqrt{\frac{\hbar}{2\rho\Omega\omega_\ell(\mathbf{q})}}|\mathbf{q}|e^{i\mathbf{q}\cdot\mathbf{r}}(a_{\mathbf{q}\ell} + a_{-\mathbf{q}\ell}^\dagger). \tag{10.48}$$

So, we have

$$\begin{aligned} H_{\text{el-ph}} &= -C_1\nabla_{\mathbf{r}}\cdot\delta\mathbf{R}(\mathbf{r}) \\ &= \sum_{\mathbf{k}'\mathbf{k}}C_1\langle\mathbf{k}'|\,\Delta\,|\mathbf{k}\rangle\,c_{\mathbf{k}'}^\dagger c_{\mathbf{k}} \\ &= -iC_1\sum_{\ell\mathbf{k}\mathbf{q}}\sqrt{\frac{\hbar}{2\rho\Omega\omega_\ell(\mathbf{q})}}|\mathbf{q}|\left(a_{\mathbf{q}\ell} + a_{-\mathbf{q}\ell}^\dagger\right)c_{\mathbf{k}+\mathbf{q}}^\dagger c_{\mathbf{k}}, \end{aligned} \tag{10.49}$$

because $\langle\mathbf{k}'|\,e^{i\mathbf{q}\cdot\mathbf{k}}\,|\mathbf{k}\rangle = \delta_{\mathbf{k}',\mathbf{k}+\mathbf{q}}$ for planewaves. Hence, we obtain

$$M_{\mathbf{q}}(\mathbf{k}\to\mathbf{k}+\mathbf{q}) = -iC_1\sqrt{\frac{\hbar}{2\rho\Omega\omega_\ell(\mathbf{q})}}|\mathbf{q}|. \tag{10.50}$$

This is the deformation-potential coupling matrix element.

We can estimate C_1 for a free electron metal, where

$$E_F = \frac{\hbar^2}{2m}\left(\frac{3\pi^2 N}{\Omega}\right)^{2/3}. \tag{10.51}$$

Since

$$\frac{1}{E_F}\times\frac{\delta E_F}{\delta\Omega} = -\frac{2}{3}\frac{1}{\Omega}, \tag{10.52}$$

we have

$$\frac{\delta E_F}{E_F} = -\frac{2}{3}\frac{\delta\Omega}{\Omega} = -\frac{2}{3}\Delta. \tag{10.53}$$

Hence, for metals,

$$\delta E = -\frac{2}{3}E_F\Delta \Rightarrow |C_1| = \frac{2}{3}E_F \sim 2 - 8\,\text{eV}. \tag{10.54}$$

For a semiconductor like silicon, if all the valence electrons are considered to be free electron-like, then E_F is of order 16 eV, and C_1 would be of order 10 eV.

10.3 Polarons

We next explore an example of the coupling of an electron to LO phonons in a polar crystal. As shown in Fig. 10.10, when a dressed electron moves in an ionically bonded crystal, the

(a)   (b)

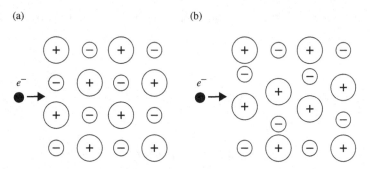

Figure 10.10 An electron moving in an ionic crystal (schematic): (a) rigid lattice, and (b) in the presence of lattice distortion.

Coulomb attraction and repulsion of the positive and negative ions with the electron will cause distortions or strains of the lattice. The displacements of the lattice caused by the distortion can be expressed in terms of phonon excitations, and the combination of the electron and the "cloud of phonons" produced is viewed together as a new quasiparticle, a polaron. This use of the term polaron, which is an electron in a polar crystal coupled to LO phonons via the Fröhlich interaction, is often extended to electrons coupled via lattice polarization in other situations and materials, such as polymers and metals.

The electron–lattice polarization lowers the electron energy. It also increases the effective mass of the electron since the electron composing the polaron is now traveling with a lattice strain. In addition, there is a change in the electron mobility since the polaron experiences scattering effects that are different from those of a free quasielectron.

Often polarons are sorted into two classes, large and small. If we define the size of a polaron ℓ_p to be an effective radius of the phonon cloud or polarization area, then large polarons are those for which $\ell_p > d$ (interatomic distance), and this indicates weak electron–phonon coupling. For small polarons, $\ell_p \lesssim d$ suggests strong coupling. Sometimes the coupling can be so strong that an electron becomes "self-trapped" in its polarization region. For large polarons and weak coupling, we are in a regime where we may use perturbation theory for the electron–phonon interaction. More sophisticated mathematical techniques are needed to solve for polaron properties for stronger coupling.

If we consider the weak coupling case, the Hamiltonian for this problem is (with the last term on the right side considered to be small)

$$
\begin{aligned}
H &= H_{\text{el}} + H_{\text{ph}} + H_{\text{el-ph}} \\
&= \sum_{\mathbf{k}} \varepsilon_{\mathbf{k}} c_{\mathbf{k}}^{\dagger} c_{\mathbf{k}} + \sum_{\mathbf{q}} (a_{\mathbf{q}}^{\dagger} a_{\mathbf{q}} + \tfrac{1}{2}) \hbar \omega_{\mathbf{q}} + \sum_{\mathbf{k'kq}} M_{\mathbf{q}} (\mathbf{k} \to \mathbf{k'}) c_{\mathbf{k'}}^{\dagger} c_{\mathbf{k}} (a_{\mathbf{q}} + a_{-\mathbf{q}}^{\dagger}).
\end{aligned} \tag{10.55}
$$

For the present case, we consider that the electron only couples to LO phonons, hence M can be written (Eq. (10.32)) as

$$
M_{\mathbf{q}} = i \frac{\hbar \omega_{\text{LO}}}{|\mathbf{q}|} \left(\frac{\hbar}{2m\omega_{\text{LO}}} \right)^{1/4} \left(\frac{4\pi \alpha}{\Omega} \right)^{1/2}, \tag{10.56}
$$

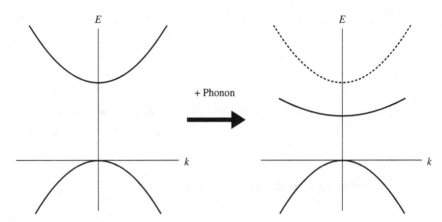

Figure 10.11 Self-energy shift for a polaron.

where the dimensionless Fröhlich coupling constant is defined as

$$\alpha = \frac{e^2}{\hbar} \left(\frac{1}{\epsilon_\infty} - \frac{1}{\epsilon_0} \right) \sqrt{\frac{m}{2\hbar\omega_{\text{LO}}}}, \tag{10.57}$$

where m is the electron band mass. Using the definitions given in Eqs. (10.56) and (10.57), we arrive at our earlier expression for M given in Eq. (10.32).

Assuming weak coupling $\alpha < 1$ allows the use of perturbation theory to compute the self-energy shift of an electron, its effective mass m^* to lowest order, and the number of phonons associated with an electron in a polaron.

The self energy arising from electron–phonon interactions lowers the energy of an electron in a band, as shown in Fig. 10.11. The expression for the self energy at $T = 0$, using second-order perturbation theory, is (for an electron in state a)

$$\Sigma_a = \Delta E_a = \sum_b \frac{|\langle a|H_{\text{el-ph}}|b\rangle|^2}{E_a - E_b}, \tag{10.58}$$

where b are all possible intermediate states and the scattering processes are illustrated in Fig. 10.12.

For the initial state a, we assume an electron is in a free electron state $|k\rangle$ with no phonons present so that the total wavefunction can be represented by $|k, 0\rangle$ and the total energy of state a is $E_a = \frac{\hbar^2 k^2}{2m}$. When the electron is scattered to an intermediate state b, the electron is in state $|\mathbf{k} - \mathbf{q}\rangle$ and one phonon is created. Hence, the total wavefunction is $a_{\mathbf{q}}^\dagger |\mathbf{k} - \mathbf{q}, 0\rangle$ and the energy is now

$$
\begin{aligned}
E_b &= \frac{\hbar^2 (\mathbf{k} - \mathbf{q})^2}{2m} + \hbar\omega_{\mathbf{q}} \\
&= \frac{\hbar^2 k^2}{2m} - \frac{\hbar^2 \mathbf{k} \cdot \mathbf{q}}{m} + \frac{\hbar^2 q^2}{2m} + \hbar\omega_{\text{LO}}
\end{aligned}
\tag{10.59}
$$

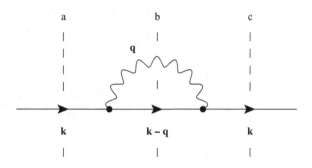

Figure 10.12 Scattering diagrams for calculating the self energy of a polaron.

and

$$\Delta E_a = \sum_b \frac{|M_{\mathbf{q}}|^2}{E_a - E_b} = -\sum_{\mathbf{q}} \frac{|M_{\mathbf{q}}|^2}{\hbar\omega_{LO} + \frac{\hbar^2 q^2}{2m} - \frac{\hbar^2 \mathbf{k} \cdot \mathbf{q}}{m}}. \tag{10.60}$$

If we denote

$$|M_{\mathbf{q}}|^2 = \frac{A}{q^2}, \tag{10.61}$$

where

$$A = (\hbar\omega_{LO})^2 \left(\frac{\hbar}{2m\omega_{LO}} \right)^{1/2} \frac{4\pi\alpha}{\Omega}, \tag{10.62}$$

and we convert the sum to an integral,

$$\sum_{\mathbf{q}} \to \frac{\Omega}{(2\pi)^3} \int 2\pi q^2 dq d\mu, \tag{10.63}$$

where $\mu = \frac{\mathbf{k} \cdot \mathbf{q}}{|\mathbf{k}||\mathbf{q}|}$, then

$$\Delta E_{\mathbf{k}} = \frac{-\Omega}{(2\pi)^3} A \int_{-1}^{1} d\mu \int_{0}^{q_{BZ}} dq \left(\frac{1}{\hbar\omega_{LO} + \frac{\hbar^2 q^2}{2m} - \frac{\hbar^2 kq\mu}{m}} \right). \tag{10.64}$$

Although this may be evaluated exactly, we can extract the main physical results by expanding for small k, letting $q_{BZ} \to \infty$, and evaluating the definite integrals to obtain

$$\Delta E = -\alpha\hbar\omega_{LO} - \frac{\hbar k^2}{2m} \left(\frac{\alpha}{6} \right) + O(k^4). \tag{10.65}$$

The energy of the electron is

$$E(\mathbf{k}) = \frac{\hbar^2 \mathbf{k}^2}{2m} + \Delta E(\mathbf{k}). \tag{10.66}$$

Table 10.1 Experimental values of α for a variety of polar crystals.[3]			
substance	α	substance	α
KCl	4.0	ZnO	0.85
KBr	3.5	PbS	0.16
AgCl	2.0	GaAs	0.06
AgBr	1.7	InSb	0.014

Combining Eqs. (10.65) and (10.66), we see that the ground-state energy is decreased by the factor $\alpha \hbar \omega_{LO}$, and the k^2-dependent part of the energy, i.e. $E_{kin}(\mathbf{k})$, has the form

$$E_{kin}(\mathbf{k}) = \frac{\hbar^2}{2m}(1 - \frac{1}{6}\alpha)k^2 = \frac{\hbar^2 k^2}{2m^*}. \tag{10.67}$$

Hence, the effective mass m^* of the polaron is

$$m^* = \frac{m}{(1 - \frac{\alpha}{6})} \approx m\left(1 + \frac{\alpha}{6}\right). \tag{10.68}$$

As expected, the polaron mass is larger than the electron mass at the bottom of the conduction band.

Another interesting calculation related to polarons is to estimate the average number of phonons coupled to an electron in a polaron at $T = 0$. The phonon number operator is given by

$$\hat{N}_{ph} = \sum_{\mathbf{q}} a_{\mathbf{q}}^{\dagger} a_{\mathbf{q}}, \tag{10.69}$$

and its average value is

$$\langle \hat{N}_{ph} \rangle = \langle \psi | \sum_{\mathbf{q}} a_{\mathbf{q}}^{\dagger} a_{\mathbf{q}} | \psi \rangle, \tag{10.70}$$

where $|\psi\rangle$ is the polaron wavefunction. For an electron wavevector $\mathbf{k}$,

$$|\psi_{\mathbf{k}}\rangle = |\mathbf{k}, 0\rangle + |\delta\psi_{\mathbf{k}}\rangle, \tag{10.71}$$

where $|\delta\psi\rangle$ is the change in the electron wavefunction $|\mathbf{k}, 0\rangle$ due to the electron–phonon interaction.

Using standard perturbation theory, to first order in M we have

$$|\delta\psi_{\mathbf{k}}\rangle = \sum_{\mathbf{q}} \frac{M_{\mathbf{q}} a_{\mathbf{q}}^{\dagger} |\mathbf{k} - \mathbf{q}, 0\rangle}{E_{\mathbf{k}} - (E_{\mathbf{k}-\mathbf{q}} + \hbar\omega_{LO})}. \tag{10.72}$$

[3] C. Kittel, *Introduction to Solid State Physics*, 8th ed. (New York, NY: Wiley, 2005).

Following the same procedure as was done in Eqs. (10.60)–(10.64), the expression for the average number of phonons given by Eq. (10.70) in $|\psi_{\mathbf{k}}\rangle$ becomes

$$\langle \hat{N}_{\text{ph}} \rangle = \sum_{\mathbf{q}} \frac{|\mathbf{M_q}|^2}{\left(\hbar \omega_{\text{LO}} + \frac{\hbar^2 q^2}{2m} - \frac{\hbar^2 kq\mu}{m} \right)^2}. \tag{10.73}$$

Converting the above sum to an integral and expanding for small k, we obtain to lowest order in k

$$\langle \hat{N}_{\text{ph}} \rangle = \frac{\Omega}{(2\pi)^2} 2A \left(\frac{2m}{\hbar^2} \right)^2 \int_0^\infty \frac{dq}{\left(\frac{2m\omega_{\text{LO}}}{\hbar} + q^2 \right)^2} = \frac{\alpha}{2}. \tag{10.74}$$

So α is, in some sense, a measure of the number of phonons traveling with an electron in a polaron.

Although the above three results, the self-energy shift, the renormalization of the effective mass, and the relationship between α and the average number of phonons "in a polaron" are calculated only for small α, they give considerable insight into the properties of this quasiparticle. As α becomes larger, this approach breaks down, but there are techniques for dealing with stronger coupling.[4] Some representative values for α in different materials are given in Table 10.1.

[4] R. P. Feynman, R. W. Hellwarth, C. K. Iddings, and P. M. Platzman, "Mobility of slow electrons in a polar crystal," *Phys. Rev.* 127(1962), 1004.

11 Dynamics of crystal electrons in a magnetic field

In general, the motion of an electron is quantized in the presence of a magnetic field. We can learn a great deal about the electronic properties of a crystalline solid by examining the behavior of the material in a magnetic field. Various magnetic field-induced phenomena have been observed and have been used as detailed probes in characterizing materials. Moreover, they have led to the discovery of new phases of matter and have been exploited in various applications. Examples include striking phenomena such as the integer and fractional quantum Hall effects, Wigner electron crystallization, and giant magnetoresistance.

Much of the behavior of electrons in a crystal with an applied magnetic field may be understood by assuming that the electrons are non-interacting fermionic quasiparticles. In this chapter, we shall consider the dynamics of crystal electrons in a magnetic field within this picture. That is, the electrons are non-interacting fermions but with masses and energy-wavevector dispersion relations that are modified by band structure, electron–electron, and electron–phonon interaction effects. We defer the discussion of the integer quantum Hall effect to Chapter 12 as part of the discussion on transport properties.

11.1 Free electrons in a uniform magnetic field and Landau levels

We first consider the simple case of free electrons in a uniform magnetic field $\mathbf{B}$. Let $\mathbf{B} = B\hat{z}$ and allow the electrons to be confined to a large box of dimension L on each side (see Fig. 11.1). Classically, we know that the electron moves freely along the z-directon, and its motion in the x–y plane is confined to cyclotron orbits. Quantum mechanically, the energy eigenvalue ε and eigenfunction ψ of the electron are given by

$$H\psi(\mathbf{r}) = \left[\frac{1}{2m} \left(\mathbf{p} - \frac{q}{c}\mathbf{A}(\mathbf{r}) \right)^2 \right] \psi(\mathbf{r}) = \varepsilon\psi(\mathbf{r}), \tag{11.1}$$

where $q = -|e|$ is the electron charge and $\mathbf{A}(\mathbf{r})$ is the vector potential, which, in the Landau gauge, is given by $\mathbf{A}(\mathbf{r}) = (0, x, 0)B$. In this gauge, $\mathbf{A}$ depends only on x. Since $\mathbf{A}$ is static and translationally invariant along the y- and z-directions, the energy and y- and z-components of the momentum of the electron are constants of motion. Equation (11.1) may be transformed to that of a one-dimensional harmonic oscillator. Substituting

$$\psi(\mathbf{r}) = \frac{e^{i(k_z z + k_y y)}}{L}\phi(x) \tag{11.2}$$

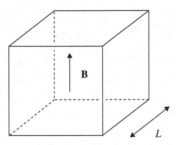

Figure 11.1 Free electrons confined in a box of dimension L in the presence of a uniform magnetic field **B**.

in Eq. (11.1) yields

$$\left[\frac{p_x^2}{2m} + \frac{q^2 B^2}{2mc^2} \left(x - \frac{\hbar c}{qB} k_y \right)^2 \right] \phi(x) = \left[\varepsilon - \frac{\hbar^2 k_z^2}{2m} \right] \phi(x). \tag{11.3}$$

The above equation corresponds to the Schrödinger equation of a simple harmonic oscillator located at position $x_0 = \frac{\hbar c}{qB} k_y$ with spring constant $\kappa = \frac{q^2 B^2}{mc^2}$. The electron energies are given by

$$\varepsilon_\lambda(k_z) = \left(\lambda + \frac{1}{2} \right) \hbar \omega_c + \frac{\hbar^2 k_z^2}{2m}, \tag{11.4}$$

with

$$\lambda = 0, 1, 2, 3, \ldots, \tag{11.5}$$

$$\omega_c = \sqrt{\frac{\kappa}{m}} = \frac{|e|B}{mc}, \tag{11.6}$$

and

$$k_z = \left(\frac{2\pi}{L} \right) n_z, \tag{11.7}$$

where n_z are integers.

The expression for ω_c has the same value as the classical cyclotron frequency. Thus, the electron behaves like a free particle parallel to **B** and performs cyclotron motion in the plane perpendicular to **B**. The free electron energy dispersion relation $\varepsilon(\mathbf{k}) = \frac{\hbar^2 k^2}{2m}$ is now changed into a series of parabolic curves as a function of k_z labeled by the quantum number λ, as illustrated in Fig. 11.2. Each of the curves is called a Landau level.[1] The Landau levels are highly degenerate because any value of k_y yields the same set of solutions in Eq. (11.3). For given λ and k_z, it can be shown that the number of degenerate states D is given by

$$D = \left(\frac{2|e|}{hc} \right) BL^2 = \frac{2BL^2}{\phi_0}. \tag{11.8}$$

---

[1] L. Landau, "Diamagnetism of metal," *Z. Phys.* 64(1930), 629.

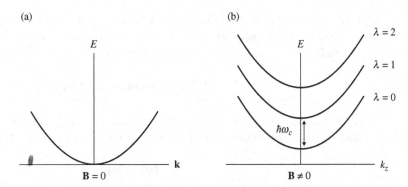

Figure 11.2 Energy vs. wavevector dispersion relation for a free electron in a magnetic field **B**: (a) **B** $= 0$ and (b) **B** $\neq 0$. The Landau-level subbands labeled by λ are separated by $\hbar w_c$, where w_c is the cyclotron frequency.

Here $\phi_0 \equiv \frac{hc}{|e|}$ is the quantum of magnetic flux, which has a value of 4.136×10^{-7}G-cm^2. Thus, D may be viewed as two times the number of magnetic fluxoids penetrating the sample. In a field of $10^3 G$ or 0.1 T and sample size of $L = 1$ cm, $D \sim 10^{10}$. The large degeneracy D here reflects the fact that an electron with a given energy and k_z classically spirals about a line parallel to the z-axis, which can have arbitrary x and y coordinates. This is the reason D is proportional to the sample area L^2.

For a two-dimensional electron gas (2DEG), k_z does not come into consideration. The degeneracy of each Landau level is given by Eq. (11.8) and the total number of occupied Landau levels is given by

$$\nu = \frac{N}{D} = n\left(\frac{\phi_0}{2B}\right), \tag{11.9}$$

where N and n are the total number of electrons and the number of electrons per unit area, respectively.

11.2 Crystal electrons in a static B-field

Many phenomena of interest involve electrons with energies near the Fermi level E_F. Under moderate magnetic fields in laboratory conditions, these electrons are in highly excited quantum states with the Landau-level index λ_F at the Fermi energy given by

$$\lambda_F \sim \frac{E_F}{\hbar \omega_c} = \frac{E_F}{\hbar \left(\frac{|e|B}{mc} \right)}. \tag{11.10}$$

For an applied field $B = 1$ Tesla and $E_F = 5$ eV, $\lambda_F \sim 10^4$. Thus, most conventional experiments are not in the extreme quantum limit, where λ_F is a small integer, and a semiclassical treatment of the phenomena should be adequate. Exceptions are phenomena such as the integer and fractional quantum Hall effects, Wigner electron crystallization, etc., which occur at high B-field and low electron-density conditions.

11.2.1 Semiclassical analysis

As discussed in Chapter 5, the dynamics of electrons in a crystal in a slowly varying applied field are basically determined by its band structure $E(\mathbf{k})$, i.e. by the Hamiltonian

$$H(\mathbf{p}) = E(\mathbf{p}), \qquad (11.11)$$

with $\mathbf{p} = \hbar \mathbf{k}$. Given $H(\mathbf{p})$, classically we obtain the Hamiltonian for $\mathbf{B} \neq 0$ by using the substitution (here e is the charge of the electron, i.e. $e = -|e|$)

$$\mathbf{P} = \mathbf{p} + \frac{|e|\mathbf{A}}{c}, \qquad (11.12)$$

where $\mathbf{p} = \frac{\hbar}{i}\nabla$ is the total momentum and $\mathbf{P}$ is defined using Eq. (11.12), which is known as the kinetic momentum of the electron, resulting in

$$H(\mathbf{P}) = E\left(\mathbf{p} + \frac{|e|\mathbf{A}}{c}\right). \qquad (11.13)$$

We can formally derive the equation of motion for an electron in the semiclassical picture. The velocity is given by

$$\mathbf{v} = \dot{\mathbf{r}} = \frac{\partial H}{\partial \mathbf{p}} = \frac{\partial E(\mathbf{P})}{\partial \mathbf{P}}, \qquad (11.14)$$

which is only a function of the band structure. (Here, for simplicity, we shall neglect the anomalous velocity term due to a possible nonzero Berry curvature of the band structure. This term, as discussed in Chapter 5, can be important for some phenomena.) The time derivative of $\mathbf{P}$ is

$$\dot{\mathbf{P}} = \dot{\mathbf{p}} + \frac{|e|}{c}\dot{\mathbf{A}} = \frac{-|e|}{c}\left(\mathbf{v}(\mathbf{P}) \times \mathbf{B}\right) \qquad (11.15)$$

and is also a function of $\mathbf{P}$ only.

Let us consider a uniform static field along the $\hat{\mathbf{z}}$-direction $\mathbf{B} = (0, 0, B)$. Equation (11.15) in Cartesian coordinates is

$$\dot{P}_x = \frac{-|e|}{c}v_y B,$$

$$\dot{P}_y = \frac{|e|}{c}v_x B, \qquad (11.16)$$

$$\dot{P}_z = 0.$$

Since $dE = \mathbf{F} \cdot \mathbf{r} = \frac{q}{c}(\mathbf{v} \times \mathbf{B}) \cdot \mathbf{v}dt$, the electron cannot gain energy from a uniform B-field. The motion of the electron in $\mathbf{P}$ space is dictated by the band structure $E(\mathbf{P})$ and constrained

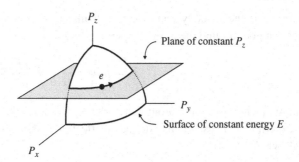

Figure 11.3 Schematic of the electron orbit in **P**-space for an electron in a magnetic field **B** along the $\hat{z}$-direction.

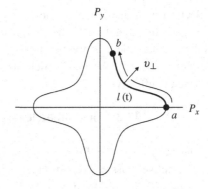

Figure 11.4 **P**-space path of an electron in a magnetic field along the $\hat{z}$-direction; $\ell(t)$ is the length traveled in time t.

to move in a path formed by the intersection of a surface of constant energy and a plane of constant P_z, as shown in Fig. 11.3. Thus, the trajectory in **P**-space is dependent on the energy of the electron and the orientation of the applied B-field. The dynamics, i.e. how fast the electron is moving in **P**-space, is however determined by the strength of the magnetic field through Eq. (11.16).

11.3 Effective mass and real-space orbits

Since many experimental studies, such as cyclotron resonance measurements, are of the magneto-transport type, it is useful to introduce the concept of an effective dynamical mass. Consider an electron moving along a closed trajectory (called a closed orbit) in **P**-space in the presence of a B-field, as shown in Fig. 11.4. Let $\ell(t)$ be the distance from points a to b along the orbit in **P**-space. Then, using Eq. (11.16),

$$\frac{d\ell}{dt} = \sqrt{\dot{P}_x^2 + \dot{P}_y^2} = \frac{|e|B}{c}\sqrt{v_x^2 + v_y^2} = \frac{|e|B}{c}v_\perp(P_z, E, \ell). \qquad (11.17)$$

Here $v_\perp$ is the velocity that is perpendicular to $\mathbf{B}$ and to the tangent of the orbit. The electron goes around the orbit in a period T with

$$T = \oint dt = \frac{c}{|e|B} \oint \frac{d\ell}{v_\perp}. \tag{11.18}$$

The frequency for going around the orbit $\omega_c^* = \frac{2\pi}{T}$ is the cyclotron frequency measured in experiment. Thus, we may define an effective mass m^* through the identification $\omega_c^* = \frac{|e|B}{m^*c}$, where

$$m^*(E, P_z) = \frac{1}{2\pi} \oint \frac{d\ell}{v_\perp(P_z, E, \ell)}. \tag{11.19}$$

For a free electron with dispersion relation $E(\mathbf{k}) = \frac{\hbar^2 k^2}{2m_e}$, the orbit is circular and one obtains $m^* = m_e$. In general, m^* depends on the band structure, electron energy, and direction of the magnetic field. Moreover, if we go beyond an independent picture, many-body interactions will modify the quasiparticle energy dispersion, giving rise to corrections to the band structure effective mass.

11.3.1 A geometric expression for m^*

Let $A(E, P_z)$ be the area of the $\mathbf{P}$-space orbital formed by $E(\mathbf{P}) = $ constant and $P_z = $ constant. The effective mass m^* of Eq. (11.19) in fact is related to $A(E, P_z)$. Consider $\delta\mathbf{n}(P_x, P_y) = (\delta P_x, \delta P_z)$ the vector normal to the constant energy surface due to a change in E (see Fig. 11.5). The change in area of the orbit as $E \to E + \delta E$ is given by the area of a strip around the orbit, $\delta A = \oint |\delta\mathbf{n}| d\ell$, because $\dot{P}_z = 0$. Since $\delta E = v_\perp \delta n$, we have

$$\delta A = \oint \delta n \, d\ell = \oint \frac{\delta E d\ell}{v_\perp}. \tag{11.20}$$

Comparing Eqs. (11.19) and (11.20), we have the relation

$$m^*(E, P_z) = \frac{1}{2\pi} \frac{\partial A(E, P_z)}{\partial E}. \tag{11.21}$$

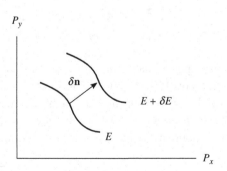

Figure 11.5   Electron orbital change in **P**-space as the energy changes from E to $E + \delta E$.

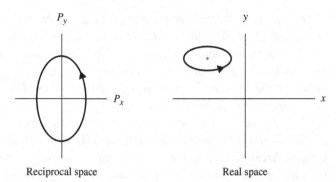

P_y

P_x

y

x

Reciprocal space

Real space

Figure 11.6   Electron in a magnetic field along the $\hat{z}$-direction. (Left) Orbit in **P**-space and (right) orbit in real space.

A measurement of the cyclotron frequencies thus contains geometric information on the electron orbits.

11.3.2 Real-space orbits

Within the semiclassical approach, we can also easily obtain the motions of the carriers in real space. Integrating Eq. (11.16) with respect to time, we obtain

$$y(t) - y_0 = \frac{-c}{|e|B}P_x(t)$$

and

$$x(t) - x_0 = \frac{c}{|e|B}P_y(t). \tag{11.22}$$

The orbit in real space is centered about some position (x_0, y_0), $y(t)$ depends on $P_x(t)$, and $x(t)$ depends on $P_y(t)$ with a scaling factor of $\frac{c}{|e|B}$. To obtain the real-space orbit, one translates the P-space orbit to (x_0, y_0), rotates it by $90°$ counterclockwise, and rescales it, as seen in Fig. 11.6.

The shape of the real-space orbits is dictated by the P-space orbit. In general, it can be quite complex because of band structure (i.e. interaction of the electron with the lattice) effects. As expected, the scaling factor also tells us that the orbit in real space tightens as the strength of the magnetic field increases. We shall return to the concept of carrier orbits in metals when we discuss magnetoresistivity in the next chapter.

11.4 Quantum oscillations: periodicity in 1/B and the de Haas–van Alphen effect in metals

Many of the electronic properties of crystalline metals at low temperature T are oscillatory functions of $1/B$, where B is an applied field. Examples include the oscillatory behaviors in the resistivity $\rho(1/B)$, known as the Shubnikov–de Haas effect, and in the magnetic susceptibility $\chi(1/B)$, known as the de Haas–van Alphen effect.[2] These effects, generally called quantum oscillations, have proven to be extraordinarily powerful probes of the electronic structure, particularly the Fermi surface properties, of metals. We shall focus on the physics of the de Haas–van Alphen effect in this section. The periodic behavior in $1/B$ in other effects has a similar physical origin.

The presence of an external magnetic field causes the formation of a magnetization $\mathbf{M}(\mathbf{B})$ in metals. The magnetic susceptibility $\chi(\mathbf{B}) = \mathbf{M}(\mathbf{B})/\mathbf{B}$ at low field can be measured using different means, such as the induced torque on a sample. The oscillations in the measured magnetization or magnetic susceptibility owe their origin to the quantization of the electron orbits.

As a simple example, we consider the free electron gas. Again, taking $\mathbf{B} = B\hat{z}$, a uniform magnetic field, the electron energy eigenvalues are given by Eq. (11.4). For a given k_z, the states are quantized into Landau levels, each having a large degeneracy given by Eq. (11.8). The induced magnetic moment of the system is (in a unit volume)

$$\mathbf{M} = \sum_{\ell} f_{\ell} \langle \mathbf{m}_{\ell} \rangle, \tag{11.23}$$

where $\langle \mathbf{m}_{\ell} \rangle$ is the moment of the ℓth state and f_{ℓ} is the Fermi–Dirac distribution function

$$f_{\ell}(T) = \frac{1}{e^{\beta(\varepsilon_{\ell} - \mu)} + 1}. \tag{11.24}$$

In a magnetic field $\mathbf{B}$, as seen in Eq. (11.4), the energy of the states is $\mathbf{B}$-dependent, i.e. $\varepsilon_{\ell}(\mathbf{B})$, and

$$\langle \mathbf{m}_{\ell} \rangle = \frac{-\partial \varepsilon_{\ell}}{\partial \mathbf{B}}. \tag{11.25}$$

Equation (11.25) follows from $\delta H = -\mathbf{m} \cdot \delta \mathbf{B}$, where $\mathbf{m}$ is the dipole moment operator, and the Hellmann–Feynman theorem $\left(\frac{\partial \varepsilon_{\ell}(\zeta)}{\partial \zeta} = \langle \frac{\partial H}{\partial \zeta} \rangle_{\ell} \right)$. Equation (11.23) becomes

$$\mathbf{M} = \sum_{\ell} \frac{1}{e^{\beta(\varepsilon_{\ell} - \mu)} + 1} \left(-\frac{\partial \varepsilon_{\ell}}{\partial \mathbf{B}} \right) = -\frac{\partial \Omega}{\partial \mathbf{B}} \bigg|_{\beta, \mu}. \tag{11.26}$$

Here Ω is the usual grand potential in thermal physics for fermions,

$$\Omega = -\frac{1}{\beta} \log \mathcal{Z} = \frac{1}{\beta} \sum_{\ell} \log \left\{ 1 + e^{-\beta(\varepsilon_{\ell} - \mu)} \right\}, \tag{11.27}$$

where $\mathcal{Z}$ denotes the grand partition function.

In order to evaluate $\Omega(\mathbf{B}, \mu, \beta)$, the sum over the states in Eq. (11.27) may be converted to an integral using the density of states of the system $g(\varepsilon, \mathbf{B})$. Owing to the formation of

[2] W. J. de Haas and P. M. van Alphen, " The dependence of the susceptibility of diamagnetic metals on the field," *Proc. Koninklijke Akademie van Wetenschappen te Amsterdam* 33(1930), 1106.

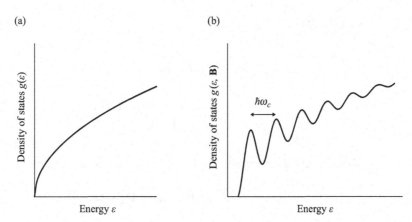

(a)   (b)

Figure 11.7 Schematic of the density of states of a free electron: (a) for $\mathbf{B} = 0$ and (b) in the presence of a uniform magnetic field with some broadening to mimic experiment.

Landau levels, the density of states of an electron gas in an applied uniform B-field no longer has the simple form $g(\varepsilon) \sim \sqrt{\varepsilon}$, but becomes one with structures that are separated by energy $\hbar\omega_c$. This is illustrated in Fig. 11.7. It is the structure of $g(\varepsilon, \mathbf{B})$ that induces structure in the physical properties of a metal as a function of B.

If we define an integrated density of state

$$G(\varepsilon, \mathbf{B}) = \int_{-\infty}^{\varepsilon} g(\varepsilon', \mathbf{B}) d\varepsilon', \qquad (11.28)$$

then the grand potential, which is

$$\Omega(\mathbf{B}, \beta, \mu) = -\frac{1}{\beta} \int d\varepsilon g(\varepsilon, \mathbf{B}) \log\left[1 + e^{-\beta(\varepsilon - \mu)}\right], \qquad (11.29)$$

may be expressed as (using integration by parts twice)

$$\Omega(\mathbf{B}, \beta, \mu) = -\int_{-\infty}^{\infty} \left[\int_{-\infty}^{\varepsilon} G(\varepsilon', \mathbf{B}) d\varepsilon'\right]\left[-\frac{\partial f(\varepsilon)}{\partial \varepsilon}\right] d\varepsilon, \qquad (11.30)$$

where $f(\varepsilon)$ is again the Fermi–Dirac distribution function. The evaluation of the grand potential is thereby reduced to the evaluation of $g(\varepsilon, \mathbf{B})$ and its integral, i.e. a counting of electronic states.

In evaluating $g(\varepsilon, \mathbf{B})$ or $G(\varepsilon, \mathbf{B})$ in three dimensions, we need to include the motion of the electron along z (i.e. the quantum number k_z) and the degeneracy of the Landau levels D as given in Eq. (11.8). The final expression needed for the integrand of

Eq. (11.30) is

$$
\int_{-\infty}^{\varepsilon} G(\varepsilon', \mathbf{B}) d\varepsilon' = a\varepsilon^{5/2} + b(\hbar\omega_c)^2 \varepsilon^{1/2}
$$

$$
- \frac{V}{\pi^4}\left(\frac{2m}{\hbar^2}\right)^{3/2}\left(\frac{\hbar\omega_c}{2}\right)^{5/2}\frac{1}{2}\sum_{n=1}^{\infty}(-1)^n\frac{\cos\left(\frac{2\pi n\varepsilon}{\hbar\omega_c} - \frac{\pi}{4}\right)}{n^{5/2}}, \quad (11.31)
$$

where a and b are constants and V is the volume. The three terms in Eq. (11.31) lead to

$$
\Omega(\mathbf{B}, \beta, \mu) = \Omega_a + \Omega_b + \Omega_c. \quad (11.32)
$$

The first term Ω_a is independent of $\mathbf{B}$ and does not contribute to $\mathbf{M} = -\frac{\partial\Omega}{\partial\mathbf{B}}\big|_{\beta,\mu}$. The second term is of the form

$$
\Omega_b(\mathbf{B}, \beta, \mu) = -b(\hbar\omega_c)^2\int_{-\infty}^{\infty} d\varepsilon\,\varepsilon^{1/2}\left(-\frac{\partial f}{\partial\varepsilon}\right) \sim -B^2. \quad (11.33)
$$

This term leads to a contribution to the magnetic moment, which is of the form

$$
\mathbf{M}_b(\mathbf{B}) = -\frac{N}{2E_\mathrm{F}}\mu_B^2\mathbf{B}, \quad (11.34)
$$

where N is the number of electrons, E_F is the Fermi energy, and μ_B the Bohr magneton in a metal is replaced with an effective Bohr magneton

$$
\mu_B^* = \frac{e\hbar}{2m^*c}, \quad (11.35)
$$

which may differ from the Bohr magneton because the effective mass m^* may be different from the electron mass in vacuum. This term gives a diamagnetic response due to the minus sign appearing in Eq. (11.34), and is called the Landau diamagnetism. If $m^* = m_e$, then Eq. (11.34) in fact is numerically equal to $-1/3$ of the Pauli paramagnetic moment due to the spins of the electrons. The physical origin of Landau diamagnetism arises from the electrons orbiting to screen the applied $\mathbf{B}$-field and has nothing to do with the spins of the electrons.

The third term in Eq. (11.32) leads to another contribution to the magnetic moment $\mathbf{M}_c(\mathbf{B}, \beta, \mu) = -\frac{\partial\Omega_c}{\partial\mathbf{B}}\big|_{\beta,\mu}$. This may be evaluated to yield the dominant contributions

$$
\mathbf{M}_c(\mathbf{B}, \beta, \mu) = \frac{3\pi}{2}\frac{N\mu_B^*}{E_\mathrm{F}\beta}\left(\frac{E_\mathrm{F}}{\mu_B^*B}\right)^{1/2}\hat{\mathbf{B}}\left[\sum_{n=1}^{\infty}\frac{(-1)^{n+1}\sin\left(\frac{\pi E_\mathrm{F}n}{\mu_B^*B} - \frac{\pi}{4}\right)}{n^{1/2}\sinh\left(\frac{\pi^2 n}{\mu_B^*B\beta}\right)}\right]. \quad (11.36)
$$

The total magnetic moment $\mathbf{M} = \mathbf{M}_b + \mathbf{M}_c$ is hence a function of T and has an oscillatory component from $\mathbf{M}_c$ that is periodic in $1/B$.

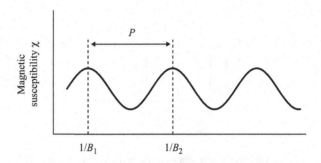

Figure 11.8 Schematic of the magnetic susceptibility χ of a free electron gas as a function of the inverse of the applied magnetic field B.

At high temperature, i.e. $k_B T \gg \hbar \omega_c = 2\mu_B^* B$,

$$\frac{1}{\sinh\left(\frac{\pi^2 k_B T n}{\mu_B^* B}\right)} \sim e^{-\frac{\pi^2 k_B T n}{\mu_B^* B}}. \tag{11.37}$$

Thus, the contribution of M_c to M is exponentially small. The oscillatory component of M can only be observed at low temperature in the regime of $k_B T \leq 2\mu_B^* B$. In this limit, the terms in the sum in Eq. (11.36) are on the order of 1 and

$$\frac{M_c}{M_b} \sim \left(\frac{k_B T}{\mu_B^* B}\right)\left(\frac{E_F}{\mu_B^* B}\right)^{1/2} \gg 1, \tag{11.38}$$

showing that the oscillatory component can be quite large at low temperature.

Taking the largest term $n = 1$ in Eq. (11.36) and defining $K = \frac{\pi E_F}{\mu_B^*}$, we have

$$M_c(\mathbf{B}) \sim \frac{1}{\sqrt{B}} \sin\left(\frac{K}{B} - \frac{\pi}{4}\right), \tag{11.39}$$

which is oscillatory as a function of $1/B$. In a de Haas–van Alphen experiment, $\chi = M/B$ will also be an oscillatory function in $1/B$ with period $P = \frac{1}{B_1} - \frac{1}{B_2} = \frac{2\mu_B^*}{E_F}$, as shown in Fig. 11.8.

For the free electron case, $E_F = \frac{\hbar^2 k_F^2}{2m^*}$. Thus, we may rewrite P in terms of the Fermi surface cross-section area, $A_{FS} = \pi k_F^2$, and obtain

$$P = \frac{\left(2\pi e/\hbar c\right)}{A_{FS}} = \frac{4\pi^2}{\phi_0 A_{FS}}, \tag{11.40}$$

where ϕ_0 is the flux quantum. Equation (11.40) can be shown (see below) to hold for any arbitrary Fermi surface if one replaces A_{FS} with the extremal cross-section areas of the Fermi surface.

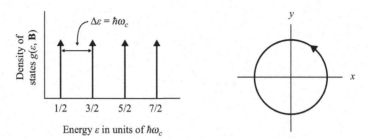

Figure 11.9 (Left) Density of states of a two-dimensional electron gas in a perpendicular magnetic field. (Right) Real-space orbit of an electron in a Landau level.

This oscillatory phenomenon has its physical origin from (1) the quantization of the electron orbits in a magnetic field, (2) the fact that adding a quantum of flux inside a fixed orbit does not change any physical measurable properties, and (3) the factor $-\partial f/\partial \varepsilon$ in Eq. (11.30), only orbits of energy within $k_B T$ of the Fermi level contribute to the grand potential Ω and hence M. Under a change of magnetic field, the quantum states within $k_B T$ of E_F will change, but will map into a set that is identical to the original set after a quantum of magnetic flux is either added or subtracted from the enclosure of the orbit. This can be seen more clearly in two dimensions. In a uniform magnetic field, the density of states of a two-dimensional electron system is a series of delta functions and the electrons undergo a closed orbit motion in real space, as illustrated in Fig. 11.9. For quantum mechanical systems, if $\{E_0, \varphi_0\}$ are solutions to the Hamiltonian

$$H_0 = \frac{\left(\mathbf{P} - \frac{e}{c}\mathbf{A}_0\right)^2}{2m} + V(\mathbf{r}), \tag{11.41}$$

then, if $\mathbf{B}_0 \to \mathbf{B}_0 + \delta\mathbf{B}$ (with the corresponding change $\mathbf{A} = \mathbf{A}_0 + \delta\mathbf{A}$), it can be shown that

$$\varphi(\mathbf{r}) = e^{\frac{ie}{\hbar c}\int^{\mathbf{r}} \delta\mathbf{A}\cdot d\boldsymbol{\ell}} \varphi_0(\mathbf{r}) \tag{11.42}$$

satisfies the Schrödinger equation with the new $H = \frac{\left(\mathbf{P} - \frac{e}{c}\mathbf{A}\right)^2}{2m} + V(\mathbf{r})$, and is a solution of the same energy if

$$\frac{e}{\hbar c}\oint \delta\mathbf{A}\cdot d\boldsymbol{\ell} = 2\pi n \quad (n = \text{integer}), \tag{11.43}$$

owing to the requirement of single-valueness for ϕ as it goes around the real-space orbit. The line integral of Eq. (11.43) may be converted to a surface integral resulting in the condition that

$$\delta B A_{\text{orbit}} = n\phi_0, \tag{11.44}$$

where A_{orbit} is the area of the real-space orbit. This result implies that the states near E_F and the properties of the system determined by these states will duplicate themselves if B is

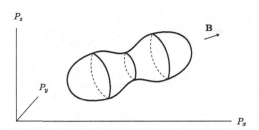

Figure 11.10 Different orbits and therefore cross-sectional areas of the orbits for electrons on the Fermi surface for a given orientation of the applied **B**-field.

changed by an amount that adds or removes a flux quantum in the orbit. This is the general principle of gauging away the effect of an integer quantum of flux in a loop system.

Since the states of importance in the de Haas–van Alphen effect are those at the Fermi energy, this gives the condition for the periodicity in $1/B$ as (using Eq. (11.44))

$$P = \frac{1}{B - \delta B} - \frac{1}{B} = \frac{\delta B}{B^2} = \frac{\phi_0}{B^2 A_{\text{orbit}}}. \tag{11.45}$$

For the states at the Fermi level, the real-space orbital area is related to the P-space orbital area by $\hbar^2 A_{\text{FS}} = (\frac{e^2 B^2}{c^2}) A_{\text{orbit}}$, giving rise to

$$P = \frac{\phi_0}{B^2 \left(\frac{\hbar^2 c^2}{e^2} \frac{1}{B^2} A_{\text{FS}} \right)} = \frac{4\pi^2}{\phi_0 A_{\text{FS}}}, \tag{11.46}$$

which is the same as Eq. (11.40).

In real materials, the Fermi surface may be quite complex. For a given orientation of **B**, each cross-sectional area will contribute a frequency of oscillations, as seen in Fig. 11.10. However, only the extremal cross sections will constructively add to provide experimental signals. Since $P = \frac{1}{B} - \frac{1}{B + \Delta B} = \frac{2\mu_B^*}{E_F}$, for most metals P is very small, on the order of 10^{-9}/G. Thus, one needs a very high field, typically $B \sim 10T$, to perform de Haas–van Alphen measurements.

The sinh term in Eq. (11.36) provides further information since it is temperature dependent. Even at low temperature, usually only the $n = 1$ term is observable. By varying the temperature at a fixed magnetic field, we may measure the combination $\frac{\pi^2 k_B T}{\mu_B^* B}$. Such measurements yield μ_B^* and hence information on m^*.

A careful de Haas–van Alphen experiment will provide detailed information on the cross-sectional area of the Fermi surface of a metal and on the effective mass of the carriers. Moreover, since $\Omega(\mathbf{B}, \mu, \beta)$ is oscillatory in $1/B$, we expect other physical quantities derived from Ω to show similar oscillatory behavior. The conductivity of a metal also exhibits oscillations as a function of $1/B$. This is called the Shubnikov–de Haas effect.

12 Fundamentals of transport phenomena in solids

The electronic and thermal transport behaviors are among the most important properties of a material, being fundamental to their understanding and characterization and central in their applications. We have discussed topics such as electrical conductivity in previous chapters. In this chapter, we develop the fundamentals of transport phenomena. The discussions focus on the important physical processes and make use of mostly kinetic theory and semiclassical formalisms.[1]

Topics to be presented include elementary treatments of magnetoresistance, the conventional Hall and integer quantum Hall effects, and various transport processes and phenomena in materials employing the Boltzmann equation formalism.

12.1 Elementary treatment of magnetoresistance and the Hall effect

An applied magnetic field in general changes the electrical conduction of a metal or semiconductor. This phenomenon of magnetoresistance has led to fundamental discoveries (such as the integer and fractional quantum Hall effects), and a number of important technologies such as magnetic recording devices.

The experimental setup for measuring magnetoresistance usually takes the arrangement shown schematically in Fig. 12.1. In this standard geometry, the sample takes on the form of a bar (often called a Hall bar) with current drawn along its length (the $\hat{x}$-direction) and a magnetic field applied transverse to its flat face (the $\hat{z}$-direction). The interesting experiments are those carried out at low temperature T and high magnetic field $\mathbf{B}$ on a pure sample. These conditions correspond to large cyclotron frequency ω_c and long collision time τ. That is,

$$\omega_c \tau = \left(\frac{|e|B}{mc} \right) \tau \gg 1. \tag{12.1}$$

The carriers go through many orbits before a collision; the details of the collision processes, as discussed below, are suppressed and the details of the electronic structure, in particular the Fermi surface, are exhibited.

---

[1] Some pioneering papers on the quantum theory of transport in metals include: W. Kohn and J. M. Luttinger, "Quantum theory of electrical transport phenomena," *Phys. Rev.* 108(1957), 590; J. M. Luttinger and W. Kohn, "Quantum theory of electrical transport phenomena. II," *Phys. Rev.* 109(1958), 1892; and I. M. Lifshitz, "Quantum theory of the electrical conductivity of metals in a magnetic field," *Soviet Phys. JETP* 5(1957), 1227.

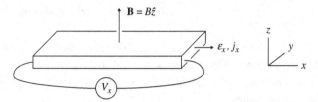

Figure 12.1 Schematic of a standard experimental setup for a magnetoresistance measurement. This is known as a Hall bar geometry.

General experimental findings for the magnetoresistance with a transverse **B** field include:

I. *Weak fields*. The fractional change in resistance at low field tends to increase as B^2. That is,

$$\frac{\Delta R}{R} = \frac{R(\mathbf{B}) - R(0)}{R(0)} \sim (\omega_c \tau)^2 \sim B^2, \tag{12.2}$$

where R (sometimes denoted as R_{xx}) is given by V_x/I_x and I_x is the total current through the Hall bar. (See Fig. 12.1.)

II. *Strong fields*.

(a) Saturation. $R(\mathbf{B})$ goes to a constant as B increases for all orientations of **B** with respect to the crystalline directions. This is found for metals with a closed Fermi surface, e.g. In, Al, Na, Li, etc.

(b) No saturation. $R(\mathbf{B})$ keeps increasing as **B** increases. This occurs in crystals with equal numbers of electron and hole carriers, e.g. Bi, Sb, W, Mo, etc.

(c) Selected saturation directions. $R(\mathbf{B})$ is unsaturated for **B** oriented along some crystalline directions, but not for all orientations. This occurs in some metals, e.g. Cu, Ag, Au, Mg, Zn, etc.

The magnetoresistance has been used successfully to study the topology of the Fermi surface of both simple and complex metals. The above-mentioned differences in the observed behavior of the magnetoresistance may be understood from elementary considerations of charge particles moving in electric and magnetic fields within the formalism of crystalline electron dynamics, which we developed in Chapter 5. The basic physics is that the Lorentz force,

$$\mathbf{F} = \frac{e}{c}\mathbf{v} \times \mathbf{B}, \tag{12.3}$$

deflects the carriers with charge e from their trajectory, and thus makes it less effective for the electric field to impart momentum to the charge carriers. Classically, for a pure sample with no collisions, the particle would drift in the direction perpendicular to the applied electric $\mathcal{E}$ and magnetic **B** fields with a drift velocity $v_d = \frac{c\mathcal{E}}{B}$.

12.1.1 One-carrier-type model

We first consider the case of a metal with a single type of carrier that is characterized by an isotropic effective mass m^* and a constant collision (or relaxation) time τ. This model corresponds to an idealized simple metal with a spherical Fermi surface and weak scattering centers.

The equation of motion for the carrier is given by

$$m^* \left(\dot{\mathbf{v}} + \frac{1}{\tau} \mathbf{v} \right) = e \left(\boldsymbol{\mathcal{E}} + \frac{1}{c} \mathbf{v} \times \mathbf{B} \right), \tag{12.4}$$

where $\boldsymbol{\mathcal{E}}$ and $\mathbf{B}$ are uniform, static, and applied electric and magnetic fields, respectively. At steady state, $\dot{\mathbf{v}} = 0$, Eq. (12.4) reduces to

$$\mathbf{v} = \frac{e\tau}{m^*} \left(\boldsymbol{\mathcal{E}} + \frac{1}{c} \mathbf{v} \times \mathbf{B} \right). \tag{12.5}$$

Let $\mathbf{B} = B\hat{z}$, and defining $\mu = e\tau/m^*$ and $\xi = eB\tau/m^*c$, the components of Eq. (12.5) are

$$\begin{aligned} v_z &= \mu \mathcal{E}_z, \\ v_x &= \frac{\mu}{1+\xi^2} \left(\mathcal{E}_x + \xi \mathcal{E}_y \right), \\ v_y &= \frac{\mu}{1+\xi^2} \left(\mathcal{E}_y - \xi \mathcal{E}_x \right). \end{aligned} \tag{12.6}$$

Here we take e as the carrier charge. The quantity μ is the mobility of the system in the absence of the **B**-field. Thus, the **B**-field reduces the effective mobility of the carriers along directions that are perpendicular to the direction of the **B**-field.

The conductivity tensor $\overset{\leftrightarrow}{\sigma}$ is given through the relation

$$\mathbf{j} = \overset{\leftrightarrow}{\sigma} \boldsymbol{\mathcal{E}}, \tag{12.7}$$

where $\mathbf{j}$ is the current density. Since $\mathbf{j} = ne\mathbf{v}$, where n is the carrier density, we obtain in this elementary treatment for the conductivity tensor

$$\overset{\leftrightarrow}{\sigma} = \frac{ne\mu}{1+\xi^2} \begin{pmatrix} 1 & \xi & 0 \\ -\xi & 1 & 0 \\ 0 & 0 & 1+\xi^2 \end{pmatrix}, \tag{12.8}$$

and the resistivity tensor

$$\overset{\leftrightarrow}{\rho} = \frac{1}{ne\mu} \begin{pmatrix} 1 & -\xi & 0 \\ \xi & 1 & 0 \\ 0 & 0 & 1 \end{pmatrix}, \tag{12.9}$$

by taking the inverse of $\overset{\leftrightarrow}{\sigma}$.

In the standard experimental geometry shown in Fig. 12.1, $j_y = j_z = 0$. The measured quantities satisfy the relation

$$\begin{pmatrix} \mathcal{E}_x \\ \mathcal{E}_y \\ \mathcal{E}_z \end{pmatrix} = \frac{1}{ne\mu} \begin{pmatrix} 1 & -\xi & 0 \\ \xi & 1 & 0 \\ 0 & 0 & 1 \end{pmatrix} \begin{pmatrix} j_x \\ 0 \\ 0 \end{pmatrix}, \tag{12.10}$$

which yields

$$\mathcal{E}_x = \frac{1}{ne\mu} j_x. \tag{12.11}$$

Eq. (12.11) shows that the longitudinal resistance $\rho_{xx} = \mathcal{E}_x/j_x = \frac{1}{ne\mu} = \frac{m^*}{ne^2\tau}$ is independent of B. The absence of magnetoresistance in this model is the result of an electric field generated along the y-direction in Fig. 12.1, called the Hall field $\mathcal{E}_y$, which just balances the Lorentz force from the magnetic field.[2] From Eqs. (12.10) and (12.11), the Hall field is

$$\mathcal{E}_y = \frac{\xi}{ne\mu} j_x = \xi \mathcal{E}_x. \tag{12.12}$$

This result arises from having one relaxation time τ and one kind of carrier. In this case, we can determine the density and the charge of the carriers by measuring the transverse or Hall resistivity ρ_{yx}. From Eq. (12.12),

$$\rho_{yx} = \frac{\mathcal{E}_y}{j_x} = \frac{\xi}{ne\mu} = \frac{\left(\frac{eB}{m^*c}\right)\tau}{ne\left(\frac{e\tau}{m^*}\right)} = \frac{B}{nec}. \tag{12.13}$$

The material parameters m^* and τ do not enter the Hall resistivity, leaving a simple relation that yields the density and sign of the charge of the carriers within this simple model.

In general, there will be magnetic field dependence on ρ_{xx}, and the Hall resistivity ρ_{yx} may be more complex than Eq. (12.13). The complexities arise because there are, in general, many carrier types in a metal and the Hall field $\mathcal{E}_y$ may not balance the Lorentz force on each carrier, resulting in a magnetoresistance in ρ_{xx}.

12.1.2 Two-carrier-type model

The simplest model to illustrate the physics of magnetoresistance in ρ_{xx} is to assume two carrier types in the metal. In this case, at steady state, the velocity of each of the carrier types, $i = 1$ or 2, is given by

$$\mathbf{v}_i = \left(\frac{e_i\tau_i}{m_i^*}\right)\mathcal{E} + \frac{e_i\tau_i}{m_i^*c}\mathbf{v}_i \times \mathbf{B}. \tag{12.14}$$

---

[2] This was discovered by E. H. Hall in 1879, 18 years before the discovery of the electron: E. H. Hall, "On a new action of the magnet on electric currents," *Amer. J. Math.* 2(1879), 287.

The current density $\mathbf{j}$ is

$$\mathbf{j} = n_1 e_1 \mathbf{v}_1 + n_2 e_2 \mathbf{v}_2 = (\overleftrightarrow{\sigma}_1 + \overleftrightarrow{\sigma}_2) \cdot \boldsymbol{\mathcal{E}}. \tag{12.15}$$

In the standard geometry with $\mathbf{B} = B\hat{z}$, we may drop the z-components of $\overleftrightarrow{\sigma}$, since, as seen in Eq. (12.10), it is not relevant to the analysis. Equation (12.15) yields

$$\overleftrightarrow{\sigma} = \overleftrightarrow{\sigma}_1 + \overleftrightarrow{\sigma}_2 = \begin{pmatrix} \sigma_{xx} & \sigma_{xy} \\ -\sigma_{xy} & \sigma_{yy} \end{pmatrix}, \tag{12.16}$$

with

$$\sigma_{xx}^{(i)} = \sigma_{yy}^{(i)} = \frac{n_i e_i^2 \tau_i}{m_i^* \left(1 + \xi_i^2\right)} = \frac{\sigma_0^{(i)}}{1 + \beta_i^2 B^2} \tag{12.17}$$

and

$$\sigma_{xy}^{(i)} = \frac{\sigma_0^{(i)} \beta_i B}{1 + \beta_i^2 B^2}. \tag{12.18}$$

Here we have used the notation $\sigma_0^{(i)} = \frac{n_i e_i^2 \tau_i}{m_i^*}$ and $\beta_i = \frac{e\tau_i}{m_i^* c}$. The resistivity tensor is given by

$$\overleftrightarrow{\rho} = \frac{1}{\sigma_{xx}\sigma_{yy} + \sigma_{xy}^2} \begin{pmatrix} \sigma_{yy} & -\sigma_{xy} \\ \sigma_{xy} & \sigma_{xx} \end{pmatrix}. \tag{12.19}$$

The explicit expression for ρ_{xx} is given by (using $\sigma_{xx} = \sigma_{yy}$)

$$\rho_{xx} = \frac{\sigma_{yy}^{(1)} + \sigma_{yy}^{(2)}}{\left(\sigma_{xx}^{(1)} + \sigma_{xx}^{(2)}\right)^2 + \left(\sigma_{xy}^{(1)} + \sigma_{xy}^{(2)}\right)^2}. \tag{12.20}$$

We may evaluate the change in ρ_{xx} due to the magnetic field and obtain

$$\frac{\Delta\rho_{xx}}{\rho_{xx}} = \frac{\rho_{xx}(B) - \rho_{xx}(0)}{\rho_{xx}(0)} = \frac{\sigma_0^{(1)}\sigma_0^{(2)}(\beta_1 - \beta_2)^2 B^2}{\left(\sigma_0^{(1)} + \sigma_0^{(2)}\right)^2 + B^2\left(\beta_2\sigma_0^{(1)} + \beta_1\sigma_0^{(2)}\right)^2}. \tag{12.21}$$

Equation (12.21) exhibits the experimental behavior observed in many metals. At weak field, the first term in the denominator in Eq. (12.21) dominates. Hence, we have $\Delta\rho_{xx}(B)/\rho_{xx}(0) \sim B^2$ and the change in resistivity is positive. At high field, the B^2 dependence of the numerator and that of the denominator cancel each other and $\Delta\rho_{xx}(B)/\rho_{xx}(0)$ saturates to a constant value.

The phenomenon of saturation may be understood as follows. At high field, as $B \to \infty$, Eq. (12.17) gives $\sigma_{xx} \to 0$. This does not mean that j_x is zero, because

$$j_x = \sigma_{xx}\mathcal{E}_x + \sigma_{xy}\mathcal{E}_y. \tag{12.22}$$

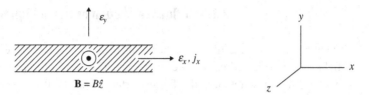

Figure 12.2 Schematic of current density j_x, and electric and magnetic field components in a Hall experiment.

In the large B-field limit, although σ_{xx} vanishes, the existence of an electric field along the x-direction $\mathcal{E}_x$ induces a Hall field $\mathcal{E}_y$ along the y-direction, to annul any Hall current that may be generated along the y-direction. In turn, $\mathcal{E}_y$ generates a Hall current in the x-direction. The fact that $j_y = 0$ leads to $\mathcal{E}_y = -\frac{\sigma_{yx}}{\sigma_{yy}}\mathcal{E}_x$. Thus, the current along the wire, as given in Eq. (12.22), in the large B limit, is due to the Hall current generated by $\mathcal{E}_y$, and the current is proportional to $\mathcal{E}_x$ but independent of B, leading to the phenomenon of saturation.

In the high B-field limit, the Hall conductivity σ_{xy} simplifies in the two-carrier-type model. Since $\sigma_{xy} = \sigma_{xy}^{(1)} + \sigma_{xy}^{(2)}$ and $1 + \xi_i^2 \to \xi_i^2$ as $B \to \infty$, we have from Eq. (12.18)

$$\sigma_{xy}^{(i)} \to \frac{n_i e^2 \tau_i^2 \xi_i}{m_i^* \xi_i^2} = \frac{n_i e_i c}{B} \tag{12.23}$$

and

$$\sigma_{xy} = \frac{(n_1 e_1 + n_2 e_2)c}{B}, \tag{12.24}$$

independent of m_i^* and τ_i.

For a compensated metal in which there are equal numbers of electrons and holes, we have $n_1 = n_2$ and $e_1 = -e_2$. Examples of this kind of system are the semimetals Bi and Sb. For such metals, Eq. (12.24) gives $\sigma_{xy} = 0$. This implies that there is no Hall field $\mathcal{E}_y$ to contribute a current to j_x in Eq. (12.22) and from Eq. (12.20),

$$\rho_{xx}(B) = \frac{\sigma_{yy}}{\sigma_{xx}^2 + \sigma_{xy}^2} \sim B^2. \tag{12.25}$$

Physically, because one cannot establish a Hall field, it becomes increasingly hard for the field $\mathcal{E}_x$ to impart momentum to the carriers. The magnetoresistance does not saturate if there are equal numbers of holes and electrons, explaining the experimental observation stated in II(b) of Section 12.1.

In fact, we can get the condition for unsaturated magnetoresistance directly from Eq. (12.21). Notice that in the denominator of Eq. (12.21),

$$\beta_2 \sigma_0^1 + \beta_1 \sigma_0^2 = \frac{\tau_1 \tau_2 e_1 e_2 (n_1 e_1 + n_2 e_2)}{m_1 m_2 c}.$$

When we have $(n_1 e_1 + n_2 e_2) = 0$, there will be no saturation of magnetoresistance in the high B-field. Such a condition can be achieved in the case where $n_1 = n_2$ and $e_1 = -e_2$.

12.1.3 Influence of open orbits in high field

What remains to be explained is behavior II(c), i.e. the phenomenon of saturation at high field only in selected crystalline directions. This kind of behavior occurs in metals with open orbits for states near the Fermi surface. That is, the intersection of the surface of constant energy and the plane of constant P_z does not form a loop, but joins continuously to the next Brillouin zone indefinitely in the repeated zone scheme.

Let us consider a metal with an open orbit along the P_x-direction. Such an orbit can create a current in real space along the y-direction, as illustrated in Fig. 12.3. This arises from Eq. (11.16) in the standard geometry for $\mathbf{B} = B\hat{z}$, i.e.

$$\dot{P}_y = -\frac{eB\dot{x}}{c} \tag{12.26}$$

and

$$\dot{P}_x = \frac{eB\dot{y}}{c}. \tag{12.27}$$

If an electric field is applied along the y-direction, there is an imbalance in the occupation of orbits in the upper plane versus the lower plane in Fig. 12.3(a), leading to a net current along the y-direction in Fig. 12.3(b). Thus, for open orbits along P_x, there will be an extra contribution to σ_{yy}.

From the discussion in Chapter 11, we can show that in the presence of a uniform electric field $\mathcal{E}$ in addition to the $\mathbf{B}$-field, the projection of the real-space orbit in a plane perpendicular to $\mathbf{B}$ is

$$\mathbf{r}_\perp(t) - \mathbf{r}_\perp(0) = \frac{-\hbar c}{eB}\hat{\mathbf{B}} \times [\mathbf{k}(t) - \mathbf{k}(0)] + \mathbf{v}_d t, \tag{12.28}$$

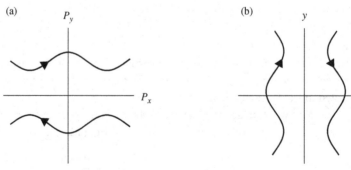

(a) P_y P_x (b) y x

Reciprocal space open orbit   Real space open orbit

Figure 12.3 Sketch of an open orbit in reciprocal space (a) and the corresponding orbit in real space (b). The applied magnetic field is along the $\hat{z}$-direction.

where $\mathbf{v}_d = c\frac{\mathcal{E}}{B}(\hat{\mathcal{E}} \times \hat{\mathbf{B}})$ is the drift term and $\mathbf{k}(t)$ is the orbital motion given by

$$\hbar\dot{\mathbf{k}} = -\frac{e}{c}\frac{1}{\hbar}\frac{\partial \bar{E}(\mathbf{k})}{\partial \mathbf{k}} \times \mathbf{B}, \tag{12.29}$$

with $\bar{E}(\mathbf{k}) = E(\mathbf{k}) - \hbar\mathbf{k} \cdot \mathbf{v}_d$. But $\hbar\mathbf{k} \cdot \mathbf{v}_d < \hbar\frac{1}{a_B}c\frac{\mathcal{E}}{B} = \left(\frac{e^2}{a_B}\right)\left(\frac{e\mathcal{E}a_B}{\hbar\omega_c}\right)$ in general is negligibly small in laboratory conditions and we can just take $\bar{E}(\mathbf{k})$ as equal to $E(\mathbf{k})$. (Here a_B is the Bohr radius.)

In calculating the current density $\mathbf{j} = ne\mathbf{v}$, the velocity $\mathbf{v}$ should be the drift velocity of the electron acquired after collision averaged over all states, i.e.

$$\langle \mathbf{v} \rangle = \left\langle \frac{\mathbf{r}_\perp(0) - \mathbf{r}_\perp(-\tau)}{\tau} \right\rangle = \left\langle -\frac{\hbar c}{eB}\hat{\mathbf{B}} \times \left[\frac{\mathbf{k}(0) - \mathbf{k}(-\tau)}{\tau}\right] \right\rangle + \tau\langle \mathbf{v}_d \rangle. \tag{12.30}$$

For closed orbits, $\Delta\mathbf{k} = \mathbf{k}(0) - \mathbf{k}(-\tau)$ is bounded in the high field limit of $\omega_c\tau \gg 1$. The dominant term in Eq. (12.30) is $\langle \mathbf{v}_d \rangle$. This is not the case for open orbits, however.

In the high B-field limit, we can then associate an extra conductivity σ_{yy}^0 from the open orbits along P_x, which is independent of B in the form of

$$\sigma_{yy}^0 = sne\mu, \tag{12.31}$$

where s is a constant. The expression in Eq. (12.31) is independent of B because the first term on the right-hand side of Eq. (12.30) is independent of B since $\Delta k(t) \sim B$. The conductivity tensor $\overleftrightarrow{\sigma}$ in the high field limit takes on the form

$$\overleftrightarrow{\sigma} = ne\mu \begin{pmatrix} \xi^{-2} & \xi^{-1} \\ -\xi^{-1} & s \end{pmatrix}, \tag{12.32}$$

including the contribution from the open orbit along $\mathbf{P}_x$. For σ_{yy}, we have dropped the term ξ^{-2} in comparison to s, since ξ^{-2} vanishes as B^{-2}.

In the standard geometry of drawing a current along the x-direction, $j_y = 0$ dictates that $\mathcal{E}_y$ is now scaled like $\frac{1}{B}\mathcal{E}_x$ and becomes vanishingly small at large B. The resistivity along the x-direction takes on the form, for $\xi \gg 1$,

$$\rho_{xx}(B) = \frac{\xi^2}{ne\mu}\left(\frac{s}{s+1}\right) \sim B^2, \tag{12.33}$$

which shows no saturation. However, if the open orbit is along P_y, then, for $\xi \gg 1$,

$$\overleftrightarrow{\sigma} = ne\mu \begin{pmatrix} s & \xi^{-1} \\ -\xi^{-1} & \xi^{-2} \end{pmatrix} \tag{12.34}$$

and

$$\rho_{xx} = \frac{1}{ne\mu}\left(\frac{1}{1+s}\right). \tag{12.35}$$

Therefore, for this orientation of the open orbit, the resistance saturates at high field. For an open orbit in a general direction and $\xi \gg 1$, we have, with s_1 and s_2 constants,

$$\overleftrightarrow{\sigma} = ne\mu \begin{pmatrix} s_1 & \xi^{-1} \\ -\xi^{-1} & s_2 \end{pmatrix} \tag{12.36}$$

and

$$\rho_{xx} = \frac{1}{ne\mu} \left(\frac{s_2}{s_1 s_2 + \xi^{-2}} \right),$$

which shows saturation.

The analysis here shows that, in general, the magnetoresistance saturates except when the open orbit carries current almost precisely parallel to the y-direction (i.e. the open orbit in **P**-space is parallel to x). The existence of open orbits is the key to understanding the behavior of selected saturation directions described above as behavior II(c). Good examples of materials that exhibit this kind of behavior are the noble metals, such as Cu and Ag.

Figure 12.4 depicts the Fermi surface of copper. There are many kinds of orbits that can be formed by intersection of the Fermi surface with a plane perpendicular to the direction of the magnetic field. Since the Fermi surface is connected to the neighboring Brillouin zones, open orbits are possible for the appropriate orientation of the **B**-field. Hence, the magnetoresistance shows non-saturation along selected orientations of the **B**-field with respect to the crystal axes.

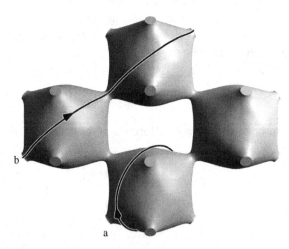

b

a

Figure 12.4   Fermi surface of copper in the repeat zone scheme, showing both closed and open orbits.

12.2 The integer quantum Hall effect

The integer quantum Hall effect (IQHE) was an unexpected discovery made by K. von Klitzing, G. Dorda, and M. Pepper in 1980,[3] and von Klitzing was awarded the 1985 Nobel Prize in Physics for the discovery. The IQHE is the manifestation of a new phase of matter of a two-dimensional electron gas (2DEG) system in the quantum limit of low carrier density, low temperature, and high magnetic field. It is an early example of the so-called topological insulators.[4] The IQHE has been used to make the most precise measurement of the fine structure constant $\alpha = \frac{e^2}{\hbar c}$ and to set a new standard for resistance. In this section, we describe the phenomenon, the experimental setup and results, and the physical origin of the effect.

12.2.1 The phenomenon

The IQHE occurs in a 2DEG. As was discussed in the previous section, the magnetoconductivity in the standard geometry is given by (from Eq. (12.8))

$$\overset{\leftrightarrow}{\sigma} = \frac{ne^2\tau}{m\left(1 + (\omega_c\tau)^2\right)} \begin{pmatrix} 1 & -\omega_c\tau \\ \omega_c\tau & 1 \end{pmatrix} \tag{12.37}$$

and

$$\overset{\leftrightarrow}{\rho} = \begin{pmatrix} \rho_0 & -\frac{B}{nec} \\ \frac{B}{nec} & \rho_0 \end{pmatrix}, \tag{12.38}$$

with $\rho_0 = \frac{m}{ne^2\tau}$. (For notation simplicity, we shall use m to denote the carrier effective mass in this section.) In the quantum limit, in which the experimental conditions are such that only a few Landau levels are occupied, it is found that the measured quantities, as a function of $\frac{n}{B}$ over large ranges, unlike those given by Eqs. (12.37) and (12.38), take the form

$$\overset{\leftrightarrow}{\sigma} = \begin{pmatrix} 0 & -\frac{\nu e^2}{h} \\ \frac{\nu e^2}{h} & 0 \end{pmatrix} \tag{12.39}$$

and

$$\overset{\leftrightarrow}{\rho} = \begin{pmatrix} 0 & \frac{h}{\nu e^2} \\ -\frac{h}{\nu e^2} & 0 \end{pmatrix}, \tag{12.40}$$

where ν is a small integer corresponding to the number of occupied Landau levels, as shown in Fig. 12.5.

[3] K. von Klitzing, G. Dorda, and M. Pepper, "New method for high-accuracy determination of the fine-structure constant based on quantized Hall resistance," *Phys. Rev. Lett.* 45(1980), 494.

[4] See, for example, the review by M. Z. Hasan and C. L. Kane, "Colloquium: Topological insulators," *Rev. Mod. Phys.* 82(2010), 3045.

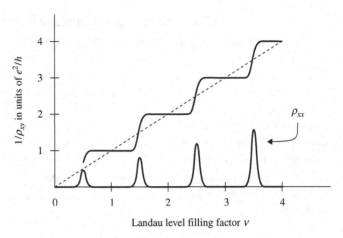

Figure 12.5 A sketch of the Hall conductivity and the longitudinal resistivity of a 2DEG with impurities, illustrating the IQHE at low Landau-level filling factor ν.

This is an extraordinary result in that the Hall plateaus in Fig. 12.5 are found to be quantized to $\frac{e^2}{h}$, the fundamental unit of conductance, to better than one part in 10^9. Since c, the speed of light, is an exactly defined quantity, the IQHE plateau value provides one of the most accurate measurements of the fine structure constant $\frac{e^2}{\hbar c}$. It also gives a new and convenient standard for resistance. Moreover, the diagonal resistance in the IQHE regime is exactly zero, except in narrow transition regions between plateaus in σ_{xy}. Thus, there is dissipationless current flow along the Hall bar, i.e. no voltage drop in the Hall plateau regions in Fig. 12.5.

12.2.2 Experimental setup

The typical setup for an IQHE experiment is schematically shown in Fig. 12.6. A 2DEG is created at a metal-oxide-semiconductor (MOS) interface or at a semiconductor hetero-junction. Contacts are made in the standard Hall bar geometry (of length L and width W) to measure the current flow and voltage drop along and across the Hall bar. The resistance along the Hall bar is

$$R_L = \frac{V_L}{I} = \frac{\mathcal{E} \times L}{j \times W} = \rho_{xx} \cdot \left(\frac{L}{W} \right), \tag{12.41}$$

and the Hall resistance is

$$R_H = \frac{V_H}{I} = \frac{\mathcal{E}_y W}{j_x W} = \frac{\mathcal{E}_y}{j_x} = \rho_{yx}. \tag{12.42}$$

In two-dimensional (2D) systems, the dimensions of resistance are the same as those of resistivity. Furthermore, for the Hall resistance ρ_{yx} (or ρ_H), the geometric size factors drop out of Eq. (12.42), making precise geometry-independent measurements possible.

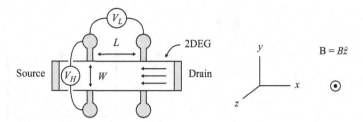

Figure 12.6 Sketch of the experimental setup for an IQHE study.

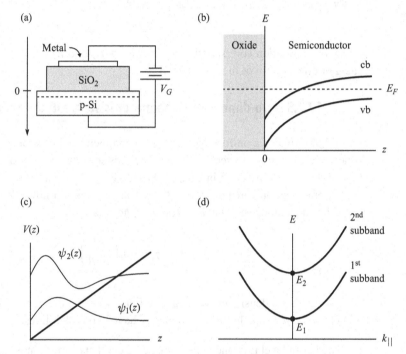

Figure 12.7 Schematic of the sample used in the IQHE study: (a) a MOS junction, (b) band structure of a p-doped semiconductor near the oxide-semiconductor interface, (c) potential $V(z)$ and subband states at the oxide-semiconductor interface, (d) subband dispersions as a function of the wavevector parallel to the oxide-semiconductor interface ($k_{\parallel}$) in absence of a magnetic field.

The original experiment by von Klitzing, Dorda, and Pepper was performed on an MOS device (see Fig. 12.7(a)). By applying a gate voltage across the interface, a layer of electrons is accumulated at the interface on the p-doped silicon side by bending the conduction band of Si below the Fermi level E_F. These electrons see a confining potential $V(z)$ along the z-direction perpendicular to the interface forming discrete subband states (see Fig. 12.7(c)), which are freely propagating in the $x-y$ plane. Each of the subbands gives rise to a 2D flat density of states. At sufficiently low temperature T and electron density n, only the lowest subband is occupied, making the system a 2DEG. The conditions

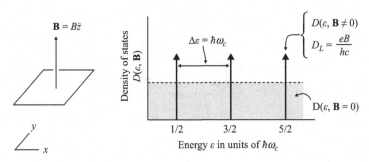

Figure 12.8 Density of states of a 2DEG in a perpendicular magnetic field.

for 2DEG formation correspond to $k_B T < E_2 - E_1$ and $E_F < E_2 - E_1$, where E_i is the energy of the bottom of the ith subband.

12.2.3 Two-dimensional magnetoresistance in the quantum limit

For a 2DEG under a uniform perpendicular magnetic field, its density of states per unit area $D(\varepsilon)$ transforms to a series of delta functions separated by $\hbar\omega_c$ due to the formation of Landau levels, as discussed in Chapter 11 (see Fig. 12.8). Each of the delta functions has a degeneracy per unit area of $D_L = \frac{|e|B}{hc}$ (we neglect spin for now). The number of Landau levels occupied at density n at low $T(k_B T \ll \hbar\omega_c)$ is

$$\nu = \frac{n}{D_L} = n\left(\frac{hc}{|e|B}\right). \tag{12.43}$$

As the electron density or the magnetic field is varied, the Fermi level will move and ν will change. From the discussion so far, we have two possibilities.

(a) The Fermi level is in one of the Landau levels. This corresponds to an arbitrary density. In this case, $\overleftrightarrow{\rho}$ is given by Eq. (12.38), and both ρ_{xx} and ρ_{xy} behave as expected for a traditional metal.

(b) The Fermi level is in between Landau levels, i.e. ν is an integer. Then there is a gap in the excitation spectrum, which means that there is no scattering that can dissipate energy. Therefore, τ now goes to infinity, $\rho_{xx} \to 0$, and $\rho_{xy} = -\frac{B}{nec} = \frac{h}{\nu e^2}$, resulting in the expressions for $\overleftrightarrow{\rho}$ and $\overleftrightarrow{\sigma}$ in Eqs. (12.40) and (12.39), respectively.

The above naive analysis, however, has serious problems in explaining the experimental data. (i) Why are the plateaus in ρ_{xy} so extended as a function of $\frac{n}{B}$? The scenario in case (b) above should occur only at singular points on the $\frac{n}{B}$-axis. (ii) Why is the quantization so perfect and not dependent on details of sample quality and geometry? The answer to (i) involves the effects of impurities and imperfections in the 2DEG, and the answer to (ii) involves knowing that the electron charge is quantized and may be understood in terms of a gauge argument.

12.2.4 Effects of random impurities and defects

Owing to various defects in or nearby the 2DEG, the potential seen by an electron has a randomly fluctuating component, as illustrated in Fig. 12.9. The random potential fluctuations lead to a broadening of the Landau-level density of states as well as the existence of localized states. Classically, a charged particle in a magnetic field can only move along paths defined by iso-energy contours on the energy surface. For a large-sized sample, virtually all states are trapped by local potential fluctuations that are either above or below the mean value. Only states with energy virtually exactly at the mean potential can propagate across the sample. Thus, as shown schematically in Fig. 12.9, only states at the original Landau-level energies are extended, and all other states with energy between Landau levels are localized states.

Since localized states do not carry current as the carrier density is varied, the conductivity σ_{xy} of the system does not change as long as E_F remains in the mobility gap (defined as the energy region over which the states are localized) between the Landau levels. These constant σ_{xy} regimes correspond to the plateaus in the ρ_{xy} vs. $\frac{n}{B}$ curve. When E_F moves through the extended state portion of the density of states, ρ_{xy} makes a transition from one plateau to the next. What remains to be explained is why σ_{xy} is exactly an integer times e^2/h.

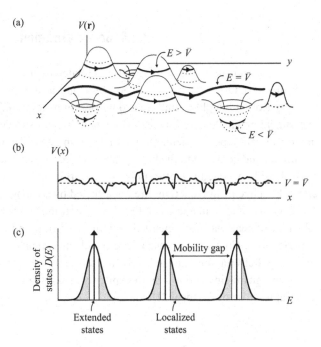

Figure 12.9 Potential fluctuations at an oxide-semiconductor interface: (a) schematic of potential fluctuations and electron orbits in real space, (b) potential fluctuations along a given direction x, (c) density of states in the presence of random potential fluctuations.

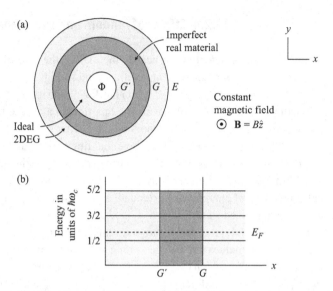

Figure 12.10 (a) A disk of a 2DEG with a hole in its center, having a solenoid of flux Φ through the hole, and a uniform applied magnetic field $B = B\hat{z}$. (b) Energy of Landau levels. G is the boundary separating the real material from the ideal 2DEG, G' is another boundary, and E is the edge of the disk.

12.2.5 Gauge argument

The exact quantization of σ_{xy} in the IQHE is related to the fact that the charge of an electron is quantized, and it may be understood by using a very general and eloquent argument made by Robert Laughlin[5] and refined by Bertrand Halperin.[6] Let us consider a gedanken experiment in which there is a disk with a hole in it, as shown in Fig. 12.10. The **B**-field is in the z-direction perpendicular to the plane of the disk. There is a variable needle of flux Φ through and confined to the hole. This flux Φ can nevertheless affect those electrons in the disk with extended wavefunctions that encircle the hole through the Aharonov–Bohm effect. Localized wavefunctions are not affected by a change in Φ.

It is convenient to imagine that the disk is partitioned into three parts. The disk consists of a ring of real material "guarded" by an inner ring and an outer ring that are free of imperfections. Consider the case where the electron density n is such that ν is an integer, so E_F is in between the Landau levels for the guard rings, as illustrated in Fig. 12.10. The use of the guard rings allows us to explicitly relate the conducting properties of the real material to those of the ideal 2DEG.

[5] R. B. Laughlin, "Quantized Hall conductivity in two dimensions," *Phys. Rev. B* 23(1981), 5632.

[6] B. I. Halperin, "Quantized Hall conductance, current-carrying edge states, and the existence of extended states in a two-dimensional disordered potential," *Phys. Rev. B* 25(1982), 2185.

Now imagine that we adiabatically change the value of Φ. There is an electric field generated through Faraday's law.

$$\frac{1}{c}\frac{d\Phi}{dt} = \frac{1}{c}\int d\mathbf{A}\cdot\frac{\partial\mathbf{B}}{\partial t} = \oint_C d\boldsymbol{\ell}\cdot\mathcal{E}_y = \oint_C d\ell\rho_{yx}j_x, \tag{12.44}$$

where C is any contour enclosing the hole, and locally we define x to be the radial direction and y to be along the tangential or the direction of the electric field $\mathcal{E}$. If there is a Hall current density perpendicular to $\mathcal{E}_y$, there will be a charge transport across the contour. Equation (12.44) can be rewritten as

$$\frac{\Delta\Phi}{c} = \frac{1}{c}\frac{d\Phi}{dt}\Delta t = \rho_H\oint_C j_\perp d\ell\,\Delta t = \rho_H\Delta Q. \tag{12.45}$$

Here ΔQ denotes the total charge transported across the contour C. Hence we have, for all $\Delta\Phi$, the relation

$$\rho_H = \frac{1}{c}\frac{\Delta\Phi}{\Delta Q}. \tag{12.46}$$

In particular, Eq. (12.46) holds for $\Delta\Phi = \Phi_0$, where Φ_0 is the flux quantum $\frac{hc}{e}$. But after a change of Φ by one flux quantum, the system is the same as before, using the same argument as in the case of our discussion of the de Haas–van Alphen effect in Chapter 11. This implies that ΔQ must be an integer times e, or

$$\rho_H = \frac{h}{e^2}\times\left(\frac{1}{\text{integer}}\right). \tag{12.47}$$

Now we further consider the loop to be in one of the guard rings, say the outer GE-ring in Fig. 12.10. The electrons transported across EG must have come from the edge E and been transferred into the impure region at boundary G. In the Landau gauge, as discussed in the previous chapter, the states in a Landau level correspond to one-dimensional harmonic oscillator states located at different locations along the x-direction at location $x_0 = \frac{c\hbar}{eB}k_y$, and have the wavefunction

$$\psi(x,y) = e^{ik_y y}\varphi(x), \tag{12.48}$$

with $k_y = \frac{2\pi}{L}p$, where p is an integer and L is the circumference along the disk. At the physical edge E, the increase in the potential raises the energy of the states within a Landau level located near the edge (specified by the label p) and gives rise to edge states that span the Fermi level for each occupied Landau level, as shown in Fig. 12.11. The way the current flows across the guard ring as the flux Φ changes is such that a particle in the state labeled p adiabatically transforms into one in state $p-1$, and state $p-1$ to state $p-2$, and so on. After a change of one flux quantum, the states have adiabatically transformed to the same set of wavefunctions and energies as before, since an integer number of the flux quantum can be removed by a gauge transformation. However, it is evident from Fig. 12.11 that one

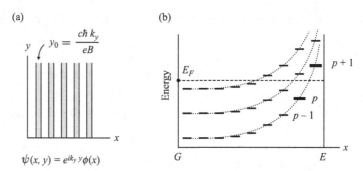

Figure 12.11 (a) Wavefunction of the electronic states in the Landau gauge. (b) Energy of the Landau-level states localized at different locations near the disk edge E.

electron is carried through the region by each occupied Landau level, since all states are shifted to the left by one position, and therefore

$$\rho_H = \frac{h}{e^2 \nu},\tag{12.49}$$

where ν is the number of occupied Landau levels. Now, assuming that there is no dissipation in the impure region GG', i.e. E_F is in the mobility gap, there are two possibilities: (1) charges pile up at the interface and we have an insulator, or (2) charges transport through and we have the IQHE.

12.3 The Boltzmann equation formalism and transport in real materials

In this section, we develop a Boltzmann equation formalism for transport. It is based on a semiclassical approach within the one-electron or quasiparticle framework. Nevertheless, the formalism is very powerful for treating various transport processes of crystal electrons in nonzero electric field $\mathcal{E}$, magnetic field $\mathbf{B}$, and temperature gradient ∇T with impurity and other scattering centers.

The basic assumptions, aside from the use of one-electron theory, are that electron collisions are taken as random, uncorrelated events, and that the form of the nonequilibrium electron distribution has little effect on the rate or distribution of electrons emerging from the collision. These assumptions are generally valid if weak fields and weak scattering centers are involved. The utility of the formulation lies in its versatility in treating different phenomena in clear physical terms and in incorporating realistic material parameters into the calculations.

12.3.1 The electron distribution function $f(\mathbf{k}, \mathbf{r}, t)$

A central concept in treating transport phenomena in the Boltzmann equation formalism is the electron distribution function $f(\mathbf{k}, \mathbf{r}, t)$. It is defined (for a paramagnetic system) through the relation

$$dN = \frac{2}{(2\pi)^3} f(\mathbf{k}, \mathbf{r}) d\mathbf{k}\, d\mathbf{r}, \tag{12.50}$$

where dN is the number of electrons in a small volume $d\mathbf{r}d\mathbf{k}$ in phase space centered around the point $(\mathbf{r}, \mathbf{k})$. From the above definition, for a homogeneous sample in equilibrium,

$$f^0(\mathbf{k}, \mathbf{r}) = n_{\mathbf{k}} = \frac{1}{e^{\beta(\mathcal{E}_{\mathbf{k}} - \mu)} + 1}, \tag{12.51}$$

which is the Fermi–Dirac distribution with $\beta = \frac{1}{k_B T}$, and μ is the chemical potential. This can be seen from the fact that the number of particles with $\mathbf{k}$ in volume $d\mathbf{k}$ is

$$dN_k = \frac{2V}{(2\pi)^3} \int_{\mathbf{k}}^{\mathbf{k}+d\mathbf{k}} n_{\mathbf{k}'} d\mathbf{k}' = \frac{2V}{(2\pi)^3} n_{\mathbf{k}} d\mathbf{k}, \tag{12.52}$$

and the fraction in volume $d\mathbf{r}$ is $\frac{d\mathbf{r}}{V}$. Hence, $dN = \frac{2}{(2\pi)^3} n_{\mathbf{k}} d\mathbf{k} d\mathbf{r}$, which gives $f^0(\mathbf{k}, \mathbf{r}) = n_{\mathbf{k}}$ for a homogeneous system.

The goal of transport studies is to determine $f(\mathbf{k}, \mathbf{r}, T, \mathcal{E}, \mathbf{B}, t)$ for a perturbed system and to obtain its physical properties, such as various response functions, from f.

Since f is a function of both $\mathbf{k}$ and $\mathbf{r}$, it is a concept that is valid only in a semiclassical sense. The uncertainty principle prevents a precise determination of both $\mathbf{k}$ and $\mathbf{r}$ of a particle. This distribution function is meaningful only if it is defined over large enough $d\mathbf{r}$ such that it contains enough electrons to give a good representation of the phenomena of interest. For example, if n is the density of the electrons, the volume $d\mathbf{r}$ should be significantly larger than $\frac{1}{n}$. On the other hand, $d\mathbf{r}$ has to be small enough so that the physical properties of the system within the small volume do not vary appreciatively. Thus, if λ is the wavelength of the perturbation (e.g. $\mathcal{E}, \mathbf{B}$, or ∇T), then we would need to have $dx \ll \lambda$. Finally, we need $d\mathbf{k}$ such that the uncertainty principle $\Delta k_x \Delta x \gtrsim 1$ is satisfied.

12.3.2 Equation of motion for $f(\mathbf{k}, \mathbf{r}, t)$

We start with the premise that the time rate of change of $f(\mathbf{k}, \mathbf{r}, t)$ may be written as

$$\frac{\partial f(\mathbf{k}, \mathbf{r}, t)}{\partial t} = \left(\frac{\partial f}{\partial t}\right)_{\text{drift}} + \left(\frac{\partial f}{\partial t}\right)_{\text{coll}}. \tag{12.53}$$

Here the first term on the right-hand side (RHS) of Eq. (12.53) is taken to be the effects of externally applied forces on $\frac{\partial f}{\partial t}$ and the second term on the RHS is from collision processes in the system. Equation (12.53) is an approximation because the drift and collision terms are treated separately. One can imagine physical conditions for which the collision term

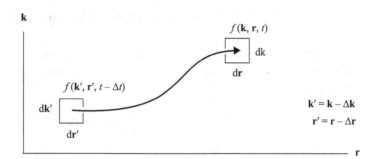

Figure 12.12   Phase space trajectory of the distribution function $f(\mathbf{k}, \mathbf{r}, \mathbf{t})$.

depends intimately on the applied fields. Equation (12.53) would not be valid for some classes of problems, e.g. very strong applied fields or strong collisions.

The drift term may be obtained by considering the time evolution of $f(\mathbf{k}, \mathbf{r}, t)$ in phase space, as depicted in Fig. 12.12. We consider a small volume $d\mathbf{k}'d\mathbf{r}'$ around $(\mathbf{k}', \mathbf{r}')$. This volume will evolve over Δt to $d\mathbf{k}\,d\mathbf{r}$ around $(\mathbf{k}, \mathbf{r})$ with $\mathbf{k} = \mathbf{k}' + \Delta\mathbf{k}$ and $\mathbf{r} = \mathbf{r}' + \Delta\mathbf{r}$. From classical mechanics of dynamical systems, we know that

$$d\mathbf{k}'d\mathbf{r}' = d\mathbf{k}\,d\mathbf{r}. \tag{12.54}$$

The distribution of particles near $\mathbf{k}', \mathbf{r}'$ at time $t' = t - \Delta t$ will drift over to near $(\mathbf{k}, \mathbf{r})$ at time t. That is,

$$f(\mathbf{k}, \mathbf{r}, t)d\mathbf{k}d\mathbf{r} = f(\mathbf{k}', \mathbf{r}', t - \Delta t)d\mathbf{k}'d\mathbf{r}'. \tag{12.55}$$

Making use of Eq. (12.54) and performing a first-order Taylor series expansion of $f(\mathbf{k}', \mathbf{r}', t')$ with respect to $\Delta\mathbf{r}$, $\Delta\mathbf{k}$, and Δt, we obtain

$$f(\mathbf{k}, \mathbf{r}, t) = f(\mathbf{k}, \mathbf{r}, t) - \frac{\partial f}{\partial \mathbf{k}} \cdot \Delta\mathbf{k} - \frac{\partial f}{\partial \mathbf{r}} \cdot \Delta\mathbf{r} - \frac{\partial f}{\partial t} \cdot \Delta t. \tag{12.56}$$

Collecting terms and labeling $\left(\frac{\partial f}{\partial t}\right)_{\text{drift}}$ as the rate of change due to drift processes, we have

$$\left(\frac{\partial f}{\partial t}\right)_{\text{drift}} = -\frac{\partial f}{\partial \mathbf{k}} \cdot \frac{d\mathbf{k}}{dt} - \frac{\partial f}{\partial \mathbf{r}} \cdot \frac{d\mathbf{r}}{dt}. \tag{12.57}$$

Since $\frac{\hbar d\mathbf{k}}{dt}$ is the force on the particle and $\frac{d\mathbf{r}}{dt}$ is the velocity $\mathbf{v}$, Eq. (12.57) may be rewritten as

$$\left(\frac{\partial f}{\partial t}\right)_{\text{drift}} = -\left[\frac{\partial f}{\partial \mathbf{k}} \cdot \frac{e}{\hbar}\left(\mathcal{E} + \frac{\mathbf{v}}{c} \times \mathbf{B}\right) + \frac{\partial f}{\partial \mathbf{r}} \cdot \mathbf{v}\right], \tag{12.58}$$

where $\mathcal{E}$ and $\mathbf{B}$ are external electric and magnetic fields, respectively. The first term on the RHS of the above equation comes from changes induced by the applied fields, and the second term is the diffusion term.

The collision term in Eq. (12.53) describes the changes in $f(\mathbf{k}, \mathbf{r}, t)$ due to scattering processes. If we denote $W_{\mathbf{k}',\mathbf{k}}$ to be the probability per unit time for a particle in state $\mathbf{k}'$ to scatter into state $\mathbf{k}$, then

$$\left(\frac{\partial f}{\partial t}\right)_{\text{coll}} = \int d\mathbf{k}' \left\{ W_{\mathbf{k}'\mathbf{k}}\left[f(\mathbf{k}', \mathbf{r})\left(1 - f(\mathbf{k}, \mathbf{r})\right)\right] - W_{\mathbf{k}\mathbf{k}'}\left[f(\mathbf{k}, \mathbf{r})\left(1 - f(\mathbf{k}', \mathbf{r})\right)\right] \right\}. \quad (12.59)$$

If "microscopic reversibility" holds for the scattering processes under consideration, $W_{\mathbf{k}'\mathbf{k}} = W_{\mathbf{k}\mathbf{k}'}$. Further, if one makes the usual assumptions that there are uniform scattering centers and that $W_{\mathbf{k}\mathbf{k}'}$ is independent of $\mathbf{r}$, then Eq. (12.59) simplifies to

$$\left(\frac{\partial f(\mathbf{k}, \mathbf{r}, t)}{\partial t}\right)_{\text{coll}} = \int d\mathbf{k}' \, W_{\mathbf{k}\mathbf{k}'}\left[f(\mathbf{k}', \mathbf{r}) - f(\mathbf{k}, \mathbf{r})\right]. \quad (12.60)$$

Adding the two contributions, we have

$$\frac{\partial f(\mathbf{k}, \mathbf{r}, t)}{\partial t} = -\left[\frac{\partial f}{\partial \mathbf{k}} \cdot \frac{e}{\hbar}\left(\mathcal{E} + \frac{\mathbf{v}}{c} \times \mathbf{B}\right) + \frac{\partial f}{\partial \mathbf{r}} \cdot \mathbf{v}\right] + \int d\mathbf{k}' \, W_{\mathbf{k}\mathbf{k}'}\left[f(\mathbf{k}', \mathbf{r}) - f(\mathbf{k}, \mathbf{r})\right]. \quad (12.61)$$

This equation is called the Boltzmann equation. It is an integro-differential equation, which in general is difficult to solve. However, as we see below, if we did have the solution $f(\mathbf{k}, \mathbf{r}, t)$, then we would be able to arrive at many of the transport properties of materials. For example, the electrical current density $\mathbf{j}(\mathbf{r}, t)$ is the number of charge carriers crossing a unit area per unit time. Since $e\mathbf{v}(\mathbf{k})\left[\frac{2}{(2\pi)^3}f(\mathbf{k}, \mathbf{r}, t)d\mathbf{k}\right]$ is the charge flux generated by carriers with wavenumber $\mathbf{k}$, we have

$$\mathbf{j}(\mathbf{r}, t) = \frac{2e}{(2\pi)^3} \int d\mathbf{k} \, \mathbf{v}(\mathbf{k}) f(\mathbf{k}, \mathbf{r}, t). \quad (12.62)$$

Similarly, the energy current density $\mathbf{Q}(\mathbf{r}, t)$ of the electrons is given by

$$\mathbf{Q}(\mathbf{r}, t) = \frac{2}{(2\pi)^3} \int d\mathbf{k} \, E(\mathbf{k}) \mathbf{v}(\mathbf{k}) f(\mathbf{k}, \mathbf{r}, t), \quad (12.63)$$

where $E(\mathbf{k})$ is the energy of the particle.

For crystals, we need to extend the formalism to include band structure effects. The distribution function is generalized to $f_n(\mathbf{k}, \mathbf{r}, t)$, with n the band index, $W_{\mathbf{k}\mathbf{k}'}$ goes to $W_{\mathbf{k}\mathbf{k}'}^{nn'}$, allowing for interband scattering, and $E(\mathbf{k})$ is replaced by $E_n(\mathbf{k})$ and $\mathbf{v}(\mathbf{k})$ by $\mathbf{v}_n(\mathbf{k}) = \frac{1}{\hbar}\frac{\partial E_n(\mathbf{k})}{\partial \mathbf{k}}$ plus a Berry curvature term if it is nonzero.

12.3.3 Steady-state transport

Many experiments and applications are concerned with the steady-state response of the system. At steady state, physical quantities such as transport coefficients and currents do not change with time. Thus, $f(\mathbf{k}, \mathbf{r}, t)$ in a stationary state satisfies

$$\frac{\partial f}{\partial t} = 0 = \left(\frac{\partial f}{\partial t}\right)_{\text{drift}} + \left(\frac{\partial f}{\partial t}\right)_{\text{coll}}, \quad (12.64)$$

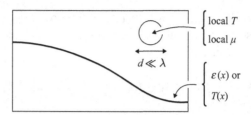

Figure 12.13 Slow spatial variation of applied field $\mathcal{E}$ or temperature T allows definition of local temperature T and chemical potential μ.

which says that the drift processes must balance the collision processes at steady state. Therefore, from Eq. (12.61), we have

$$\left[\mathbf{v}(\mathbf{k}) \cdot \frac{\partial}{\partial \mathbf{r}} + \frac{e}{\hbar} \left(\mathcal{E} + \frac{\mathbf{v}(\mathbf{k})}{c} \times \mathbf{B} \right) \cdot \frac{\partial}{\partial \mathbf{k}} \right] f(\mathbf{k}, \mathbf{r}, t) = \left(\frac{\partial f}{\partial t} \right)_{\text{coll}}. \tag{12.65}$$

This is the equation we need to solve in order to determine the steady-state transport properties of materials. Material-specific parameters come in through $\mathbf{v}(\mathbf{k})$ and $\left(\frac{\partial f}{\partial t} \right)_{\text{coll}}$.

Let us first consider the simplest case, which is a spatial homogeneous (i.e. uniform impurity or defect density) system at equilibrium with $\mathbf{B} = 0$, $\mathcal{E} = 0$, and $\nabla T = 0$. Both the $\frac{\partial}{\partial \mathbf{r}}$ and $\frac{\partial}{\partial \mathbf{k}}$ terms on the left-hand side (LHS) of Eq. (12.65) are zero and f goes to f^0 given by Eq. (12.51). To satisfy Eq. (12.65), we must have $\left(\frac{\partial f^0}{\partial t} \right)_{\text{coll}} = 0$, or from Eq. (12.60),

$$\int d\mathbf{k}' \, W_{\mathbf{k}\mathbf{k}'} \left[f^0(\mathbf{k}') - f^0(\mathbf{k}) \right] = 0. \tag{12.66}$$

We now introduce the concept of local equilibrium. For slowly varying perturbations with variation length λ, it is useful to define a local temperature $T(\mathbf{r})$ and local chemical potential $\mu(\mathbf{r})$ over a region of size d with $d \ll \lambda$, although there may be an overall variation in T and μ over the whole sample, as illustrated in Fig. 12.13.

In the local region at $\mathbf{r}$, we may write

$$f = f^{\ell 0} + f^{(1)} \tag{12.67}$$

with

$$f^{\ell 0}(\mathbf{k}, \mathbf{r}) = \frac{1}{e^{\beta(\mathbf{r})[E(\mathbf{k}) - \mu(\mathbf{r})]} + 1} \tag{12.68}$$

and

$$\left(\frac{\partial f^{\ell 0}}{\partial t} \right)_{\text{coll}} = 0 \tag{12.69}$$

from Eqs. (12.51) and (12.66). Since $\left(\frac{\partial f^{\ell 0}}{\partial t} \right)_{\text{coll}}$ does not contribute to currents, we may then focus on $f^{(1)}$ in the analysis.

At steady state, the Boltzmann equation in terms of $f = f^{\ell 0} + f^{(1)}$ becomes

$$\left[\mathbf{v(k)} \cdot \frac{\partial}{\partial \mathbf{r}} + \frac{e}{\hbar}\left(\mathcal{E} + \frac{\mathbf{v(k)}}{c} \times \mathbf{B} \right) \cdot \frac{\partial}{\partial \mathbf{k}} \right] (f^0 + f^{(1)}) = \left(\frac{\partial f^{(1)}}{\partial t} \right)_{\text{coll}}. \tag{12.70}$$

The above equation may be rewritten, dropping terms not linear in fields or $f^{(1)}$, to obtain the following linearized Boltzmann equation (LBE):

$$\mathbf{v(k)} \cdot \frac{\partial f^0}{\partial \mathbf{r}} + \frac{e}{\hbar}\mathcal{E} \cdot \frac{\partial f^0}{\partial \mathbf{k}} = \left(\frac{\partial f^{(1)}}{\partial t} \right)_{\text{coll}} - \frac{e}{\hbar c}\left(\mathbf{v(k)} \times \mathbf{B} \right) \cdot \frac{\partial}{\partial \mathbf{k}} f^{(1)} - \mathbf{v(k)} \cdot \frac{\partial}{\partial \mathbf{r}} f^{(1)}. \tag{12.71}$$

In the above two expressions, f^0 is understood to be $f^{\ell 0}$ for notational simplicity. This LBE is valid for steady-state and low-field conditions. However, even this linearized equation is in general difficult to solve.

12.3.4 Relaxation time approximation

The collision term is usually the hardest term to determine in Eq. (12.71). Since $\left(\frac{\partial f}{\partial t} \right)_{\text{coll}}$ has the dimensions of $[f]/[\text{time}]$ and $\left(\frac{\partial f}{\partial t} \right)_{\text{coll}} \to 0$ as $f \to f^0$, the simplest approximation one can make would be to take $\left(\frac{\partial f}{\partial t} \right)_{\text{coll}}$ to be equal to $-f^{(1)}$ divided by some characteristic collision time $\tau_{\mathbf{k}}$. Also, for a sample with uniform density of scatterers, $\left(\frac{\partial f}{\partial t} \right)_{\text{coll}}$ should be independent of $\mathbf{r}$. That is,

$$\left(\frac{\partial f}{\partial t}(\mathbf{k}, \mathbf{r}) \right)_{\text{coll}} = -\frac{f(\mathbf{k}, \mathbf{r}) - f^0(\mathbf{k}, \mathbf{r})}{\tau_{\mathbf{k}}} = -\frac{f^{(1)}(\mathbf{k})}{\tau_{\mathbf{k}}} \tag{12.72}$$

for a homogeneous system. The minus sign takes into account the physics that f would go back to f^0 with the removal of the perturbation. $\tau_{\mathbf{k}}$ is called the relaxation time.

In the absence of any external field, if f is deviated from its equilibrium value (i.e. $f(t=0) = f^0 + A$), then

$$\frac{\partial f}{\partial t} = \left(\frac{\partial f}{\partial t} \right)_{\text{drift}} + \left(\frac{\partial f}{\partial t} \right)_{\text{coll}} = -\frac{f^{(1)}}{\tau_{\mathbf{k}}}, \tag{12.73}$$

since the drift term is zero, leading to a solution of

$$f^{(1)}(\mathbf{k}, t) = Ae^{-t/\tau_{\mathbf{k}}} \tag{12.74}$$

or

$$f(\mathbf{k}, t) = f^0(\mathbf{k}) + Ae^{-t/\tau_{\mathbf{k}}}. \tag{12.75}$$

The distribution function returns to its equilibrium value in the time scale of $\tau_{\mathbf{k}}$. Thus, τ gives the characteristic time for a disturbed system to relax back to equilibrium after the perturbation is removed.

Within the relaxation time approximation and the assumption of uniform scattering centers and external field, the LBE (Eq. (12.71)) may be further simplified to

$$\mathbf{v}(\mathbf{k}) \cdot \frac{\partial f^0}{\partial \mathbf{r}} + \frac{e}{\hbar} \boldsymbol{\mathcal{E}} \cdot \frac{\partial f^0}{\partial \mathbf{k}} = \left(\frac{\partial f^{(1)}}{\partial t} \right)_{\mathrm{coll}} - \frac{e}{\hbar c} \left(\mathbf{v}(\mathbf{k}) \times \mathbf{B} \right) \cdot \frac{\partial f^{(1)}}{\partial \mathbf{k}}. \tag{12.76}$$

(The last term on the RHS of Eq. (12.71) has been neglected with the assumption of uniformity, i.e., the spatial variation of $f^{(0)}$ is large compared to that of $f^{(1)}$.) Noting that f^0 is the local Fermi–Dirac distribution function, we have

$$\frac{\partial f^0}{\partial \mathbf{r}} = \frac{1}{\beta} \left(\frac{\partial f^0}{\partial E} \right) \frac{\partial}{\partial \mathbf{r}} (\beta(E - \mu)), \tag{12.77}$$

$$\mathbf{v}(\mathbf{k}) \cdot \frac{\partial f^0}{\partial \mathbf{r}} = \frac{1}{\beta} \left(\frac{\partial f^0}{\partial E} \right) \mathbf{v}(\mathbf{k}) \cdot \frac{\partial}{\partial \mathbf{r}} (\beta(E - \mu)), \tag{12.78}$$

$$\frac{e}{\hbar} \boldsymbol{\mathcal{E}} \cdot \frac{\partial f^0}{\partial \mathbf{k}} = \frac{e}{\hbar} \boldsymbol{\mathcal{E}} \cdot \frac{\partial f^0}{\partial E} \frac{\partial E}{\partial \mathbf{k}} = e\boldsymbol{\mathcal{E}} \cdot \mathbf{v}(\mathbf{k}) \left(\frac{\partial f^0}{\partial E} \right). \tag{12.79}$$

In this form, we may single out the most rapidly changing term on the LHS of Eq. (12.76), which is $\frac{\partial f^0}{\partial E}$. For the RHS of Eq. (12.76), we define

$$f^{(1)} \equiv - \left(\frac{\partial f^0}{\partial E} \right) \phi(\mathbf{k}, \mathbf{r}), \tag{12.80}$$

then

$$\left(\frac{\partial f^{(1)}}{\partial t} \right)_{\mathrm{coll}} = - \frac{f^{(1)}}{\tau_{\mathbf{k}}} = \left(\frac{\partial f^0}{\partial E} \right) \frac{\phi(\mathbf{k}, \mathbf{r})}{\tau_{\mathbf{k}}} \tag{12.81}$$

in the relaxation time model, and the second term on the RHS of Eq. (12.76) is

$$- \frac{e}{\hbar c} \left(\mathbf{v}(\mathbf{k}) \times \mathbf{B} \right) \cdot \frac{\partial f^{(1)}}{\partial \mathbf{k}} = \left(\frac{\partial f^0}{\partial E} \right) \frac{e}{\hbar c} \left(\mathbf{v}(\mathbf{k}) \times \mathbf{B} \right) \cdot \frac{\partial \phi}{\partial \mathbf{k}}. \tag{12.82}$$

Now all the terms in the LBE have the factor $-\frac{\partial f^0}{\partial E}$ (which is $\sim \delta(E - \mu)$ at low T), and we obtain, for the LBE in the relaxation time model,

$$T(\mathbf{r})\mathbf{v}(\mathbf{k}) \cdot \frac{\partial}{\partial \mathbf{r}} \left(\frac{E(\mathbf{k}) - \mu(\mathbf{r})}{T(\mathbf{r})} \right) + e\boldsymbol{\mathcal{E}} \cdot \mathbf{v}(\mathbf{k}) = \frac{e}{\hbar c} \left(\mathbf{v}(\mathbf{k}) \times \mathbf{B} \right) \cdot \frac{\partial}{\partial \mathbf{k}} \phi(\mathbf{k}, \mathbf{r}) + \frac{\phi(\mathbf{k}, \mathbf{r})}{\tau_{\mathbf{k}}}. \tag{12.83}$$

From the solution ϕ, we may obtain $f^{(1)}$ for the determination of transport properties of materials.

12.4 Electrical and thermal transport with the linearized Boltzmann equation

12.4.1 Isothermal electrical conductivity

Consider a sample under a uniform electric field $\mathcal{E}$ in thermal equilibrium such that the temperature T and chemical potential μ are independent of $\mathbf{r}$. If $\mathcal{E}$ is small, we may use the LBE to obtain f and $\mathbf{j}$, and hence the electrical conductivity tensor $\overleftrightarrow{\sigma}$. In the relaxation time model, since T and μ are constants and $\mathbf{B} = 0$, Eq. (12.83) yields

$$\phi(\mathbf{k}) = e\mathcal{E} \cdot \mathbf{v}(\mathbf{k})\tau_{\mathbf{k}}, \tag{12.84}$$

and

$$f^{(1)} = -\frac{\partial f^0}{\partial E}\phi(\mathbf{k}) = -\frac{\partial f^0}{\partial E}e\mathcal{E} \cdot \mathbf{v}(\mathbf{k})\tau_{\mathbf{k}}. \tag{12.85}$$

Now, from Eq. (12.62) and the fact that f^0 carries no current, we have

$$\mathbf{j} = \frac{2}{(2\pi)^3}e\int d\mathbf{k}\,\mathbf{v}(\mathbf{k})f^{(1)} = \frac{e}{4\pi^3}\int d\mathbf{k}\,\mathbf{v}(\mathbf{k})\left(-\frac{\partial f^0}{\partial E}\right)e\mathcal{E}\cdot\mathbf{v}(\mathbf{k})\tau_{\mathbf{k}}. \tag{12.86}$$

Using $\mathbf{j} = \overleftrightarrow{\sigma}\cdot\mathcal{E}$, we have

$$\overleftrightarrow{\sigma} = \frac{e^2}{4\pi^3}\int d\mathbf{k}\,\mathbf{v}(\mathbf{k})\mathbf{v}(\mathbf{k})\left(-\frac{\partial f^0}{\partial E}\right)\tau_{\mathbf{k}}. \tag{12.87}$$

This is the linearized Boltzmann equation result for the dc conductivity of metals within the relaxation time approximation. The quantum mechanics and material-specific parameters are contained in $\mathbf{v}(\mathbf{k})$ and $\tau_{\mathbf{k}}(T)$. We note that, in general, τ is $\mathbf{k}$ and temperature dependent. Since $-\frac{\partial f^0}{\partial E}$ is a very sharp function near the Fermi energy, the Fermi surface geometry and the properties of the electronic states near the Fermi surface determine $\overleftrightarrow{\sigma}$. It is a consequence of the Pauli exclusion principle. The conductivity is dependent on temperature T through $\tau_{\mathbf{k}}(T)$ and the factor $-\frac{\partial f^0}{\partial E}(T)$. There are, of course, many processes that contribute to $\tau_{\mathbf{k}}(T)$, including defect, electron–phonon, and electron–electron scatterings.

From Eq. (12.85), we may write

$$f(\mathbf{k}) = f^0(\mathbf{k}) + f^{(1)}(\mathbf{k}) = f^0(\mathbf{k}) - \frac{\partial f^0}{\partial E}e\mathcal{E}\cdot\mathbf{v}(\mathbf{k})\tau_{\mathbf{k}}, \tag{12.88}$$

and, using $\mathbf{v}(\mathbf{k}) = \frac{1}{\hbar}\frac{\partial E(\mathbf{k})}{\partial \mathbf{k}}$, we have

$$f(\mathbf{k}) = f^0(\mathbf{k}) - \frac{\partial f^0}{\partial \mathbf{k}}\cdot\frac{e\mathcal{E}}{\hbar}\tau_{\mathbf{k}}$$

$$\approx f^0\left(\mathbf{k} - \frac{e\tau_{\mathbf{k}}}{\hbar}\mathcal{E}\right). \tag{12.89}$$

If $\tau_{\mathbf{k}} = \tau$, i.e. $\tau_{\mathbf{k}}$ is independent of $\mathbf{k}$, then in Eq. (12.89) we have a distribution f as if the whole Fermi volume has been shifted by the amount $\frac{e\tau}{\hbar}\mathcal{E}$ in $\mathbf{k}$-space, a picture commonly presented in elementary discussion of transport.

The temperature dependence of the electrical resistivity of bulk metals due to phonons may be understood from Eq. (12.87) through $\tau_{\mathbf{k}}(T)$. Using Fermi's golden rule, with the electron–phonon interaction matrix elements discussed in Chapter 10, both the Bose factor

$$\langle n_{\lambda\mathbf{q}}(T)\rangle = \frac{1}{e^{\hbar\omega_\lambda(\mathbf{q})/k_{\mathrm{B}}T} - 1} \tag{12.90}$$

and the Fermi factor go into the determination of $\tau_{\mathbf{k}}(T)$. At high temperature $(T > T_{\mathrm{D}})$, the scattering rate $(\frac{1}{\tau})$ is proportional to the number of thermally excited phonons, which is given by

$$\langle n_{\lambda\mathbf{q}}(T)\rangle \sim \frac{k_{\mathrm{B}}T}{\hbar\omega_\lambda(\mathbf{q})}. \tag{12.91}$$

This simple argument together with Eq. (12.87) leads to the commonly observed result that, at high temperature, $\sigma(T) \sim 1/T$ and $\rho(T) \sim T$ for metals.

At very low temperature, it can be shown that, because of both the Bose and Fermi factors, only phonons with $\hbar\omega < k_{\mathrm{B}}T$ are effective in scattering electrons. This restriction leads to $\tau_{\mathbf{k}}(T) \sim T^{-5}$ and an electrical resistivity $\rho \sim T^5$, the so-called Bloch T^5 law. However, for most metals, scatterings at low T will be dominated by defects, impurities, etc., which give rise to a T-independent contribution to the resistivity. Therefore, at low T, the electrical resistivity or conductivity of normal metals saturates to a sample-dependent constant value.

12.4.2 Thermo-electric transport

We now consider the case of a sample in a uniform electric field $\mathcal{E}$ with two ends at different temperature T. There will be a position dependence on the local temperature $T(\mathbf{r})$ and the local chemical potential $\mu(\mathbf{r})$. Under these conditions, Eq. (12.83) becomes

$$T\mathbf{v}(\mathbf{k}) \cdot \frac{\partial}{\partial\mathbf{r}}\left(\frac{E(\mathbf{k}) - \mu}{T}\right) + e\mathcal{E}\cdot\mathbf{v}(\mathbf{k}) = \frac{e}{\hbar c}(\mathbf{v}(\mathbf{k})\times\mathbf{B})\cdot\frac{\partial}{\partial\mathbf{k}}\phi(\mathbf{k},\mathbf{r}) + \frac{\phi(\mathbf{k},\mathbf{r})}{\tau_{\mathbf{k}}}. \tag{12.92}$$

Since $\mathbf{B} = 0$, $\mathcal{E} \neq 0$, and $\nabla T \neq 0$, Eq. (12.92) reduces to

$$\phi(\mathbf{k},\mathbf{r}) = \tau_{\mathbf{k}}\mathbf{v}(\mathbf{k})\cdot\left[T\left(-\frac{E(\mathbf{k})}{T^2}\nabla T - \nabla\left(\frac{\mu}{T}\right)\right) + e\mathcal{E}\right]. \tag{12.93}$$

Using Eqs. (12.62) and (12.63), we have, for the current density,

$$\mathbf{j} = \frac{1}{4\pi^3}\int d\mathbf{k}\left(-\frac{\partial^0}{\partial E}\right)\phi(\mathbf{k})\mathbf{v}(\mathbf{k})e, \tag{12.94}$$

and for the energy flux,

$$\mathbf{Q} = \frac{1}{4\pi^3} \int d\mathbf{k} \left(-\frac{\partial f^0}{\partial E} \right) \phi(\mathbf{k}) \mathbf{v}(\mathbf{k}) E(\mathbf{k}). \tag{12.95}$$

To bring out explicitly the material parameters that will determine the transport properties, let us define the following quantities:

$$\overset{\leftrightarrow}{K}^{(n)} \equiv \frac{1}{4\pi^3} \int d\mathbf{k} \tau_\mathbf{k} \mathbf{v}(\mathbf{k}) \mathbf{v}(\mathbf{k}) E^{n-1}(\mathbf{k}) \left(-\frac{\partial f^0}{\partial E} \right). \tag{12.96}$$

The $\overset{\leftrightarrow}{K}^{(n)}$ are symmetric tensors for $\mathbf{B} = 0$ and are basically determined by the electronic structure and scattering properties of the electronic states near the Fermi surface of the material. They are related to another material parameter $\overset{\leftrightarrow}{R}(E)$ defined by

$$\overset{\leftrightarrow}{R}(E) \equiv \frac{1}{4\pi^3} \int d\mathbf{k} \tau_\mathbf{k} \mathbf{v}(\mathbf{k}) \mathbf{v}(\mathbf{k}) \delta \Big(E(\mathbf{k}) - E \Big) \tag{12.97}$$

through the relation

$$\overset{\leftrightarrow}{K} = \int \overset{\leftrightarrow}{R}(E) E^{n-1} \left(-\frac{\partial f^0}{\partial E} \right) dE. \tag{12.98}$$

In terms of the $\overset{\leftrightarrow}{K}$'s,

$$\mathbf{j} = e\overset{\leftrightarrow}{K}^{(1)} \left[e\boldsymbol{\mathcal{E}} - T\nabla\left(\frac{\mu}{T}\right) \right] + e\overset{\leftrightarrow}{K}^{(2)} \left[-\frac{1}{T}\nabla T \right] \tag{12.99}$$

and

$$\mathbf{Q} = \overset{\leftrightarrow}{K}^{(2)} \left[e\boldsymbol{\mathcal{E}} - T\nabla\left(\frac{\mu}{T}\right) \right] + \overset{\leftrightarrow}{K}^{(3)} \left[-\frac{1}{T}\nabla T \right]. \tag{12.100}$$

From Eqs. (12.99) and (12.100), we see that there is a close connection between $\mathbf{j}$ and $\mathbf{Q}$, and that ∇T gives rise to a current.

12.4.3 Heat current

The heat current density may be formulated through the thermodynamic relation that a change in heat is given by

$$Tds = dE - \mu dN. \tag{12.101}$$

The heat current density $\mathbf{u}$ is then given by

$$\mathbf{u} = \frac{2}{(2\pi)^3} \int f_\mathbf{k}^{(1)} \left[E(\mathbf{k}) - \mu \right] \mathbf{v}(\mathbf{k}) \, d\mathbf{k}. \tag{12.102}$$

If we rewrite Eq. (12.93) as

$$\phi(\mathbf{k}, \mathbf{r}) = \tau_{\mathbf{k}} \mathbf{v}(\mathbf{k}) \cdot \left\{ e\left(\mathcal{E} - \frac{1}{e}\nabla\mu\right) + \left(\frac{E(\mathbf{k}) - \mu}{T}\right)(-\nabla T) \right\} \tag{12.103}$$

and define another set of material parameters

$$\overset{\leftrightarrow}{L}^{(n)} \equiv \frac{1}{4\pi^3} \int d\mathbf{k} \tau_{\mathbf{k}} \mathbf{v}(\mathbf{k})\mathbf{v}(\mathbf{k})(E(\mathbf{k}) - \mu)^n \left(-\frac{\partial f^0}{\partial E}\right), \tag{12.104}$$

then we have

$$\mathbf{j} = e\overset{\leftrightarrow}{L}^{(0)} \cdot \left[e\left(\mathcal{E} - \frac{1}{e}\nabla\mu\right) \right] + e\overset{\leftrightarrow}{L}^{(1)} \left[-\frac{1}{T}\nabla T \right] \tag{12.105}$$

and

$$\mathbf{u} = \overset{\leftrightarrow}{L}^{(1)} \cdot \left[e\left(\mathcal{E} - \frac{1}{e}\nabla\mu\right) \right] + \overset{\leftrightarrow}{L}^{(2)} \left[-\frac{1}{T}.\nabla T \right]. \tag{12.106}$$

We note here that one often drops the term $\nabla\mu$ in the above two equations and replaces $\mathcal{E}$ with an "observed" electric field $\mathcal{E}'$, which contains the added term from the gradient in the chemical potential μ caused by the temperature gradient.

12.4.4 Thermal conductivity

The thermal conductivity of a material due to the electrons is not given by $\overset{\leftrightarrow}{L}^{(2)}/T$, but through the definition of relating how much heat flows through the system under a temperature difference, with the boundary condition that there is no electrical current flow. That is, the thermal conductivity tensor $\overset{\leftrightarrow}{K}_{th}$ is defined using

$$\mathbf{u} = -\overset{\leftrightarrow}{K}_{th}\nabla T, \tag{12.107}$$

with $\mathbf{j} = 0$.

Eq. (12.105) shows that, with $\nabla T \neq 0$, having $\mathbf{j} = 0$ implies that an electric field is produced even if there is no applied electric field. Defining $\mathcal{E}' \equiv \mathcal{E} - \frac{1}{e}\nabla\mu, \mathbf{j} = 0$ requires that

$$0 = e\overset{\leftrightarrow}{L}^{(0)} \cdot e\mathcal{E}' + e\overset{\leftrightarrow}{L}^{(1)} \cdot \left(-\frac{1}{T}\nabla T\right). \tag{12.108}$$

Substituting $\mathcal{E}'$ from Eq. (12.108) into Eq. (12.106), we have

$$\overset{\leftrightarrow}{K}_{th} = \frac{1}{T}[\overset{\leftrightarrow}{L}^{(2)} - \overset{\leftrightarrow}{L}^{(1)}(\overset{\leftrightarrow}{L}^{(0)})^{-1}\overset{\leftrightarrow}{L}^{(1)}] \tag{12.109}$$

as the thermal conductivity tensor.

From the definition $\overset{\leftrightarrow}{R}(E)$ (see Eq. (12.97)), we also have the relation

$$\overset{\leftrightarrow}{L}^{(n)} = \int \overset{\leftrightarrow}{R}(E)(E - \mu)^n \left(-\frac{\partial f^0}{\partial E} \right) dE. \tag{12.110}$$

Since at low T $(k_B T \ll E_F)$, $-\frac{\partial f^0}{\partial E}$ is a sharp, symmetric function around μ, we have

$$\overset{\leftrightarrow}{L}^{(0)} = \overset{\leftrightarrow}{R}(\mu), \tag{12.111}$$

$$\overset{\leftrightarrow}{L}^{(1)} = \frac{1}{3}\pi^2 (k_B T)^2 \left[\frac{\partial}{\partial E} L^{(0)}(E) \right]_{E=\mu}, \tag{12.112}$$

and

$$\overset{\leftrightarrow}{L}^{(2)} = \frac{1}{3}\pi^2 (k_B T)^2 \overset{\leftrightarrow}{R}(\mu). \tag{12.113}$$

Thus, at low temperature, from Eq. (12.109),

$$\overset{\leftrightarrow}{K}_{th} = \frac{1}{T}\frac{\pi^2}{3}(k_B T)^2 \overset{\leftrightarrow}{R}(\mu). \tag{12.114}$$

Comparing the expression for $\overset{\leftrightarrow}{\sigma}$ in Eq. (12.87) with the definition of $\overset{\leftrightarrow}{R}(\mu)$ in Eq. (12.97), the electrical conductivity tensor may also be expressed as

$$\overset{\leftrightarrow}{\sigma} = e\mathbf{R}(\mu). \tag{12.115}$$

Equations (12.114) and (12.115) lead to the relation

$$\overset{\leftrightarrow}{K}_{th} = \frac{\pi^2}{3e^2} k_B^2 T \overset{\leftrightarrow}{\sigma}. \tag{12.116}$$

The ratio

$$\frac{K_{th}}{T\sigma} = \frac{\pi^2}{3}\left(\frac{k_B}{e} \right)^2 \tag{12.117}$$

is called the Lorentz ratio and is equal to 2.72×10^{-13} esu/deg^2. This is known as the Wiedemann–Franz law and was discovered in 1853.[7] In comparison with experiment, the difficulty is in subtracting out the lattice contribution to the thermal conductivity.

12.4.5 Examples of thermo-electric effect

The discussion above clearly shows that there are strong interaction effects between electrical current and thermal current, due to the electrons. To make it clearer, Eqs. (12.105)

---

[7] R. Franz and G. Wiedemann, "Ueber die Wärme-Leitungsfähigkeit der Metalle," *Ann. Physik* 165(1853), 497.

and (12.106) may be rewritten in the form

$$\mathbf{j} = e^2 \overset{\leftrightarrow(0)}{L} \cdot \boldsymbol{\mathcal{E}}' + \frac{e}{T} \overset{\leftrightarrow(1)}{L} (-\nabla T), \tag{12.118}$$

$$\mathbf{u} = e \overset{\leftrightarrow(1)}{L} \cdot \boldsymbol{\mathcal{E}}' + \frac{1}{T} \overset{\leftrightarrow(2)}{L} (-\nabla T). \tag{12.119}$$

Several well-known effects with important applications result from the interplay between Eqs. (12.118) and (12.119).

A sample with different temperatures at each end in an open circuit, in general, will generate a voltage across it. This is seen by setting $\mathbf{j} = 0$ in Eq. (12.118), resulting in

$$\boldsymbol{\mathcal{E}}' = \frac{1}{eT}\left[\left(\overset{\leftrightarrow(0)}{L}\right)^{-1}\overset{\leftrightarrow(1)}{L}\right]\nabla T = Q\nabla T. \tag{12.120}$$

Q is called the thermo-electric power. For $k_{\mathrm{B}}T \ll E_{\mathrm{F}}$, Eqs. (12.111), (12.112), and (12.115) show that

$$Q \sim \frac{T}{e\sigma} \frac{\partial \sigma(E)}{\partial E}\bigg|_{\mu}. \tag{12.121}$$

The so-called Seebeck effect is observed when two different materials (A and B) of different Q are arranged in an open circuit, as shown in Fig. 12.14(a). A voltage is generated when junctions 1 and 2 are at different temperatures. The electromotive force around the circuit is

$$V = \int_0^1 \mathcal{E}_{\mathrm{B}}' dx + \int_1^2 \mathcal{E}_{\mathrm{A}}' dx + \int_2^0 \mathcal{E}_{\mathrm{B}}' dx = (Q_{\mathrm{B}} - Q_{\mathrm{A}})(T_1 - T_2). \tag{12.122}$$

The voltage V is a function of the difference in temperature of the two junctions and of the difference in the thermo-electric power Q.

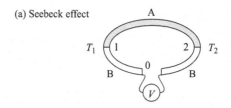

(a) Seebeck effect

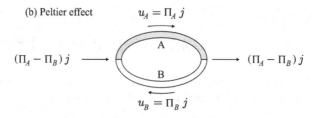

(b) Peltier effect

The Peltier effect occurs if a closed-circuit geometry is used, but with no initial temperature difference and with a driven current density j, as illustrated in Fig. 12.14(b). Under the initial conditions of $\nabla T = 0$ and $j \neq 0$, Eqs. (12.118) and (12.119) lead to

$$u = eL^{(1)}\mathcal{E}' \tag{12.123}$$

and

$$j = e^2 L^{(0)}\mathcal{E}'. \tag{12.124}$$

For a fixed and finite j, the heat current in each material is given by

$$u = \frac{1}{e}\left[L^{(0)}\right]^{-1} L^{(1)} j = \Pi j, \tag{12.125}$$

where Π is called the Peltier coefficient. The heat current is proportional to the electric current, but is also dependent on the Peltier coefficient $\Pi = \frac{1}{e}[L^{(0)}]^{-1}L^{(1)}$, which is material specific. Since j is the same in both segments, for different materials A and B, the heat current is different in the two branches. This leads to a warming of one junction and a cooling of the other. For example, the Peltier effect is used in solid-state refrigeration (see Fig. 12.14(b)).

Part III Problems

III.1. **Light absorption in InSb**. For InSb ($m_c^* = 0.014m_0$, $m_v^* = 0.4m_0$, and $E_{\text{gap}} = 0.18\text{eV}$), how much energy from a 0.50 eV photon is converted into free electron kinetic energy, and how much is converted into free hole energy? If the photon momentum is included in the equation for conservation of momentum, then by what percentage do the electron and hole momenta differ?

III.2. **Exciton**. Use the effective mass theory for the following.

    (a) Calculate the binding energy for the lowest-energy exciton at the absorption edge in GaAs. (Please look up the band structure parameters for GaAs in any standard semiconductors handbook.)

    (b) Can an exciton exist at an indirect optical absorption edge? (Give reasons and discussion.)

III.3. **Normal incidence**. Derive an expression for the reflection coefficient for normal incidence as a function of the real and imaginary parts of (a) the dielectric function ϵ and (b) n and k.

III.4. **Transverse displacement vector**. If we use the Coulomb gauge, $\nabla \cdot \boldsymbol{A} = 0$, and ignore all the external (free) charge distribution ρ^{ext}, free charge current $\boldsymbol{J}^{\text{ext}}$, and magnetization current $\nabla \times \boldsymbol{M}$ in the system, we can prove that the displacement vector $\boldsymbol{D}_G(\boldsymbol{q}; \omega)$ only has transverse components. We write down all the equations (including the Maxwell equations, the Coulomb gauge, and the external charge conditions) as below,

$$\rho^{\text{ext}} = 0, \tag{III.1}$$

$$\boldsymbol{J}^{\text{ext}} = \boldsymbol{0}, \tag{III.2}$$

$$\boldsymbol{J}^{\text{ind}} = \frac{\partial \boldsymbol{P}}{\partial t}, \tag{III.3}$$

$$\rho^{\text{ind}} = -\nabla \cdot \boldsymbol{P}, \tag{III.4}$$

$$\boldsymbol{E} = -\nabla\phi - \frac{1}{c}\frac{\partial \boldsymbol{A}}{\partial t}, \tag{III.5}$$

$$\nabla^2\phi = -4\pi\rho, \tag{III.6}$$

$$\nabla^2\boldsymbol{A} = -\frac{4\pi}{c}\boldsymbol{J} + \frac{1}{c^2}\frac{\partial \boldsymbol{A}}{\partial t} + \frac{1}{c}\frac{\partial}{\partial t}\nabla\phi, \tag{III.7}$$

$$\nabla \cdot \boldsymbol{A} = 0. \tag{III.8}$$

Take the Fourier transform of the displacement vector $D(x, t)$,

$$D(x; \omega) = \int dt\, e^{i\omega t} D(x, t), \tag{III.9}$$

$$D_G(q; \omega) = \int dx\, e^{-i(q+G)\cdot x} D(x; \omega). \tag{III.10}$$

Define the dielectric tensor as

$$D_G(q, \omega) = \epsilon_0 E_G(q, \omega) + P_G(q, \omega) = \sum_{G'} \epsilon_{GG'} E_{G'}(q, \omega), \tag{III.11}$$

and the displacement vector can be split into a transverse part and a longitudinal part:

$$D_G \doteq D_G^L + D_G^T = \left[\frac{q+G}{|q+G|} \cdot D_G\right] \frac{q+G}{|q+G|} + \left[D_G - D_G^L\right], \tag{III.12}$$

where we have omitted the (q, ω) index, which is the same for all the quantities. Try your best to prove that $D_G^L = 0$, and therefore $D_G = D_G^T$. This is why in the literature we often encounter the transverse dielectric function.

III.5. **Optical properties of simple metals**.

(a) Prove that energy conservation and momentum conservation make it impossible for a free electron to absorb a photon: (Hint: if you use the non-relativistic dispersion relation $\varepsilon = p^2/2m$ for the electronic energy, you will find that such absorption is possible, but only at an electronic energy so high [on the scale of mc^2] that the non-relativistic approximation is not valid. It is therefore necessary to use the relativistic dispersion relation, $\varepsilon = (p^2c^2 + m^2c^4)^{1/2}$, to prove that the absorption is impossible at any electronic energy.)

(b) Using the results from the Drude model, and the expression for the reflectivity being

$$R(\omega) = \frac{(1-n)^2 + k^2}{(1+n)^2 + k^2}, \tag{III.13}$$

examine the behavior of the reflectivity of a simple metal when $\omega\tau \gg 1$. Show that the reflectivity is near unity below the plasma frequency, and that

$$R(\omega) = \left(\frac{\omega_p^2}{4\omega^2}\right)^2 \tag{III.14}$$

when $\omega \gg \omega_p$.

III.6. **Indirect optical transition**. Using the virtual band model shown in Fig. III.1, calculate for an *indirect* optical transition (phonon-assisted processes) the dependence of $\varepsilon_2(\omega)$ on the temperature T and on the energy difference of the photon energy $\hbar\omega$ from the indirect bandgap energy E_g (that is, fill in the steps missing in the text discussion). You may assume that each band has a parabolic dispersion near its extremum.

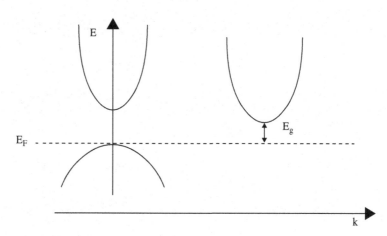

Figure III.1   Band structure with an indirect bandgap.

III.7. **Hydrogenic excitons with anisotropic perturbation**. Consider a direct gap semi-conductor with anisotropic effective masses such that $m_\perp^* \neq m_\parallel^*$ for both electrons and holes. Determine the energy levels for the $1s$, $2s$, $2p_0$, and $2p_{\pm 1}$ states of a hydrogenic exciton if the anisotropy is small.

III.8. **Two-phonon scattering**. Recall that the basic laws governing neutron scattering from phonons could be deduced from conservation of energy and crystal momentum, without detailed quantum mechanical calculation. Generalize this argument to include processes in which neutrons incident upon a crystal participate in the creation or destruction of two phonons. Assume a three-dimensional crystal.

(a) Enumerate the number of distinct ways in which such a process may occur.

(b) Show that two-phonon processes do not produce sharp peaks in the number of scattered neutrons as a function of neutron energy change during the scattering process for fixed observation angle, and that they may therefore be distinguished from one-phonon processes.

III.9. **Polarons in a covalent semiconductor**.

(a) Compute the self energy of an electron due to electron–phonon interaction for an electron at the bottom of the conduction band (with $m^* = 0.5 m_e$) in a co-valent semiconductor. Assume only interaction with the longitudinal acoustic phonons, and use the Debye approximation to the phonon dispersion curve. Use as the zero-order state $|0\rangle = c_k^\dagger |vac\rangle$ with $k = 0$. Compute only to lowest order in $m v_s / \hbar q_D$ or $m v_s^2 / \hbar \omega_D$, where v_s is the sound velocity and q_D the De-bye wavevector. Evaluate the shift in energy of the electron using a deformation potential of 10 eV, an atomic mass $M = 5 \times 10^4 m_e$, and $\omega_D = 0.02$eV.

(b) What is the expectation value of the total number of phonons $\langle N \rangle$ in the corresponding first-order state?

(c) Suppose now that the effective mass is $m^* = 100 m_e$. What is $\langle N \rangle$? Is the above theory still reasonable?

III.10. **Phonon population.** Show explicitly that the average number of phonons in a polaron is given by

$$\sum_q \frac{|M_q|^2}{\left[\hbar\omega + \frac{\hbar^2 q^2}{2m} - \frac{\hbar^2 k \cdot q}{m}\right]^2}. \tag{III.15}$$

III.11. **Numerical study of a polaron.** For a polaron with Fröhlich coupling constant $\alpha = 0.5$ in an ionic crystal with conduction band mass $m_c = 0.6m$ and longitudinal optical phonon frequency $\omega = 1.6 \times 10^{13}$ sec^{-1}, give the numerical (and sign) values for:
(a) the self-energy shift in eV;
(b) the effective mass m^* in terms of the free electron mass m;
(c) the number of phonons in the polaron.

III.12. **Free electron gas.** For a free electron gas, calculate
(a) the paramagnetic spin susceptibility;
(b) the thermal conductivity;
(c) the electrical conductivity;
(d) the Wiedemann–Franz ratio.

III.13. **Electron conductance in a semiconductor.** Consider a semiconductor at $T = 0$ with N donors per unit volume, each with an electron bound to it (and no electrons in the conduction band). The low-frequency optical conductivity is nonzero due to the possibility of excitation of the bound electrons into the conduction band. Compute the corresponding $\text{Re}[\sigma_{ij}(\omega)]$. (Hint: appropriately modify the formula derived in the text [to include localized states] for the imaginary part of the dielectric function.)

The bound wavefunctions may be taken as $\psi_d(r) = Ae^{-\alpha r}$ (i.e. 1s functions); the conduction band wavefunctions may be taken as planewaves orthogonalized to the donor states (which do not overlap). The donor binding energy is $\hbar^2\alpha^2/2m$. The form of the result is simpler if written in terms of ω, α, and the magnitude of the wavenumber of the final state k. Thus, $\hbar^2 k^2/2m + \hbar^2\alpha^2/2m = \hbar\omega$. Sketch the result for $\sigma(\omega)$.

III.14. **Electrons in a magnetic field.** Consider an electron on the Fermi surface of Na metal with initial motion in the x–y plane.
(a) What is the orbital radius in an applied magnetic field of 10^4 G?
(b) How is the area of the orbit in real space related to the k-space orbit?
(c) If we define a vector operator $\hat{k} \doteq \frac{1}{\hbar}\left(\hat{p} - \frac{e}{c}A(\hat{r})\right)$, show that

$$\hat{k} \times \hat{k} = i\frac{eB}{\hbar c}. \tag{III.16}$$

III.15. **Magnetic properties of materials in classical physics.** Consider classical electrons in a magnetic field using statistical mechanics.
(a) Write down the classical partition function Z for N particles. The particles interact with each other according to an arbitrary potential U that depends only on their positions.
(b) Show how a spatially uniform static magnetic field is to be included in the partition function.

(c) Show finally that with a change of integration variables, the magnetic field can be eliminated from the partition function and therefore cannot affect any thermodynamic variable.

III.16. **Magnetoresistivity**. Consider the magnetoresistance of the two-carrier-type problem for all values of the magnetic field, but assuming $e_1 = -e_2$, $m_1 = m_2$, and $\tau_1 = \tau_2$. Show that, for the standard geometry (in the notation defined in the text),

$$j_x = (n_1 + n_2)e\mu \frac{1}{1 + \xi^2}\left[1 + \frac{(n_1 - n_2)^2}{(n_1 + n_2)^2}\xi^2\right]E_x. \tag{III.17}$$

III.17. **Landau levels**. Solve exactly the Schrödinger equation for a free electron in a uniform magnetic field along the z-direction. Solve it first in Cartesian coordinates with the gauge

$$A_x = 0,$$
$$A_y = Bx, \tag{III.18}$$
$$A_z = 0,$$

and then in cylindrical coordinates with the gauge

$$A_x = -\frac{1}{2}By,$$
$$A_y = \frac{1}{2}Bx, \tag{III.19}$$
$$A_z = 0.$$

Show that the energy states are discrete for each fixed value of p_z, and that energy separation is in accord with the cyclotron formula. Comment on the apparent difference in the wavefunctions of the two gauges used.

III.18. **Bohr–Sommerfeld quantization**. Use the Bohr–Sommerfeld quantization formula to show that the oscillations of M in the de Haas–van Alphen effect have a $1/B$ period that gives the maximal or minimal cross-sectional area of the Fermi surface perpendicular to B.

III.19. **Cyclotron orbit**. Consider the problem of an extremal cyclotron orbit in the k_x–k_y plane of the Fermi surface for a non-isotropic solid. The equation for the Fermi energy is $E = (Ak_x^2 + Bk_y^2)$. Apply a magnetic field along z and describe the rate at which k changes during an orbit when A and B are unequal. Describe the orbit performed in real space for this situation.

III.20. **Conduction band electrons in a magnetic field**. Show that the appropriate Hamiltonian for a Ge conduction band electron (which has an ellipsoidal energy surface in k-space) in a magnetic field is

$$H = \frac{p_y^2}{2m_t} + \frac{1}{2}\left(\frac{eB}{c}\right)^2\left[\frac{\sin^2\theta}{m_l} + \frac{\cos^2\theta}{m_t}\right]y'^2 + A, \tag{III.20}$$

where θ, m_t, and m_l are related to the direction of the magnetic field and the effective mass tensor of the the conduction electron, respectively. Give the explicit expressions for y' and A.

III.21. **Sommerfeld expansion.** In the text, we present the expressions of $\overleftrightarrow{L}^n$, $n = 0, 1, 2$ in the free electron case. Now let us use the Sommerfeld expansion to derive these expressions. First consider a generic integral of the form

$$I = \int_0^\infty d\epsilon f(\epsilon) g(\epsilon), \tag{III.21}$$

where $f(\epsilon) = 1/(e^{\beta(\epsilon - \mu)} + 1)$ is the Fermi–Dirac distribution function. Many ensemble-averaged physical properties can be reduced to this form. $g(\epsilon)$ is supposed to be a smooth function and always finite, which is presumed for physical quantities, such as energy, velocity, etc. Integrate by parts

$$I = \int_0^\infty d\epsilon \frac{df(\epsilon)}{d\epsilon} G(\epsilon), \tag{III.22}$$

where $G(\epsilon)$ is defined as

$$G(\epsilon) = -\int_0^\epsilon dx g(x). \tag{III.23}$$

We know that $df(\epsilon)/d\epsilon$ is strongly peaked at $\epsilon = \mu$ and vanishes elsewhere. Expand $G(\epsilon)$ near $\epsilon = \mu$:

$$G(\epsilon) = G(\mu) + \left.\frac{\partial G}{\partial \epsilon}\right|_{\epsilon = \mu} (\epsilon - \mu) + \frac{1}{2}\left.\frac{\partial^2 G}{\partial \epsilon^2}\right|_{\epsilon = \mu} (\epsilon - \mu)^2. \tag{III.24}$$

Define a series of integrals

$$I_n = -\int_0^\infty d\epsilon (\epsilon - \mu)^n \frac{df(x)}{dx}. \tag{III.25}$$

(a) Evaluate I_0.

(b) Evaluate I_1 and I_2. (Hint: you can change the lower bound of the integral to $-\infty$.)

III.22. **Mott formula for thermopower.** Derive (using the Boltzmann equation formalism within the relaxation time approximation, or by other means if you prefer) the expression for the thermopower (denoted as Q in the text; sometimes it is called the Seebeck coefficient S) of a system, showing that it is in the form

$$S = \frac{\pi^2}{3} \frac{k_B^2 T}{e} \frac{\sigma'}{\sigma}, \tag{III.26}$$

where σ is the electrical conductivity and σ' is its energy derivative.

PART IV

MANY-BODY EFFECTS, SUPERCONDUCTIVITY, MAGNETISM, AND LOWER-DIMENSIONAL SYSTEMS

Using many-body techniques

In Chapter 6, the many-electron problem was discussed and electron–electron interactions were considered using the standard approaches of many-body quantum theory. The quasi-particle and collective excitations of the interacting system are expected to emerge as elementary excitations of the interacting system, and their properties should dominate the spectra associated with various response functions. Since the elementary excitations are emergent properties, which can be viewed in terms of excitations or particles that can be created or destroyed, it is convenient to use techniques associated with the fields of quantum electrodynamics (QED) and statistical physics. The essential point is that in QED, because relativistic effects are included, particles can be created and destroyed, and the formalism accounts for the changes associated with these events.

For condensed matter systems, we are not usually probing the systems in energy ranges where relativistic effects are important. However, the techniques of QED are useful if they are applied to cases where elementary excitations are created or destroyed. In this chapter, we focus on recipes for calculations of this kind and refer the reader to books in this area that provide the proofs and more complete discussions.

13.1 General formalism

The first step in setting up the formalism for a many-electron- or many-phonon-type study is to rely on second quantization techniques. The approach, which is described in many texts on quantum mechanics, consists in replacing operators in the usual quantum theory, such as the kinetic energy, that act on the coordinate variables in a wavefunction by other operators that operate on a wavefunction describing the number of particles in a given state. These new operators change the number of particles in specific states, and the result is the same as would be obtained when using ordinary quantum theory techniques. So second quantization is a convenient method, which although in principle is not required for the study of solids, allows efficient calculations and provides new insights. Analogies like the one comparing electron–hole excitations by photons with Dirac's electron–positron theory, and viewing a collection of phonons as a Bose system, are very useful. Because particles can be created and destroyed at will, the grand canonical ensemble approach of statistical physics is employed. Another very convenient aspect of second quantization is that the task of symmetrizing wavefunctions for many-body systems is achieved through commutation relations. The operators we need to deal with most are the kinetic energy T, the one-body potential V, and the two-body potential U. For N electrons as an example, these are given by

$$T = \sum_{i=1}^{N} \frac{\mathbf{p}_i^2}{2m} = \sum_{i=1}^{N} t_i, \qquad (13.1)$$

$$V = \sum_{i=1}^{N} v(\mathbf{r}_i), \qquad (13.2)$$

and

$$U = \frac{1}{2} \sum_{ij}' u(\mathbf{r}_i, \mathbf{r}_j), \qquad (13.3)$$

where $\mathbf{p}_i$ and $\mathbf{r}_i$ are the momenta and coordinates of individual electrons, v and u can be viewed as the electron–ion and electron–electron Coulomb potentials, respectively, and the prime on the sum for u signifies $i \neq j$.

The form of the many-body wavefunction we use was introduced in Chapter 2:

$$|\psi(n)\rangle = |n_1, n_2, n_3, \ldots\rangle, \qquad (13.4)$$

where the n_s variable measures the number of particles in the single-particle state s. The wavefunction is assumed to be normalized, and the operators acting on it create and destroy particles in a specific state. (Here we will sometimes use the term "state" for a single-particle state or single-particle orbital.) For bosons, the creation and destruction operators for state s are $a_s^\dagger$ and a_s. For fermions, they are $c_s^\dagger$ and c_s. The application of these operators is illustrated in Eq. (13.5), using bosons as an example:

$$a_s \psi(n_1, n_2, \ldots, n_s \ldots) = \sqrt{n_s}\ \psi(n_1, n_2, \ldots, n_s - 1, \ldots),$$
$$a_s^\dagger \psi(n_1, n_2, \ldots, n_s \ldots) = \sqrt{n_s + 1}\ \psi(n_1, n_2, \ldots, n_s + 1, \ldots). \qquad (13.5)$$

The number operator for the sth orbital is $\hat{n}_s = a_s^\dagger a_s$, since subsequent operation of a_s followed by $a_s^\dagger$ yields

$$a_s^\dagger a_s \psi(n_1, n_2, \ldots, n_s, \ldots) = n_s \psi(n_1, n_2, \ldots, n_s, \ldots), \qquad (13.6)$$

which gives the number of particles in the s-state as n_s. The symmetry constraints on the full many-body state for bosons are guaranteed by using the commutation relations

$$[a_s, a_{s'}]_- = 0,$$
$$[a_s^\dagger, a_{s'}^\dagger]_- = 0, \qquad (13.7)$$
$$[a_s, a_{s'}^\dagger]_- = \delta_{ss'}.$$

The operators given in Eqs. (13.1), (13.2), and (13.3) can now be expressed in terms of a_s and $a_s^\dagger$ in the following way. For the one-body Hamiltonian $H_1 = T + V =$

$\sum_{i=1}^{N} h(\mathbf{r}_i, \mathbf{k}_i, \sigma_i)$ where σ is a spin coordinate, we have

$$H_1|\psi(n)\rangle = \sum_{ss'} \langle s'|h|s\rangle a_{s'}^{\dagger} a_s |\psi(n)\rangle, \qquad (13.8)$$

with

$$\langle s'|h|s\rangle = \int \phi_{s'}^* h \phi_s d\mathbf{r}, \qquad (13.9)$$

where ϕ is a single-particle orbital and a sum over spins is implicit. The one-particle wavefunctions ϕ_s are assumed to be a complete orthonormal set

$$\langle \phi_s|\phi_{s'}\rangle = \int \phi_s^* \phi_{s'} d\mathbf{r} = \delta_{ss'}, \qquad (13.10)$$

where a spin sum is assumed and the s and s' labels can be replaced by wavevectors such as $\mathbf{k}$ and $\mathbf{k}'$ to label states. The two-body operator u given in Eq. (13.3) can be represented by

$$U|\psi(n)\rangle = \frac{1}{2} \sum_{\substack{jk \\ \ell m}} \langle jk|u|\ell m\rangle a_j^{\dagger} a_k^{\dagger} a_m a_\ell |\psi(n)\rangle, \qquad (13.11)$$

where we denote

$$\langle jk|u|\ell m\rangle = \int \phi_j^*(\mathbf{r}_1)\phi_k^*(\mathbf{r}_2)u(\mathbf{r}_1,\mathbf{r}_2)\phi_\ell(\mathbf{r}_1)\phi_m(\mathbf{r}_2)d\mathbf{r}_1 d\mathbf{r}_2. \qquad (13.12)$$

We now introduce a more compact formulation involving wave field or field operators that are related to the amplitudes for the creation and destruction of particles in space. These are defined as

$$\psi(\mathbf{r}) = \sum_s \phi_s(\mathbf{r})a_s \qquad (13.13)$$

and

$$\psi^{\dagger}(\mathbf{r}) = \sum_s \phi_s^*(\mathbf{r})a_s^{\dagger}.$$

They satisfy the commutation relations

$$\left[\psi(\mathbf{r}), \psi(\mathbf{r}')\right]_- = \left[\psi^{\dagger}(\mathbf{r}), \psi^{\dagger}(\mathbf{r}')\right]_- = 0 \qquad (13.14)$$

and

$$\left[\psi(\mathbf{r}), \psi^{\dagger}(\mathbf{r}')\right]_- = \delta(\mathbf{r} - \mathbf{r}').$$

For a one-particle operator like the kinetic energy

$$T|\psi(n)\rangle = \left(\int \psi^\dagger(\mathbf{r})\, t\psi(\mathbf{r})\, d\mathbf{r}\right)|\psi(n)\rangle, \qquad (13.15)$$

the density operator is given by

$$\rho(\mathbf{r}) = \psi^\dagger(\mathbf{r})\psi(\mathbf{r}) \qquad (13.16)$$

and the number operator of particles is

$$N = \int \rho(\mathbf{r})\, d\mathbf{r} = \sum_{ss'} a_{s'}^\dagger a_s \int \phi_{s'}^* \phi_s\, d\mathbf{r} = \sum_s a_s^\dagger a_s. \qquad (13.17)$$

It is often convenient to deal with the Fourier transform of the density operator $\rho(\mathbf{r})$. This is given by

$$\rho_\mathbf{q} = \int e^{i\mathbf{q}\cdot\mathbf{r}}\rho(\mathbf{r})\, d\mathbf{r} = \sum_{ss'} a_{s'}^\dagger a_s \int e^{i\mathbf{q}\cdot\mathbf{r}}\phi_{s'}^*\phi_s\, d\mathbf{r}. \qquad (13.18)$$

For the case of planewaves for the single-particle states,

$$\phi_\mathbf{k}(\mathbf{r}) = \frac{1}{\sqrt{\Omega}}e^{i\mathbf{kr}}. \qquad (13.19)$$

Since

$$\frac{1}{\Omega}\int e^{i\mathbf{q}\cdot\mathbf{r}}e^{-i\mathbf{k}'\cdot\mathbf{r}}e^{i\mathbf{k}\cdot\mathbf{r}}\, d\mathbf{r} = \delta_{\mathbf{k}',\mathbf{k}+\mathbf{q}}, \qquad (13.20)$$

then

$$\rho_\mathbf{q} = \sum_\mathbf{k} a_{\mathbf{k}+\mathbf{q}}^\dagger a_\mathbf{k}, \qquad (13.21)$$

where Ω is the crystal volume.

Hence, for bosons, the total Hamiltonian becomes

$$H = \int \psi^\dagger(\mathbf{r})\langle\mathbf{r}|h|\mathbf{r}\rangle\psi(\mathbf{r})\, d\mathbf{r} + \frac{1}{2}\int \psi^\dagger(\mathbf{r}_1)\psi^\dagger(\mathbf{r}_2)u(\mathbf{r}_1,\mathbf{r}_2)\psi(\mathbf{r}_1)\psi(\mathbf{r}_2)\, d\mathbf{r}_1 d\mathbf{r}_2. \qquad (13.22)$$

The above description in terms of field operators can be extended to describe Fermion systems where $n_s = 0$ or $n_s = 1$. This is done using "+" anticommutation relations appropriate for the fermions

$$[c_s, c_{s'}]_+ = [c_s^\dagger, c_{s'}^\dagger]_+ = 0,$$
$$[c_s, c_{s'}^\dagger]_+ = \delta_{ss'}, \qquad (13.23)$$

$$\left[\psi(\mathbf{r}), \psi(\mathbf{r}')\right]_+ = \left[\psi^\dagger(\mathbf{r}), \psi^\dagger(\mathbf{r}')\right]_+ = 0,$$
$$\left[\psi(\mathbf{r}), \psi^\dagger(\mathbf{r}')\right]_+ = \delta(\mathbf{r} - \mathbf{r}').$$

(13.24)

Everything follows as before, but now we include the spin labels σ and σ' to write the total Hamiltonian for fermions:

$$H = \sum_{\mathbf{k}\mathbf{k}'\sigma\sigma'} \langle \mathbf{k}'\sigma' | h | \mathbf{k}\sigma \rangle \, c^\dagger_{\mathbf{k}'\sigma'} c_{\mathbf{k}\sigma} + \frac{1}{2} \sum_{jk\ell m \sigma\sigma'} \langle jk | u_{\sigma\sigma'}(\mathbf{r}_1, \mathbf{r}_2) | \ell m \rangle \, c^\dagger_{j\sigma} c^\dagger_{k\sigma'} c_{m\sigma'} c_{\ell\sigma}.$$

(13.25)

There is a choice of quantum mechanical representations for expressing and using the wave field operators. The choice taken here is the Heisenberg representation, where the many-body wavefunctions are independent of time and the operators are time dependent. The transformation from the Schrödinger to the Heisenberg picture can be made by applying a unitary transformation to the Schrödinger wavefunction to make it independent of time. Another useful formalism is the interaction representation, where the perturbing part of a Hamiltonian determines the time evolution of the wavefunction, while the time dependence of the operators is governed by the unperturbed, usually solvable, part of the Hamiltonian. However, as stated above, we will restrict ourselves to the Heisenberg picture.

13.2 Interacting Green's functions

We next introduce Green's functions, which are also called propagators. The underlying motivation for this approach goes back to the discussions presented in Chapter 1. Our goal is to perturb a system, let it evolve, and study its responses. We do not want, and would have difficulty using, all the coordinates and momenta of all the particles in a solid after it has been probed. The most studied Green's function is the so-called one-particle Green's function, which can be used to obtain the spin density, charge density, momentum distribution, quasiparticle excitation spectrum, and, for some cases like a fermion system interacting via a two-body potential, it can give the ground-state energy of the system. One can also define a two-particle Green's function that can be used to obtain properties such as the electrical conductivity, optical properties, magnetic susceptibility, and many non-equilibrium properties. Here we will focus on the one-particle Green's function.

The underlying idea is to set up a bare state exactly, perturb it, and treat the excitations of the perturbed system as quasiparticles or collective excitations with finite lifetimes. We begin by introducing some of the definitions and mathematical properties of this approach, and return later to discuss its physical relevance and applications.

Assuming that we are dealing with spin-independent potentials, the one-particle Green's function for an electron is defined as

$$G(\mathbf{r}, t; \mathbf{r}', t') = -i \langle 0 | T[\psi(\mathbf{r}, t) \, \psi^\dagger(\mathbf{r}', t')] | 0 \rangle,$$

(13.26)

where $|0\rangle$ is the exact ground-state wavefunction of the system in the Heisenberg representation. The field operators ψ and $\psi^\dagger$ are also in the Heisenberg representation. They create or destroy a particle at $(\mathbf{r}, t)$. The Wick time-ordering operator T acts on the operators in the brackets [ ] by ordering them so that earlier times are on the right. When operators change their position, one has to ensure that the sign changes resulting from anticommutation are included when dealing with fermions. Therefore,

$$
\begin{aligned}
G(\mathbf{r}, t; \mathbf{r}', t') &= -i\langle \psi(\mathbf{r}, t)\, \psi^\dagger(\mathbf{r}', t')\rangle \quad (t > t') \\
&= +i\langle \psi^\dagger(\mathbf{r}', t')\, \psi(\mathbf{r}, t)\rangle \quad (t \le t'),
\end{aligned}
\tag{13.27}
$$

where $\langle\ \rangle$ represent a quantum mechanical average with respect to the ground state. The case where $t = t'$ is absorbed into the $< t'$ case by convention. Assuming translational and time invariance, it is expected that G will be a function of $\mathbf{r} - \mathbf{r}'$ and $t - t'$. Hence we replace these differences by $\mathbf{r}$ and t.

The Fourier transform $G(\mathbf{p}, \omega)$ of $G(\mathbf{r}, t)$ is useful for many calculations, as is the spatially transformed version $G(\mathbf{p}, t)$. These are given by (setting $\hbar = 1$)

$$
G(\mathbf{p}, \omega) = \int e^{-i(\mathbf{p}\cdot\mathbf{r} - \omega t)} G(\mathbf{r}, t)\, d\mathbf{r}\, dt
\tag{13.28}
$$

and

$$
G(\mathbf{p}, t) = \int e^{-i\mathbf{p}\cdot\mathbf{r}} G(\mathbf{r}, t)\, d\mathbf{r}.
\tag{13.29}
$$

The inverse Fourier transform for unit volume is

$$
G(\mathbf{r}, t) = \int e^{i(\mathbf{p}\cdot\mathbf{r} - \omega t)} G(\mathbf{p}, \omega) \frac{d\mathbf{p}\, d\omega}{(2\pi)^4}.
\tag{13.30}
$$

For planewave orbitals, the field operator is $\psi(\mathbf{r}, t) = \frac{1}{\sqrt{\Omega}} \sum_\mathbf{p} c_\mathbf{p}(t) e^{i\mathbf{p}\cdot\mathbf{r}}$, which yields

$$
G(\mathbf{p}, t) = -i\langle T[c_\mathbf{p}(t)\, c_\mathbf{p}^\dagger(0)]\rangle.
\tag{13.31}
$$

A useful example is the free electron gas. In the ground state, all planewave states are occupied up to the Fermi momentum p_F by spin-up and spin-down electrons. If we focus only on up spins and suppress the spin index,

$$
H = \sum_p \varepsilon_p c_\mathbf{p}^\dagger c_\mathbf{p},
\tag{13.32}
$$

where $\varepsilon_p = p^2/2m$. The momentum- and time-dependent non-interacting free electron Green's function is then (using $c_\mathbf{p}(t) = e^{iHt} c_\mathbf{p}(0) e^{-iHt}$)

$$
\begin{aligned}
G_0(\mathbf{p}, t) &= -i\langle T[c_\mathbf{p}(t)\, c_\mathbf{p}^\dagger(0)]\rangle \\
&= -i\langle c_\mathbf{p} c_\mathbf{p}^\dagger\rangle e^{-i\varepsilon_p t} \quad (t > 0) \\
&= i\langle c_\mathbf{p}^\dagger c_\mathbf{p}\rangle e^{-i\varepsilon_p t} \quad (t \le 0).
\end{aligned}
\tag{13.33}
$$

The number operator in momentum space is $n_{\mathbf{p}} = c_{\mathbf{p}}^{\dagger} c_{\mathbf{p}}$ and $n_{\mathbf{p}} = 1$ for $p < p_{\mathrm{F}}$. Hence,

$$
\begin{aligned}
G_0(\mathbf{p}, t) &= -i(1 - n_p) \, e^{-i\varepsilon_p t} \quad (t > 0) \\
&= i n_p e^{-i\varepsilon_p t} \quad (t \le 0).
\end{aligned}
\tag{13.34}
$$

The associated Green's function in momentum and frequency space is

$$
G_0(\mathbf{p}, \omega) = \frac{1}{\omega - \varepsilon_p + i\eta(p - p_{\mathrm{F}})},
\tag{13.35}
$$

where η is an infinitesimal positive parameter.

We now consider the interacting system where we expect a propagating electron-like excitation to have a finite lifetime, a shift in energy, and an effective mass because of its interaction with other electrons and other elementary excitations. If we consider only the interacting electron gas, it is convenient to explore a spectral representation of the Green's function $G(\mathbf{p}, t)$ and later obtain $G(\mathbf{p}, \omega)$ from $G(\mathbf{p}, t)$.

We begin by assuming that we have a ground-state wavefunction $|0\rangle = \left|\psi_0^N\right\rangle$ for an N-electron system with ground-state energy E_0^N and a complete set of states $\left|\psi_m^{N\pm1}\right\rangle$, to represent eigenstates of the Hamiltonian H for the $N \pm 1$ electron system with possible excited states labeled by m. Using this nomenclature,

$$
\begin{aligned}
G(\mathbf{p}, t) &= -i\left\langle \psi_0^N \right| T\left[c_{\mathbf{p}}(t) \, c_{\mathbf{p}}^{\dagger}(0) \right] \left| \psi_0^N \right\rangle \\
&= -i\left\langle \psi_0^N \right| e^{iHt} c_{\mathbf{p}} e^{-iHt} c_{\mathbf{p}}^{\dagger} \left| \psi_0^N \right\rangle \quad (t > 0) \\
&= i\left\langle \psi_0^N \right| c_{\mathbf{p}}^{\dagger} e^{iHt} c_{\mathbf{p}} e^{-iHt} \left| \psi_0^N \right\rangle \quad (t \le 0),
\end{aligned}
\tag{13.36}
$$

where the explicit time-dependence of the operators in the Heisenberg representation is included. If we now insert a complete set of eigenstates and operate on the ground-state wavefunction with e^{iHt} and e^{-iHt}, we have, for $t > 0$,

$$
G(\mathbf{p}, t) = -i \sum_{mm'} e^{iE_0^N t} \left\langle \psi_0^N \right| c_{\mathbf{p}} \left| \psi_m^{N+1} \right\rangle \left\langle \psi_m^{N+1} \right| e^{-iHt} \left| \psi_{m'}^{N+1} \right\rangle \left\langle \psi_{m'}^{N+1} \right| c_{\mathbf{p}}^{\dagger} \left| \psi_0^N \right\rangle.
\tag{13.37}
$$

Since

$$
\left\langle \psi_m^{N+1} \right| e^{-iHt} \left| \psi_{m'}^{N+1} \right\rangle = e^{-iE_m^{N+1} t} \delta_{mm'},
\tag{13.38}
$$

and if we denote

$$
\begin{aligned}
\left(c_{\mathbf{p}}^{\dagger} \right)_{m0} &\equiv \left\langle \psi_m^{N+1} \right| c_{\mathbf{p}}^{\dagger} \left| \psi_0^N \right\rangle, \\
\left(c_{\mathbf{p}} \right)_{m0} &\equiv \left\langle \psi_m^{N-1} \right| c_{\mathbf{p}} \left| \psi_0^N \right\rangle,
\end{aligned}
\tag{13.39}
$$

then

$$
\begin{aligned}
G(\mathbf{p}, t) &= -i \sum_m \left| \left(c_{\mathbf{p}}^{\dagger} \right)_{m0} \right|^2 e^{-i(E_m^{N+1} - E_0^N)t} \quad (t > 0) \\
&= i \sum_m |(c_{\mathbf{p}})_{m0}|^2 e^{i(E_m^{N-1} - E_0^N)t} \quad (t < 0).
\end{aligned}
\tag{13.40}
$$

It is convenient to separate the energy terms as follows:

$$E_m^{N+1} - E_0^N = \left(E_0^{N+1} - E_0^N\right) + \left(E_m^{N+1} - E_0^{N+1}\right), \tag{13.41}$$

where the first term is the ground-state energy difference between a system of N and $N+1$ particles, which is just the chemical potential μ_N. The second term is the energy required to excite the $N+1$ particle system from the ground state to the m excited state. If we assume (for a large system)

$$\mu_N \approx \mu_{N-1} \equiv \mu \tag{13.42}$$

and

$$\begin{aligned}
\omega_m^{N+1} &= E_m^{N+1} - E_0^N - \mu, \\
\omega_m^{N-1} &= E_0^N - E_m^{N-1} - \mu,
\end{aligned} \tag{13.43}$$

where $\omega_m^{N+1} > 0$ and $\omega_m^{N-1} < 0$, then

$$\begin{aligned}
G(\mathbf{p}, t) &= -i \sum_m |(c_p^\dagger)_{m0}|^2 e^{-i(\omega_m^{N+1}+\mu)t} \quad (t > 0) \\
&= i \sum_m |(c_{\mathbf{p}})_{m0}|^2 e^{-i(\omega_m^{N-1}+\mu)t} \quad (t < 0),
\end{aligned} \tag{13.44}$$

which resembles the form of $G_0(\mathbf{p}, t)$.

We now introduce the spectral weight function

$$A(\mathbf{p}, \omega) = \sum_m |(c_{\mathbf{p}}^\dagger)_{m0}|^2 \delta\left(\omega - \omega_m^{N+1}\right) + \sum_m |(c_{\mathbf{p}})_{m0}|^2 \delta\left(\omega - \omega_m^{N-1}\right). \tag{13.45}$$

If we write

$$G(\mathbf{p}, \omega_0) = \int_{-\infty}^{\infty} \frac{A(\mathbf{p}, \omega)\, d\omega}{\omega_0 - \omega - \mu + i\omega\eta} \tag{13.46}$$

for $\eta > 0$, then

$$G(\mathbf{p}, \omega_0) = \int_{-\infty}^{\infty} \left\{ \frac{\sum_m |(c_{\mathbf{p}}^\dagger)_{m0}|^2 \delta\left(\omega - \omega_m^{N+1}\right)}{\omega_0 - \omega - \mu + i\omega\eta} + \frac{\sum_m |(c_{\mathbf{p}})_{m0}|^2 \delta\left(\omega - \omega_m^{N-1}\right)}{\omega_0 - \omega - \mu + i\omega\eta} \right\} d\omega, \tag{13.47}$$

which can be viewed as a weighted average of non-interacting G_0's, with $\omega > 0$ in the first term and $\omega < 0$ in the second. By taking the ω_0 Fourier transform of $G(\mathbf{p}, \omega_0)$, the Green's function $G_0(\mathbf{p}, t)$ can be obtained. The non-interacting spectral weight function is given by

$$A_0(\mathbf{p}, \omega) = \delta(\omega - (\varepsilon_p - \mu)). \tag{13.48}$$

Some easily calculated properties of $A(\mathbf{p}, \omega)$ are

$$A(\mathbf{p}, \omega) = A^*(\mathbf{p}, \omega), \tag{13.49}$$

$$\int_{-\infty}^{\infty} A(\mathbf{p}, \omega)d\omega = 1, \tag{13.50}$$

and

$$A(\mathbf{p}, \omega - \mu) = -\mathrm{sgn}\,(\omega - \mu)\,\mathrm{Im}G(\mathbf{p}, \omega). \tag{13.51}$$

Equation (13.51) can be used to obtain the dispersion relations between the real and imaginary parts of $G(\mathbf{p}, \omega)$.

We now describe the physical interpretations and some uses of the various forms of the one-particle Green's function. For the real-space G case, where $\mathbf{r} = \mathbf{r}'$ and $t = t + \delta t$, with $\delta t = 0^+$, we have

$$G(\mathbf{r}, t; \mathbf{r}, t + \delta t) = i\,\langle 0|\,\psi^\dagger(\mathbf{r}, t)\,\psi(\mathbf{r}, t)\,|0\rangle = i\,\langle 0|\,\rho(\mathbf{r}, t)\,|0\rangle. \tag{13.52}$$

So the equal time real-space G gives the expectation value of the density in the ground state. For unequal times, assuming $t > t'$, then the real-space G becomes (using Eq. (13.26))

$$iG(\mathbf{r}, t; \mathbf{r}', t') = \langle 0|\,\psi(\mathbf{r}, t)\,\psi^\dagger(\mathbf{r}', t')\,|0\rangle, \tag{13.53}$$

which gives the probability amplitude of a particle added at $\mathbf{r}'$ and time t' propagates to $\mathbf{r}$ at time t (not necessarily the same particle, since we have identical quantum particles). Similarly, $iG(\mathbf{r}, t; \mathbf{r}', t')$ gives the same probability amplitude for holes if $t < t'$.

For the momentum space G's, we have for equal times $t = 0^-$,

$$-iG(\mathbf{p}, t) = \langle 0|\,c_\mathbf{p}^\dagger c_\mathbf{p}\,|0\rangle = \langle 0|\,n_\mathbf{p}\,|0\rangle, \tag{13.54}$$

so this quantity gives the momentum distribution, which is the number of particles in the ground state per unit volume in a unit volume element of momentum space about $\mathbf{p}$. For unequal times,

$$iG(\mathbf{p}, t) = \langle 0|\,c_\mathbf{p}(t)\,c_\mathbf{p}^\dagger(0)\,|0\rangle \quad (t > 0) \tag{13.55}$$

gives the probability amplitude for a particle with momentum $\mathbf{p}$ added to the interacting system in the ground state at $t = 0$, and then a particle is found in state $\mathbf{p}$ at $t > 0$.

For $G_0(\mathbf{p}, \omega)$, given in Eq. (13.35), there is a pole at the excitation energy and the entire "spectral strength" of this excitation is connected with this pole. In anticipation of developing a lifetime Γ, we rewrite Eq. (13.48) using a standard expression for the δ-function:

$$A_0(\mathbf{p}, \omega) = \lim_{\Gamma \to 0} \frac{1}{2\pi}\left[\frac{i}{\omega - (\varepsilon_\mathbf{p} - \mu) + i\Gamma} - \frac{i}{\omega - (\varepsilon_\mathbf{p} - \mu) - i\Gamma}\right]. \tag{13.56}$$

Comparing this expression with the one in Eq. (13.48), we can make the following observations and derive the expression for the interacting Green's functions.

The interactions change $A(\mathbf{p}, \omega)$ from a δ-function, and these changes are governed by the values of matrix elements of the operators such as $(c_{\mathbf{p}}^{\dagger})_{m0}$. If the interactions are not too strong, we expect a strong peak in $A(\mathbf{p}, \omega)$ "near" the real axis at $\omega = \varepsilon_{\mathbf{p}} - \mu$. The lifetime effects anticipated in Eq. (13.56), which tend to zero in the non-interacting case, are now finite, and the pole at $\varepsilon_{\mathbf{p}} - \mu$ moves to $\tilde{\varepsilon}_{\mathbf{p}} - \mu - i\Gamma_{\mathbf{p}}$ with an overall strength of $Z_{\mathbf{p}}$. By comparing Eqs. (13.34) and (13.44), we expect and can argue for expressions having the following form:

$$G(\mathbf{p}, t) = -iZ_{\mathbf{p}} e^{-i\tilde{\varepsilon}_{\mathbf{p}} t} e^{-\Gamma_{\mathbf{p}} t} + \text{corrections}, \tag{13.57}$$

$$A(\mathbf{p}, \omega) = \frac{iZ_{\mathbf{p}}/2\pi}{\omega - \left[(\tilde{\varepsilon}_{\mathbf{p}} - \mu) - i\Gamma_{\mathbf{p}}\right]} + \text{complex conjugate}, \tag{13.58}$$

$$G(\mathbf{p}, \omega) = \frac{iZ_{\mathbf{p}}}{\omega - \tilde{\varepsilon}_{\mathbf{p}}} + i\Gamma_{\mathbf{p}} + \text{ small corrections}. \tag{13.59}$$

This is in essence the quasiparticle approximation for electron Green's functions in the case of an interacting system. There are cases where several poles are present in the spectral function. The one closest to the real axis with the longest lifetime (smallest $\Gamma_{\mathbf{p}}$) will dominate. The underlying physics displayed by the Green's function approach is that interactions spread out the δ-function since $c_{\mathbf{p}}^{\dagger} |0\rangle$ is not an eigenstate and the strength of the quasiparticle peak is related to the matrix element of $(c_{\mathbf{p}}^{\dagger})_{m0}$. The total amplitude of the function is spread out in frequency, usually leaving behind a fairly strong peak. In general, the spectral weight function has two peaks. One is at $\omega > 0$, and is related to the probability of adding an electron, and one is at $\omega < 0$, and is related to the probability of adding a hole. In general, $A(\mathbf{p}, \omega)$ is a complicated function except for the feature of having two Lorentzian peaks at some specific ω near μ. Also, there can sometimes be satellites in $A(\mathbf{p}, \omega)$ away from the quasiparticle peak due to interactions with other excitations such as plasmons and phonons.

Landau[1] exploited this property in his theory of Fermi liquids, which has features in common with the quasiparticle description implied in Eq. (13.57). That is, there is an elementary excitation of strength $Z_{\mathbf{p}}$ having energy $\tilde{\varepsilon}_{\mathbf{p}}$ and energy width $\Gamma_{\mathbf{p}}$. In the limit of $Z_{\mathbf{p}} \to 1, \tilde{\varepsilon}_{\mathbf{p}} \to \varepsilon_{\mathbf{p}}$, and $\Gamma_{\mathbf{p}} \to 0$, the Green's function becomes that of a free particle. In this picture, the excitations look like "dressed" versions of the free particles.

The Green's function theory for phonons can be developed in the same manner as for electrons. Using definitions from previous chapters, the displacement of an atom can be expanded in phonon coordinates and in the creation and destruction operators $a_{-\mathbf{q}\alpha}^{\dagger}$ and

[1] L. D. Landau, "The theory of a Fermi liquid", *Soviet Phys. JETP* 3(1956), 920.

$a_{\mathbf{q}\alpha}$:

$$\delta \mathbf{R}_{\ell a} = \sum_{\mathbf{q}\alpha} \left(\frac{\hbar}{2MN\omega_{\mathbf{q}\alpha}} \right)^{\frac{1}{2}} \hat{\epsilon}_{\mathbf{q}\alpha}^{a} e^{i\mathbf{q}\cdot\mathbf{R}_{\ell}^{a}} \left(a_{\mathbf{q}\alpha} + a_{-\mathbf{q}\alpha}^{\dagger} \right),$$

(13.60)

where N is the number of cells per unit volume, M is the ionic mass per cell, ℓ is the lattice vector index to locate a cell, a gives the atomic site in the cell, α is the branch index, $\omega_{\mathbf{q}\alpha}$ is the phonon frequency for wavevector $\mathbf{q}$ and α, and $\hat{\epsilon}_{\mathbf{q}\alpha}^{a}$ is the phonon polarization unit vector.

We define the wave field operator

$$\phi_{\alpha}(\mathbf{r}, t) = \sum_{\mathbf{q}} \phi_{\mathbf{q}\alpha}(t) e^{-i\mathbf{q}\cdot\mathbf{r}},$$

(13.61)

where

$$\phi_{\mathbf{q}\alpha} = a_{\mathbf{q}\alpha} + a_{-\mathbf{q}\alpha}^{\dagger},$$

(13.62)

and the one-particle Green's function for phonons is

$$\begin{aligned}
D_{\alpha}(\mathbf{r}, t; \mathbf{r}'t') &= -i \langle 0| T\big(\phi_{\alpha}(\mathbf{r}, t) \phi_{\alpha}^{\dagger}(\mathbf{r}', t')\big) |0\rangle \\
&= -i \langle \phi_{\alpha}(\mathbf{r}, t) \phi_{\alpha}^{\dagger}(\mathbf{r}', t') \rangle \quad (t > t') \\
&= -i \langle \phi_{\alpha}^{\dagger}(\mathbf{r}', t') \phi_{\alpha}(\mathbf{r}, t) \rangle \quad (t \leq t').
\end{aligned}$$

(13.63)

Following our discussion of electrons for a homogeneous system, we let $(t - t') \to t$ and $(\mathbf{r} - \mathbf{r}') \to \mathbf{r}$, the Fourier-transformed phonon Green's function becomes

$$D_{\alpha}(\mathbf{q}, t) = -i \langle 0| T\left[\phi_{\mathbf{q}\alpha}(t) \phi_{\mathbf{q}\alpha}^{\dagger}(0) \right] |0\rangle,$$

(13.64)

and the spectral function is

$$B_{\alpha}(\mathbf{q}, \omega) = \sum_{m} |\langle m|\phi_{\mathbf{q}\alpha}^{\dagger}|0\rangle|^2 \delta(\omega - \omega_m^{N+1}) - \sum_{m} |\langle m|\phi_{\mathbf{q}\alpha}|0\rangle|^2 \delta(\omega + \omega_m^{N-1}),$$

(13.65)

where $\omega_m^N = E_m^N - E_0^N$ is the excitation energy. For bare phonons,

$$B_{0\alpha}(\mathbf{q}, \omega) = \delta(|\omega| - \omega_{\mathbf{q}\alpha})\text{sgn}(\omega).$$

(13.66)

The frequency-transformed Green's function becomes

$$D_{\alpha}(\mathbf{q}, \omega_0) = \int_{-\infty}^{\infty} \frac{B_{\alpha}(\mathbf{q}, \omega)\mathrm{d}\omega}{\omega_0 - \omega + i\omega\eta}$$

(13.67)

(with $\eta = 0^+$), and, for a system with inversion $B_{\alpha}(\mathbf{q}, \omega) = B_{\alpha}(\mathbf{q}, -\omega)$,

$$D_{\alpha}(\mathbf{q}, \omega_0) = \int_{0}^{\infty} \frac{B_{\alpha}(\mathbf{q}, \omega)2\omega\mathrm{d}\omega}{\omega_0^2 - \omega^2 + i\eta}.$$

(13.68)

For the non-interaction "bare" case,

$$D_{0\alpha}(\mathbf{q}, \omega_0) = \frac{2\omega_{\mathbf{q}\alpha}}{\omega_0^2 - \omega_{\mathbf{q}\alpha}^2 + i\eta}. \tag{13.69}$$

13.3 Feynman diagrams and many-body perturbation theory techniques

Our next task is to use a perturbation scheme to calculate the changes in the non-interacting electron and phonon Green's functions when interactions are present. This is done most often using the technique of Feynman[2] diagrams. The proofs are given in texts on field theory or many-body physics. Here we discuss the resulting recipes and follow the more or less standard description.

In this approach the electron $iG_0(\mathbf{p}, \omega)$ is represented by a solid line, $iD_{0\alpha}(\mathbf{q}, \omega)$ by a wavy line, and the Coulomb interaction $-iv(\mathbf{q}, \omega)$ by a dashed line. The one-body potential $-iU$ is represented by a dotted line, and $-ig_{\mathbf{q}\alpha}$, which represents the coupling constant for electron–phonon scattering, is represented by a dot. Momenta, frequencies, and spins are conserved at each vertex.

After an appropriate diagram is drawn for a physical process, all the factors are multiplied together and a factor of $(-1)^n$ is included, where n is the number of closed electron lines or loops in the graph. An integration is next done over all free internal momenta and frequencies, together with a sum over phonon polarizations α and internal line-spin indices. Different graphs describing the same process are added together.

It is instructive to choose specific graphs to illustrate the method. If we consider the possible interactions for an electron in an electron gas, the lowest-order term (the Hartree term), in which the electron interacts with all the other electrons, is canceled by the background positive charge of the lattice. Therefore, the first-order diagram shown in Fig. 13.1 representing this interaction is pictured as an electron entering from the left and the line contributes $iG_0(p)$. The integration is done over the free internal momentum q. There are no loops and no spin sum since there are no internal lines. Hence, this graph represents the exchange term in Chapter 6, which is given by

$$iG_0(p) \left[\frac{d\mathbf{q}d\omega}{(2\pi)^4}(-i)v(q)iG_0(p - q) \right] iG_0(p), \tag{13.70}$$

where we make use of the notation that the p and q variables have a frequency component, i.e. $p \to (\mathbf{p}, \omega)$, $q \to (\mathbf{q}, \omega)$, and s denotes the spin index.

If we take into account the screening that can be described by having the photon creating and absorbing virtual electron–hole pairs, there is a loop and free internal lines requiring a

---

[2] R. P. Feynman, "Space-time approach to quantum electrodynamics," *Phys. Rev.* 76(1949), 769.

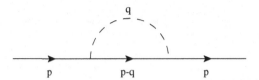

Figure 13.1 Lowest-order graph for an electron interacting via the Coulomb interaction in an electron gas. This is called the lowest-order exchange contribution.

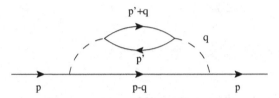

Figure 13.2 Virtual electron–hole processes (the bubble) causing screening of the exchange interaction in Fig. 13.1.

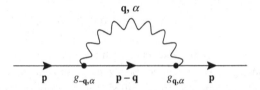

Figure 13.3 The lowest-order contribution to the electron self energy arising from the emission followed by the absorption of a phonon.

spin sum. The diagram is shown in Fig. 13.2, and the contribution is

$$iG_0(p)\left[\sum_s(-1)\int\frac{d\mathbf{q}d\omega}{(2\pi)^4}\int\frac{d\mathbf{p}'d\omega'}{(2\pi)^4}iG_0(p-q)iG_0(p')iG_0(p'+q)(iv(q))^2\right]iG_0(p).$$

(13.71)

The graph for the contribution of electron–phonon scattering is shown in Fig. 13.3, and the associated expression is

$$iG_0(p)\left[\sum_\alpha\int\frac{d\mathbf{q}d\omega}{(2\pi)^4}|g_{q\alpha}|^2D_{0\alpha}(q)G_0(p-q)\right]iG_0(p).$$

(13.72)

When calculating the integrals, the integrations over the frequencies are usually not difficult to do using the calculus of residues, while the integrations over the momenta require knowledge of the matrix elements.

If we denote the Green's function appropriate for an interacting system $iG(p)$ as a double line in analogy with the single line for $iG_0(p)$, we can add contributions such as those shown in Figs. 13.1 and 13.3. We will consider screening effects such as those represented

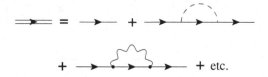

Figure 13.4   Contributions to the interacting Green's function.

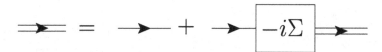

Figure 13.5   Diagrams defining the irreducible self energy Σ.

in Fig. 13.2 later. The scheme is illustrated graphically in Fig. 13.4. In all cases, for the interactions, there is a bare $iG_0(p)$ line coming in from the left and leaving from the right. If we denote the various contributions as A, B, C,..., then Fig. 13.4 can be described as

$$iG(p) = iG_0(p) + iG_0(p) [A + B + C \ldots] iG(p). \tag{13.73}$$

If the term in square brackets contains all graphs with no external lines entering or leaving and only graphs that cannot be separated into two unconnected pieces by cutting a single line, then the bracket term is called the irreducible self energy $\Sigma(p)$ with a multiplying factor $-i$. To sum an infinite number of repeated graphs, we use the so-called Dyson approach, shown in Fig. 13.5, which gives

$$iG(p) = iG_0(p) + iG_0(p) (-i\Sigma(p)) iG(p), \tag{13.74}$$

so that

$$\frac{1}{G(p)} = \frac{1}{G_0(p)} - \Sigma(p), \tag{13.75}$$

which is often called Dyson's equation.

This approach to summing an infinite contribution in perturbation theory is powerful. A trivial similar example to illustrate the summing of an infinite series is

$$
\begin{aligned}
x &= 1 + \tfrac{1}{2} + \tfrac{1}{4} + \tfrac{1}{8} + \cdots, \\
x &= 1 + \tfrac{1}{2}(1 + \tfrac{1}{2} + \tfrac{1}{8} + \cdots) = 1 + \tfrac{1}{2}(x), \\
x &= 2.
\end{aligned}
\tag{13.76}
$$

Using the explicit form of G_0 (Eq. (13.35)) in Eq. (13.75), we have

$$\frac{1}{G(\mathbf{p}, \omega)} = \omega - \varepsilon_{\mathbf{p}} - \Sigma(\mathbf{p}, \omega) + i\eta(|\mathbf{p}| - p_{\mathrm{F}}) \tag{13.77}$$

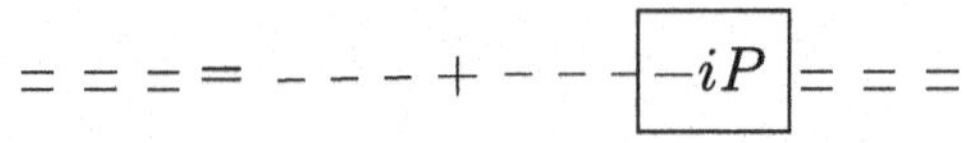

Figure 13.6 Diagrams defining the polarization function P.

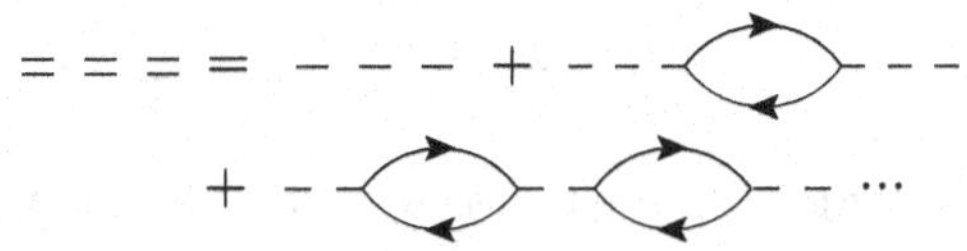

Figure 13.7 The screened Coulomb interaction expressed in terms of a series of virtual electron–hole processes causing screening.

and

$$G(\mathbf{p}, \omega) = \frac{1}{\omega - \varepsilon_{\mathbf{p}} - \Sigma(\mathbf{p}, \omega) + i\eta(|\mathbf{p}| - p_F)}. \tag{13.78}$$

Therefore, we can calculate $\Sigma(\mathbf{p}, \omega)$ to any order of interactions and use it to obtain $G(\mathbf{p}, \omega)$. This approach often gives satisfactory results even when a straightforward series expansion in powers of interaction strengths does not. We note that $\Sigma(\mathbf{p}, \omega)$ in general is a complex function with real and imaginary parts. The poles of the analytic continuation of G into the lower half-plane give the energy E_p of the electron excitation and the damping rate:

$$E_{\mathbf{p}} = \varepsilon_{\mathbf{p}} + \mathrm{Re}\,\Sigma(\mathbf{p}, E_{\mathbf{p}} + i\Gamma_{\mathbf{p}}), \tag{13.79}$$

$$\Gamma_{\mathbf{p}} = \mathrm{Im}\,\Sigma(\mathbf{p}, E_{\mathbf{p}} + i\Gamma_{\mathbf{p}}). \tag{13.80}$$

The similarity to the discussion associated with the quasiparticle description is noted.

The Dyson[3] approach can also be applied to the interaction potentials between particles. For example, if we represent the screened Coulomb interaction as $-iW(\mathbf{q}, \omega)$ and $-iP$ as the polarization component to be defined below, then Fig. 13.6 can be represented by

$$-iW(\mathbf{q}, \omega) = -iv(\mathbf{q}) + \big(-iv(\mathbf{q})\big)\big(-iP(\mathbf{q}, \omega)\big)\big(-iW(\mathbf{q}, \omega)\big). \tag{13.81}$$

In lowest order, we can take one "bubble" as shown in Fig. 13.2 and add an infinite number of them as shown in Fig. 13.7. These can be summed as shown in Fig. 13.8.

This limitation to one bubble is equivalent to the random phase approximation (RPA). Therefore, Fig. 13.8 yields

$$W^{\mathrm{RPA}}(\mathbf{q}, \omega) = \frac{v(\mathbf{q})}{1 + v(\mathbf{q})P^{\mathrm{RPA}}(\mathbf{q}, \omega)}, \tag{13.82}$$

[3] F. J. Dyson, "The S matrix in quantum electrodynamics," *Phys. Rev.* 75(1949), 1736.

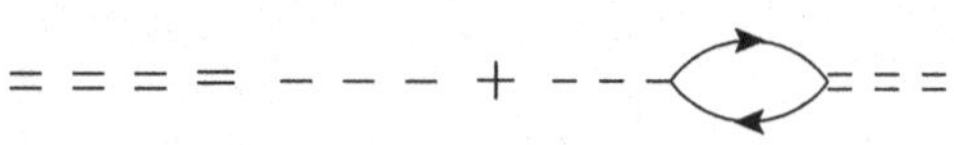

Figure 13.8   A Dyson approach for summing the contributions shown in Fig. 13.7.

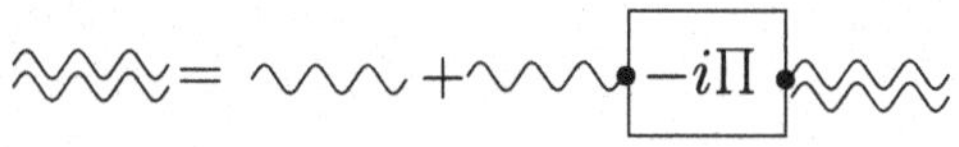

Figure 13.9   The renormalized phonon Green's function expressed as a Dyson sum.

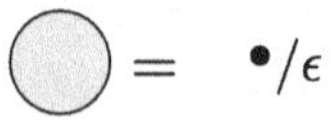

Figure 13.10   Schematic for the screening of the electron–phonon matrix element.

where we have used Eq. (13.81).

We now identify the dielectric function with the denominator in Eq. (13.82) (which is of the same form discussed in Chapter 8):

$$\epsilon^{\mathrm{RPA}}(\mathbf{q}, \omega) = 1 + v(\mathbf{q}) P^{\mathrm{RPA}}(\mathbf{q}, \omega), \tag{13.83}$$

where

$$P^{\mathrm{RPA}}(\mathbf{q}, \omega) = 2i \sum \int \frac{d\mathbf{p}\, d\omega_0}{(2\pi)^4} G_0(\mathbf{p} + \mathbf{q}, \omega + \omega_0) G_0(\mathbf{p}, \omega_0), \tag{13.84}$$

and we have used Eq. (13.71). Although there are other ways to compute $P^{\mathrm{RPA}}(\mathbf{q}, \omega)$, this approach can be systematically extended to include higher-order contributions.

The Dyson approach can be used to renormalize the phonon Green's function and screen the electron–phonon matrix element as shown in the graphs in Figs. 13.9 and 13.10, which gives the results

$$D_\alpha(\mathbf{q}, \omega) = \frac{2\omega_{\mathbf{q}\alpha}}{\omega^2 - \omega_{\mathbf{q}\alpha}^2 - 2\omega_{\mathbf{q}\alpha}\pi_\alpha(\mathbf{q}, \omega) + i\delta}, \tag{13.85}$$

where $\delta \to 0^+$. For lowest order (RPA screening),

$$\pi_\alpha(\mathbf{r}, \omega) = -g_{\mathbf{q}\alpha} \frac{1}{\epsilon^{\mathrm{RPA}}(\mathbf{q}, \omega)}, \tag{13.86}$$

and the electron–phonon matrix element can be screened $g \to \tilde{g}$ by the dielectric function as shown in Fig. 13.10.

At this point, the total electron self energy arising from Coulomb and phonon scattering can be written as

$$\Sigma(\mathbf{p}, \omega) = i \int \frac{d\mathbf{q}\, d\omega}{(2\pi)^4} G_0(\mathbf{p} + \mathbf{q}, \omega + \omega_0) \Big[W(\mathbf{q}, \omega_0) + \sum_\alpha |g_{\mathbf{q}\alpha}|^2 D_\alpha(\mathbf{q}, \omega_0) \Big]. \tag{13.87}$$

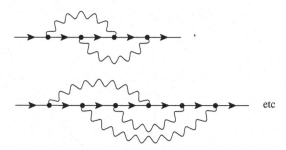

Figure 13.11 Graphs contributing to the vertex function Γ.

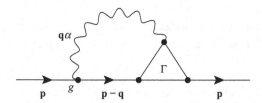

Figure 13.12 The electron self energy arising from the emission and absorption of a phonon including vertex corrections.

Keeping the first term in the bracket for the electron self energy is known as the GW approximation, discussed in Chapter 7.

We now consider one more aspect of summing electron–phonon graphs, by considering graphs such as those shown in Fig. 13.11. In these graphs, there is an electron line entering and leaving the cluster of phonon scatterings as shown in Fig. 13.12. The diagram in Fig. 13.12 resembles the diagram in Fig. 13.3, except that at one of the electron–phonon vertices g is replaced by a Γ. This procedure is called a vertex correction, to account for diagrams of the type shown in Fig. 13.11. Therefore, the phonon contribution to the total electron self energy given in Eq. (13.87) can be modified to include vertex corrections to the phonon contribution to give

$$\Sigma^{\text{ph}}(\mathbf{p}, \omega) = i \sum_{\alpha} \int \frac{d\mathbf{q}\,d\omega_0}{(2\pi)^4} g_{\mathbf{q}\alpha} G_0(\mathbf{p} + \mathbf{q}, \omega + \omega_0) \Gamma(\mathbf{p}, \mathbf{q}) D(\mathbf{q}, \omega_0). \qquad (13.88)$$

Because it can be shown that graphs of the type shown in Fig. 13.11 usually contribute with an expansion parameter proportional to the square root of the electron-to-ion mass ratio, they are neglected for many standard calculations. This approach is referred to as the Migdal[4] approximation. It essentially involves replacing Γ with $\tilde{g}$. This approximation is sometimes difficult to justify, especially in cases where electrons are scattered between parallel faces on a Fermi surface.

The usual calculation for Σ^{ph} is to put $\mathbf{p}$ at p_{F} and apply the Migdal approximation. For this case, it can be shown that the electron dispersion curve near E_{F} is modified with a

---

[4] A. B. Migdal, "Interaction between electrons and lattice vibrations in a normal metal," *Soviet Phys. JETP* 7(1958), 996.

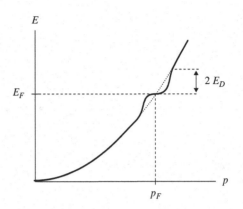

Figure 13.13 Illustration of the electron–phonon "wrinkle" at the Fermi energy resulting in an enhanced electronic effective mass arising from electron–phonon interactions, with E_D the Debye energy.

"wrinkle" over an energy width on the order of the phonon energy or Debye energy E_D, as shown in Fig. 13.13. This is interpreted in terms of an effective mass

$$m^* = m_b(1 + \lambda), \tag{13.89}$$

where m_b is the band mass and λ is the electron–phonon coupling parameter, which can be calculated from the self energy using Eq. (13.88).

For the BCS (Bardeen, Cooper, and Schrieffer) theory of superconductivity, the attractive part of the electron pairing is viewed in terms of the exchange of phonons between electrons. Since this coupling also involves λ, it is sometimes instructive to compare the λ obtained by measuring the heat capacity (and hence m^*) of a metal to the one extracted by using superconductivity data, such as the transition temperature. A third experimental source of λ is to extract it from the temperature-dependent resistivity, since this involves the scattering of electrons by phonons. The first two cases involve the emission and absorption of virtual phonons, whereas the latter involves the emission or absorption of real phonons. Therefore, they are not exactly the same quantity, especially when extracted from transport measurements. However, in general, values obtained in all three cases are in reasonable agreement.

To conclude this chapter, it is appropriate to state that the graphical approach to calculate many-body processes in solids is an important and standard tool at this time. It allows a systematic approach to evaluate physical properties, and it is useful for understanding the role of higher-order processes in a perturbation expansion.

14 Superconductivity

14.1 Brief discussion of the experimental background

Superconductivity is one of the most studied areas of condensed matter physics. The discovery by H. Kamerlingh-Onnes[1] in 1911 demonstrated that, at 4.3 K, Hg had zero resistance and remained in this state at temperatures below the transition temperature T_c (Fig. 14.1). It appeared that the solid had made a transition to a new state, and this visualization of the superconducting transition as a change of state remained. The transition was found to be very sharp with $\Delta T \sim 10^{-2}$ K, and no crystallographic changes were noted.

Further tests of various materials yielded new superconductors, and it became clear that non-metals were not superconducting, while metals with higher room-temperature resistivities tended to be more likely candidates for superconductivity. Another feature of superconductors was that at finite frequencies ω, the resistance $\rho(\omega)$ remained zero up to a critical value $\omega = \omega_c$, where $\hbar\omega_c$ became known as the optical superconducting energy gap (Fig. 14.2). The relation $\hbar\omega_c \approx 3.5 k_B T_c$ was established experimentally and later derived theoretically for a large class of superconductors.

The maximum T_c for materials increased at an average rate of 1 K every three years until the mid 1980s when it was around 25 K. In 1986, Bednorz and Müller[2] found superconductivity in a copper oxide material in the 50 K range, and, soon afterward, Chu and collaborators[3] achieved T_c's above liquid nitrogen temperature and, in 1996, achieved T_c's around 160 K. At this time, this is the highest T_c achieved in this class of materials.

An important early discovery was the existence of persistent currents, where supercurrents, once generated, persist in a superconducting ring for as long as the ring is kept below T_c. A magnetic field is induced in the hole of the ring and the system is very stable. Theoretical estimates of the time for the decay of the currents at $T \ll T_c$ are enormously long, perhaps of the order of the age of the universe. Another important magnetic field effect is the destruction of superconductivity in the presence of a magnetic field above a value called the critical field H_c (Fig. 14.3). The critical field is temperature dependent, rising from $H_c(T_c) = 0$ to a maximum $H_c(T = 0)$, which for superconductors like Hg is only around several hundred G. It can be orders of magnitude larger for other superconductors.

[1] H. Kamerlingh–Onnes, "The superconductivity of mercury," *Leiden Comm.*, Nos. 122 and 124(1911).

[2] J. G. Bednorz and K. A. Müller, "Possible high T_c superconductivity in the Ba–La–Cu–O system," *Z. Phys. B* 64(1986), 189.

[3] L. Gao, Y. Xue, F. Chen, Q. Xiong, R. L. Meng, D. Ramirez, and C. W. Chu, "Superconductivity up to 164 K in HgBa$_2$Ca$_{m-1}$Cu$_m$O$_{2m+2+\delta}$ ($m = 1, 2$, and 3) under quasihydrostatic pressures," *Phys. Rev. B* 50(1994), 4260.

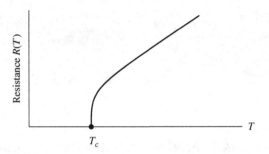

Figure 14.1 First (1911) superconductor, mercury, $T_c = 4.3$ K.

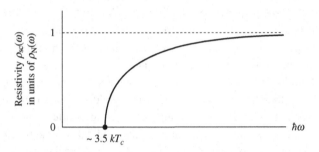

Figure 14.2 Resistivity of a superconductor ρ_{SC} as a function of frequency divided by the normal state resistivity ρ_N.

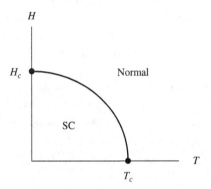

Figure 14.3 Temperature dependence of the critical field for a superconductor like mercury.

Below H_c, the so-called type I superconductors expel magnetic flux. This discovery by Meissner and Ochsenfeld[4] in 1933 (Fig. 14.4) was extremely important for understanding superconductors. A superconductor is a perfect diamagnet. Therefore, using Maxwell's

---

[4] W. Meissner and R. Ochsenfeld, "A new effect in penetration of superconductors," *Die Naturwissenschaften* 21(1933), 787.

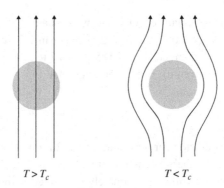

$T > T_c$ $T < T_c$

Figure 14.4   The Meissner effect. Magnetic flux is expelled when the system is superconducting.

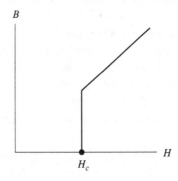

Figure 14.5   The magnetic field in a superconductor like mercury.

equation

$$\mathbf{B} = \mathbf{H} + 4\pi\mathbf{M}, \qquad (14.1)$$

and the observation (Fig. 14.5) that $B = 0$ inside a superconductor implies that

$$\mathbf{M} = -\frac{1}{4\pi}\mathbf{H},$$

or

$$\chi \equiv \frac{M}{H} = -\frac{1}{4\pi}. \qquad (14.2)$$

Since the energy density needed to expel flux is $\frac{H^2}{8\pi}$, the energy difference between the superconducting and normal states is $\frac{1}{8\pi}H_c^2(T)$. This energy, for $T = 0$, is the "condensation energy" between the normal and superconducting states.

Another important experiment, which had a great deal of influence on the search for understanding of the underlying mechanism of superconductivity, is the isotope effect.[5] The T_c of Hg changed from 4.185 K to 4.146 K when the isotopic Hg mass changed from 199.5 to 203.4 (amu). The observation that

$$T_c \sim M^{-\alpha}, \tag{14.3}$$

where M is the nuclear mass and $\alpha \approx 1/2$ for many superconductors, led to the suggestion that, since the Debye temperature $T_D \sim M^{-1/2}$, then T_c/T_D is constant and the lattice vibrations of a solid are intimately connected with superconductivity. Later experiments found deviations from $\alpha = 1/2$, but the observation that T_c is related to the ionic mass led to the discovery of an electron–phonon mechanism for superconductivity. This discovery is also consistent with the observation that bad conductors were good superconductors, since the higher resistivities in the normal state were caused by enhanced scattering arising from stronger electron–phonon couplings.

Another very important experimental result, which gave significant insight into the nature of superconductivity, was the measurement of the temperature-dependent heat capacity at constant volume $C_V(T)$. As the temperature is lowered, the linear dependence on temperature resulting from excitations of "normal" electrons changed with a jump at $T = T_c$ and an exponential behavior for the heat capacity

$$C_V \sim \exp\left(-1.76\frac{T_c}{T}\right). \tag{14.4}$$

This jump and exponential factor (Fig. 14.6) added credence to the superconducting energy gap concept. Using the standard thermodynamic relationships between C_v and the entropy, it is easy to show that not only was there a condensation energy when passing into the superconducting state, but that this state was more ordered.

Two other early experimental results showed the effects of increased electronic order: (i) the energy gap and special correlations between the "superconducting" electrons were deduced from the optical properties, which showed a "gap" in the spectrum, and (ii) the exponential decay of the temperature dependence of the ultrasonic attenuation below T_c (Fig. 14.7) is also caused by the energy gap and special correlation effects.

The process of testing for a new superconducting material using just the resistivity was sometimes clouded by the problem that filaments could carry the superconducting current. Hence, an impurity phase or a specific structural phase of a sample could mask an experiment that was trying to establish superconductivity as a property of the bulk sample. Of course, if T_c were very high, the direction of the investigation would turn to establishing what the "smaller" component was. Similarly, a superconducting skin on a non-superconductor could exclude flux. There are safeguards, like measurements of C_V,

---

[5] E. Maxwell, "Isotope effect in the superconductivity of mercury," *Phys. Rev.* 78(1950), 477; C. A. Reynolds, B. Serin, W. H. Wright, and L. B. Nesbitt, "Superconductivity of isotopes of mercury," *Phys. Rev.* 78(1950), 487.

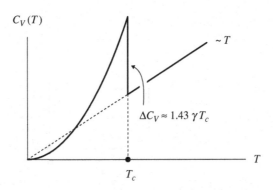

Figure 14.6 Heat capacity of a superconductor as a function of temperature near T_c. Here γ is the linear coefficient of the normal state electronic specific heat.

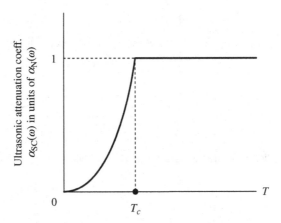

Figure 14.7 Acoustic attenuation as a function of temperature in a superconductor compared to the normal state.

which give the volume of the sample that is superconducting and responsible for the expulsion of flux. However, caution needs to be the watchword when a new superconductor is reported.

The geometry of a sample can affect the measured value of the critical magnetic field H_c. If H is along a needle-shaped sample, one gets a good measurement of H_c, but for a sphere, or a flat or needle-shaped object with the field perpendicular to the flat side, there can be domains of normal and superconducting material. This is called the intermediate state.

Superconductors are classified according to their magnetic properties. Assuming no geometric effects coming from "intermediate state" phenomena, if a superconductor becomes normal for $H > H_c$ and is a perfect diamagnet for $H < H_c$, then it is a type I superconductor (Fig. 14.8). If the above is not true, then it is a type II superconductor. Essentially, magnetic flux penetrates a type II superconductor on a microscopic length scale (unlike

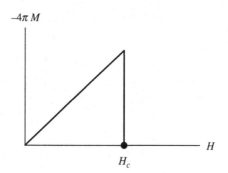

Figure 14.8   Magnetic behavior of a type I superconductor.

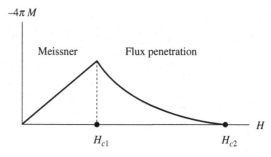

Figure 14.9   Magnetic behavior of a type II superconductor, which does not trap flux. The magnetic behavior is reversible.

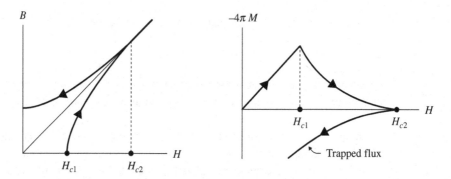

Figure 14.10   Magnetic behavior of a type II superconductor with flux trapping sites.

the intermediate state domains) at a magnetic field H_{c1}, and the superconductivity remains until an upper critical field H_{c2} is reached. This state is the mixed or vortex state for superconductors (Figs. 14.9 and 14.10).

14.2 Theories of superconductivity

The thermodynamic properties, zero resistivity, and the Meissner effect were the pillars on which the early theories were built. The London[6] theory expressed the current density within a skin depth of the surface of a superconductor in terms of a magnetic vector potential, and the Gorter–Casimir[7] model envisioned a two-fluid description of superconductors with normal and superconducting "fluids" of electrons, with a ratio depending on temperature. In 1950, Ginzburg and Landau[8] (GL) introduced a phenomenological theory with a complex order parameter that described the macroscopic properties of superconductors very well. And in 1957, Bardeen, Cooper, and Schrieffer[9] (BCS) produced a microscopic description of superconductivity that explains all the properties for a broad class of superconductors. Since 1957, refinements of BCS theory and extensions have allowed the derivation of the GL equations from microscopic theory, and these developments have resulted in detailed and predictive calculations of superconducting properties.

14.2.1 London theory

The London theory explains the Meissner effect and many of the electrodynamic properties of superconductors. We begin by considering electrons in a magnetic field where the kinetic momentum of an electron is given by $\mathbf{p} - \frac{e}{c}\mathbf{A} = m\mathbf{v}$, and the current density for N electrons is

$$\mathbf{j}(\mathbf{r}) = \sum_{\ell=1}^{N} \int \left\{ \frac{e\hbar}{2mi}\left[\psi^*\nabla_\ell\psi - \psi\nabla_\ell\psi^*\right] - \frac{e^2}{mc}\mathbf{A}(\mathbf{r}_\ell)\psi^*\psi \right\} \times \delta(\mathbf{r} - \mathbf{r}_\ell)d\mathbf{r}_1,\ldots,d\mathbf{r}_N,$$

(14.5)

where $\psi(\mathbf{r}_1,\ldots,\mathbf{r}_N)$ is the many-electron wavefunction. The first term produces paramagnetism, while the second term is the diamagnetic contribution. London theory added the assumption that the wavefunction is rigid with respect to a transverse vector potential $\mathbf{A}$, where $\nabla \cdot \mathbf{A} = 0$. Therefore, in a superconductor, the paramagnetic term vanishes, and we have

$$\mathbf{j}(\mathbf{r}) = -\frac{e^2}{mc}n_s\mathbf{A}.$$

(14.6)

This is the London equation, where n_s is the number density of superconducting electrons, unlike Ohm's law for normal electrons where

$$\mathbf{j} = \sigma\mathbf{E}.$$

(14.7)

[6] F. London and H. London, "The electromagnetic equations of the supraconductor," *Proc. R. Soc. Lond. A* 149(1935), 75.

[7] C. J. Gorter and H. Casimir, "On supraconductivity," *Physica* 1(1934), 305.

[8] V. L. Ginzburg and L. D. Landau, "On the theory of superconductivity," *Soviet Phys. JETP* 20(1950), 1064.

[9] J. Bardeen, L. N. Cooper, and J. R. Schrieffer, "Theory of superconductivity," *Phys. Rev.* 108(1957), 1175.

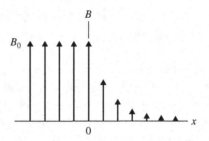

Penetration of a magnetic field into a superconductor; $x = 0$ is the metal surface. The decay length is λ_L.

The London equation connects the current density and the magnetic vector potential, and this leads to the Meissner effect,

$$B(x) = B_0 e^{-x/\lambda_L}, \tag{14.8}$$

where λ_L is the London penetration depth and x is along the surface normal of a superconducting sample into the bulk.

Starting with Maxwell's equations, together with Eq. (14.6), we obtain

$$\nabla \times \mathbf{B} = \frac{4\pi}{c}\mathbf{j} = -\frac{4\pi n_s e^2}{mc^2}\mathbf{A}. \tag{14.9}$$

Using

$$\nabla \times (\nabla \times \mathbf{B}) = \nabla(\nabla \cdot \mathbf{B}) - \nabla^2\mathbf{B}, \tag{14.10}$$

we have

$$\nabla^2\mathbf{B} = \frac{\mathbf{B}}{\lambda_L^2}. \tag{14.11}$$

This then leads to an expression for the temperature-dependent London penetration depth and plasma frequency associated with the superconducting electrons

$$\lambda_L^2(T) = \frac{mc^2}{4\pi e^2 n_s(T)} = \frac{c^2}{\omega_p^2(T)}. \tag{14.12}$$

Typically, $\lambda_L \sim 500$ Å (Fig. 14.11).

We note that this result is different from the skin effect for an ordinary metal, where

$$\nabla \times \mathbf{B} = \frac{4\pi}{c}\mathbf{j} = \frac{4\pi}{c}\sigma\mathbf{E} \tag{14.13}$$

leads to

$$\nabla \times (\nabla \times \mathbf{B}) = \frac{4\pi\sigma}{c}(\nabla \times \mathbf{E}), \tag{14.14}$$

which yields

$$-\nabla^2 \mathbf{B} = -\frac{4\pi\sigma}{c^2}\frac{\partial \mathbf{B}}{\partial t}, \qquad (14.15)$$

$$k^2 = i\frac{4\pi\omega\sigma}{c^2}, \qquad (14.16)$$

$$\lambda_{\text{skin}} \sim \frac{1}{\sqrt{\omega\sigma}}, \qquad (14.17)$$

where $\lambda_{\text{skin}} \to \infty$ as $\omega \to 0$.

To gain some insight into the nature of the superconducting state, it is instructive to compare the excitation spectrum of a normal metal with that of a superconductor. For a normal metal, if we examine the energy difference between the lowest-energy states with $N+1$ and N electrons, we get

$$E_{N+1} - E_N = \frac{dE}{dN} \sim \frac{E_{\text{F}}}{N} = \frac{10^1}{10^{22}} = 10^{-21}\, eV \ll k_{\text{B}}T \qquad (14.18)$$

for almost any imaginable temperature (for a macroscopic solid). This situation is shown in Fig. 14.12. If we view the system "as a whole," the picture is shown in Fig. 14.13, where the first excited state is microscopically higher in energy than the ground state. The ground state is a filled Fermi sea, while the excited state has excited electrons and holes.

For superconductors, the ground state and the first excited state are separated by a macroscopic gap Δ (2Δ is previously called $\hbar\omega_c$ in an optical measurement that creates an electron and hole, which is $\sim kT_c$ (Fig. 14.14)).

The gap comes from electron–electron interactions, which suggests a special stability, and there is a phase transition at $T = T_c$. The gap manifests itself strongly at the Fermi energy. The electronic band structure given by $E(\mathbf{k})$ will still be essentially the same, except for wavevectors close to the Fermi wavevector k_{F}, as shown in Fig. 14.15.

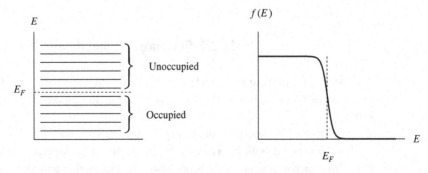

Figure 14.12 Energy level diagram of normal electron gas, where E_{F} is the Fermi level. The smearing at E_{F} in the distribution function $f(E)$ is assumed here to arise from finite temperature effects.

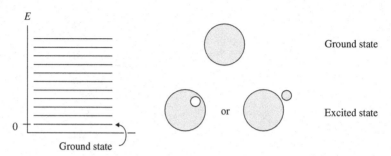

Figure 14.13 Quasiparticle excited states of a normal electron gas. The ground state is represented by a shaded circle symbolizing a Fermi sphere. The small circles are electrons (shaded) and holes (unshaded) for an excited state.

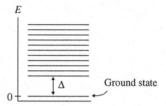

Figure 14.14 Quasiparticle excitation spectrum of a superconductor, where Δ is the superconducting energy gap.

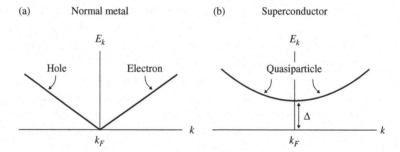

Figure 14.15 Dispersion curves for a normal metal and a superconductor for wavevectors **k** near the Fermi wavevector k_F.

14.2.2 Ginzburg–Landau theory

The theory of Ginzburg and Landau (GL) was developed as an extension of the London theory to take into account the spatial dependence of the superconducting electron density $n_s(\mathbf{r})$.

Although the GL theory set out to explore the variations of $n_s(\mathbf{r})$ to treat problems such as those connected with interfaces between normal and superconducting materials and thin film superconductors, the theory also encompassed thermodynamic properties and type II superconductivity, and had many applications. In addition, the two equations that resulted have the two important lengths for superconductors, the penetration depth and the coherence length. The first is connected to the variation of the magnetic field near a superconducting/normal interface, and the latter is associated with the variation in density of the superconducting electrons (actually the order parameter, as discussed below). The GL equations were derived to be valid near T_c, although some extensions have been made to other regimes. After the BCS theory was developed, the GL theory was derived from the microscopic theory.

An important observation by Pippard[10] based on detailed measurements of the London penetration depth demonstrated the need for a second characteristic length in addition to λ_L to understand the behavior of superconductors. The generalization of the London theory by Pippard was to allow the London equation to include non-local behavior. Instead of $\mathbf{j}(\mathbf{r})$ depending on $\mathbf{A}(\mathbf{r})$, $\mathbf{j}(\mathbf{r})$ was expressed as a function of $\mathbf{A}(\mathbf{r}')$, where the influence on $\mathbf{j}(\mathbf{r})$ within a range

$$|\mathbf{r} - \mathbf{r}'| < \xi_0 \tag{14.19}$$

was important. The length ξ_0 is a coherence length, which is roughly the length over which an electron "knows" that it is superconducting. Since experimentally it was shown that the penetration depth depended on the normal state mean free path ℓ, this implied that the non-local effects depended on ℓ. Similar non-local and mean free path effects are seen in normal metals, and Ohm's law was generalized to include these effects. A similar analysis by Pippard for superconductors led to the expression

$$\mathbf{j}_s(\mathbf{r}) = -\frac{3n_s e^2}{4\pi \xi_0 mc} \int \frac{\mathbf{R}\left[\mathbf{R} \cdot \mathbf{A}(\mathbf{r}')\right] e^{-R/\xi} d^3 r'}{R^4}, \tag{14.20}$$

where $\mathbf{R} = \mathbf{r} - \mathbf{r}'$, and

$$\frac{1}{\xi} = \frac{1}{\xi_0} + \frac{1}{\alpha \ell}, \tag{14.21}$$

where α is an empirical constant of order unity. For short ℓ (the dirty limit), $\xi \to \ell$ and the effective penetration depth increases in agreement with experiment. To obtain an estimate for ξ_0, Pippard used an uncertainty principle argument,

$$\xi_0 \sim \Delta x \sim \frac{\hbar}{\Delta p} \sim \frac{\hbar v_F}{\Delta E} \sim \frac{\hbar v_F}{k_B T_c}. \tag{14.22}$$

The subsequent microscopic derivations of ξ_0 give similar results. Equation (14.20) reduces to a simpler expression for the case where $\mathbf{A}(\mathbf{r})$ varies slowly over a coherence length

$$\mathbf{j}_s(\mathbf{r}) = -\frac{c}{4\pi} \frac{1}{\lambda_L^2} \frac{\xi}{\xi_0} \mathbf{A}(\mathbf{r}), \tag{14.23}$$

[10] A. B. Pippard, "An experimental and theoretical study of the relation between magnetic field and current in a superconductor," *Proc. R. Soc. Lond. A* 216(1953), 547.

so the effective penetration depth is

$$\lambda_{\mathrm{eff}} = \lambda_{\mathrm{L}}\sqrt{1 + \xi_0/\ell} \tag{14.24}$$

if $\alpha \approx 1$, and therefore the measured increases in this quantity are explained.

In their macroscopic derivation, Ginzburg and Laudau introduce an effective complex wavefunction $\psi(\mathbf{r})$ that describes the condensed superconducting electrons at a given temperature with

$$|\psi|^2 = \frac{n_s(T)}{n}. \tag{14.25}$$

Here, $n_s(T)$ is equal to the density of electrons condensed into the superconducting state, and n = total number of electrons. This approach follows the logic of the Gorter–Casimir two-fluid model, where

$$|\psi|^2 = 1 \text{ at } T = 0$$
$$= 0 \text{ at } T > T_c. \tag{14.26}$$

However, in GL theory, ψ is treated as a complex order parameter, which is determined at each $\mathbf{r}$ by minimizing the free energy. To derive the GL equations, we begin by considering the homogeneous case. The free energy density difference (per unit volume) $f(\psi, T) = f_{\mathrm{S}} - f_{\mathrm{N}}$ can be taken to be the change in free energy between the superconducting and normal phases. We assume that f can be expanded in $|\psi|^2$ as Landau had done previously to study other phase transitions. Near T_c, we expand f as follows:

$$f(\psi, T) = a(T)|\psi|^2 + \frac{1}{2}b(T)|\psi|^4 + \cdots \tag{14.27}$$

for $|\psi|^2 \ll 1$ (near the phase transition). Only even powers are included. The linear term in ψ is excluded because f is real and a physical quantity, and therefore f cannot depend on overall phase. Odd powers of $|\psi|$ are excluded because f would not be analytic as $\psi \to 0$. Following Landau's[11] theory of second-order phase transitions, and assuming the superconducting phase transition is second order, we keep the first two terms in the expansion. The coefficient $b(T)$ for the $|\psi|^4$-term should be positive for all T; if not, the free energy minimum would occur for an arbitrarily large value of $|\psi|^2$. If $a > 0$, the minimum of the free energy would occur at $|\psi|^2 = 0$, which corresponds to the normal state. If $a < 0$, we have a minimum at a finite $|\psi|$, as shown in Fig. 14.16(a). Therefore, $a(T)$ must change sign at $T = T_c$. Expanding $a(T)$, it should have the form

$$a(T) = \alpha \left(1 - \frac{T}{T_c}\right) + \cdots \tag{14.28}$$

---

[11] L. D. Landau, "On the theory of phase transitions," *Soviet Phys. JETP* 7(1937), 627.

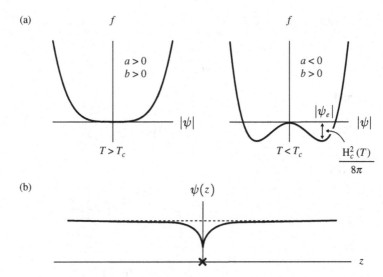

Figure 14.16 (a) Free energy density difference versus order parameter. (b) Perturbation of the position-dependent order parameter in a small region near $z = 0$.

For the homogeneous case, we denote the equilibrium value of ψ as ψ_e. By minimizing f, we get

$$\frac{\partial f}{\partial |\psi|^2} = 0 = a(T) + b(T)|\psi_e|^2, \tag{14.29}$$

$$|\psi_e|^2 = -\frac{a(T)}{b(T)}, \tag{14.30}$$

and

$$f_e(T) = -\frac{1}{2}\frac{a^2(T)}{b(T)}. \tag{14.31}$$

Hence, one can expect $n_s(T) \propto (1 - T/T_c)$, and $a(T)$ and $b(T)$ can be determined from experiment. Recall that

$$f_e(T) = f_S(T) - f_N(T) = -\frac{H_c^2(T)}{8\pi}, \tag{14.32}$$

hence

$$\frac{1}{2}\frac{a^2(T)}{b(T)} = \frac{H_c^2(T)}{8\pi}. \tag{14.33}$$

Also, the London penetration depth is related to $n_s(T)$:

$$\lambda^2(T) \propto \frac{1}{n_s(T)}. \tag{14.34}$$

Therefore, we have

$$\frac{\lambda^2(0)}{\lambda^2(T)} = \frac{n_s(T)}{n} = |\psi_e(T)|^2 = -\frac{a(T)}{b(T)} \tag{14.35}$$

and

$$b(T) = -a(T)\frac{\lambda^2(T)}{\lambda^2(0)}. \tag{14.36}$$

So, for $T < T_c$,

$$a(T) = -\frac{H_c^2(T)}{4\pi}\frac{\lambda^2(T)}{\lambda^2(0)} \tag{14.37}$$

and

$$b(T) = \frac{H_c^2(T)}{4\pi}\frac{\lambda^4(T)}{\lambda^4(0)}. \tag{14.38}$$

Therefore, under these conditions, $f(T)$ can be expressed as a function of experimentally measured quantities.

If we now consider the inhomogeneous case, then the order parameter has a spatial-dependent phase

$$\psi(\mathbf{r}) = |\psi(\mathbf{r})|e^{i\theta(\mathbf{r})}, \tag{14.39}$$

and it may not be uniform in space, resulting in additional terms in f as shown in Fig. 14.16(b). Both the kinetic energy density associated with $\nabla\psi(\mathbf{r})$ arising from the wiggles in the many-body wavefunction describing n_s and the potential energy density, which is influenced by the variation of $n_s(\mathbf{r})$ in a region surrounding the point in question, contribute changes. Ginzburg and Landau suggested that if $|\psi|^2$ varies slowly in space, it should be sufficient to keep the leading term in $|\nabla\psi|^2$.

In a magnetic field, they assumed that the free energy has the following additional terms:

$$\int \frac{n^*}{2m^*}\left|\frac{\hbar}{i}\nabla\psi + \frac{e^*}{c}\mathbf{A}(\mathbf{r})\psi(\mathbf{r})\right|^2 d^3r + \int \frac{1}{8\pi}B^2(\mathbf{r})d^3r. \tag{14.40}$$

Later, from microscopic theory, it was found that $n^* = \frac{n}{2}$, $e^* = 2|e|$, and $m^* = 2m$. Thus, the total free energy is

$$F(\psi(\mathbf{r}), T, \mathbf{A}(\mathbf{r}), \ldots) = \int \left(a(T)|\psi|^2 + \frac{1}{2}b(T)|\psi|^4\right)d\mathbf{r}$$

$$+ \int \frac{n^*}{2m^*}\left|\frac{\hbar}{i}\nabla\psi + \frac{e^*}{c}\mathbf{A}(\mathbf{r})\psi(\mathbf{r})\right|^2 d\mathbf{r}$$

$$+ \int \frac{1}{8\pi}B^2(\mathbf{r})d\mathbf{r}. \tag{14.41}$$

Minimizing F with respect to ψ^*, $\left(\frac{\delta F}{\delta \psi^*} = 0\right)$, one obtains the following Euler–Lagrange equation:

$$-\frac{\hbar^2}{2m^*}\left[\nabla + \frac{ie^*}{\hbar c}\mathbf{A}(\mathbf{r})\right]^2 \psi(\mathbf{r}) - \frac{H_c^2(T)}{4\pi n^*}\frac{\lambda^2(T)}{\lambda^2(0)}\left[1 - \frac{\lambda^2(T)}{\lambda^2(0)}|\psi(\mathbf{r})|^2\right]\psi(\mathbf{r}) = 0. \quad (14.42)$$

This is the first Ginzburg–Landau equation (GL1). It is a Schrödinger-like equation, but it is nonlinear in $\psi(\mathbf{r})$. The nonlinear term acts like a repulsive potential of ψ on itself, tending to favor wavefunctions that are spread out and as uniform as possible. Writing

$$\psi(\mathbf{r}) = |\psi(\mathbf{r})|e^{i\theta(\mathbf{r})} \quad (14.43)$$

and noting that

$$\psi^*\nabla\psi = |\psi|\nabla|\psi| + |\psi|^2 i\nabla\theta,$$
$$\psi\nabla\psi^* = |\psi|\nabla|\psi| - |\psi|^2 i\nabla\theta, \quad (14.44)$$

we can rewrite the integrand in the first term in Eq. (14.40) as

$$\frac{n^*}{2m^*}\left|\frac{\hbar}{i}\nabla\psi + \frac{e^*}{c}\mathbf{A}(\mathbf{r})\psi(\mathbf{r})\right|^2 = \frac{n^*}{2m^*}\hbar^2(\nabla|\psi|)^2 + \frac{n^*}{2m^*}\left(\hbar\nabla\theta + \frac{e^*}{c}\mathbf{A}\right)^2|\psi|^2. \quad (14.45)$$

The first term on the right-hand side is the extra energy associated with the spatial variations of the order parameter ψ. The second term on the right-hand side is the kinetic energy of the supercurrent in a gauge-invariant form $\left(\text{cf. } m^*\mathbf{v}_s = \mathbf{p}_s + \frac{e^*\mathbf{A}}{c}\right)$.

Minimizing $F(\psi, T, \mathbf{A})$ with respect to $\mathbf{A}(\mathbf{r})$, $\left(\frac{\delta F}{\delta \mathbf{A}} = 0\right)$, and using $\nabla \times \mathbf{B} = \frac{4\pi\mathbf{j}}{c}$ and $\mathbf{B}\cdot\delta\mathbf{B} = \mathbf{B}\cdot(\nabla \times \delta\mathbf{A})$, one arrives at the second Ginzburg–Landau equation (GL2),

$$\begin{aligned}
\mathbf{j}_s(\mathbf{r}) &= -\frac{n^*}{m^*c}|\psi(\mathbf{r})|^2(e^*)^2\mathbf{A}(\mathbf{r}) + \frac{in^*e^*\hbar}{2m^*}[\psi^*\nabla\psi - \psi\nabla\psi^*] \\
&= -\frac{n^*e^*}{m^*}\left(\hbar\nabla\theta(\mathbf{r}) + \frac{e^*}{c}\mathbf{A}\right)|\psi(\mathbf{r})|^2 \\
&= (-e^*)n^*|\psi|^2\mathbf{v}_s.
\end{aligned} \quad (14.46)$$

The first and second Ginzburg–Landau equations, in conjunction with Maxwell's equation $\nabla \times (\nabla \times \mathbf{A}) = \frac{4\pi\mathbf{J}}{c}$, lead to two nonlinear differential equations that determine the functions $\psi(\mathbf{r})$ and $\mathbf{A}(\mathbf{r})$.

We note that the boundary condition for an insulator–superconductor interface is (since there is no current into the insulator)

$$\left(\frac{\hbar}{i}\nabla + \frac{e^*}{c}\mathbf{A}(\mathbf{r})\right)\psi\bigg|_{\text{normal}} = 0, \quad (14.47)$$

and, for a metal–superconductor interface, is

$$\left(\frac{\hbar}{i}\nabla + \frac{e^*}{c}\mathbf{A}(\mathbf{r})\right)\psi\bigg|_{\text{normal}} = \frac{i}{b}\psi, \quad (14.48)$$

where b is a material-dependent constant. Also, in the weak field case and using $\nabla \times (\nabla \times \mathbf{A}) = \nabla(\nabla \cdot \mathbf{A}) - \nabla^2\mathbf{A}$, we recover the London equation

$$\nabla^2\mathbf{A} = \frac{|\psi|^2\mathbf{A}}{\lambda_L(0)}. \tag{14.49}$$

The GL theory contains another length. For $\mathbf{A} = 0$ and ψ uniform, the GL1 equation becomes

$$1 - \frac{\lambda^2(T)}{\lambda^2(0)}|\psi_e|^2 = 0, \tag{14.50}$$

which implies that

$$|\psi_e|^2 = \frac{\lambda^2(0)}{\lambda^2(T)}. \tag{14.51}$$

If ψ is perturbed slightly (as in Fig. 14.16(b)) in a plane at $z = 0$, then

$$\psi(z) = \psi_e + \delta\psi(z).$$

Since $A = 0$ and $\delta\psi$ is small, we have (from GL1)

$$\delta\psi(z) = \delta\psi_0 \exp\left(-\frac{z\sqrt{2}}{\xi}\right). \tag{14.52}$$

If we define ξ through the equation

$$\frac{\hbar^2}{2m^*}\frac{1}{\xi^2} = \frac{H_c^2(T)\lambda^2(T)}{4\pi n^*\lambda^2(0)}, \tag{14.53}$$

then

$$\xi(T) = \sqrt{\frac{2\pi n^*\hbar^2}{m^*H_c^2(T)}}\frac{\lambda(0)}{\lambda(T)}. \tag{14.54}$$

The temperature dependence of the coherence length ξ can be examined by referring to experimental properties. If we define t as

$$t = \frac{T}{T_c}, \tag{14.55}$$

then from Gorter–Casimir theory and experimentally,

$$\lambda^{-2} \propto 1 - t^4$$

and

$$H_c \propto 1 - t^2, \tag{14.56}$$

while GL theory gives

$$\lambda^{-2} \propto a(T) \propto (1 - t). \tag{14.57}$$

So, for small t, the T dependence of the coherence length ξ is

$$\xi \approx \frac{\xi_0}{\sqrt{1 - t}}. \tag{14.58}$$

From rough energetic arguments introduced earlier, ξ_0 can be estimated as

$$\xi_0 \equiv \frac{\hbar v_F}{2\Delta}, \tag{14.59}$$

where the energy gap Δ is assumed proportional to T_c, so

$$\xi_0 \sim \frac{1}{T_c}. \tag{14.60}$$

Numerically, ξ_0 for ordinary superconductors is of the order of microns. The other important length in the GL theory is the penetration depth of the magnetic field

$$\lambda = \frac{c}{\omega_p(T)}. \tag{14.61}$$

The relative magnitude of these two lengths determines the characteristics of many of the electromagnetic properties of the superconductor. The dimensionless parameter κ is defined as

$$\kappa = \frac{\lambda(T)}{\xi(T)} \approx \frac{\lambda(0)}{\xi(0)}. \tag{14.62}$$

As we will show later, the condition $\kappa \gg 1$ arises for dirty samples and type II superconductors. There are two different regimes of particular interest for superconductivity: small κ representing type I superconductors and large κ representing type II superconductors.

Type II superconductivity. To illustrate how κ, which is a material-dependent parameter, describes superconducting properties, we examine the free energy density difference inside a superconductor in an external magnetic field H:

$$f(H) = -\frac{H_c^2}{8\pi} + \frac{H^2}{8\pi}. \tag{14.63}$$

Near the surface of the superconductor, but inside, there is an energy lowering of $-\frac{\lambda H^2}{8\pi}$ per unit area, because we relax the expulsion condition, whereas there is an energy loss of $\frac{\xi H_c^2}{8\pi}$ per unit area, because there are fewer electrons in the condensate. Therefore, the free energy difference for the whole sample becomes

$$F_S - F_N = \left(-\frac{H_c^2}{8\pi} + \frac{H^2}{8\pi} \right) V + \left(\frac{\xi H_c^2}{8\pi} - \frac{\lambda H^2}{8\pi} \right) A, \tag{14.64}$$

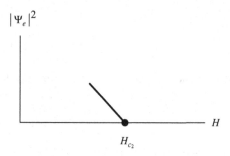

Figure 14.17 The order parameter of a type II superconductor near the critical field H_{c_2}.

where V and A denote volume and surface area, respectively. The second term (involving A) is the surface energy, which can be positive or negative.

For case 1 ($\kappa \ll 1 \Rightarrow \lambda \ll \xi$), the surface energy is > 0 and the system tries to minimize the interface area. This results in flux exclusion and the Meissner effect (type I superconductivity). For case 2 ($\kappa \gg 1 \Rightarrow \lambda \gg \xi$), the surface energy can be < 0 at sufficiently large H. In this case, the system allows flux penetration to increase the superconducting normal interface and this leads to an incomplete Meissner effect (type II superconductivity). In the next section, it will be shown that a superconductor is of type II if $k > \frac{1}{\sqrt{2}}$. For that case, the M vs. H curve is shown in Fig. 14.9, and the order parameter field dependence near H_{c_2} is shown in Fig. 14.17.

To explore the behavior of a type II superconductor[12] in a magnetic field, we examine the nucleation of superconductivity in bulk samples. We define H_{c_2} to be the highest field at which superconductivity can exist in the interior of a large sample in a decreasing external field, and apply the GL theory to determine H_{c_2}. We begin by rewriting the GL equations in terms of dimensionless quantities:

$$
\begin{aligned}
\psi' &= \psi/\psi_e, \\
H'^2 &= H^2/(2H_{cb}^2), \\
\mathbf{A}' &= \mathbf{A}/(2H_{cb}\lambda),
\end{aligned}
\tag{14.65}
$$

and

$$
\nabla' = \lambda \nabla.
\tag{14.66}
$$

The GL parameter κ becomes

$$
\kappa = \frac{\sqrt{2}e^* H_{cb}\lambda^2}{\hbar c} = \frac{\lambda}{\xi},
\tag{14.67}
$$

[12] A. A. Abrikosov, "On the magnetic properties of superconductors of the second group," *Soviet Phys. JETP* 6(1957), 489.

where $\lambda = \frac{c}{\omega_p}$, as before, and H_{cb} is the bulk thermodynamic critical field. For simplicity, we drop all primes, which gives the two GL equations as

$$\left(i\frac{\nabla}{\kappa} + \mathbf{A}\right)^2 \psi = \psi - |\psi|^2 \psi \qquad (14.68)$$

and

$$\nabla \times (\nabla \times \mathbf{A}) = -|\psi|^2 \mathbf{A} + \frac{1}{2i\kappa}\left(\psi^*\nabla\psi - \psi\nabla\psi^*\right). \qquad (14.69)$$

For high fields where ψ is small, we can linearize GL1 to give

$$\left(i\frac{\nabla}{\kappa} + \mathbf{A}\right)^2 \psi = \psi \qquad (14.70)$$

or

$$(i\nabla + \kappa\mathbf{A})^2 \psi = \kappa^2\psi. \qquad (14.71)$$

If we now ask for the highest field for which this equation has a non-trivial solution, we can determine the κ for superconductivity above H_{cb}. In the Landau gauge, $\mathbf{A} = (0, H_0 x, 0)$, $H = \nabla \times A = H_0\hat{z}$, and the linearized GL1 resembles a Schrödinger equation,

$$\left(-\nabla^2 + i2\kappa H_0 x\frac{\partial}{\partial y} + \kappa^2 H_0^2 x^2\right)\psi = \kappa^2\psi. \qquad (14.72)$$

Since the effective potential depends only on x, we may write

$$\psi(x,y,z) = e^{ik_y y}e^{ik_z z}f(x). \qquad (14.73)$$

This gives

$$-\frac{\partial^2}{\partial x^2}f(x) + (\kappa H_0)^2(x - x_0)^2 f(x) = (\kappa^2 - k_z^2)f(x), \qquad (14.74)$$

where $x_0 = \frac{k_y}{\kappa H_0}$. This is a harmonic oscillator equation and the solutions for κ are

$$\kappa^2 - k_z^2 = \left(n + \frac{1}{2}\right)2\kappa H_0 \qquad (14.75)$$

or

$$H_0 = \frac{\kappa^2 - k_z^2}{(2n + 1)\kappa}, \qquad (14.76)$$

where $n = 0, 1, 2, \ldots$ The largest H_0 corresponds to $k_z = 0$ and $n = 0$. Hence, the critical field is $H_0 = \kappa$ in dimensionless units. In physical units, the critical field is therefore

$$\sqrt{2}H_{cb}\kappa = H_{c2} \qquad (14.77)$$

and the corresponding wavefunction is

$$f(x) = \exp\left(-\frac{\kappa^2(x-x_0)^2}{2}\right). \tag{14.78}$$

We note that the factor $e^{ik_y y}$ shifts the location of the local minimum of the effective potential. This is not important in the present calculation, but would be if one were to deal with superconductivity near the surface of a finite sample or in a vortex lattice where one is constructing a space-filling solution rather than a localized one.

The above result illustrates that the value of $\kappa = \frac{1}{\sqrt{2}}$ separates the materials for which

$$H_{c2} > H_{cb} \quad \text{(type II superconductivity)}$$

and those for which

$$H_{c2} < H_{cb} \quad \text{(type I superconductivity).} \tag{14.79}$$

Examining the solutions of the GL equations in more detail for a type II superconductor shows that for $H_{cb} < H < H_{c2}$, the complete solution for ψ is in fact a lattice of normal regions at $T = 0$, known as the Abrikosov lattice. This refers to Abrikosov's solution of a square lattice of normal regions. A subsequent, more accurate solution shows that a triangular lattice of normal regions is more stable. The normal region is called a vortex. Each normal vortex contains a single quantum unit of magnetic flux $\phi_0 = \frac{hc}{2e}$. The factor of 2 in the denominator comes from Cooper pairing, so $e^* = 2e$. We can argue for this result by observing that the negative interface energy between the normal and superconducting states will maximize the number of normal regions and the smallest possible flux per vortex line.

Now consider a loop c around a vortex that is well within the superconducting region, so that there is no magnetic field penetration and therefore no current. Along this loop, $\psi(\mathbf{r}) = |\psi|e^{i\theta(\mathbf{r})}$ and $v_s = \frac{1}{m*}(\hbar\nabla\theta + \frac{e^*}{c}\mathbf{A})$. We have

$$\oint_c d\boldsymbol{\ell} \cdot \left(\hbar\nabla\theta + \frac{e^*}{c}\mathbf{A}\right) = 0 \tag{14.80}$$

and the flux enclosed

$$\begin{aligned}
\phi &= \oint_c \mathbf{A} \cdot d\boldsymbol{\ell} = -\frac{c}{e^*}\oint d\boldsymbol{\ell} \cdot \hbar\nabla\theta \\
&= \frac{\hbar c}{e^*}\oint \nabla\theta \cdot d\boldsymbol{\ell} = \frac{\hbar c}{2e}(2\pi n),
\end{aligned} \tag{14.81}$$

where $n = 1, 2, \ldots$ from the single-valueness of the order parameter ψ. Equation (14.81), together with the condition for the smallest value of flux in each vortex, gives $\phi = hc/2e = \phi_0$. In scanning tunneling spectroscopy, one can measure the trapped localized quasiparticle states in a vortex core. See Fig. 14.18.

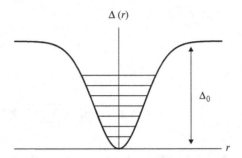

$\Delta(r)$

Δ_0

r

Figure 14.18 The behavior of the energy gap near a vortex core. The horizontal lines represent the energies associated with localized normal electron states in the core.

14.2.3 Microscopic theory

The Cooper pair. The wavefunction of the superconducting phase is qualitatively different from that of the normal phase in that the electrons are strongly correlated. In particular, pair correlations dominate and this pairing concept is essential to BCS theory. To illustrate the pairing concept, we consider the pairing problem posed by Leon Cooper.[13] He showed that the normal state is unstable for an arbitrarily weak attractive interaction between electrons in the following sense. By choosing two electrons from the non-interacting Fermi sea and allowing them to occupy orbitals above E_{F} and interact through an attractive potential, he showed that these two electrons can be in a lower-energy state. We begin with two electrons in two planewave states:

$$\phi_{\mathbf{k}_1}(\mathbf{r}_1) = \frac{1}{\sqrt{\Omega}} e^{i\mathbf{k}_1 \cdot \mathbf{r}_1}, \ \phi_{\mathbf{k}_2}(\mathbf{r}_2) = \frac{1}{\sqrt{\Omega}} e^{i\mathbf{k}_2 \cdot \mathbf{r}_2}. \tag{14.82}$$

By allowing an attractive interaction between the electrons, we can express the pair wavefunction as

$$\psi_{\mathrm{pair}}(\mathbf{r}_1, \mathbf{r}_2) = \sum_{\mathbf{k}_1, \mathbf{k}_2} a(\mathbf{k}_1, \mathbf{k}_2) \phi_{\mathbf{k}_1}(\mathbf{r}_1) \phi_{\mathbf{k}_2}(\mathbf{r}_2). \tag{14.83}$$

By changing to center of mass and relative coordinates, we have

$$\mathbf{K} = \mathbf{k}_1 + \mathbf{k}_2,$$
$$\mathbf{R} = \frac{\mathbf{r}_1 + \mathbf{r}_2}{2},$$
$$\mathbf{k} = \frac{\mathbf{k}_1 - \mathbf{k}_2}{2}, \tag{14.84}$$
$$\mathbf{r} = \mathbf{r}_1 - \mathbf{r}_2,$$

[13] L. N. Cooper, "Bound electron pairs in a degenerate Fermi gas," *Phys. Rev.* 104(1956), 1189.

then

$$\mathbf{k}_1 \cdot \mathbf{r}_1 + \mathbf{k}_2 \cdot \mathbf{r}_2 = \mathbf{K} \cdot \mathbf{R} + \mathbf{k} \cdot \mathbf{r} \tag{14.85}$$

and

$$\psi_{\text{pair}}(\mathbf{r}_1, \mathbf{r}_2) = \sum_{\mathbf{k},\mathbf{K}} a(\mathbf{k}, \mathbf{K}) e^{i\mathbf{K}\cdot\mathbf{R}} e^{i\mathbf{k}\cdot\mathbf{r}}. \tag{14.86}$$

As in exciton formation, translational invariance implies only a single $\mathbf{K}$ in the sum above, and for the lowest-energy state, $\mathbf{K} = 0$, yielding

$$\psi_{\text{pair}}(\mathbf{r}_1, \mathbf{r}_2) = \sum_{\mathbf{k}} a(\mathbf{k}) e^{i\mathbf{k}\cdot\mathbf{r}_1} e^{-i\mathbf{k}\cdot\mathbf{r}_2}, \tag{14.87}$$

where $a(\mathbf{k}) = 0$ if $\mathbf{k} < \mathbf{k}_F$ because this is a degenerate Fermi system. The resulting equation to solve is (measuring E from $2\,E_F$)

$$\left(-\frac{\hbar^2}{2m} \left(\nabla_1^2 + \nabla_2^2 \right) + V(\mathbf{r}_1, \mathbf{r}_2) \right) \psi(\mathbf{r}_1, \mathbf{r}_2) = (E + 2E_F) \psi(\mathbf{r}_1, \mathbf{r}_2). \tag{14.88}$$

If we assume $V(\mathbf{r}_1, \mathbf{r}_2) = V(\mathbf{r})$ and refer the energy to that of two particles on the Fermi surface, then Eq. (14.88) becomes

$$-\frac{\hbar^2}{m} \psi''(\mathbf{r}) + V(\mathbf{r}) \psi(\mathbf{r}) = (E + 2E_F) \psi(\mathbf{r}). \tag{14.89}$$

Using Eq. (14.87) and defining

$$\varepsilon_{\mathbf{k}} = \frac{\hbar^2 k^2}{2m} - E_F, \quad V_{\mathbf{k}\mathbf{k}'} = \langle \mathbf{k} | V | \mathbf{k}' \rangle, \tag{14.90}$$

we arrive at the Bethe–Goldstone equation

$$\sum_{\mathbf{k}'}' V_{\mathbf{k}\mathbf{k}'} a(\mathbf{k}) = (E - 2\varepsilon_{\mathbf{k}}) a(\mathbf{k}), \tag{14.91}$$

where the summation is over all $\mathbf{k}' \neq \mathbf{k}$. This describes the scattering of a pair $(\mathbf{k}, -\mathbf{k}) \to (\mathbf{k}', -\mathbf{k}')$, as shown in Fig. 14.19.

Figure 14.19   Feynman diagram for pair scattering, involving an arbitrary interaction represented by a dashed line.

If $V_{kk'}$ is attractive, we can expect that the electrons may bind in pairs. Cooper chose $V_{kk'}$ to be attractive and constant over an energy range from zero to some energy cutoff E_D measured from the Fermi energy E_F:

$$V_{kk'} = \begin{cases} -\dfrac{V}{\Omega} & \varepsilon_k, \ \varepsilon_{k'} < E_D, \\ 0 & \text{otherwise.} \end{cases} \tag{14.92}$$

The Bethe–Goldstone equation then becomes

$$(E - 2\varepsilon_k)\, a(\mathbf{k}) = -\frac{V}{\Omega} \sum_{k'}{}' a(\mathbf{k}'), \tag{14.93}$$

and

$$a(\mathbf{k}) = \frac{c}{E - 2\varepsilon_k} \tag{14.94}$$

for all $|\mathbf{k}| > k_F$. Here c is the constant corresponding to the right hand side of Eq. (14.93). This gives a self-consistent condition on E and c,

$$c = -\frac{V}{\Omega} \sum_{k'} \frac{c}{E - 2\varepsilon_{k'}}, \tag{14.95}$$

or

$$1 = V \int_0^{E_D} \frac{N_\uparrow(\varepsilon)\, d\varepsilon}{2\varepsilon - E}, \tag{14.96}$$

where $N_\uparrow(\varepsilon)$ is the density of states for one spin type. If $E_D \ll E_F$, the electronic density of states for a single spin is $N_\uparrow(\varepsilon) \approx N_\uparrow(0)$. Therefore, we have (denoting $N_\uparrow(0) = N(0)$)

$$1 = \frac{N(0)V}{2} \int_0^{E_D} \frac{d\varepsilon}{\varepsilon - E/2}. \tag{14.97}$$

By integrating, we obtain

$$E = -\frac{2E_D}{\exp\left(\frac{2}{N(0)V}\right) - 1}. \tag{14.98}$$

For weak coupling, $N(0)V \ll 1$, we obtain

$$E = -2E_D e^{\left(-\frac{2}{N(0)V}\right)}. \tag{14.99}$$

For strong coupling, $N(0)V \gg 1$, we obtain

$$E = -E_D N(0)V. \tag{14.100}$$

Since $E < 0$, we have binding and the normal state is unstable for an arbitrarily weak attractive interaction. The electrons are expected to form Cooper pairs. We note that if we let $x = \frac{N(0)V}{2}$, the solution $\sim e^{-1/x}$ is non-analytic as $x \to 0$. Hence, it is difficult to obtain the above result with ordinary perturbation approaches.

More generally, we can examine the pairing interactions between electrons, starting first with the repulsive Coulomb interaction. Expressing the pairing potential as

$$V_{\mathbf{kk'}} = \frac{1}{\Omega} \int V(\mathbf{r}) e^{i(\mathbf{k'}-\mathbf{k})\cdot\mathbf{r}} d\mathbf{r}, \tag{14.101}$$

for the Coulomb interaction,

$$V^C(\mathbf{r}) = \frac{e^2}{r}, \tag{14.102}$$

and, for $\mathbf{k'} - \mathbf{k} = \mathbf{q}$,

$$V^C_{\mathbf{kk'}} = V_{\mathbf{q}} = \frac{4\pi e^2}{\Omega q^2}, \tag{14.103}$$

which is repulsive and corresponds to pair scattering, as shown in Fig. 14.20.

The Coulomb interaction is mediated by photons and is basically instantaneous ($v = c$), but a screened interaction is not. Here we need to add the effects of dielectric screening and obtain (Fig. 14.21)

$$V_{\mathbf{q}} = \frac{4\pi e^2}{\Omega q^2 \epsilon(\mathbf{q}, \omega)}. \tag{14.104}$$

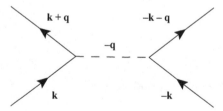

Figure 14.20   Coulomb pair scattering. The dashed line represents the propagation of a photon.

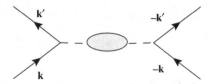

Figure 14.21   Screened Coulomb interaction. The shaded area represents the screening of the Coulomb interaction, which can be implemented using a dielectric function.

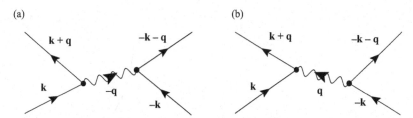

Figure 14.22 Two types of phonon-mediated interaction: (a) emission of a phonon with wavevector $-\mathbf{q}$ and (b) absorption of a phonon with wavevector $\mathbf{q}$.

An attractive phonon-mediated interaction may be obtained using second-order perturbation theory in the electron–phonon interaction V_{ep} (Fig. 14.22).

If we assume

$$V^{\text{Phonon-induced}}_{\substack{\mathbf{k},\mathbf{k}' \\ \mathbf{k}'=\mathbf{k}+\mathbf{q}}} = V^{\text{Phonon-induced}}_{\mathbf{q}}, \tag{14.105}$$

then we can get the form of this phonon-induced pairing interaction. Starting with the free electron states and $\mathbf{k}' = -\mathbf{k} - \mathbf{q}$

$$E_i = 2\varepsilon_{\mathbf{k}}, \quad E_f = 2\varepsilon_{\mathbf{k}'}, \tag{14.106}$$

we can envision two intermediate states $j = a, b$ (Fig. 14.22), and using perturbation theory,

$$\langle i| V^{\text{Phonon-induced}} |f\rangle = \frac{1}{2} \sum_j \left[\frac{\langle i| V_{ep} |j\rangle \langle j| V_{ep} |f\rangle}{E_f - E_j} + \frac{\langle i| V_{ep} |j\rangle \langle j| V_{ep} |f\rangle}{E_i - E_j} \right], \tag{14.107}$$

and energy denominators (using $\omega_{\mathbf{q}} = \omega_{-\mathbf{q}}$),

$$E_f - E_a = \varepsilon_{\mathbf{k}'} - \varepsilon_{\mathbf{k}} - \hbar\omega_{\mathbf{q}},$$
$$E_f - E_b = \varepsilon_{\mathbf{k}'} - \varepsilon_{\mathbf{k}} - \hbar\omega_{\mathbf{q}},$$
$$E_i - E_a = \varepsilon_{\mathbf{k}} - \varepsilon_{\mathbf{k}'} - \hbar\omega_{\mathbf{q}},$$
$$E_i - E_b = \varepsilon_{\mathbf{k}} - \varepsilon_{\mathbf{k}'} - \hbar\omega_{\mathbf{q}}, \tag{14.108}$$

we obtain

$$V^{\text{Phonon-induced}}_{\mathbf{q}} = |V^{ep}_{\mathbf{q}}|^2 \left[\frac{1}{(\varepsilon_{\mathbf{k}'} - \varepsilon_{\mathbf{k}}) - \hbar\omega_{\mathbf{q}}} - \frac{1}{(\varepsilon_{\mathbf{k}'} - \varepsilon_{\mathbf{k}}) + \hbar\omega_{\mathbf{q}}} \right]$$
$$= \frac{|V^{ep}_{\mathbf{q}}|^2 2\hbar\omega_{\mathbf{q}}}{(\varepsilon_{\mathbf{k}'} - \varepsilon_{\mathbf{k}})^2 - (\hbar\omega_{\mathbf{q}})^2}, \tag{14.109}$$

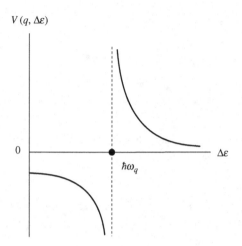

$V(q, \Delta\varepsilon)$

$\Delta\varepsilon$

0

$\hbar\omega_q$

Figure 14.23 The energy dependence of the Bardeen–Pines interaction as a function of energy transfer $\Delta\varepsilon$.

where $\langle i| V_{ep} |j\rangle \equiv V_{\mathbf{q}}^{ep}$ for a specific phonon with $\mathbf{q}$ and $\omega_{\mathbf{q}}$. This is the Bardeen–Pines[14] interaction (Fig. 14.23), which Cooper modeled with a constant attractive interaction.

The BCS theory. In 1957, using the pairing concept along with the assumption that the phonon-induced attractive interaction is the mechanism that causes pair formation, Bardeen, Cooper, and Schrieffer developed their highly successful BCS theory of superconductivity, starting with a general Hamiltonian of the form

$$H = \sum_{\mathbf{k}} \varepsilon_{\mathbf{k}} c_{\mathbf{k}}^{\dagger} c_{\mathbf{k}} + \sum_{\mathbf{k}\mathbf{k}'\mathbf{q}} V(\mathbf{k}, \mathbf{k}', \mathbf{k} - \mathbf{q}, \mathbf{k}' + \mathbf{q}) c_{\mathbf{k}'+\mathbf{q}}^{\dagger} c_{\mathbf{k}-\mathbf{q}}^{\dagger} c_{\mathbf{k}} c_{\mathbf{k}'}, \qquad (14.110)$$

where $\varepsilon_{\mathbf{k}}$ is referenced to the Fermi level, and the spin indices are omitted for convenience at this point (see Figs. 14.21 and 14.22). The $c_{\mathbf{k}}^{\dagger}$ and $c_{\mathbf{k}}$ are the creation and destruction operators for electrons.

To solve the Hamiltonian, we may expand the solution in terms of many-body basis functions or Slater determinants of the form

$$|n_1, n_2, \ldots, n_{\ell}, \ldots\rangle = \ldots (c_{\ell}^{\dagger})^{n_{\ell}} \ldots (c_2^{\dagger})^{n_2} (c_1^{\dagger})^{n_1} |0\rangle, \qquad (14.111)$$

with $n_{\ell} = 1$ or 0, and $\sum_{\ell} n_{\ell} = N$.

For the superconducting state,

$$|\Phi\rangle = \sum_{\alpha} A_{\alpha} |\{n\}_{\alpha}\rangle, \qquad (14.112)$$

we want a coherent state with only the subset in the sum (i.e. one of the $\{n\}_{\alpha}$) having the form

$$|n_1, n_1'; n_2, n_2'; \ldots\rangle \qquad (14.113)$$

[14] J. Bardeen and D. Pines, "Electron–phonon interaction in metals," *Phys. Rev.* 99(1955), 1140.

with $n_1 = n_1', n_2 = n_2', \ldots$ to represent the formation of Cooper pairs.

The reason is that for fermions,

$$c_j \left| n_1, \ldots, n_j, \ldots \right\rangle = \theta_j n_j \left| n_1, \ldots, 1 - n_j, \ldots \right\rangle,$$

$$c_j^\dagger \left| n_1, \ldots, n_j, \ldots \right\rangle = \theta_j (1 - n_j) \left| n_1, \ldots, 1 - n_j, \ldots \right\rangle,$$

$$\theta_j = (-1)^{v_j}, \quad v_j = \sum_{l=1}^{j-1} n_l,$$

$$(14.114)$$

that is, in operations with $c_j, c_j^\dagger$, there occurs multiplication by ± 1 according to whether there are even or odd numbers of occupied states that precede the state j in the ordering of states that have been adopted. As a consequence, the matrix $H_{\alpha\beta}$ has off-diagonal elements that are random in sign. To get a coherent state of low energy, we need the off-diagonal elements of the Hamiltonian to have the same sign and be negative. This can be done by working with a subset of basis functions, for which the matrix elements $H_{\alpha\beta}$ of the interaction are always negative. This property is realized for V negative if the Bloch states are always occupied only in pairs, as in Eq. (14.113). In addition, the interaction conserves wavevectors because of translation invariance, so we will consider all pairs to have the same total momentum

$$\mathbf{k}_j + \mathbf{k}_j' = \mathbf{K}. \tag{14.115}$$

If $\mathbf{K}$ is zero, the pair state is $(\mathbf{k}, -\mathbf{k})$. Since the energy for attractive interaction will usually be lower for an antiparallel $(\uparrow\downarrow)$ pair than for a pair of parallel spin, we assume that the total spin state is zero.

For the ground state within this framework, we may work with the pair subspace using a truncated Hamiltonian (the so-called BCS-reduced Hamiltonian)

$$H_{\text{red}} = \sum_{\mathbf{k}} \varepsilon_{\mathbf{k}} (c_{\mathbf{k}\uparrow}^\dagger c_{\mathbf{k}\uparrow} + c_{-\mathbf{k}\downarrow}^\dagger c_{-\mathbf{k}\downarrow}) + \sum_{\mathbf{k}\mathbf{k}'} V_{\mathbf{k}\mathbf{k}'} c_{\mathbf{k}'\uparrow}^\dagger c_{-\mathbf{k}'\downarrow}^\dagger c_{-\mathbf{k}\downarrow} c_{\mathbf{k}\uparrow}. \tag{14.116}$$

The Hamiltonian is different from the one in the original BCS paper by a constant term $\sum_{\mathbf{k}<k_F} |\varepsilon_{\mathbf{k}}|$ because here we used the excitation energy $\varepsilon_{\mathbf{k}} = \text{energy} - E_F$. We can also look at $\delta H = H - H_{\text{red}}$ and do perturbation theory on the solutions of H_{red} to consider effects that are neglected by H_{red}. If we define the pair operators as

$$b_{\mathbf{k}}^\dagger = c_{\mathbf{k}\uparrow}^\dagger c_{-\mathbf{k}\downarrow}^\dagger,$$

$$b_{\mathbf{k}} = c_{-\mathbf{k}\downarrow} c_{\mathbf{k}\uparrow}, \tag{14.117}$$

then in this pair subspace, for the ground state, there is no pair state $\mathbf{k} \uparrow, -\mathbf{k} \downarrow$ occupied by a single electron.

Since in the pair subspace $n_{\mathbf{k}\uparrow} = n_{-\mathbf{k}\downarrow}$ and $n_{\mathbf{k}\uparrow} + n_{-\mathbf{k}\downarrow} = 2b_{\mathbf{k}}^\dagger b_{\mathbf{k}}$, for the ground state we arrive at

$$H_{\text{red}}^G = \sum_{\mathbf{k}} 2\varepsilon_{\mathbf{k}} b_{\mathbf{k}}^\dagger b_{\mathbf{k}} + \sum_{\mathbf{k}\mathbf{k}'} V_{\mathbf{k}\mathbf{k}'} b_{\mathbf{k}}^\dagger b_{\mathbf{k}'}. \tag{14.118}$$

The commutation relations for the pair operators are

$$[b_{\mathbf{k}}, b_{\mathbf{k}'}^{\dagger}] = b_{\mathbf{k}} b_{\mathbf{k}'}^{\dagger} - b_{\mathbf{k}'}^{\dagger} b_{\mathbf{k}} = 0 \quad (\text{for } \mathbf{k} \neq \mathbf{k}'), \tag{14.119}$$

$$[b_{\mathbf{k}}, b_{\mathbf{k}}^{\dagger}] = b_{\mathbf{k}} b_{\mathbf{k}}^{\dagger} - b_{\mathbf{k}}^{\dagger} b_{\mathbf{k}} = 1 - (n_{\mathbf{k}\uparrow} + n_{-\mathbf{k}\downarrow}), \tag{14.120}$$

and

$$[b_{\mathbf{k}}, b_{\mathbf{k}'}] = [b_{\mathbf{k}}^{\dagger}, b_{\mathbf{k}'}^{\dagger}] = 0. \tag{14.121}$$

This is similar, but not the same as for bosons, since in that case $[\mathbf{b_k}, \mathbf{b_k^{\dagger}}] = 1$. We can think of the pair operator as satisfying Bose–Einstein statistics for $\mathbf{k} \neq \mathbf{k}'$ and satisfying the Pauli principle $\mathbf{b_k^{\dagger 2}} = 0 = \mathbf{b_k^2}$ for $\mathbf{k} = \mathbf{k}'$. In terms of the pair operators, we may write the ground-state wavefunction as

$$|\Phi_{\mathrm{N}}\rangle = \prod_{\mathbf{k} < \mathbf{k}_{\mathrm{F}}} b_{\mathbf{k}}^{\dagger} |0\rangle \tag{14.122}$$

for the normal state. For the superconducting ground state, we seek a coherent superposition of pair configurations with the general form

$$|\Phi\rangle = \prod_{\text{all pair config.}} g(\mathbf{k}_l, \ldots, \mathbf{k}_m) b_{\mathbf{k}_{\ell}}^{\dagger}, \ldots, b_{\mathbf{k}_m}^{\dagger} |0\rangle. \tag{14.123}$$

Since the number of terms is $\sim N! \sim 10^{10^{20}}$, this appears intractable.

To simplify this problem, Bardeen, Cooper, and Schrieffer chose a mean-field theory, in which the occupancy of each $\mathbf{k}$ orbital is taken to depend only on the average occupancy of other orbitals. This relaxes the constraint of the total number of particles being N, since occupancies are treated only statistically. Within this approximation, BCS theory takes for the ground state

$$|\Phi\rangle = \prod_{\mathbf{k}} (u_{\mathbf{k}} + v_{\mathbf{k}} b_{\mathbf{k}}^{\dagger}) |0\rangle, \tag{14.124}$$

where $u_{\mathbf{k}}$ and $v_{\mathbf{k}}$ are variational parameters, which we will take to be real for now. Hence for this approximation, electrons only appear in pairs, with $u_{\mathbf{k}}$ as the probability amplitude for an empty pair orbital $(\mathbf{k}, -\mathbf{k})$, and $v_{\mathbf{k}}$ is the probability amplitude for an occupied pair orbital $(\mathbf{k}, -\mathbf{k})$.

We now proceed to find the values of $u_{\mathbf{k}}$ and $v_{\mathbf{k}}$ that will minimize the energy, but subject to the condition of fixing the expectation value of the number of electrons at N. To normalize the wavefunction,

$$
\begin{aligned}
\langle \Phi \,|\, \Phi \rangle &= \langle 0| \prod_{\mathbf{k}'} (u_{\mathbf{k}'} + v_{\mathbf{k}'} b_{\mathbf{k}'}) \prod_{\mathbf{k}} (u_{\mathbf{k}} + v_{\mathbf{k}} b_{\mathbf{k}}^{\dagger}) \,|0\rangle \\
&= \langle 0| \prod_{\mathbf{k}} (u_{\mathbf{k}} + v_{\mathbf{k}} b_{\mathbf{k}})(u_{\mathbf{k}} + v_{\mathbf{k}} b_{\mathbf{k}}^{\dagger}) \,|0\rangle \\
&= \langle 0| \prod_{\mathbf{k}} (u_{\mathbf{k}}^2 + u_{\mathbf{k}} v_{\mathbf{k}} b_{\mathbf{k}} + u_{\mathbf{k}} v_{\mathbf{k}} b_{\mathbf{k}}^{\dagger} + v_{\mathbf{k}}^2 b_{\mathbf{k}} b_{\mathbf{k}}^{\dagger}) \,|0\rangle \\
&= \langle 0| \prod_{\mathbf{k}} (u_{\mathbf{k}}^2 + v_{\mathbf{k}}^2) \,|0\rangle = 1,
\end{aligned}
\tag{14.125}
$$

which dictates that

$$
u_{\mathbf{k}}{}^2 + v_{\mathbf{k}}{}^2 = 1,
\tag{14.126}
$$

where we have used

$$
[b_{\mathbf{k}}, b_{\mathbf{k}}^{\dagger}] = b_{\mathbf{k}} b_{\mathbf{k}}^{\dagger} - b_{\mathbf{k}}^{\dagger} b_{\mathbf{k}} = 1 - (n_{\mathbf{k}\uparrow} + n_{-\mathbf{k}\downarrow}).
\tag{14.127}
$$

We now work in a grand canonical ensemble and obtain the u's and v's through minimization of the energy. We use

$$
H_{\text{red}}^{G} = H_{\text{KE}} + H_{\text{int}}
\tag{14.128}
$$

and minimize

$$
\langle H_{\text{red}} - \mu N_{\text{op}} \rangle,
\tag{14.129}
$$

where μ is the chemical potential, $N_{\text{op}} = 2 \sum_{\mathbf{k}} b_{\mathbf{k}}^{\dagger} b_{\mathbf{k}}$, and

$$
H_{\text{KE}} = \sum_{\mathbf{k}} 2\varepsilon_{\mathbf{k}} b_{\mathbf{k}}^{\dagger} b_{\mathbf{k}}.
\tag{14.130}
$$

The results are

$$
\begin{aligned}
\langle \phi| H_{\text{KE}} |\phi \rangle &= 2 \sum_{\mathbf{k}} \langle 0| (u_{\mathbf{k}} + v_{\mathbf{k}} b_{\mathbf{k}}) b_{\mathbf{k}}^{\dagger} b_{\mathbf{k}} (u_{\mathbf{k}} + v_{\mathbf{k}} b_{\mathbf{k}}^{\dagger}) \,|0\rangle \, \varepsilon_{\mathbf{k}} \\
&= 2 \sum_{\mathbf{k}} v_{\mathbf{k}}^2 \, \varepsilon_{\mathbf{k}} \, \langle 0| b_{\mathbf{k}} b_{\mathbf{k}}^{\dagger} b_{\mathbf{k}} b_{\mathbf{k}}^{\dagger} \,|0\rangle = 2 \sum_{\mathbf{k}} v_{\mathbf{k}}^2 \, \varepsilon_{\mathbf{k}},
\end{aligned}
\tag{14.131}
$$

$$
\begin{aligned}
\langle \phi| H_{\text{int}} |\phi \rangle &= \sum_{\mathbf{k},\mathbf{k}'} V_{\mathbf{k}\mathbf{k}'} \langle 0| (u_{\mathbf{k}} + v_{\mathbf{k}} b_{\mathbf{k}})(u_{\mathbf{k}'} + v_{\mathbf{k}'} b_{\mathbf{k}'}) b_{\mathbf{k}}^{\dagger} b_{\mathbf{k}'} (u_{\mathbf{k}'} + v_{\mathbf{k}'} b_{\mathbf{k}'}^{\dagger})(u_{\mathbf{k}} + v_{\mathbf{k}} b_{\mathbf{k}}^{\dagger}) \,|0\rangle \\
&= \sum_{\mathbf{k}\mathbf{k}'} V_{\mathbf{k}\mathbf{k}'} u_{\mathbf{k}} v_{\mathbf{k}} u_{\mathbf{k}'} v_{\mathbf{k}},
\end{aligned}
\tag{14.132}
$$

where $u_{\mathbf{k}} v_{\mathbf{k}'}$ is the probability amplitude that $\mathbf{k}'$ is full but $\mathbf{k}$ is empty and $u_{\mathbf{k}'} v_{\mathbf{k}}$ is the probability amplitude that $\mathbf{k}$ is full but $\mathbf{k}'$ is empty. The constraint on the number of particles is

$$\langle \mu N_{\text{op}} \rangle = 2 \sum_{\mathbf{k}} \mu v_{\mathbf{k}}^2. \tag{14.133}$$

Therefore

$$\left\langle H_{\text{red}}^G - \mu N_{\text{op}} \right\rangle = 2 \sum_{\mathbf{k}} v_{\mathbf{k}}^2 \varepsilon_{\mathbf{k}} + \sum_{\mathbf{k},\mathbf{k}'} V_{\mathbf{k}\mathbf{k}'} u_{\mathbf{k}} v_{\mathbf{k}'} u_{\mathbf{k}'} v_{\mathbf{k}} - 2 \sum_{\mathbf{k}} \mu v_{\mathbf{k}}^2. \tag{14.134}$$

We now need to minimize with respect to $v_{\mathbf{k}}$. We first change variables so that

$$\zeta_{\mathbf{k}} = \varepsilon_{\mathbf{k}} - \mu \tag{14.135}$$

and (since $u^2 + v^2 = 1$)

$$v_{\mathbf{k}} = \cos \theta_{\mathbf{k}} = \sqrt{\frac{1}{2}(1 + \cos 2\theta_{\mathbf{k}})},$$

$$u_{\mathbf{k}} = \sin \theta_{\mathbf{k}} = \sqrt{\frac{1}{2}(1 - \cos 2\theta_{\mathbf{k}})}, \tag{14.136}$$

then

$$\langle H_{\text{red}}^G - \mu N_{\text{op}} \rangle = 2 \sum_{\mathbf{k}} \zeta_{\mathbf{k}} \cos^2 \theta_{\mathbf{k}} + \sum_{\mathbf{k},\mathbf{k}'} V_{\mathbf{k}\mathbf{k}'} \sin \theta_{\mathbf{k}} \cos \theta_{\mathbf{k}} \sin \theta_{\mathbf{k}'} \cos \theta_{\mathbf{k}'}$$

$$= 2 \sum_{\mathbf{k}} \zeta_{\mathbf{k}} \cos^2 \theta_{\mathbf{k}} + \frac{1}{4} \sum_{\mathbf{k},\mathbf{k}'} V_{\mathbf{k}\mathbf{k}'} \sin 2\theta_{\mathbf{k}} \sin 2\theta_{\mathbf{k}'}. \tag{14.137}$$

In this form, we can do the minimization with respect to $\theta_{\mathbf{k}}$, and we obtain

$$0 = -4 \zeta_{\mathbf{k}} \cos \theta_{\mathbf{k}} \sin \theta_{\mathbf{k}} + \sum_{\mathbf{k}'} V_{\mathbf{k}\mathbf{k}'} \cos 2\theta_{\mathbf{k}} \sin 2\theta_{\mathbf{k}'}. \tag{14.138}$$

Using

$$\cos \theta_{\mathbf{k}} \sin \theta_{\mathbf{k}} = \frac{1}{2} \sin 2\theta_{\mathbf{k}}, \tag{14.139}$$

we have

$$\tan 2\theta_{\mathbf{k}} = \frac{1}{2\zeta_{\mathbf{k}}} \sum_{\mathbf{k}'} V_{\mathbf{k}\mathbf{k}'} \sin 2\theta_{\mathbf{k}'}. \tag{14.140}$$

Defining

$$\tan 2\theta_{\mathbf{k}} \equiv \frac{-\Delta_{\mathbf{k}}}{\zeta_{\mathbf{k}}},$$

$$E_{\mathbf{k}} = \sqrt{\zeta_{\mathbf{k}}^2 + \Delta_{\mathbf{k}}^2}, \tag{14.141}$$

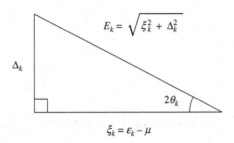

Figure 14.24 Definition of θ_k in terms of E, ξ, and Δ.

and

$$\sin 2\theta_{\mathbf{k}} = \frac{\Delta_{\mathbf{k}}}{\sqrt{\xi_{\mathbf{k}}^2 + \Delta_{\mathbf{k}}^2}} = \frac{\Delta_{\mathbf{k}}}{\xi_{\mathbf{k}}} = 2u_{\mathbf{k}}v_{\mathbf{k}} \qquad (14.142)$$

from the definition of $\theta_{\mathbf{k}}$ (see Fig. 14.24), we arrive at the famous BCS gap equation

$$\Delta_{\mathbf{k}} = -\frac{1}{2}\sum_{\mathbf{k}'} V_{\mathbf{k}\mathbf{k}'}\frac{\Delta_{\mathbf{k}'}}{E_{\mathbf{k}'}}. \qquad (14.143)$$

As discussed below,

$$E_{\mathbf{k}} = \sqrt{(\varepsilon_{\mathbf{k}} - \mu)^2 + \Delta_{\mathbf{k}}^2} \qquad (14.144)$$

is the energy for creating a quasiparticle of momentum $\mathbf{k}$ in the superconducting state with $\Delta_{\mathbf{k}}$ as the energy gap. In general, the energy gap is $\mathbf{k}$-dependent. The probability amplitudes become

$$u_{\mathbf{k}}^2 = \sin^2\theta_{\mathbf{k}} = \frac{1}{2}\left(1 + \frac{\varepsilon_{\mathbf{k}} - \mu}{E_{\mathbf{k}}}\right), \qquad (14.145)$$

$$v_{\mathbf{k}}^2 = \cos^2\theta_{\mathbf{k}} = \frac{1}{2}\left(1 - \frac{\varepsilon_{\mathbf{k}} - \mu}{E_{\mathbf{k}}}\right), \qquad (14.146)$$

and

$$u_{\mathbf{k}}v_{\mathbf{k}} = \frac{\Delta_{\mathbf{k}}}{2E_{\mathbf{k}}}. \qquad (14.147)$$

The aim of the procedure is to simultaneously solve the gap equation and the constraint condition

$$\langle \phi | N_{\mathrm{op}} | \phi \rangle = 2\sum_{\mathbf{k}} v_{\mathbf{k}}^2 = N \qquad (14.148)$$

to determine $\Delta_{\mathbf{k}}$ and μ.

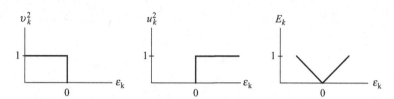

Figure 14.25 Amplitudes v_k^2 and u_k^2 and energy dispersion relation for the normal state.

Since $\varepsilon_{\mathbf{k}}$ is the energy relative to E_F in the normal state, μ is the shift of the chemical potential between the normal and superconducting states. In general, $\mu \approx 0$ because of electron–hole symmetry, and we will assume this here. Therefore, $\xi_{\mathbf{k}} \to \varepsilon_{\mathbf{k}}$. If there is a relative phase between $u_{\mathbf{k}}$ and $v_{\mathbf{k}}$, then we have

$$|u_{\mathbf{k}}| \, |v_{\mathbf{k}}| \, e^{i\phi_{\mathbf{k}}} = \frac{\Delta_{\mathbf{k}}}{2E_{\mathbf{k}}}, \tag{14.149}$$

where $\phi_{\mathbf{k}}$ is the phase of $\Delta_{\mathbf{k}}$. The solution of the BCS gap equation ($T = 0$) for the normal state where $V_{\mathbf{kk'}} = 0$ is (see Fig. 14.25)

$$\Delta_{\mathbf{k}} = 0 = \Delta_{\mathbf{k'}}, \tag{14.150}$$

for $\varepsilon_{\mathbf{k}} > 0$,

$$v_{\mathbf{k}}^2 = \frac{1}{2}\left(1 - \frac{\varepsilon_{\mathbf{k}}}{|\varepsilon_{\mathbf{k}}|}\right) = 0,$$
$$u_{\mathbf{k}}^2 = \frac{1}{2}\left(1 + \frac{\varepsilon_{\mathbf{k}}}{|\varepsilon_{\mathbf{k}}|}\right) = 1, \tag{14.151}$$

and for $\varepsilon_{\mathbf{k}} < 0$,

$$v_{\mathbf{k}}^2 = \frac{1}{2}\left(1 - \frac{\varepsilon_{\mathbf{k}}}{|\varepsilon_{\mathbf{k}}|}\right) = 1,$$
$$u_{\mathbf{k}}^2 = \frac{1}{2}\left(1 + \frac{\varepsilon_{\mathbf{k}}}{|\varepsilon_{\mathbf{k}}|}\right) = 0, \tag{14.152}$$

and

$$E_N = 2 \sum_{\mathbf{k}} \varepsilon_{\mathbf{k}} v_{\mathbf{k}}^2. \tag{14.153}$$

For the superconducting state, $\Delta_{\mathbf{k}} \neq 0$ (see Fig. 14.26), and

$$E_S = 2 \sum_{\mathbf{k}} \xi_{\mathbf{k}} v_{\mathbf{k}}^2 + \sum_{\mathbf{kk'}} V_{\mathbf{kk'}} u_{\mathbf{k}} v_{\mathbf{k}} u_{\mathbf{k'}} v_{\mathbf{k'}}. \tag{14.154}$$

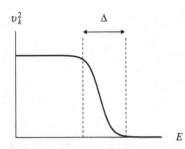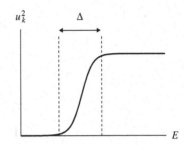

Figure 14.26 Amplitudes v_k^2 and u_k^2 as a function of energy for the superconducting state.

The BCS gap equation and thermodynamics. Starting with the gap equation,

$$\Delta_{\mathbf{k}} = -\frac{1}{2} \sum_{\mathbf{k}'} V_{\mathbf{k}\mathbf{k}'} \frac{\Delta_{\mathbf{k}'}}{E_{\mathbf{k}'}}$$

and

$$E_{\mathbf{k}'} = \sqrt{\xi_{\mathbf{k}'}^2 + \Delta_{\mathbf{k}'}^2}, \tag{14.155}$$

where $V_{\mathbf{k}\mathbf{k}'}$ has two parts corresponding to the Coulomb and phonon-induced interactions, to solve the gap equation, Bardeen, Cooper, and Schrieffer introduced a model similar in form to that used by Cooper for the one-pair problem. If we use the Bardeen–Pines interaction for the phonon-mediated part

$$V_{\mathbf{k}\mathbf{k}'}^{BP} = \frac{2\hbar\omega_{\mathbf{q}} |V_{\mathbf{q}}^{ep}|^2}{(\varepsilon_{\mathbf{k}} - \varepsilon_{\mathbf{k}'})^2 - (\hbar\omega_{\mathbf{q}})^2}, \tag{14.156}$$

the BCS model assumes a constant attractive potential with a cutoff at the Debye energy E_D as an approximation for this attractive interaction. For the Coulomb interaction V_c, BCS also takes a constant potential with at cutoff at E_D. There is no real justification for taking E_D as a cutoff for V_c. These approximations yield

$$V_{\mathbf{k}\mathbf{k}'} = V_{\mathbf{k}\mathbf{k}'}^{ph} + V_{\mathbf{k}\mathbf{k}'}^c = \begin{cases} -\frac{V}{\Omega} & |\varepsilon_{\mathbf{k}}|, |\varepsilon_{\mathbf{k}'}| < E_D, \\ 0 & \text{otherwise}, \end{cases} \tag{14.157}$$

$$\Delta_{\mathbf{k}} = \frac{V}{2} \int_{|\varepsilon| < \varepsilon_D} \frac{\Delta_{\mathbf{k}'}}{E_{\mathbf{k}'}} N_\uparrow (\varepsilon_{\mathbf{k}'}) \, d\varepsilon_{\mathbf{k}'}. \tag{14.158}$$

Assuming $N_\uparrow(\varepsilon_{\mathbf{k}'}) = N(0)$, the trial solution

$$\Delta (\varepsilon_{\mathbf{k}}) = \begin{cases} \Delta = \text{constant} & |\varepsilon_{\mathbf{k}}| < E_D, \\ 0 & \text{otherwise}, \end{cases} \tag{14.159}$$

yields

$$\Delta = \frac{V}{2} N(0) \Delta \int_{-E_{\mathrm{D}}}^{E_{\mathrm{D}}} \frac{d\varepsilon}{\sqrt{\varepsilon^2 + \Delta^2}}, \tag{14.160}$$

or

$$\frac{1}{N(0)V} = \int_0^{E_{\mathrm{D}}} \frac{d\varepsilon}{\sqrt{\varepsilon^2 + \Delta^2}} = \sinh^{-1}\left(\frac{E_{\mathrm{D}}}{\Delta}\right), \tag{14.161}$$

and

$$\Delta = \frac{E_{\mathrm{D}}}{\sinh\left(\frac{1}{N(0)V}\right)}. \tag{14.162}$$

For $N(0)V \ll 1$ (i.e. weak coupling), we obtain

$$\Delta \approx 2E_{\mathrm{D}} e^{-\frac{1}{N(0)V}}. \tag{14.163}$$

Using this result, we can now calculate the condensation energy. The ground-state energies are

$$E_{\mathrm{S}} = \langle \phi | H_{\mathrm{red}}^{G} - \mu N_{\mathrm{op}} | \phi \rangle = 2 \sum_{\mathbf{k}} \xi_{\mathbf{k}} v_{\mathbf{k}}^2 + \sum_{\mathbf{k}\ell} V_{\mathbf{k}\ell} u_{\mathbf{k}} v_{\mathbf{k}} u_{\ell} v_{\ell}$$

$$= \sum_{\mathbf{k}} \left(\xi_{\mathbf{k}} - \frac{\xi_{\mathbf{k}}^2}{E_{\mathbf{k}}} \right) - \sum_{\ell} \Delta_{\ell} u_{\ell} v_{\ell}$$

$$= \sum_{\mathbf{k}} \left(\xi_{\mathbf{k}} - \frac{\xi_{\mathbf{k}}^2}{E_{\mathbf{k}}} \right) - \frac{1}{2} \sum_{\mathbf{k}} \frac{\Delta_{\mathbf{k}}^2}{E_{\mathbf{k}}} = \sum_{\mathbf{k}} \left[\left(\varepsilon_{\mathbf{k}} - \frac{\xi_{\mathbf{k}}^2}{E_{\mathbf{k}}} \right) - \frac{\Delta_{\mathbf{k}}^2}{2E_{\mathbf{k}}} \right], \tag{14.164}$$

and

$$E_{\mathrm{N}} = \sum_{|\mathbf{k}| < \mathbf{k}_{\mathrm{F}}} 2\xi_{\mathbf{k}}. \tag{14.165}$$

If we integrate over a small energy range of $E_{\mathrm{F}} - E_{\mathrm{D}}$ to $E_{\mathrm{F}} + E_{\mathrm{D}}$, then (letting $\mu = 0$)

$$E_{\mathrm{N}} - E_{\mathrm{S}} = N(0) \int_{-E_{\mathrm{D}}}^{E_{\mathrm{D}}} d\varepsilon \left(\frac{\varepsilon^2 + \frac{1}{2}\Delta^2}{\sqrt{\varepsilon^2 + \Delta^2}} - |\varepsilon| \right) = -N(0) E_{\mathrm{D}}^2 \left(1 - \sqrt{1 + \left(\frac{\Delta}{E_{\mathrm{D}}}\right)^2} \right), \tag{14.166}$$

and for small Δ,

$$E_{\mathrm{N}} - E_{\mathrm{S}} = \frac{1}{2} N(0) \Delta^2. \tag{14.167}$$

Since thermodynamics gives

$$E_{\mathrm{N}} - E_{\mathrm{S}} = \frac{H_c^2(T=0)}{8\pi} = \frac{N(0)\Delta^2}{2}, \tag{14.168}$$

then

$$H_c(0) = 2\sqrt{\pi N(0)}\,\Delta(0). \tag{14.169}$$

Therefore, we get H_c from Δ and $N(0)$. Also, because of Eq. (14.163) and the fact that $E_{\mathrm{D}} \propto M^{-1/2}$, there should be an isotope effect for both Δ and H_c as one changes the mass of an ion in the system.

We are now in a position to explore quasiparticle excitations of a superconductor by using

$$H_{\mathrm{red}} = \sum_{\mathbf{k}} \varepsilon_{\mathbf{k}}(c_{\mathbf{k}\uparrow}^{\dagger}c_{\mathbf{k}\uparrow} + c_{\mathbf{k}\downarrow}^{\dagger}c_{\mathbf{k}\downarrow}) + \sum_{\mathbf{k}\mathbf{k}'} V_{\mathbf{k}\mathbf{k}'} b_{\mathbf{k}}^{\dagger} b_{\mathbf{k}'}, \tag{14.170}$$

and defining the quasiparticle (QP) energy as

$$\text{QP energy} = E(N+1) - E(N)$$

$$= \text{total change of energy when an external electron}$$

$$\text{is added to the system.}$$

$$= \delta W_{\ell} + \xi_{\ell}. \tag{14.171}$$

If we add an electron $\ell \uparrow$ (with $-\ell \downarrow$ empty), the one effect of this process is to block the pair state $(\ell \uparrow, -\ell \downarrow)$ from participating in the pairing interactions because, since $-\ell \downarrow$ is empty, the electron in $\ell \uparrow$ cannot be scattered out of this state. Starting with the energy

$$W = \sum_{\mathbf{k}} 2\xi_{\mathbf{k}} v_{\mathbf{k}}^2 + \sum_{\mathbf{k}\mathbf{k}'} V_{\mathbf{k}\mathbf{k}'} u_{\mathbf{k}} v_{\mathbf{k}} u_{\mathbf{k}'} v_{\mathbf{k}'}, \tag{14.172}$$

deleting the pairing $(\ell \uparrow, -\ell \downarrow)$ from the condensate yields

$$E_{N+1}^c - E_N^c = \delta W_{\ell} = -2\xi_{\ell} v_{\ell}^2 - 2\left[\sum_{\mathbf{k}'} V_{\ell\mathbf{k}'} u_{\mathbf{k}'} v_{\mathbf{k}'}\right] u_{\ell} v_{\ell}$$

$$= -2\xi_{\ell} v_{\ell}^2 + 2\Delta_{\ell} u_{\ell} v_{\ell}$$

$$= -2\xi_{\ell} v_{\ell}^2 + \frac{\Delta_{\ell}^2}{E_{\ell}}. \tag{14.173}$$

Since the gap is $\Delta_{\mathbf{k}} = 2E_{\mathbf{k}} u_{\mathbf{k}} v_{\mathbf{k}}$, the energy of the added particle is ξ_l, and hence the quasiparticle energy is

$$\Delta E = \xi_{\ell} + \delta W_{\ell} = \xi_{\ell}(1 - 2v_{\ell}^2) + \frac{\Delta_{\ell}^2}{E_{\ell}} = \xi_{\ell}\frac{\xi_{\ell}}{E_{\ell}} + \frac{\Delta_{\ell}^2}{E_{\ell}} = E_{\ell}. \tag{14.174}$$

Bogoliubov–Valatin transformation. The BCS variational wavefunction is physically transparent and good for the ground-state properties. However, for studying excitations and finite temperature effects, it is convenient to consider another self-consistent field approach.[15] We begin by letting

$$c_{-\mathbf{k}\downarrow}c_{\mathbf{k}\uparrow} = \langle c_{-\mathbf{k}\downarrow}c_{\mathbf{k}\uparrow}\rangle + (c_{-\mathbf{k}\downarrow}c_{\mathbf{k}\uparrow} - \langle c_{-\mathbf{k}\downarrow}c_{\mathbf{k}\uparrow}\rangle) \tag{14.175}$$

or

$$b_{\mathbf{k}} = \langle b_{\mathbf{k}}\rangle + (b_{\mathbf{k}} - \langle b_{\mathbf{k}}\rangle), \tag{14.176}$$

where $\langle b_{\mathbf{k}}\rangle$ is determined self-consistently and the second term in Eq. (14.175) is assumed to be a small quantity. We will neglect quantities that are bilinear in the small fluctuation term. So we have

$$b_{\mathbf{k}}^{\dagger}b_{\ell} = b_{\mathbf{k}}^{\dagger}\langle b_{\ell}\rangle + \langle b_{\mathbf{k}}\rangle^* b_{\ell} - \langle b_{\mathbf{k}}\rangle^*\langle b_{\ell}\rangle. \tag{14.177}$$

The reduced Hamiltonian becomes

$$H_M = \sum_{\mathbf{k}\sigma}\xi_{\mathbf{k}}c_{\mathbf{k}\sigma}^+ c_{\mathbf{k}\sigma} + \sum_{\mathbf{k}\ell}V_{\mathbf{k}\ell}(b_{\mathbf{k}}^{\dagger}\langle b_{\ell}\rangle + \langle b_{\mathbf{k}}\rangle^* b_{\ell} - \langle b_{\mathbf{k}}\rangle^*\langle b_{\ell}\rangle). \tag{14.178}$$

By defining $\Delta_{\mathbf{k}} \equiv -\sum_{\ell}V_{\mathbf{k}\ell}\langle b_{\ell}\rangle$, we have

$$H_M = \sum_{\mathbf{k}\sigma}\xi_{\mathbf{k}}c_{\mathbf{k}\sigma}^{\dagger}c_{\mathbf{k}\sigma} - \sum_{\mathbf{k}}(\Delta_{\mathbf{k}}b_{\mathbf{k}}^{\dagger} + \Delta_{\mathbf{k}}^* b_{\mathbf{k}} - \Delta_{\mathbf{k}}\langle b_{\mathbf{k}}\rangle^*). \tag{14.179}$$

The interaction term is a sum of terms, each proportional to the pair of operators corresponding to the partners in a pair.

This Hamiltonian may be diagonalized by a suitable linear transformation called the Bogoliubov–Valatin transformation by defining

$$\gamma_{\mathbf{k}0}^{\dagger} = u_{\mathbf{k}}^* c_{\mathbf{k}\uparrow}^{\dagger} - v_{\mathbf{k}}^* c_{-\mathbf{k}\downarrow} \quad \text{(which puts an electron in } \mathbf{k}\uparrow), \tag{14.180}$$

and

$$\gamma_{\mathbf{k}1}^{\dagger} = u_{\mathbf{k}}^* c_{-\mathbf{k}\downarrow}^{\dagger} + v_{\mathbf{k}}^* c_{\mathbf{k}\uparrow} \quad \text{(which puts an electron in } -\mathbf{k}\downarrow). \tag{14.181}$$

[15] N. N. Bogoliubov, V. V. Tolmachev, and D. V. Shirkov, *A New Method in the Theory of Superconductivity*, (New York, NY: Consultants Bureau, Inc. 1959).

Then

$$H_M = \sum_{\mathbf{k}} \xi_{\mathbf{k}}[(|u_{\mathbf{k}}|^2 - |v_{\mathbf{k}}|^2)(\gamma_{\mathbf{k}0}^\dagger \gamma_{\mathbf{k}0} + \gamma_{\mathbf{k}1}^\dagger \gamma_{\mathbf{k}1}) + 2|v_{\mathbf{k}}|^2 + 2u_{\mathbf{k}}^* v_{\mathbf{k}}^* \gamma_{\mathbf{k}1} \gamma_{\mathbf{k}0} + 2u_{\mathbf{k}} v_{\mathbf{k}} \gamma_{\mathbf{k}0}^\dagger \gamma_{\mathbf{k}1}^\dagger]$$

$$+ \sum_{\mathbf{k}} (\Delta_{\mathbf{k}} u_{\mathbf{k}} v_{\mathbf{k}}^* + \Delta_{\mathbf{k}}^* u_{\mathbf{k}}^* v_{\mathbf{k}})(\gamma_{\mathbf{k}0}^\dagger \gamma_{\mathbf{k}0} + \gamma_{\mathbf{k}1}^\dagger \gamma_{\mathbf{k}1} - 1)$$

$$+ \sum_{\mathbf{k}} \left[(\Delta_{\mathbf{k}} v_{\mathbf{k}}^{*2} - \Delta_{\mathbf{k}}^* u_{\mathbf{k}}^{*2}) \gamma_{\mathbf{k}1} \gamma_{\mathbf{k}0} + (\Delta_{\mathbf{k}}^* v_{\mathbf{k}}^2 - \Delta_{\mathbf{k}} u_{\mathbf{k}}^2) \gamma_{\mathbf{k}0}^\dagger \gamma_{\mathbf{k}1}^\dagger + \Delta_{\mathbf{k}} \langle b_{\mathbf{k}} \rangle^* \right].$$

$$(14.182)$$

If we choose $u_{\mathbf{k}}$ and $v_{\mathbf{k}}$ so that the coefficients for $\gamma_{\mathbf{k}1} \gamma_{\mathbf{k}0}$ and $\gamma_{\mathbf{k}0}^\dagger \gamma_{\mathbf{k}1}^\dagger$ vanish, i.e.

$$2\xi_{\mathbf{k}} u_{\mathbf{k}} v_{\mathbf{k}} + (\Delta_{\mathbf{k}}^* v_{\mathbf{k}}^2 - \Delta_{\mathbf{k}} u_{\mathbf{k}}^2) = 0, \tag{14.183}$$

then the Hamiltonian is diagonalized. Multiplying by $\frac{\Delta_{\mathbf{k}}^*}{u_{\mathbf{k}}^2}$ through Eq. (14.183), we have

$$2\xi_{\mathbf{k}} \Delta_{\mathbf{k}}^* \frac{v_{\mathbf{k}}}{u_{\mathbf{k}}} + (\Delta_{\mathbf{k}}^*)^2 \frac{v_{\mathbf{k}}^2}{u_{\mathbf{k}}^2} - |\Delta_{\mathbf{k}}|^2 = 0 \tag{14.184}$$

and

$$\Delta_{\mathbf{k}}^* \frac{v_{\mathbf{k}}}{u_{\mathbf{k}}} = \sqrt{\xi_{\mathbf{k}}^2 + |\Delta_{\mathbf{k}}|^2} - \xi_{\mathbf{k}} \equiv E_{\mathbf{k}} - \xi_{\mathbf{k}}. \tag{14.185}$$

By choosing the positive sign of the square root to get the minimum energy and using $u_{\mathbf{k}}^2 + v_{\mathbf{k}}^2 = 1$ and $|\frac{v_{\mathbf{k}}}{u_{\mathbf{k}}}| = \frac{(E_{\mathbf{k}} - \xi_{\mathbf{k}})}{|\Delta_{\mathbf{k}}|}$, we obtain

$$|v_{\mathbf{k}}|^2 = 1 - |u_{\mathbf{k}}|^2 = \frac{1}{2} \left(1 - \frac{\xi_{\mathbf{k}}}{E_{\mathbf{k}}} \right). \tag{14.186}$$

With u and v given by Eq. (14.186), we can calculate the excitation energies and the gap. The Hamiltonian in Eq. (14.182) becomes

$$H_M = \sum_{\mathbf{k}} (\xi_{\mathbf{k}} - E_{\mathbf{k}} + \Delta_{\mathbf{k}} \langle b_{\mathbf{k}} \rangle^*) + \sum_{\mathbf{k}} E_{\mathbf{k}} (\gamma_{\mathbf{k}0}^\dagger \gamma_{\mathbf{k}0} + \gamma_{\mathbf{k}1}^\dagger \gamma_{\mathbf{k}1}), \tag{14.187}$$

where $\gamma_{\mathbf{k}}$ describes the elementary quasiparticle excitation of the system with wavevector $\mathbf{k}$ and energy $E_{\mathbf{k}}$. Self-consistency requires

$$\Delta_{\mathbf{k}} \equiv -\sum_{\ell} V_{\mathbf{k}\ell} \langle c_{-\ell\downarrow} c_{\ell\downarrow} \rangle \tag{14.188}$$

and, from Eqs. (14.180) and (14.181),

$$\Delta_{\mathbf{k}} = -\sum_{\ell} V_{\mathbf{k}\ell} u_{\ell}^* v_{\ell} \left\langle 1 - \gamma_{\ell 0}^\dagger \gamma_{\ell 0} - \gamma_{\ell 1}^\dagger \gamma_{\ell 1} \right\rangle. \tag{14.189}$$

Using

$$\Delta_{\mathbf{k}} v_{\mathbf{k}}^* = u_{\mathbf{k}}^* (E_{\mathbf{k}} - \zeta_{\mathbf{k}}) \tag{14.190}$$

yields

$$\Delta_{\mathbf{k}} v_{\mathbf{k}}^* u_{\mathbf{k}} = |u_{\mathbf{k}}|^2 (E_{\mathbf{k}} - \zeta_{\mathbf{k}}) = \frac{1}{2} \frac{\Delta_{\mathbf{k}}^2}{E_{\mathbf{k}}}. \tag{14.191}$$

Hence,

$$H_M = E_G + \sum_{\mathbf{k}} E_{\mathbf{k}} (\gamma_{\mathbf{k}0}^\dagger \gamma_{\mathbf{k}0} + \gamma_{\mathbf{k}1}^\dagger \gamma_{\mathbf{k}1}), \tag{14.192}$$

with

$$E_{\mathbf{k}} = \sqrt{\varepsilon_{\mathbf{k}}^2 + |\Delta_{\mathbf{k}}|^2}, \tag{14.193}$$

where E_G is the ground-state energy, which one can show is the same as the BCS result given before.

Equations (14.189) and (14.191) give the BCS gap equation at finite temperature

$$\Delta_{\mathbf{k}} = -\sum_{\ell} V_{\mathbf{k}\ell} \frac{\Delta_\ell}{2E_\ell} \left\langle 1 - \gamma_{\mathbf{k}0}^\dagger \gamma_{\mathbf{k}0} - \gamma_{\mathbf{k}1}^\dagger \gamma_{\mathbf{k}1} \right\rangle. \tag{14.194}$$

At $T = 0$, $\langle \gamma_{\mathbf{k}_0}^\dagger \gamma_{\mathbf{k}_0} \rangle = 0$ and $\langle \gamma_{\mathbf{k}_1}^\dagger \gamma_{\mathbf{k}_1} \rangle = 0$, and we get back the same BCS gap equation as before.

At finite temperature, since the γ's are fermion operators, the probability of having excitations in thermal equilibrium is given by the usual Fermi function

$$\langle \gamma_{\mathbf{k}}^\dagger \gamma_{\mathbf{k}} \rangle = f(E_{\mathbf{k}}) = \left(e^{\beta E_{\mathbf{k}}} + 1 \right)^{-1}, \tag{14.195}$$

where $\beta = \frac{1}{k_B T}$. Thus,

$$\left\langle 1 - \gamma_{\mathbf{k}0}^\dagger \gamma_{\mathbf{k}0} - \gamma_{\mathbf{k}1}^\dagger \gamma_{\mathbf{k}1} \right\rangle = 1 - 2f(E_{\mathbf{k}}) = \tanh \left(\frac{\beta E_{\mathbf{k}}}{2} \right), \tag{14.196}$$

and the gap equation becomes

$$\Delta_{\mathbf{k}} = -\sum_{\ell} V_{\mathbf{k}\ell} \frac{\Delta_\ell}{2E_\ell} \tanh \frac{\beta E_\ell}{2}. \tag{14.197}$$

Because thermally excited electrons block pair states participating in the condensate, this reduces the effective scattering strength through the $\tanh \left(\frac{\beta E_\ell}{2} \right)$ factor.

If we use the BCS model,

$$\begin{aligned} V_{\mathbf{k}\ell} &= -V/\Omega \quad \text{for } |\zeta_{\mathbf{k}}|, |\zeta_\ell| < E_D, \\ \Delta_{\mathbf{k}} &= \Delta_\ell = \Delta, \qquad E < E_D, \end{aligned} \tag{14.198}$$

the gap equation becomes

$$\frac{1}{V} = \frac{1}{2} \sum_{\mathbf{k}} \frac{1}{E_{\mathbf{k}}} \tanh \frac{\beta E_{\mathbf{k}}}{2}, \tag{14.199}$$

where the sum is over all states with $E < E_D$ and we can determine the transition temperature T_c. Since as $T \to T_c$, $\Delta(T) \to 0$, and $E_{\mathbf{k}} \to |\xi_{\mathbf{k}}|$, the gap equation for the BCS model becomes

$$\frac{1}{N(0)V} = \int_0^{\xi_D} \frac{\tanh(\frac{\beta\xi}{2})d\xi}{E}, \tag{14.200}$$

and

$$\frac{1}{N(0)V} = \int_0^{\beta_c E_D/2} \frac{\tanh(x)dx}{x} = \ln(1.13\beta_c E_D), \tag{14.201}$$

resulting in

$$k_B T_c = 1.13 E_D e^{-\frac{1}{N(0)V}}. \tag{14.202}$$

Recalling $\Delta(0) = 2E_D e^{-\frac{1}{N(0)V}}$, we have

$$2\Delta(0) = 3.52 k_B T_c. \tag{14.203}$$

At this point, we can also calculate the temperature dependence of $\Delta(T)$ from

$$\frac{1}{N(0)V} = \int_0^{E_D} \frac{\tanh\left[\frac{1}{2}\beta\sqrt{\xi^2 + \Delta^2(T)}\right]}{\sqrt{\xi^2 + \Delta^2(T)}} d\xi, \tag{14.204}$$

where $\Delta(T)$ can be computed numerically. For $\frac{E_D}{k_B T_c} \gg 1$, the ratio $\frac{\Delta(T)}{\Delta(0)}$ is a universal function of $\frac{T}{T_c}$, as shown in Fig. 14.27.

Once we know $\Delta(T)$ from the gap equation, we know the T-dependent quasiparticle energies and can calculate thermodynamic properties. The energies $E_{\mathbf{k}}(T)$ determine the occupation numbers for excitations of wavevector $\mathbf{k}$,

$$f_{\mathbf{k}} = \frac{1}{1 + e^{\beta E_{\mathbf{k}}(T)}}, \tag{14.205}$$

which goes in the electronic entropy for a gas of quasiparticles

$$S_{es} = -2k_B \sum_{\mathbf{k}} \left[(1 - f_{\mathbf{k}}) \ln(1 - f_{\mathbf{k}}) + f_{\mathbf{k}} \ln f_{\mathbf{k}}\right] \tag{14.206}$$

and the electronic heat capacity

$$C_{es} = T\frac{dS_{es}}{dT} = -\beta\frac{dS_{es}}{d\beta}. \tag{14.207}$$

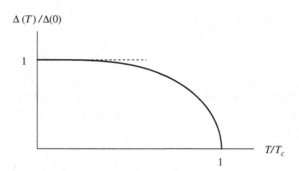

Figure 14.27 Temperature dependence of Δ.

After some algebra, we get

$$C_{es} = 2\beta k_B \sum_{\mathbf{k}} \left(-\frac{\partial f}{\partial E_{\mathbf{k}}}\right) \left(E_{\mathbf{k}}^2 + \frac{1}{2}\beta \frac{d(\Delta_{\mathbf{k}}^2)}{d\beta}\right), \tag{14.208}$$

with

$$\frac{\partial f}{\partial E_{\mathbf{k}}} = -\left(\frac{1}{1+e^{\beta E_{\mathbf{k}}}}\right)^2 e^{\beta E_{\mathbf{k}}} \beta. \tag{14.209}$$

For $T \ll T_c$, where Δ is much greater than $k_B T$,

$$C_{es} \sim e^{-\alpha/T}. \tag{14.210}$$

Here α is a constant. This is true only for $T \ll T_c$ and for the so-called s-wave super-conductors for which $\Delta(\mathbf{k}) \geq \Delta_0$, where Δ_0 is a nonzero constant. If not, then the low T behavior of C_{es} may be different. In the $T \to T_c$ limit,

$$\Delta(T \to T_c) \to 0 \tag{14.211}$$

and

$$E_{\mathbf{k}} \to |\xi_{\mathbf{k}}|, \tag{14.212}$$

and the first term on the right-hand side (RHS) in Eq. (14.208) becomes

$$C_{en} = \gamma T = \frac{2\pi^2}{3} N(0) k_B^2 T. \tag{14.213}$$

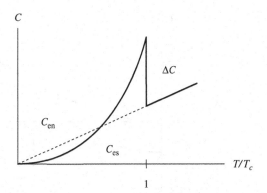

Figure 14.28 Temperature dependence of the specific heat near $T = T_c$ within the BCS model.

The second term on the RHS in Eq. (14.208) is finite below T_c, where $\frac{d\Delta^2}{dT}$ is large, but is 0 above T_c, giving rise to a discontinuity at T_c. The size of the discontinuity is

$$\Delta C = (C_{es} - C_{en})|_{T_c} = N(0)k_B\beta^2 \left(\frac{d\Delta^2}{d\beta}\right) \int_{-\infty}^{\infty} \left(-\frac{\partial f}{\partial |\xi|}\right) d\xi$$

$$= N(0) \left(-\frac{d\Delta^2}{dT}\right)\bigg|_{T_c}. \qquad (14.214)$$

Within the BCS model,

$$\frac{\Delta C}{C_{es}} = 1.43. \qquad (14.215)$$

See Fig. 14.28.

We can obtain the internal energy U since

$$U(T) = \int_0^T C(T)dT + \text{ constant.} \qquad (14.216)$$

Hence

$$U_{es}(T) + \int_T^{T_c} C_{es}(T)dT = U_{en}(T_c), \qquad (14.217)$$

and (see Fig. 14.29)

$$U_{es}(T) = U_{en}(T_c) - \int_T^{T_c} C_{es}(T)dT$$

$$= U_{en}(0) + \frac{1}{2}\gamma T_c^2 - \int_T^{T_c} C_{es}(T)dT. \qquad (14.218)$$

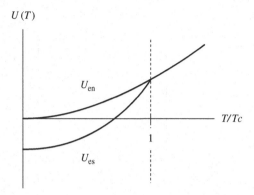

$U(T)$

U_{en}

T/T_C

1

U_{es}

Figure 14.29 Internal energy in the normal and superconducting states as a function of temperature.

For the entropy, we have

$$C_{es}(T) = T\frac{dS}{dT} \tag{14.219}$$

and

$$S_{es}(T) = \int_0^T \frac{C_{es}(T')}{T'}dT'.$$

See Fig. 14.30. For the free energy, we have

$$F_{es}(T) = U_{es}(T) - TS_{es}(T). \tag{14.220}$$

The thermodynamic critical magnetic field is given by

$$\frac{H_c^2(T)}{8\pi} = F_{en}(T) - F_{es}(T). \tag{14.221}$$

Therefore, by measuring $C_{es}(T)$, one obtains many thermodynamic properties of superconductors. Because the transition is second order, S, U, and F are continuous at T_c. Moreover,
the slope of F_{es} joins continuously (Fig. 14.31) to that of F_{en} at T_c, since $\frac{\partial F}{\partial T} = -S$.
We next explore the quasiparticle spectrum using the canonical transformation operators
for the BCS-reduced Hamiltonian. Starting with

$$\gamma_{\mathbf{k}0}^\dagger = u_\mathbf{k}^* c_{\mathbf{k}\uparrow}^\dagger - v_\mathbf{k}^* c_{-\mathbf{k}\downarrow} \tag{14.222}$$

and

$$\gamma_{\mathbf{k}1}^\dagger = u_\mathbf{k}^* c_{-\mathbf{k}\downarrow}^\dagger + v_\mathbf{k}^* c_{\mathbf{k}\uparrow}, \tag{14.223}$$

we have

$$\gamma_{\mathbf{k}0}\left|\phi_G\right\rangle = \gamma_{\mathbf{k}1}\left|\phi_G\right\rangle = 0, \tag{14.224}$$

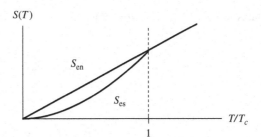

Figure 14.30 Entropy of a normal and a superconducting metal.

since

$$\gamma_{\mathbf{k}0}\,|\phi_G\rangle = (u_{\mathbf{k}}c_{\mathbf{k}\uparrow} - v_{\mathbf{k}}c^{\dagger}_{-\mathbf{k}\downarrow})\prod_{\ell}(u_{\ell} + v_{\ell}c^{+}_{\ell\uparrow}c^{+}_{-\ell\downarrow})\,|0\rangle, \qquad (14.225)$$

and when $\gamma_{\mathbf{k}0}$ acts on the $\ell = k$ term in the product of the right hand side of Eq. (14.225) gives

$$u^2_{\mathbf{k}}c_{\mathbf{k}\uparrow} + u_{\mathbf{k}}v_{\mathbf{k}}c_{\mathbf{k}\uparrow}c^{\dagger}_{\mathbf{k}\uparrow}c^{\dagger}_{-\mathbf{k}\downarrow} - v_{\mathbf{k}}u_{\mathbf{k}}c^{\dagger}_{-\mathbf{k}\downarrow} - v^2_{\mathbf{k}}c^{\dagger}_{-\mathbf{k}\downarrow}c^{\dagger}_{\mathbf{k}\uparrow}c^{\dagger}_{-\mathbf{k}\downarrow}, \qquad (14.226)$$

which yields zero when it acts on the vacuum. The same holds for $\gamma_{\mathbf{k}1}\,|\phi_G\rangle = 0$.

Next we examine an excited state

$$\gamma^{\dagger}_{\mathbf{k}0}\,|\phi_G\rangle = (|u_{\mathbf{k}}|^2 c^{\dagger}_{\mathbf{k}\uparrow} + u^{*}_{\mathbf{k}}v_{\mathbf{k}}c^{\dagger}_{\mathbf{k}\uparrow}c^{\dagger}_{\mathbf{k}\uparrow}c^{\dagger}_{-\mathbf{k}\downarrow}$$
$$- v^{*}_{\mathbf{k}}u_{\mathbf{k}}c_{-\mathbf{k}\downarrow} - |v_{\mathbf{k}}|^2 c_{-\mathbf{k}\downarrow}c^{\dagger}_{\mathbf{k}\uparrow}c^{\dagger}_{-\mathbf{k}\downarrow})\prod_{\ell\neq k}(u_{\ell} + v_{\ell}c^{\dagger}_{\ell\uparrow}c^{\dagger}_{-\ell\downarrow})\,|0\rangle, \qquad (14.227)$$

which yields (after some algebra)

$$\gamma^{\dagger}_{\mathbf{k}0}\,|\phi_G\rangle = c^{\dagger}_{\mathbf{k}\uparrow}\prod_{\ell\neq k}(u_{\ell} + v_{\ell}c^{\dagger}_{\ell\uparrow}c^{\dagger}_{-\ell\downarrow})\,|0\rangle, \qquad (14.228)$$

and similarly we have

$$\gamma^{\dagger}_{\mathbf{k}1}\,|\phi_G\rangle = c^{\dagger}_{-\mathbf{k}\downarrow}\prod_{\ell\neq k}(u_{\ell} + v_{\ell}c^{\dagger}_{\ell\uparrow}c^{\dagger}_{-\ell\downarrow})\,|0\rangle. \qquad (14.229)$$

These correspond to excitations in which a single electron is in one of the orbitals of the pair $(\mathbf{k}\uparrow, -\mathbf{k}\downarrow)$, while leaving the other orbital of the pair empty. Let N and N' be the average numbers of electrons in the ground and excited states respectively, then the excitation changes N by $\delta N = u^2_{\mathbf{k}} - v^2_{\mathbf{k}}$, which varies from -1 to 1 ($\delta N = N' - N = 1 - 2v^2_{\mathbf{k}} = u^2_{\mathbf{k}} - v^2_{\mathbf{k}}$) as $\zeta_{\mathbf{k}}$ goes from $-\infty$ to ∞.

Since there is a one-to-one correspondence between the $c^{\dagger}_{\mathbf{k}}$ of the normal metal with the $\gamma^{\dagger}_{\mathbf{k}}$ of the superconducting state, we can obtain the superconducting quasiparticle density of states $N_S(E)$ by equating

$$N_S(E)dE = N_n(\zeta)d\xi. \qquad (14.230)$$

For states near E_F, $N_n(\xi) = N(0)$, thus

$$\frac{N_S(E)}{N(0)} = \frac{d\xi}{dE}.$$ (14.231)

Also, since

$$E^2 = \xi^2 + \Delta^2$$ (14.232)

and

$$EdE = \xi d\xi,$$ (14.233)

then

$$\frac{N_S(E)}{N(0)} = \frac{d\xi}{dE} = \left\{ \begin{array}{ll} \frac{E}{\sqrt{E^2 - \Delta^2}} & (E > \Delta) \\ 0 & (E < \Delta) \end{array} \right\}.$$ (14.234)

States with a definitive number of electrons may in fact be constructed using

$$|\phi_G^N\rangle = \int_0^{2\pi} d\phi e^{-i\frac{N}{2}\phi} \prod_\ell (|u_\ell| + |v_\ell|e^{i\phi}b_\ell^\dagger) |0\rangle$$ (14.235)

and

$$\gamma_{\mathbf{k}0}^\dagger |\phi_G^N\rangle = c_{\mathbf{k}\uparrow}^\dagger \int_0^{2\pi} d\phi e^{-i\frac{N'}{2}\phi} \prod_{\ell \neq \mathbf{k}} (|u_\ell| + |v_\ell|e^{i\phi}b_\ell^\dagger) |0\rangle,$$ (14.236)

where $N' = N$ if an electron is added and $N' = N - 2$ if an electron is removed. To avoid the need for an explicit wavefunction, an operator S was introduced by Josephson for destroying a Cooper pair, where S has the eigenvalue of $e^{i\phi}$ for the BCS state. We obtain the following:

$$\gamma_{e\mathbf{k}0}^\dagger = u_{\mathbf{k}}^* c_{\mathbf{k}\uparrow}^\dagger - v_{\mathbf{k}}^* S^\dagger c_{-\mathbf{k}\downarrow} \quad \text{(putting a quasiparticle in } \mathbf{k} \uparrow\text{)},$$ (14.237)

$$\gamma_{h\mathbf{k}0}^\dagger = u_{\mathbf{k}}^* S c_{\mathbf{k}\uparrow}^\dagger - v_{\mathbf{k}}^* c_{-\mathbf{k}\downarrow} \quad \text{(putting a quasihole in } \mathbf{k} \uparrow\text{)},$$ (14.238)

$$\gamma_{e\mathbf{k}1}^\dagger = u_{\mathbf{k}}^* c_{-\mathbf{k}\downarrow}^\dagger + v_{\mathbf{k}}^* S^\dagger c_{\mathbf{k}\uparrow} \quad \text{(putting a quasiparticle in } -\mathbf{k} \downarrow\text{)},$$ (14.239)

$$\gamma_{h\mathbf{k}1}^\dagger = u_{\mathbf{k}}^* S c_{-\mathbf{k}\downarrow}^\dagger + v_{\mathbf{k}}^* c_{\mathbf{k}\uparrow} \quad \text{(putting a quasihole in } -\mathbf{k} \downarrow\text{)}.$$ (14.240)

We note that $\gamma_{h\mathbf{k}}^\dagger = S\gamma_{e\mathbf{k}}$, and that creating a hole-like excitation is equivalent to annihilating a pair and creating an electron-like excitation. The energy dispersion of the quasiparticle states, the density of quasiparticle states, and the relationships among $E_{\mathbf{k}}$ and $\xi_{\mathbf{k}}$ are shown in Figs. 14.32 and 14.33.

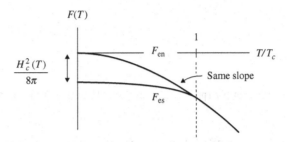

Figure 14.31 Free energy of a normal and a superconducting metal as a function of temperature.

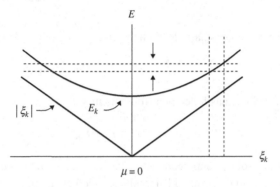

Figure 14.32 Energy dispersion of superconducting quasiparticle states and normal states $\xi_{\mathbf{k}}$.

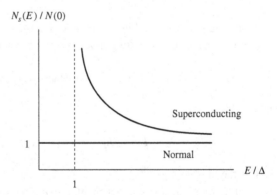

Figure 14.33 Density of quasiparticle states as a function of energy E, within the BCS model.

14.3 Superconducting quasiparticle tunneling

In dealing with tunneling[16] processes that transfer electrons from one system to another, it is useful to carefully examine the chemical potentials explicitly, since they will be different

---

[16] I. Giaever, "Energy gap in superconductors measured by electron tunneling," *Phys. Rev. Lett.* 5(1960), 147; B. D. Josephson, "Possible new effects in superconductive tunneling," *Phys. Lett.* 1 (1962), 251.

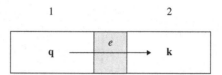

Figure 14.34 Quasiparticle tunneling process between two metallic systems through an insulating barrier.

for different conductors maintained at a finite bias voltage. Previously, we had

$$H = \mu N_{\text{op}} + E_G + \sum_{\mathbf{k},i} E_{\mathbf{k}} \gamma_{\mathbf{k}i}^{\dagger} \gamma_{\mathbf{k}i}. \tag{14.241}$$

If we add a quasiparticle, then the change in energy (δE) to the system is (for injection)

$$\delta E = E_{e\mathbf{k}} = (E_{\mathbf{k}} + \mu). \tag{14.242}$$

If we add a quasihole, then (for extraction)

$$\delta E = E_{h\mathbf{k}} = (E_{\mathbf{k}} - \mu). \tag{14.243}$$

For an isolated superconductor, the simplest number-conserving excitation consists of a hole and an electron with a total excitation energy of

$$(E_{\mathbf{k}} + \mu) + (E_{\mathbf{k}'} - \mu) = E_{\mathbf{k}} + E_{\mathbf{k}'} \geq 2\Delta. \tag{14.244}$$

In a tunneling process (see Fig. 14.34 and Fig. 14.35), an electron with $\mathbf{k}$ is transferred from system 1 to system 2 resulting in $\mathbf{k}'$ conserving energy, hence

$$(E_{\mathbf{k}1} - \mu_1) + (E_{\mathbf{k}'2} + \mu_2) = 0, \tag{14.245}$$

and this yields

$$E_{\mathbf{k}1} + E_{\mathbf{k}'2} = (\mu_1 - \mu_2) = -eV, \tag{14.246}$$

where V is the bias voltage.

By far the most detailed experimental examination of the density of states (see Fig. 14.33) is provided by electron tunneling. Using Fig. 14.34, we write the tunneling Hamiltonian as

$$H_T = \sum_{\sigma \mathbf{k} \mathbf{q}} T_{\mathbf{k}\mathbf{q}} c_{\mathbf{k}\sigma}^{\dagger} c_{\mathbf{q}\sigma} + \text{c.c.}, \tag{14.247}$$

where $\mathbf{k}$ refers to metal 2 and $\mathbf{q}$ refers to metal 1. No spin flip occurs since we assume there are no magnetic perturbations, and we note that momentum is not conserved if the interface is rough. The quantity T may be calculated using the overlap of the wavefunctions from the two metals in the insulating region where the tunneling occurs, and we will assume that it is a constant in the following analysis. The transition probability, and hence the current, is given by Fermi's golden rule:

$$W_{1 \to 2} = \frac{2\pi}{\hbar} \sum_f |\langle f|H_T|i\rangle|^2 \, \delta(E_i - E_f)$$

$$= \frac{2\pi}{\hbar} \sum_{\alpha,\beta} \left| \langle \alpha_1| \langle \beta_2| \sum_{\mathbf{kq}\sigma} Tc^\dagger_{\mathbf{k}\sigma} c_{\mathbf{q}\sigma} |G_1\rangle |G_2\rangle \right|^2 \delta(E_\alpha + E_\beta + eV), \qquad (14.248)$$

where $|G\rangle$ represents the ground state, while $|\alpha\rangle$ and $|\beta\rangle$ are excited states of the conductors, and we have made use of the energy conservation condition of Eq. (14.246). Here we have excluded the coherent processes that occur in Josephson tunneling. We can then write

$$\delta(E_\alpha + E_\beta + eV) = \int \delta(E_\alpha - E')\delta(E_\beta + (E' + eV))dE', \qquad (14.249)$$

and, from Eq. (14.248), we have for the current

$$I(V) \sim |T|^2 \int N_1^-(E)N_2^+(-eV - E)dE, \qquad (14.250)$$

where, for the added electron in conductor 2,

$$N_2^+(E) = \sum_{\mathbf{k},\beta} \left| \langle \beta_2| c^\dagger_{\mathbf{k}\sigma} |G_2\rangle \right|^2 \delta(E_\beta - E), \qquad (14.251)$$

and, for the subtracted electron in conductor 1,

$$N_1^-(E) = \sum_{\mathbf{q},\alpha} \left| \langle \alpha_1| c_{\mathbf{q}\sigma} |G_1\rangle \right|^2 \delta(E_\alpha - E). \qquad (14.252)$$

We note that these are not densities of state because the matrix elements in Eqs. (14.251) and (14.252) are not unity, i.e. $|\langle\rangle| \neq 1$. We must consider N_2^+ and N_1^- in more detail. We may write, for small bias voltage V,

$$c^\dagger_{\mathbf{k}\uparrow} = \underbrace{u_\mathbf{k}\gamma^\dagger_{e\mathbf{k}0}}_{①} + \underbrace{v^*_\mathbf{k}\gamma_{h\mathbf{k}1}}_{②} \qquad (14.253)$$

and consider an electron transferred by H_T into a state $\mathbf{k} \uparrow$. There are two terms in Eq. (14.253) corresponding to two different energy processes ($E_\mathbf{k} + \mu$ and $E_\mathbf{k} - \mu$):

$$\sum_\beta |\langle\ \rangle_2|^2 \, (\to \mathbf{k})$$

$$= \begin{cases} ① \ |u_\mathbf{k}|^2 \, (1 - f(E_\mathbf{k})) & \text{(adding a quasielectron if state } \mathbf{k} \uparrow \text{ is empty)} \\ ② \ |v_\mathbf{k}|^2 f(E_\mathbf{k}) & \text{(destroying a quasihole if state } \mathbf{k} \uparrow \text{ is occupied).} \end{cases}$$
$$(14.254)$$

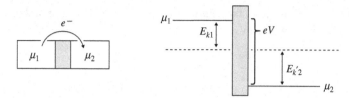

Figure 14.35 Schematic junction diagram (left) and energy diagram (right) for the tunneling process.

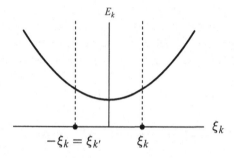

Figure 14.36 Two states $\mathbf{k}$ and $\mathbf{k}'$ with the same quasiparticle energy $E_\mathbf{k} = E_{\mathbf{k}'}$.

However, for every $\mathbf{k}$ channel, there is another $\mathbf{k}'$ channel with $\xi_{\mathbf{k}'} = -\xi_\mathbf{k}$ with exactly the same E (see Fig. 14.36). Hence, for the first term in Eq. (14.253),

$$\sum_\beta |\langle \, \rangle_2|^2 (\to \mathbf{k} \text{ and } \mathbf{k}') = u_\mathbf{k}^2 (1 - f(E_\mathbf{k})) + u_{\mathbf{k}'}^2 (1 - f(E_{\mathbf{k}'}))$$

$$= \left(u_\mathbf{k}^2 + v_\mathbf{k}^2 \right) (1 - f(E_\mathbf{k})) = 1 - f(E_\mathbf{k}), \qquad (14.255)$$

and, for the second term in Eq. (14.253),

$$\sum_\beta |\langle \, \rangle_2|^2 (\to \mathbf{k} \text{ and } \mathbf{k}') = f(E_\mathbf{k}). \qquad (14.256)$$

The coherence factor drops out in both processes when considering the effects of both $\mathbf{k}$ and $\mathbf{k}'$. This makes it possible to use the so-called semiconductor model for the tunneling process. We create a branch for the process of destroying a quasihole by using the convention of negative energy $E' \equiv -|E|$, and write $f(E) = 1 - f(-|E|)$ for the second term. See Fig. 14.37. Hence

$$N_2^+ : N_2(eV + E)\left[1 - f(eV + E)\right] \quad \text{for all energy } -\infty < E < \infty, \qquad (14.257)$$

$$N_1^- : N_1(E)f(E) \qquad\qquad\qquad \text{for all energy } -\infty < E < \infty, \qquad (14.258)$$

where N_1 and N_2 are now the densities of state of conductor 1 and conductor 2, respectively.

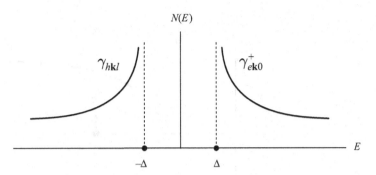

Figure 14.37 Semiconductor model density of quasiparticle states in tunneling analysis.

Within the independent-particle approximation (i.e. neglecting possible interference between $\xi_{\mathbf{k}} = -\xi_{\mathbf{k}'}$ states), tunneling between metal 1 and metal 2 is characterized by

$$I_{1\to2} = A \int_{-\infty}^{\infty} |T|^2 N_1(E) f(E) N_2(eV+E) \left[1 - f(eV+E)\right] dE, \qquad (14.259)$$

where A is a constant, and the net current is

$$I = I_{1\to2} + I_{2\to1} = A|T|^2 \int_{-\infty}^{\infty} N_1(E) N_2(eV+E) \left[f(E) - f(eV+E)\right] dE. \qquad (14.260)$$

If both metals are normal, then (making use of the properties of f and Fig. 14.38)

$$I_{nn} = A|T|^2 N_1(0) N_2(0) \int_{-\infty}^{\infty} \left[f(E) - f(eV+E)\right] dE$$

$$= A|T|^2 N_1(0) N_2(0) eV \equiv G_{nn} V, \qquad (14.261)$$

where G_{nn} is the conductance. This gives the usual ohmic behavior and is independent of temperature. See Fig. 14.39. For the case of an interface between a normal metal and a superconductor, the density of states at $V = 0$ is shown in Fig. 14.40. The current at finite V is given by

$$I_{ns} = A|T|^2 N_1(0) \int_{-\infty}^{\infty} N_{2S}(E) \left[f(E) - f(eV+E)\right] dE \qquad (14.262)$$

$$= \frac{G_{nn}}{e} \int_{-\infty}^{\infty} \frac{N_{2S}(E)}{N_2(0)} \left[f(E) - f(eV+E)\right] dE, \qquad (14.263)$$

where $N_{2S}(E)$ is the quasiparticle density of states of metal 2 in the superconducting phase.

At $T = 0$, no current will flow until $|eV| = \Delta$ (see Fig. 14.39), since the chemical potential difference must provide enough energy to create an excitation in the superconductor. The magnitude of the current is independent of the sign of V because hole and electron excitations have equal energies. For $T > 0$, the presence of thermally excited quasiparticles

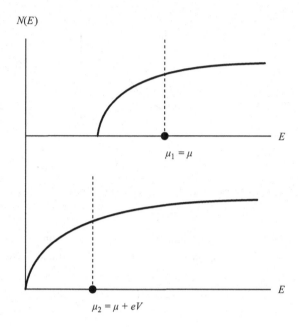

Figure 14.38   Density of states for normal metal–normal metal junction at bias voltage V.

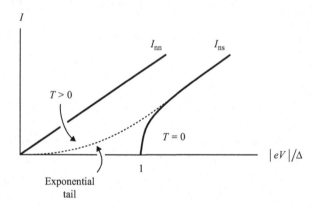

Figure 14.39 I vs. V curves of normal metal–normal metal junction and normal metal-superconductor junction.

allows them to tunnel at lower voltage. So we have

$$G_{ns}(V) \equiv \frac{dI_{ns}}{dV} = G_{nn} \int_{-\infty}^{\infty} \frac{N_{2S}(E)}{N_2(0)} \left[-\frac{\partial f(eV + E)}{\partial (eV)} \right] dE. \qquad (14.264)$$

As $T \to 0$, $-\frac{\partial f(eV+E)}{\partial(eV)}$ is a bell-shaped weighting function peaked at $E = |eV|$ with width $\sim k_{\mathrm{B}}T$ having unit area under the curve. (See Fig. 14.41.) Therefore, for $T \to 0$,

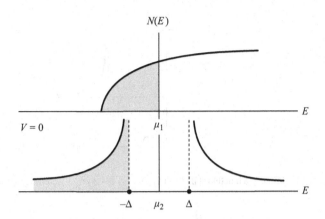

Figure 14.40 Density of states for normal metal–superconductor junction at zero bias voltage.

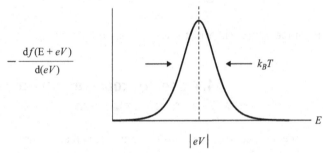

Figure 14.41 $-\frac{df(eV+E)}{d(eV)}$ as a function of energy.

we have

$$G_{ns}(V)|_{T=0} = \left.\frac{dI_{ns}}{dV}\right|_{T=0} = G_{nn}\frac{N_{2S}(e|V|)}{N_2(0)} \tag{14.265}$$

or

$$\frac{N_{2S}(e|V|)}{N_2(0)} = \frac{1}{G_{nn}}\frac{dI_{ns}}{dV}. \tag{14.266}$$

(See Fig. 14.42.) Also, one can show that

$$\left.\frac{G_{ns}(T)}{G_{nn}}\right|_{V=0} = \left(\frac{2\pi\Delta}{k_B T}\right)^{\frac{1}{2}} e^{-\frac{\Delta}{k_B T}}, \tag{14.267}$$

arising from the fact that, for $k_B T \ll \Delta$, the number of quasiparticles is exponentially dependent on T at small V. Hence, tunneling allows a detailed measurement of the superconducting gap. For tunneling between superconductors, a similar analysis yields the values of the gaps for both superconductors.

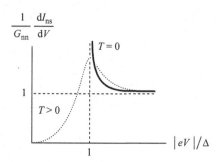

Figure 14.42 $\frac{1}{G_{nn}}\frac{dI_{ns}}{dV}$ plotted as a function of bias voltage.

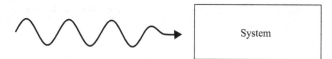

Figure 14.43   External perturbation such as light or sound on a superconducting sample.

14.4 Spectroscopies of superconductors

In general, spectroscopic measurements help obtain information about the ground and excited states of superconducting systems. Unlike the quasiparticle tunneling case, often there
are coherence effects that arise because of correlations among the electrons in the superconducting state. The experimental setup involves a time-varying external probe. (See
Fig. 14.43.) The objective of the theoretical studies of these experiments is to compute
properties by calculating the transition probabilities and coherence effects. In general, the
effects of an external perturbation for optical, nuclear magnetic resonance (NMR), acoustic
attenuation, etc. experiments on the electrons in a normal metal can be expressed in terms
of an interaction Hamiltonian of the form

$$H_{\mathrm{I}} = \sum_{\substack{\mathbf{k}\sigma' \\ \mathbf{k}'\sigma'}} B_{\mathbf{k}'\sigma',\mathbf{k}\sigma}\, c_{\mathbf{k}'\sigma'}^{\dagger} c_{\mathbf{k}\sigma}, \tag{14.268}$$

where the B's are the matrix element of the perturbing operator between the ordinary one-
electron states, the $c^{\dagger}$'s and the c's represent the creation and destruction of well-defined
normal electron excitations, and each term in this sum is independent (see Fig. 14.44).

The square of each B is proportional to a corresponding transition probability. The
measured spectrum for a normal system is given by the Fermi golden rule:

$$s(\omega) \sim \sum_{f} |\langle f| H_{\mathrm{I}} |G\rangle|^2 \delta(\hbar\omega - E_f) = \sum_{\substack{\mathbf{k}\sigma \\ \mathbf{k}'\sigma'}} \left|B_{\mathbf{k}'\sigma',\mathbf{k}\sigma}\right|^2 \delta(\hbar\omega - E\left(\mathbf{k}\sigma \to \mathbf{k}'\sigma'\right)), \tag{14.269}$$

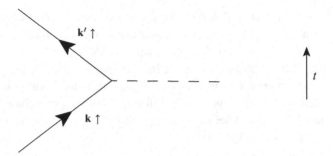

Figure 14.44   Terms in the interaction Hamiltonian between electrons and an external probe.

where $|f\rangle$ is an excited state, $E_f = E(\mathbf{k}\sigma \to \mathbf{k}'\sigma')$ is the excitation energy of state f, and $|G\rangle$ is the ground state. This is not true in the superconducting state, because it consists of a phase-coherent superposition of occupied one-electron states. As a result, there are interference terms that are not present in the expression for $s(\omega)$ for the normal state. This can be seen by expanding $c^{\dagger}_{\mathbf{k}'\uparrow}c_{\mathbf{k}\uparrow}$ in terms of the quasiparticle γ's (using Eqs. (14.222) and (14.223)),

$$c^{\dagger}_{\mathbf{k}'\uparrow}c_{\mathbf{k}\uparrow} = u_{\mathbf{k}'}u^*_{\mathbf{k}}\gamma^{\dagger}_{\mathbf{k}'0}\gamma_{\mathbf{k}0} - v^*_{\mathbf{k}'}v_{\mathbf{k}}\gamma^{\dagger}_{\mathbf{k}1}\gamma_{\mathbf{k}'1}$$
$$+ u_{\mathbf{k}'}v_{\mathbf{k}}\gamma^{\dagger}_{\mathbf{k}'0}\gamma^{\dagger}_{\mathbf{k}1} + v^*_{\mathbf{k}'}u^*_{\mathbf{k}}\gamma_{\mathbf{k}'1}\gamma_{\mathbf{k}0} \tag{14.270}$$

and

$$c^{\dagger}_{-\mathbf{k}\downarrow}c_{-\mathbf{k}'\downarrow} = -v^*_{\mathbf{k}}v_{\mathbf{k}'}\gamma^{\dagger}_{\mathbf{k}'0}\gamma_{\mathbf{k}0} + u_{\mathbf{k}}u^*_{\mathbf{k}'}\gamma^{\dagger}_{\mathbf{k}1}\gamma_{\mathbf{k}'1}$$
$$+ u_{\mathbf{k}}v_{\mathbf{k}'}\gamma^{\dagger}_{\mathbf{k}'0}\gamma^{\dagger}_{\mathbf{k}1} + v^*_{\mathbf{k}}u^*_{\mathbf{k}'}\gamma_{\mathbf{k}'1}\gamma_{\mathbf{k}0}. \tag{14.271}$$

Both of these terms lead to the same final states. Therefore, we must add the matrix elements of these two terms before squaring; that is, they add coherently in determining the transition rate. In general, this addition can be done quite readily, since one expects the coefficients $B_{\mathbf{k}'\sigma',\mathbf{k}\sigma}$ and $B_{-\mathbf{k}-\sigma,-\mathbf{k}'-\sigma'}$ to differ at most in sign because both represent processes in which

$$\Delta\mathbf{k} = \mathbf{k}' - \mathbf{k} \tag{14.272}$$

and

$$\Delta\sigma = \sigma' - \sigma. \tag{14.273}$$

Thus, we may combine them as

$$\left[(\mathbf{k}\sigma) \to (\mathbf{k}'\sigma')\right] + \left[(-\mathbf{k}' - \sigma') \to (-\mathbf{k} - \sigma)\right] = B_{\mathbf{k}'\sigma',\mathbf{k}\sigma}\left(c^{\dagger}_{\mathbf{k}'\sigma'}c_{\mathbf{k}\sigma} \pm c^{\dagger}_{-\mathbf{k}-\sigma}c_{-\mathbf{k}'-\sigma'}\right). \tag{14.274}$$

The sign $+$ or $-$ depends on the specific interaction. In Eq. (14.274), we consider the $+$ sign for electron–phonon ultrasonic attenuation experiments, since the scalar deformation potential involved in B depends only on $\Delta\mathbf{k}$ and not $\mathbf{k}$ or σ, and we consider the $-$ sign for electron–photon processes, $H_I \sim \mathbf{p} \cdot \mathbf{A}$, since this change of sign when replacing $\mathbf{k}$ by $-\mathbf{k}$ is appropriate for this case. We note that since neither interaction involves the spin, then $\sigma = \sigma'$. However, this is different if there is a spin change, as in the term $I_+ S_-$ in the hyperfine case. After some algebra, and taking u, v, and Δ to be real, the combined terms in Eq. (14.274) become

$$
B_{\mathbf{k}'\sigma',\mathbf{k}\sigma} \left[(u_{\mathbf{k}'}u_{\mathbf{k}} \mp v_{\mathbf{k}'}v_{\mathbf{k}})(\gamma^\dagger_{\mathbf{k}'\sigma'}\gamma_{\mathbf{k}\sigma} \pm \theta_{\sigma'\sigma}\gamma^\dagger_{-\mathbf{k}-\sigma}\gamma_{-\mathbf{k}'-\sigma'}) \right.
$$
$$
\left. + (v_{\mathbf{k}}u_{\mathbf{k}'} \pm u_{\mathbf{k}}v_{\mathbf{k}'})(\gamma^\dagger_{\mathbf{k}'\sigma'}\gamma^\dagger_{-\mathbf{k}-\sigma} \pm \theta_{\sigma'\sigma}\gamma_{-\mathbf{k}'-\sigma'}\gamma_{\mathbf{k}\sigma}) \right],
\tag{14.275}
$$

where

$$
\theta_{\sigma\sigma'} = \pm 1 \quad \text{for } \sigma' = \pm\sigma,
\tag{14.276}
$$

$$
\gamma_{\mathbf{k}\sigma} = \gamma_{\mathbf{k}0} \quad \text{for } \sigma = \uparrow,
\tag{14.277}
$$

$$
\gamma_{\mathbf{k}\sigma} = \gamma_{-\mathbf{k}1} \quad \text{for } \sigma = \downarrow.
\tag{14.278}
$$

We will classify the experiments using two cases: case I for the upper sign and case II for the lower sign. Case I and case II pertain to perturbations that are even or odd under time reversal of the electron states, since this interchanges the partner electrons in the Cooper pair. We note that θ in Eq. (14.275) really has no effect on the magnitude of the transition probability; it affects only the relative phase of the off-diagonal matrix element connecting disparate states. For the computation of transition probabilities, we have therefore from Eq. (14.275)

$$
|B|^2 \rightarrow (uu' \mp vv')^2 |B|^2 \quad \text{(for the scattering of a quasiparticle)}
\tag{14.279}
$$

or

$$
|B|^2 \rightarrow (vu' \pm uv')^2 |B|^2 \quad \text{(for the creation or annihilation of two quasiparticles)}.
\tag{14.280}
$$

We can write the following for the above two coherent factors:

$$
(uu' \mp vv')^2 = \frac{1}{2}\left(1 \mp \frac{\Delta^2}{EE'}\right),
\tag{14.281}
$$

and

$$
(vu' \pm uv')^2 = \frac{1}{2}\left(1 \pm \frac{\Delta^2}{EE'}\right).
\tag{14.282}
$$

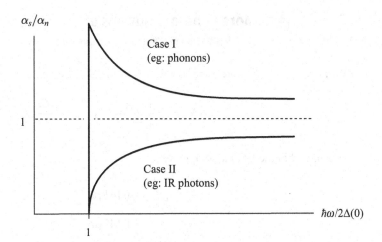

Figure 14.45 Schematic of frequency dependence of the absorption in $\alpha_s(\omega)$ by a superconductor at $T = 0$.

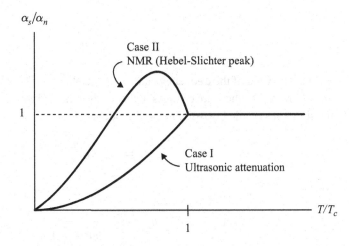

Figure 14.46 Spectroscopic response of a superconductor for $\hbar\omega \ll \Delta(0)$ where scatterings of quasiparticles dominate.

These can have very large effects in transition rates for energies E near the gap Δ. To illustrate these effects, we consider physical processes where $\hbar\omega \lesssim \Delta$. These include ultrasonic attenuation, NMR relaxation, and electromagnetic absorption. If we restrict the frequency range so that $\hbar\omega \ll \Delta(0)$, only scatterings of quasiparticles are possible. For absorption at $T = 0$, we have the creation of quasiparticle pairs only if $\hbar\omega > \Delta$. For $\hbar\omega \gg \Delta(0)$, the energies E and E' are much larger than Δ, and there is no difference between case I and case II. These effects are illustrated in Figs. 14.45 and 14.46.

14.5 More general solutions of the BCS gap equation

We now consider a more general solution of the BCS gap equation[17]

$$\Delta_{\mathbf{k}} = -\frac{1}{2} \sum_{\ell} \frac{\Delta_{\ell}}{E_{\ell}} V_{\mathbf{k}\ell} \tanh \frac{\beta E_{\ell}}{2}, \tag{14.283}$$

which, within the BCS model, gives

$$k_{\mathrm{B}} T_c = 1.14 E_{\mathrm{D}} e^{-\frac{1}{N(0)V}} \tag{14.284}$$

$$= 1.14 E_{\mathrm{D}} e^{-\frac{1}{\lambda - \mu}}, \tag{14.285}$$

where we define

$$\lambda = N(0) V^{\mathrm{ph}} \tag{14.286}$$

and

$$\mu = N(0) V^{\mathrm{C}}. \tag{14.287}$$

We will show that if the frequency cutoffs for the phonon V^{ph} and Coulomb V^{C} interactions are different, then there are changes in the results obtained by the BCS model. In particular, the simplest so-called two-square-well model gives

$$k_{\mathrm{B}} T_c = 1.14 E_{\mathrm{D}} e^{-\frac{1}{\lambda - \mu^*}}, \tag{14.288}$$

where

$$\mu^* = \frac{\mu}{1 + \mu \ln \left(\frac{E_{\mathrm{F}}}{E_{\mathrm{D}}} \right)}, \tag{14.289}$$

and this gives large changes in T_c and in the isotope effect. Although the BCS gap equation has been superseded by new formulations to be discussed later, it is still instructive to explore more general solutions of the BCS model, since these methods are also used for other formulations.

We begin with

$$V_{\mathbf{k}\ell} = V_{\mathbf{k}\ell}^{\mathrm{ph}} + V_{\mathbf{k}\ell}^{\mathrm{C}} \tag{14.290}$$

$$= -(-V_{\mathrm{ph}} - V_{\mathrm{C}}). \tag{14.291}$$

---

[17] M. L. Cohen, "Superconductivity in low-carrier-density systems: degenerate semiconductors," in *Supercon-ductivity*, ed. R. D. Parks (New York, NY: Marcel Dekker, Inc., 1969), p. 615.

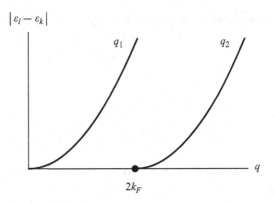

The limits q_1 and q_2 on the integral over q in the BCS gap equation.

These interactions can be complex for a real material. We first consider the simplest case going beyond the BCS model by assuming that

$$V_{\mathbf{k}\ell} = V(|\ell - \mathbf{k}|, \varepsilon_{\mathbf{k}} - \varepsilon_\ell), \tag{14.292}$$

and that Δ_ℓ and E_ℓ depend only on $|\ell|$, i.e. we have an isotropic Fermi surface. We can transform Eq. (14.283) into

$$\Delta_{\mathbf{k}} = -\frac{\Omega}{2(2\pi)^3} \int \frac{\Delta_\ell}{E_\ell} \left(\frac{\ell}{v_\ell \hbar k} \right) \tanh \left(\frac{\beta E_\ell}{2} \right) \times \left[\int_{q_1}^{q_2} q\, dq\, V(q, \varepsilon_\ell - \varepsilon_{\mathbf{k}}) \right] d\varepsilon_\ell \tag{14.293}$$

by using a Jacobian to transform the three-dimensional integral over $\mathbf{k}$ into an integral over q and ε_ℓ. (See Fig. 14.47.) Here q_1 and q_2 are determined for fixed ε_ℓ and ε_k,

$$q = \left| |\ell| - |\mathbf{k}| \right| \text{ to } |\ell| + |\mathbf{k}|. \tag{14.294}$$

With the following changes to variables ε and ε',

$$\varepsilon_k = \frac{\hbar^2 k^2}{2m} - \varepsilon_{\mathrm{F}} = \varepsilon, \tag{14.295}$$

$$\Delta_\ell = \Delta(\varepsilon_\ell) = \Delta(\varepsilon'), \tag{14.296}$$

then

$$\Delta(\varepsilon) = -\int \frac{\Delta(\varepsilon')}{E'} F(\varepsilon, \varepsilon') \tanh \left(\frac{\beta E'}{2} \right) d\varepsilon', \tag{14.297}$$

where

$$F(\varepsilon, \varepsilon') = F(\varepsilon_k, \varepsilon_\ell) = \frac{\Omega}{2(2\pi)^3} \frac{l}{\hbar k v_l} \left[\int_{||l|-|k||}^{|l|+|k|} q V(q, |\varepsilon - \varepsilon'|) dq \right]. \tag{14.298}$$

The integration limits are shown in Fig. 14.47. Another change in variables

$$\Delta(\varepsilon)\frac{k}{k_F} \equiv D(\varepsilon); \quad \frac{k}{l}F(\varepsilon,\varepsilon') \equiv K(\varepsilon,\varepsilon') \tag{14.299}$$

yields

$$D(\varepsilon) = -\int \frac{D(\varepsilon')}{E'}K(\varepsilon,\varepsilon')\tanh\frac{\beta E'}{2}d\varepsilon'. \tag{14.300}$$

As an example of a solution to Eq. (14.300), we consider the BCS or one-square-well model

$$V(q,\omega) = -\frac{V}{\Omega} \tag{14.301}$$

for $k = \ell = k_F$ (i.e. $\varepsilon, \varepsilon' = 0$),

$$K(0,0) = -\frac{\Omega}{2(2\pi)^3}\left(\frac{m}{\hbar k_F}\right)\frac{1}{\hbar}\frac{V}{\Omega}\int_0^{2k_F}qdq = -\frac{N_\uparrow(0)}{2}V, \tag{14.302}$$

$$D(0) = -\int_{-\infty}^{\infty}\frac{D(\varepsilon')}{E'}K(0,\varepsilon')\tanh\frac{\beta E'}{2}d\varepsilon'. \tag{14.303}$$

Now, in this model, $K(\varepsilon,\varepsilon') = K(0,0)$ for $-E_D < \varepsilon' < E_D$, we have

$$D(0) = \left(\frac{N(0)V}{2}\right)2\int_0^{E_D}\frac{D(\varepsilon')}{E'}\tanh\frac{\beta E'}{2}d\varepsilon'. \tag{14.304}$$

This equation, when evaluated at $T = T_c$, yields

$$k_B T_c = 1.14 E_D e^{-\frac{1}{N(0)V}}, \tag{14.305}$$

which is the BCS model solution.

For more complicated kernels like Coulomb $V^C(q,\omega) = \frac{4\pi e^2}{\Omega q^2 \epsilon(q,\omega)}$, where $\epsilon(q,\omega)$ is the Lindhard dielectric function,

$$K^C(\varepsilon,\varepsilon') = \frac{\Omega}{8\pi^2 v'\hbar}\int_{q_1}^{q_2}qV^C(q,|\varepsilon-\varepsilon'|)d\varepsilon', \tag{14.306}$$

we can numerically evaluate $K^C(\varepsilon,\varepsilon')$, which appears in Fig. 14.48, and a reasonable approximation for K^C is to assume a constant value with a cutoff at E_F.

For the phonon kernel, we use a Bardeen–Pines interaction (see Fig. 14.49),

$$V^{BP}(q,\omega) = \frac{2\hbar\omega_q|M_q|^2}{(\hbar\omega)^2 - (\hbar\omega_q)^2}, \tag{14.307}$$

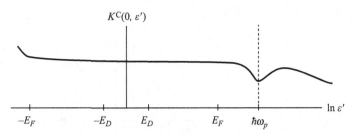

Figure 14.48 The Coulomb kernel as a function of energy ε' using a Lindhard dielectric function. Approximate values for the Fermi energy E_F, the Debye energy E_D, and the plasma energy $\hbar\omega_F$ are shown as reference energies.

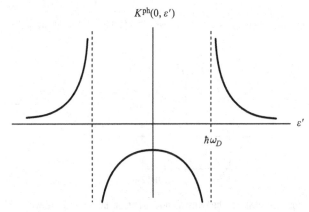

Figure 14.49 The phonon kernel as a function of energy ε' assuming a Bardeen–Pines form for the interaction.

and the corresponding kernel has the form

$$K_{\mathrm{BP}}^{ph} = \frac{\mathrm{const}}{\omega^2 - \omega_{\mathrm{D}}^2} \tag{14.308}$$

which can be modeled as a constant with a cutoff at $E_{\mathrm{D}} = \hbar\omega_{\mathrm{D}}$. The equation we now need to solve is

$$D(\varepsilon) = -\int_{-\infty}^{\infty} \frac{K(\varepsilon, \varepsilon')D(\varepsilon')}{E'} \tanh\left(\frac{\beta E'}{2}\right) d\varepsilon', \tag{14.309}$$

where

$$D(\varepsilon) = \frac{k}{k_{\mathrm{F}}}\Delta(\varepsilon_k) \tag{14.310}$$

and

$$K(\varepsilon, \varepsilon') = K^C(\varepsilon, \varepsilon') + K^{ph}(\varepsilon, \varepsilon'). \tag{14.311}$$

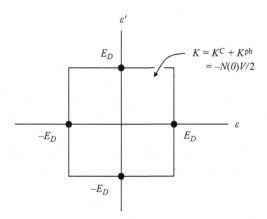

Energy plane, energy cutoffs, and kernel values for the BCS one-square-well model.

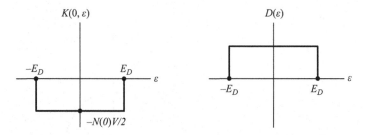

Energy dependence of the kernel K and gap D in the BCS one-square-well model.

The BCS model (Figs. 14.50 and 14.51) uses a one-square-well model with one cutoff $E_D = \hbar\omega_D$, but as we have seen, K^C extends much further to E_F. Hence, a two-square-well model is appropriate. To account for the different energy regions for ε and ε', we label the piece-wise constant regions in the $(\varepsilon, \varepsilon')$ plane to have values K_{ij} (Fig. 14.52). We now choose

$$K_{00} = -N(0)V = 2(K^C - K^{ph}) = \mu - \lambda \tag{14.312}$$

and

$$K_{01} = K_{10} = K_{11} = 0. \tag{14.313}$$

At T_c,

$$D_0 = -K_{00}D_0 \int_0^{E_D} \frac{1}{\varepsilon'} \tanh\left(\frac{\beta\varepsilon'}{2}\right) d\varepsilon' \equiv -K_{00}D_0 x, \tag{14.314}$$

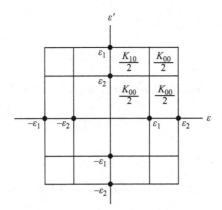

Figure 14.52 The values of the kernel in the energy plane $(\varepsilon, \varepsilon')$ for the two-square-well model.

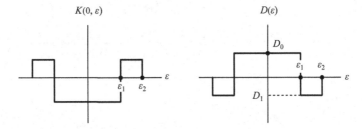

Figure 14.53 The kernel K and the energy gap D as a function of energy ε for the two-square-well model.

which leads to

$$x = \ln\left(\frac{1.14E_D}{k_B T_c}\right) = \frac{1}{K_{00}}. \tag{14.315}$$

This is the same result that BCS obtained. However, for the two-square-well model (see Figs. 14.50 and 14.53),

$$D_0 = -K_{00}D_0 \int_0^{\varepsilon_1} \frac{1}{\varepsilon'}\tanh\left(\frac{\beta\varepsilon'}{2}\right) d\varepsilon' - K_{01}D_1 \ln\left(\frac{\varepsilon_2}{\varepsilon_1}\right)$$

$$\equiv -K_{00}D_0 x - K_{01}D_1 z, \tag{14.316}$$

$$D_1 = -K_{10}D_0 x - K_{11}D_1 z. \tag{14.317}$$

Solving for x, we have

$$x = \frac{1}{-K_{00} + \frac{K_{01}K_{10}z}{K_{11}z+1}} = \ln\left(\frac{1.14\varepsilon_1}{k_B T_c}\right),$$

and

$$k_B T_c = 1.14\varepsilon_1 e^{-x}.$$ (14.318)

We can now associate the square wells with λ and μ:

$$-K_{00} = \left(|K^{ph}| - |K^C|\right) \times 2 = \lambda - \mu$$ (14.319)

and

$$K_{01} = K_{10} = K_{11} = |K^C| \times 2 = \mu.$$ (14.320)

Then,

$$x = \frac{1}{2|K^{ph}| - \frac{2K^C}{2K^C z + 1}} = \frac{1}{\lambda - \mu^*},$$ (14.321)

$$\mu^* = \frac{2K^C}{2K^C \ln\left(\frac{\varepsilon_2}{\varepsilon_1}\right) + 1} = \frac{\mu}{1 + \mu \ln\left(\frac{\varepsilon_2}{\varepsilon_1}\right)},$$ (14.322)

and

$$\lambda = 2|K^{ph}|.$$ (14.323)

We note that $\varepsilon_2 \approx E_F$, and that $\ln\left(\frac{E_F}{E_D}\right) > 1$; therefore the Coulomb interaction μ^* in Eq. (14.322) is reduced because of the frequency-dependent factors, and T_c is increased as shown in the following equations:

$$k_B T_c = 1.14 E_D e^{-\frac{1}{\lambda - \mu}} \qquad \text{(BCS model)}$$ (14.324)

and

$$k_B T_c = 1.14 E_D e^{-\frac{1}{\lambda - \mu^*}} \qquad \text{(two-square-well model).}$$ (14.325)

One of the most important experiments in the search for determining the mechanism of superconductivity is the isotope effect described earlier. The isotope effect parameter α is defined by

$$T_c \sim M^{-\alpha}$$

and

$$\alpha \equiv -\frac{d(\ln T_c)}{d(\ln M)}.$$ (14.326)

Table 14.1 Isotope effects in superconductors. $\alpha < 0.5$ for some superconducting elements[18]

Substance	α	Substance	α
Zn	0.45	Ru	~ 0.00
Sn	0.47	Os	0.15
Hg	0.50	$Nb_3 Sn$	0.08
Pb	0.49	Zr	~ 0.00

Using the BCS model,

$$\ln T_c = \text{const} + \ln E_D - N(0)V. \tag{14.327}$$

Since

$$E_D \sim \frac{1}{\sqrt{M}}, \tag{14.328}$$

this leads to $\alpha = \frac{1}{2}$.

As shown in Table 14.1, the BCS value of $\alpha = \frac{1}{2}$ is often not found, however this variation in α can be accounted for by using the two-square-well model, where

$$\ln T_c = \text{const} + \ln E_D - \frac{1}{\lambda - \mu^*}. \tag{14.329}$$

From Eq. (14.328) and

$$\frac{d\mu^*}{d\ln M} = -\frac{1}{2}\mu^{*2}, \tag{14.330}$$

we have

$$\frac{d\ln T_c}{d\ln M} = -\frac{1}{2} - \frac{1}{(\lambda - \mu^*)^2}\left(-\frac{1}{2}\mu^{*2}\right) \tag{14.331}$$

and

$$\alpha = \frac{1}{2}\left[1 - \left(\frac{\mu^*}{\lambda - \mu^*}\right)^2\right] < \frac{1}{2}. \tag{14.332}$$

Also,

$$\alpha \approx 0 \text{ if } \begin{cases} \mu^* \approx \frac{\lambda}{2} \\ \mu^* \gg \lambda. \end{cases} \tag{14.333}$$

[18] C. Kittel, *Introduction to Solid State Physics*, 8th ed. (New York, NY: Wiley, 2005)

The latter case can be ruled out, since then superconductivity would be unlikely. Experimentally, many materials show anomalous isotope effects. For example, $\alpha \approx \frac{1}{2}$ for $s - p$ metals, $\alpha < \frac{1}{2}$ for transition metals, and $\alpha \approx 0$ in some elements (Ru, Zr).

To make a rough estimate of the sizes involved, if we use a Thomas–Fermi model for μ, we have

$$\mu = \frac{0.166 r_s}{2} \ln \left| 1 + \frac{1}{0.166 r_s} \right|, \tag{14.334}$$

and note that for most metals, $2 \le r_s \le 4$, then

$$0.23 \le \mu \le 0.30, \tag{14.335}$$

and

$$0.1 \le \mu^* \le 0.13. \tag{14.336}$$

These values give reasonable estimates for T_c and α when λ is in its normal range of about 0.5.

14.6 Field theoretical methods and BCS theory

Field theoretical methods were used to improve BCS theory while retaining the fundamental concepts of pairing caused by electron–lattice interactions. Important contributions from Nambu,[19] Gorkov,[20] and Eliashberg[21] using Green's function approaches formed the basis for the modern theory. This theory has been very successful in explaining and predicting properties of a class of superconductors where it is believed that the above BCS assumptions hold. In the late 1980s, the so-called high-temperature superconductors based on copper and oxygen allowed the existence of superconductivity above liquid nitrogen. This class of superconductors and others, such as the Fe-based superconductors, appear to be different, and there is no consensus at present on whether the superconductivity in these materials is understood in terms of one consistent theory, even though many theories have been developed in this area. Hence, for our purposes, we will ignore this class of superconductors and the associated theories and simply describe some of the highlights associated with the evolution of BCS theory with phonon-induced pairing.

The Nambu–Gorkov–Eliashberg approach yielded an equation for Δ and T_c which involved a self-consistent treatment of the renormalization arising from phonons and pairing effects. The resulting kernels corresponding to those in Eqs. (14.297)–(14.311) are similar,

[19] Y. Nambu, "Quasi-particles and gauge invariance in the theory of superconductivity," *Phys. Rev.* 117(1960), 648.

[20] L. P. Gorkov, "On the energy spectrum of superconductors," *Soviet Phys. JETP* 7(1958), 505.

[21] G. M. Eliashberg, "Interactions between electrons and lattice vibrations in a superconductor," *Soviet Phys. JETP* 11(1960), 696.

except that the energies $\varepsilon, \varepsilon'$ are now renormalized to E and E', and the density of states is modified by the phonon renormalization much like in the case of polarons. McMillan[22] solved the resulting equations with a square-well model and, by fitting some data, produced the following very useful equation for T_c:

$$T_c = \frac{\Theta}{1.45} \exp\left[\frac{-1.04(1+\lambda)}{\lambda - \mu^*(1+0.62\lambda)}\right], \tag{14.337}$$

where Θ is the Debye temperature. More modern approaches deal with the so-called Eliashberg equations in a finite temperature (Matsubara) formulation. The approaches generally retain some of the notation developed by McMillan, such as

$$\lambda = 2 \int_0^\infty d\omega \frac{\alpha^2(\omega)F(\omega)}{\omega}, \tag{14.338}$$

where the spectral function $\alpha^2(\omega)F(\omega)$ for real ω (not Matsubara) is

$$\alpha^2(\omega)F(\omega) = \frac{\int_S \frac{d^2p}{v} \int_{S'} \frac{d^2p'}{(2\pi)^3 v_F'} \sum_\alpha M_{\mathbf{pp}'\alpha} \delta(\omega - \omega_{\mathbf{p-p}',\alpha})}{\int \frac{d^2p}{v_F}}, \tag{14.339}$$

where we assume scattering near the Fermi surface S for a state p to a state p' by a phonon of wavevector $p - p'$ and polarization α by the electron–phonon matrix element M. Note that the polarization index α is not the same as the function $\alpha^2(\omega)$. This expression is sometimes interpreted as $\alpha^2(\omega)$, corresponding to the electron–phonon interaction and $F(\omega)$ the phonon density of states. In the Matsubara representation, the integrals become summations over imaginary frequencies and the gap equation and spectral function become

$$\Delta(i\omega_n) = \pi k_B T \sum_m \frac{[\lambda(\omega_n - \omega_m) - \mu^*]\Delta(i\omega_m)}{\sqrt{\omega_m^2 Z^2 + \Delta(i\omega_n)^2}}, \tag{14.340}$$

where

$$\lambda(\omega_n - \omega_m) = 2 \int_0^\infty d\omega \frac{\alpha^2 F(\omega)\omega}{\omega^2 + (\omega_n - \omega_m)^2}, \tag{14.341}$$

and Z is usually taken as the renormalization in the normal state, i.e. $Z = 1 + \lambda$ and $\lambda(0) = \lambda$, as defined previously.

Equation (14.340) can be solved numerically for T_c. However, to illustrate analytical solutions, we use a specific form of the solution (without a detailed derivation) as follows. The T_c is obtained from the condition

$$\det |K_{mn}| = 0, \tag{14.342}$$

[22] W. L. McMillan, "Transition temperature of strong-coupled superconductors," *Phys. Rev.* 167(1968), 331.

where

$$K_{mn} = \lambda(m - n) + \lambda(m + n - 1) - 2\mu^* - \delta_{mn}\left(2m + 1 + \lambda(0) + 2\sum_{l=1}^{m}\lambda(l)\right),$$

(14.343)

where, from Eq. (14.341),

$$\lambda(n) = 2\int_0^\infty d\omega \frac{\alpha^2 F(\omega)\omega}{\omega^2 + (2\pi nT)^2}.$$

(14.344)

We will give two interesting solutions of Eq. (14.343), assuming $\mu^* = 0$ and an Einstein model for $\alpha^2 F(\omega)$ of the form

$$\alpha^2 F(\omega) = \frac{\lambda\omega_E}{2}\delta(\omega - \omega_E).$$

(14.345)

For large λ, we take $m = n = 0$, which yields an equation for T_c

$$K_{00} = 0,$$

(14.346)

where

$$K_{00} = \lambda(0) - \lambda(1) - (1 + \lambda(0)) = \lambda(1) - 1 = 0.$$

(14.347)

Then at T_c,

$$\lambda(1) = \frac{\lambda\omega_E^2}{\omega_E^2 + (2\pi T_c)^2}.$$

(14.348)

This gives

$$\lambda\omega_E^2 = \omega_E^2 + (2\pi T_c)^2$$

(14.349)

and

$$T_c = \frac{1}{2\pi}\sqrt{\lambda\omega_E^2}.$$

(14.350)

Allen and Dynes[23] explored the large λ limit for various $\alpha^2 F(\omega)$ and obtained

$$T_c = 0.182\sqrt{\lambda\langle\omega^2\rangle}.$$

(14.351)

Since $\frac{1}{2\pi} = 0.16$, this is close to the limit obtained for the Einstein spectrum for large λ. Unlike the McMillan equation (Eq. (14.337)), which exhibits a maximum T_c at relatively

[23] P. B. Allen and R. C. Dynes, "Transition temperature of strong-coupled superconductors reanalyzed," *Phys. Rev. B* 12(1975), 905.

small λ, the Allen–Dynes equation does not, which is encouraging for obtaining high T_c materials with the electron–phonon mechanism. The saturation of $T_c(\lambda)$ in the McMillan equation was a result of small λ approximations. The McMillan equation is appropriate for $\lambda \leq 1.5$. Another solution for Eq. (14.343) over the full range of λ, which gives the McMillan exponential dependence on λ for small λ and the Allen–Dynes square root dependence at high λ, was first described by semi-empirical methods[24] for $\mu^* = 0$ by Kresin and later for a specific spectral function by Barbee and Cohen. The latter calculation assumes a square-well spectral function

$$\alpha^2 F(\omega) = \begin{cases} \frac{\lambda}{2} & 0 < \omega < \Omega_{\max} \\ 0 & \text{otherwise.} \end{cases} \tag{14.352}$$

Hence,

$$\lambda(1) = \lambda \int_0^{\Omega_{\max}} \frac{\omega d\omega}{\omega^2 + (2\pi T_c)^2} = 1 \tag{14.353}$$

$$= \frac{\lambda}{2} \ln \left[\omega^2 + (2\pi T_c)^2 \right]\Big|_0^{\Omega_{\max}}, \tag{14.354}$$

which gives

$$T_c = \frac{\Omega_{\max}}{2\pi} \left(e^{2/\lambda} - 1 \right)^{-1/2}. \tag{14.355}$$

If we replace

$$\frac{\Omega_{\max}}{2\pi} = 0.16\Omega_{\max} = 0.25\omega_{ph}, \tag{14.356}$$

where ω_{ph} is considered to be an average phonon frequency, then

$$T_c = \frac{0.25\omega_{ph}}{(e^{2/\lambda} - 1)^{1/2}}. \tag{14.357}$$

For small λ, Eq. (14.356) yields

$$T_c = 0.25\omega_{ph}e^{-1/\lambda}, \tag{14.358}$$

which is close to the McMillan expression for this range. For large λ,

$$T_c = \frac{0.25}{\sqrt{2}}\omega_{ph}\sqrt{\lambda} = 0.18\omega_{ph}\sqrt{\lambda}, \tag{14.359}$$

and this is the result Allen and Dynes obtain. So both limits are in agreement with previous work. Numerical solutions of the Eliashberg equation for $\mu^* = 0$ give results close to those given by Eq. (14.334), hence this equation is useful over the full range of λ.

---

[24] L. C. Bourne, A. Zettl, T. W. Barbee III, and M. L. Cohen, "Complete absence of isotope effect in YBa$_2$Cu$_3$O$_7$: consequences for phonon-mediated superconductivity," *Phys. Rev. B* 36(1987), 3990; V. Z. Kresin, "Critical temperature for any strength of the electron–phonon coupling," *Bull. Am. Phys. Soc.* 32(1987), 796.

Magnetism

15.1 Background

Magnetism is one of the oldest areas of study in materials science. Almost everyone has seen the effects of ferromagnetism, and the applications of magnetism are enormous. The scientific aspects of magnetism are also extremely interesting, and this old field continues to be one of the most active in condensed matter physics. Intrinsic magnetic properties of materials are associated with orbital and spin angular momentum. This property of matter has its origin in the quantum mechanical nature of electrons and nuclei. Here we will consider only electronic properties and note that external magnetic fields can induce magnetic behavior in solids usually by changing electron spin orientation and orbital motion. Referring back to our previous discussion of Maxwell's equations, we again define the magnetic susceptibility χ as follows. Using the relation

$$\mathbf{B} = \mathbf{H} + 4\pi \mathbf{M}, \tag{15.1}$$

where $\mathbf{M}$ is the magnetic moment per unit volume, $\mathbf{B}$ is the magnetic field in the solid, and $\mathbf{H}$ is the usual "auxiliary" field associated with free currents, χ is given by

$$\chi = \frac{\mathbf{M}}{\mathbf{H}}. \tag{15.2}$$

The behavior of χ characterizes many magnetic properties of solids such as diamagnetism, paramagnetism, ferromagnetism, and antiferromagnetism.

15.2 Diamagnetism

Diamagnetism is the general property of solids that demonstrates their inclination to "repel" a magnetic field. For diamagnetism, χ is negative. We encountered this for the case of superconductors in the last chapter. For type I superconductors, $B = 0$ inside the superconductor; hence using Eq. (15.1), $\chi = -\frac{1}{4\pi}$, and a superconductor is a perfect diamagnet. Essentially, the superconducting electrons participate in a current that produces a magnetic field that opposes the applied field. We also encountered diamagnetic behavior when studying the response of normal electrons to an external magnetic field. The orbital motion of the responding electrons produced the so-called Landau diamagnetism described by

$$\chi = -\frac{(\mu_B^*)^2}{3}N(E_F), \tag{15.3}$$

where $N(E_F)$ is the density of states at the Fermi level and μ_B^* is an effective Bohr magneton with the electron mass replaced by the effective mass of the crystal electron. (See the discussion in Chapter 11.)

One can visualize some magnetic effects by considering the concepts inherent in Lenz's law, where currents are produced in a conductor resulting in magnetic fields opposite to an applied magnetic field. The classic demonstration of diamagnetism is Langevin diamagnetism, which is appropriate for atoms or ions with closed shells in which the excitation energy of electrons to higher levels is large. This approach would be suitable for a rare gas crystal or ions in an ionic crystal.

Classically, an electron in a magnetic field **B** has orbital motion with the Larmor precession frequency $\frac{eB}{2mc}$. For Z electrons, this gives a current

$$I = -\frac{Ze^2B}{4\pi mc^2}. \tag{15.4}$$

Since the magnetic moment is equal to the product of the current and the area of a loop, then, for a **B**-field along the z-axis, the average loop area is $\pi\rho^2$, where ρ^2 is the mean square distance describing the electron loop perpendicular to the z-axis. If the electronic charge is spherical and r^2 is the mean square distance of the electrons from the nucleus, then $\rho^2 = \frac{2}{3}r^2$, and the magnetic moment is

$$\mu = -\frac{Ze^2B}{4mc^2}\langle\rho^2\rangle = -\frac{Ze^2B}{6mc^2}\langle r^2\rangle. \tag{15.5}$$

Hence, assuming $\chi \ll 1$ (**H** $\approx$ **B**), the diamagnetic susceptibility for N atoms per unit volume is

$$\chi = -\frac{NZe^2}{6mc^2}\langle r^2\rangle. \tag{15.6}$$

This can be derived quantum mechanically by first-order perturbation theory of the "diamagnetic" term E_D in the energy that appears when introducing a vector potential **A** to represent the magnetic field, i.e.

$$H = \frac{1}{2m}(\mathbf{p} - \frac{e}{c}\mathbf{A})^2 \tag{15.7}$$

and

$$E_D = \langle 0|\frac{e^2}{2mc^2}A^2|0\rangle, \tag{15.8}$$

where $|0\rangle$ represents the ground-state wavefunction for each atom. This is the first-order perturbation term, which gives the same result for μ as Eq. (15.5) for a spherically symmetric system.

15.3 Paramagnetism

Pauli paramagnetism arises from the spin of unpaired electrons. Paired electrons in the singlet state have spin magnetic moments that cancel, while the unpaired electron moments can align and orient in the same direction as the applied magnetic field. For paramagnetic systems, $\chi > 0$, and these systems are attracted to an applied field. The magnetization is proportional to the applied field, and the magnitude of the paramagnetic χ is generally much larger than the diamagnetic component. In contrast, if we compare the χ's of Landau diamagnetism for a metal with the Pauli spin paramagnetism,[1] they are similar in magnitude.

To calculate the Pauli term of a non-magnetic metal, we note that a magnetic field splits the degeneracy between opposite spin electrons. For example, since spin-up electrons align along the direction of the field, a comparison of the energies of up and down electrons for a state $\mathbf{k}$ with energy $E(\mathbf{k})$ and magnetic moment μ yields

$$E_\uparrow = E(\mathbf{k}) - \mu B$$

and

$$E_\downarrow = E(\mathbf{k}) + \mu B. \tag{15.9}$$

For a single spin and no orbital moment, $\mu = \mu_B$, where μ_B is the magnetic moment of an electron. Once the chemical potential between the up and down electrons becomes equal, there will be more electrons in the up band since its energy has been lowered. Assuming an equal number of up and down electrons before the H field is applied (see Fig. 15.1), then

$$n_\uparrow - n_\downarrow = \int_0^\infty \frac{1}{2} \left[f(E - \mu_B B) - f(E + \mu_B B) \right] N(E) dE, \tag{15.10}$$

where $N(E)$ is the density of states and f is the Fermi–Dirac distribution function. Assuming $\mu_B B \ll E_F$, the resulting moment for low temperatures is

$$M = \mu_B^2 B \int_0^\infty \left(-\frac{\partial f}{\partial E} \right) N(E) dE \tag{15.11}$$

$$= \mu_B^2 B N(E_F), \tag{15.12}$$

where E_F is the Fermi energy.

Therefore, the Pauli paramagnetic spin susceptibility is

$$\chi = \mu_B^2 N(E_F), \tag{15.13}$$

which is opposite in sign and roughly three times larger in magnitude than the Landau orbital term given in Eq. (15.3).

---

[1] W. Pauli, "Uber Gasentartung und Paramagetismus," *Z. Phys.* 41(1927), 91.

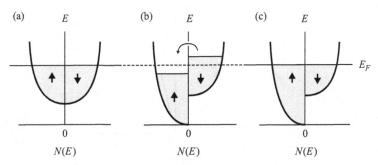

Figure 15.1 Density of states for spins (up and down) for a metal (a) with no magnetic field, (b) in a magnetic field before, and (c) after adjustment of the chemical potential.

Unlike the spin susceptibility in metals or Pauli paramagnetism for atoms, molecules, and defects with unpaired electrons, the paramagnetism of insulating solids is temperature dependent and the magnetic susceptibility obeys Curie's law

$$\chi = \frac{C}{T},$$
(15.14)

where Curie's constant C depends on the system studied. In the absence of a magnetic field, the paramagnetic moment is absent because the individual atomic moments are randomized due to thermal effects. In the presence of a magnetic field, the magnetic moments μ tend to line up with the field. Curie's law is appropriate for low fields, $\mu B \ll k_B T$. If the moments are not interacting, then one can show that

$$C = \frac{N\mu^2}{3k_B} n_{\text{eff}}^2,$$
(15.15)

where the effective number of Bohr magnetons is

$$n_{\text{eff}} = g\sqrt{J(J+1)},$$
(15.16)

and g is the Landé g-factor

$$g = 1 + \frac{J(J+1) + S(S+1) - L(L+1)}{2J(J+1)},$$
(15.17)

where J, L, and S are the total, orbital, and spin angular momenta.

To illustrate the above, we consider a case where the orbital contributions to J are small, so $J = S$. This gives $g = 2$. For $\mathbf{B}$ along the z-axis, the Zeeman splitting of the levels arises from the interaction $-\boldsymbol{\mu} \cdot \mathbf{B}$. The energy of each level is

$$E_{m_J} = -m_J g_J \mu_B B.$$
(15.18)

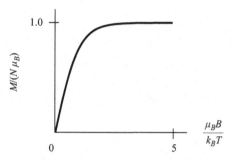

Figure 15.2 Magnetization for a system of localized spin 1/2 electrons following the Curie law.

For a single spin $m_J = \pm\frac{1}{2}$, $g_J = 2$, and the energies are $\pm\mu_B B$. For this system with two levels, the fractional populations are

$$\frac{N_\pm}{N} = \frac{e^{\pm\mu_B B/k_B T}}{e^{\mu_B B/k_B T} + e^{-\mu_B B/k_B T}} \tag{15.19}$$

which gives a magnetization for $N = N_+ + N_-$ atoms per unit volume:

$$M = (N_+ - N_-)\mu_B = N\mu_B \tanh\left(\frac{\mu_B B}{k_B T}\right) \tag{15.20}$$

(as shown in Fig. 15.2).

For $\mu_B B \ll k_B T$,

$$M = \frac{N\mu_B^2 B}{k_B T}, \tag{15.21}$$

and therefore, for this case, using Eq. (15.14),

$$C = \frac{N\mu_B^2}{k_B}. \tag{15.22}$$

More generally for an atom with total angular momentum J, there are $2J+1$ levels and the average M is

$$M = N\frac{\sum\limits_{m_J=-J}^{J} m_J g_J \mu_B e^{m_J g_J \mu_B B/k_B T}}{\sum\limits_{m_J=-J}^{J} e^{m_J g_J \mu_B B/k_B T}}, \tag{15.23}$$

and for $\mu_B B \ll k_B T$,

$$M \approx N g_J \mu_B \frac{\sum\limits_{m_J=-J}^{J} m_J(1 + m_J g_J \mu_B B/k_B T)}{\sum\limits_{m_J=-J}^{J} (1 + m_J g_J \mu_B B/k_B T)}. \tag{15.24}$$

Since $\sum_{-J}^{J} 1 = 2J + 1$, $\sum_{-J}^{J} m_J = 0$, and $\sum_{-J}^{J} m_J^2 = J(J+1)(2J+1)/3$, then

$$M = \frac{N g_J^2 \mu_B^2 B}{k_B T} \frac{\sum\limits_{m_J=-J}^{J} m_J^2}{\sum\limits_{m_J=-J}^{J} (1)}. \tag{15.25}$$

Therefore,

$$M = \frac{N g_J^2 \mu_B^2}{3 k_B T} J(J+1) B, \tag{15.26}$$

yielding

$$C = \frac{N g_J^2 \mu_B^2 J(J+1)}{3 k_B}. \tag{15.27}$$

15.4 Ferromagnetism and antiferromagnetism

As in the case of paramagnetism, ferromagnetism and antiferromagnetism both involve the interactions of unpaired electrons. Unlike paramagnetism, which occurs only as a response to an applied magnetic field, internal interactions can cause ordering of electron magnetic moments in the absence of an external magnetic field. For ferromagnets, this field tends to align moments, while for antiferromagnets, the tendency is to anti-align the moments.

Restricting ourselves first to ferromagnets such as iron, nickel, cobalt, gadolinium, and alloys containing these elements, we note that the ferromagnetic properties disappear at high temperatures when thermal disordering effects destroy the alignment. Although we will not discuss magnetic domains in ferromagnets, we note that they are important. They occur where clusters of moments gather together to form small permanent magnetic regions. These regions or domains move and can be oriented in an external magnetic field, forming a single domain and a permanent magnet.

Although it is not generally agreed that ferromagnetism is completely understood, a great deal is known about this interesting and very useful property of solids. A classical, semi-empirical approach was introduced by Pierre Weiss,[2] based on the qualitative description above. The basic concept involves a mean-field approximation, where an internal magnetic field is assumed to arise from the interaction between neighboring electrons. This internal Weiss field $\mathbf{H}_W$ is viewed as equivalent to a magnetic field and is proportional to the internal magnetization $\mathbf{M}$ with

$$\mathbf{H}_W = \lambda \mathbf{M} = N \langle \boldsymbol{\mu} \rangle, \tag{15.28}$$

[2] P. Weiss, "The hypothesis of the molecular field and the property of ferromagnetism," *J. de Phys.* 6(1907), 661.

where $\langle \boldsymbol{\mu} \rangle$ is an average magnetic moment and λ is a proportionality constant. In the mean-field approximation, this field is assumed to arise from the average environment caused by the effects of the other electrons. If we add the Weiss field to the applied field and put the sum into the expression corresponding to the Curie relation for a paramagnet, we have

$$M = \frac{C}{T}(H + \lambda M), \tag{15.29}$$

$$\chi = \frac{M}{H} = \frac{C}{T - C\lambda}, \tag{15.30}$$

and

$$\chi = \frac{C}{T - T_c}, \tag{15.31}$$

where $T_c = C\lambda$, which is called the Curie temperature.

From Eq. (15.31), which is known as the Curie–Weiss[3] law, for $T > T_c$, the system behaves as a paramagnet, but as T approaches T_c, χ has a singularity. Just as in the case of dielectric functions that were discussed previously, this signals a new order. The applied field can get vanishingly small, but the large χ assures a finite $\mathbf{M}$ and, in this case, ferromagnetism. Using Eq. (15.27), for $J = S$, we can now relate T_c and the empirical parameter λ, giving

$$\lambda = \frac{T_c}{C} = \frac{3k_B T_c}{Ng^2 \mu_B^2 S(S+1)}. \tag{15.32}$$

If we replace the applied field by the Weiss field, the spontaneous magnetization has the form (from Eq. (15.20))

$$M = N\mu_B \tanh\left(\frac{\mu_B \lambda M}{k_B T}\right), \tag{15.33}$$

as shown in Fig. 15.3.

The internal field $\mathbf{H}_W$ postulated by Weiss is also called the exchange field. It represents the effects of the exchange interaction (discussed below) between electrons arising from the fermionic symmetry restrictions on the electronic wavefunction that influence the Coulomb interaction. The dipole interaction between spins is not large enough to create the large Weiss field.

The origin of the exchange interaction is best seen by examining two atoms (a, b) and two electrons $(1, 2)$ in a molecule using the Heitler–London theory for a diatomic molecule such as diatomic hydrogen. The total wavefunction of the electrons must have the property that

$$\psi(1, 2) = -\psi(2, 1). \tag{15.34}$$

---

[3] P. Curie, "Propriétés magnétiques des crops á diverses températures," *Ann. Chim. Phys.* 5(1895), 289.

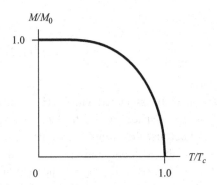

Figure 15.3 Saturation magnetization as a function of temperature for a ferromagnet within a mean field description. $M_0 = M(T = 0)$.

Assuming ψ is a product of a spatial component $\phi(r_1, r_2)$ and a spin component $\chi(s_1, s_2)$, the standard symmetric (S) and antisymmetric (A) combinations for the hybrid spatial wavefunction are

$$\phi_S(1, 2) = \frac{1}{\sqrt{2}}(\alpha_a(\mathbf{r}_1) + \alpha_b(\mathbf{r}_2)), \tag{15.35}$$

$$\phi_A(1, 2) = \frac{1}{\sqrt{2}}(\alpha_a(\mathbf{r}_1) - \alpha_b(\mathbf{r}_2)), \tag{15.36}$$

where α_a, α_b are the one-electron spatial wavefunctions for the atoms. We next construct the symmetric and antisymmetric spin wavefunctions corresponding to total spin $S = 1$ and $S = 0$, respectively.

For $S = 1$,

$$\chi_S^1 = |\uparrow\uparrow\rangle \qquad\qquad (m_S = 1), \tag{15.37}$$

$$\chi_S^0 = \frac{1}{\sqrt{2}}(|\uparrow\downarrow\rangle + |\downarrow\uparrow\rangle) \qquad (m_S = 0), \tag{15.38}$$

$$\chi_S^{-1} = |\downarrow\downarrow\rangle \qquad\qquad (m_S = -1). \tag{15.39}$$

For $S = 0$,

$$\chi_A = \frac{1}{\sqrt{2}}(|\uparrow\downarrow\rangle - |\downarrow\uparrow\rangle). \tag{15.40}$$

The total antisymmetric wavefunctions are as follows for the singlet (s) and triplet (t) states:

$$\psi_s = \phi_S(1, 2)\chi_A \tag{15.41}$$

and

$$\psi_t = \phi_A(1,2)\chi_S, \qquad (15.42)$$

for $m_S = -1, 0, 1$.

The charge density associated with $|\psi_s|^2$ is bonding-like since the antiparallel electrons can occupy the same space. For the triplet state, the Pauli principle keeps the parallel spin electrons apart because they cannot be in the same location. As a result of this avoidance, there is a node in this antibonding-like wavefunction. If we calculate the energy difference between these states, $(E_t - E_s) > 0$, since the bonding state lies lower in energy. This calculation involves taking the expectation of the Hamiltonian for the diatomic molecule.

When the two atoms are far apart, the energy of the single electrons is each E_0. The corresponding wavefunctions are non-overlapping. But the overlap integral I is not zero when the atoms are brought together and interact. In this case, the expectation value of the Hamiltonian involves the Coulomb integral V and the exchange integral U:

$$V = e^2 \int d\mathbf{r}_1 d\mathbf{r}_2 |a_a(\mathbf{r}_1)|^2 |a_b(\mathbf{r}_2)|^2 \left[\frac{1}{|\mathbf{r}_1 - \mathbf{r}_2|} - \frac{1}{|\mathbf{r}_1 - \mathbf{R}_a|} - \frac{1}{|\mathbf{r}_2 - \mathbf{R}_b|} \right] \qquad (15.43)$$

and

$$U = e^2 \int d\mathbf{r}_1 d\mathbf{r}_2 a_a^*(\mathbf{r}_1) a_b^*(\mathbf{r}_2) \left[\frac{1}{|\mathbf{r}_1 - \mathbf{r}_2|} - \frac{1}{|\mathbf{r}_1 - \mathbf{R}_b|} - \frac{1}{|\mathbf{r}_2 - \mathbf{R}_a|} \right] a_b(\mathbf{r}_1) a_a(\mathbf{r}_2), \qquad (15.44)$$

where $\mathbf{r}$ and $\mathbf{R}$ are the electron and nuclear coordinates, respectively.

Using the above definitions and the expressions for the singlet and triplet wavefunctions, the energies become (in the large separation limit)

$$E_s = 2E_0 + \frac{e^2}{R} + \frac{V + U}{1 + I^2} \qquad (15.45)$$

and

$$E_t = 2E_0 + \frac{e^2}{R} + \frac{V - U}{1 - I^2}, \qquad (15.46)$$

where $R = |\mathbf{R}_a - \mathbf{R}_b|$. Hence,

$$E_t - E_s = \frac{2(I^2 V - U)}{1 - I^4} \equiv -J. \qquad (15.47)$$

If $J < 0$, then $E_s < E_t$, implying antiferromagnetic coupling between the two atoms. The spin alignment puts symmetry restrictions on the spatial part of the wavefunction and hence on the Coulomb interaction. Therefore, the Hamiltonian for the system that operates on the spatial part of the wavefunction and yields the energies E_t and E_s can be replaced by a Hamiltonian that is spin dependent.

Expressing the Hamiltonian in the spin representation, we have

$$H = A - B\mathbf{S}_1 \cdot \mathbf{S}_2, \tag{15.48}$$

where A and B are constructed to give the energies E_t and E_s when H operates on the triplet and singlet wavefunctions, respectively. Using

$$\mathbf{S}_1 \cdot \mathbf{S}_2 = S_1^z S_2^z + \frac{1}{2}(S_1^+ S_2^- + S_1^- S_2^+), \tag{15.49}$$

where

$$S^\pm = S^x \pm iS^y, \tag{15.50}$$

we have

$$\mathbf{S}_1 \cdot \mathbf{S}_2 \chi_s = -\frac{3}{4}\chi_s \tag{15.51}$$

and

$$\mathbf{S}_1 \cdot \mathbf{S}_2 \chi_t = \frac{1}{4}\chi_t. \tag{15.52}$$

So we arrive at the following:

$$A = \frac{1}{4}(E_s + 3E_t), \tag{15.53}$$

$$B = (E_s - E_t), \tag{15.54}$$

$$H = \frac{1}{4}(E_s + 3E_t) - (E_s - E_t)\mathbf{S}_1 \cdot \mathbf{S}_2, \tag{15.55}$$

$$H\psi_s = E_s\psi_s, \tag{15.56}$$

and

$$H\psi_t = E_t\psi_t. \tag{15.57}$$

Using Eqs. (15.45), (15.46), and (15.47),

$$H = 2E_0 + \frac{e^2}{R} + \frac{V - U}{1 - I^2} + \left(\frac{1}{4} - \mathbf{S}_1 \cdot \mathbf{S}_2\right)J. \tag{15.58}$$

If we shift the zero of energy in Eq. (15.58), then we can assume a form of H that is spin dependent:

$$H = -J\mathbf{S}_1 \cdot \mathbf{S}_2, \tag{15.59}$$

where $J > 0$ yields parallel spins and $J < 0$ yields antiparallel spins.

Schematic representation of a spin wave state.

The formalism gives the physical background for the Heisenberg[4] model for a system of magnetic ions on a lattice. As has been demonstrated, the spin interactions are explained using the concept of the exchange integral and not a direct dipolar interaction. The Heisenberg Hamiltonian for a crystal has the form

$$H = -\sum_{i \neq j} J_{ij} \mathbf{S}_i \cdot \mathbf{S}_j, \tag{15.60}$$

where i and j label lattice sites and J_{ij} is assumed to be short-ranged and to act over a few lattice spacings. If we consider only nearest-neighbor interactions, then parallel or antiparallel neighbor interactions (i.e. different sign for J) are possible. Assuming a constant J describing these interactions, then one can write an expression for a magnetic moment as $g\mu_B \mathbf{S}_i$ interacting with an effective Weiss field $\mathbf{H}_W = \lambda \mathbf{M}$ from Eq. (15.28). Then, for Z nearest neighbors, the Hamiltonian is

$$H = -\sum_{i} \left[\sum_{j=1}^{Z} J \mathbf{S}_j \right] \cdot \mathbf{S}_i, \tag{15.61}$$

and J can be written in this mean-field approximation (using Eq. (15.33) at $T = 0$) as

$$J = \frac{3 k_B T_c}{2 Z S (S + 1)}. \tag{15.62}$$

Different crystal structures have been considered to refine Eq. (15.62) for specialized cases.

If we examine the ground state of a ferromagnet at $T = 0$, using the Heisenberg model for $J > 0$, all spins point in the same direction. There is a degeneracy since all directions are equal until an external perturbation breaks this symmetry. As temperature is raised, low-lying excitations called spin waves can also decrease the magnetization. The spin waves result from changes in the orientation of the spins as a function of their positions in the lattice (Fig. 15.4). In terms of collective excitations, which in this case are called magnons, these waves can be viewed as analogous to the elastic waves in solids. The description of the energy spectrum, heat capacity, and other properties resulting from excitations of magnetic systems can be achieved by using the concept of magnons, in the same manner as we used the concept of phonons in studies of elastic waves in earlier chapters.

The dispersion curve for spin waves can be obtained using models similar to the spring models discussed earlier for lattice vibrations. A quantum approach based on the Heisenberg model allows the interpretation of excited states of the spin system in terms of

[4] W. Heisenberg, "Zur Theorie des Ferromagnetismus," *Z. Phys.* 49(1928), 619.

elementary excitations. We begin with the Heisenberg model for direct exchange interactions between an atom at lattice site i and its nearest-neighbor site $j = i \pm 1$, in a chain,

$$H = -J \sum_{\langle ij \rangle}{}' \mathbf{S}_i \cdot \mathbf{S}_j. \tag{15.63}$$

We have chosen a constant $J > 0$ (ferromagnetic case) for all interactions and use the notation $\langle ij \rangle$ to denote nearest-neighbor sites and i, j run over all sites. In the ground state, all the spins are aligned in the z-direction, and we look for an excitation such as a spin reversal, described by H, which can be rewritten as (see Eq. (15.49))

$$H = -J \sum_{\langle ij \rangle}{}' \left[S_i^z S_j^z + \frac{1}{2}(S_i^+ S_j^- + S_i^- S_j^+) \right], \tag{15.64}$$

where $S^\pm = S_x \pm iS_y$.

In the ground state, the S^+ operators yield zeros when operating on the system with all spins aligned up. Equation (15.64) gives a ground-state energy

$$E_0 = -JNZS^2, \tag{15.65}$$

where N is the total number of sites and Z is the coordination number for each site (for our chain case, $Z = 2$). If we pick a specific site ℓ and reduce the spin quantum number m_ℓ on that site by 1 as discussed above (spin reversal for $S = 1/2$), then the corresponding state is

$$|\phi_\ell\rangle = S_\ell^- |\phi_0\rangle, \tag{15.66}$$

where $|\phi_0\rangle$ is the ground state with energy E_0. We perform the Holstein – Primakoff transformation by introducing Bose operation a_ℓ to simplify the Heisenberg Hamiltonian Eq. (15.64)

$$S_\ell^+ = \left(\sqrt{2S - a_\ell^\dagger a_\ell} \right) a_\ell,$$

$$S_\ell^- = a_\ell^\dagger \left(\sqrt{2S - a_\ell^\dagger a_\ell} \right),$$

$$S_\ell^z = S - a_\ell^\dagger a_\ell. \tag{15.67}$$

For large S, $\sqrt{2S - a_\ell^\dagger a_\ell} \approx \sqrt{2S}$, the Hamiltonian of Eq. (15.67) can be shown to have the form

$$H = E_0 - J \sum_{\langle ij \rangle}{}' S(a_i^\dagger a_j + a_j^\dagger a_i - a_i^\dagger a_i - a_j^\dagger a_j). \tag{15.68}$$

The derivation, based on the so-called $\frac{1}{S}$ expansion, is related to the $\frac{1}{N}$ expansions commonly used in statistical physics. The Hamiltonian (Eq. (15.68)) can be diagonalized by using a Fourier representation, where

$$a_i = \frac{1}{\sqrt{N}} \sum_{\mathbf{k}} a_{\mathbf{k}} e^{-i\mathbf{k} \cdot \mathbf{r}_i}. \tag{15.69}$$

If $\mathbf{r}_i - \mathbf{r}_j = \boldsymbol{\delta}$, the vector to the nearest neighbors, Eq. (15.68) becomes

$$
\begin{aligned}
H &= E_0 - 2JS \sum_{\mathbf{k}} \sum_{\delta} [\cos(\mathbf{k} \cdot \boldsymbol{\delta}) - 1] \, a_{\mathbf{k}}^{\dagger} a_{\mathbf{k}} \\
&= E_0 + \sum_{\mathbf{k}} n_{\mathbf{k}} \hbar\omega_{\mathbf{k}},
\end{aligned}
\tag{15.70}
$$

where the excited state energy is given by

$$\hbar\omega_{\mathbf{k}} = 2SJ \sum_{\delta} [1 - \cos(\mathbf{k} \cdot \boldsymbol{\delta})]. \tag{15.71}$$

Equation (15.70) describes a Hamiltonian yielding a ground-state energy E_0 where all the spins are aligned plus an excited state term describing a state with $n_{\mathbf{k}}$ magnons having energy $\omega_{\mathbf{k}}$ excited above the ground state. The expression for $\omega_{\mathbf{k}}$ given in Eq. (15.71) describes the spin wave dispersion relation (see Fig. 15.5). This spin-state contribution to the energy at temperature T can be written as

$$E = \sum_{\mathbf{k}} \frac{\hbar\omega_{\mathbf{k}}}{e^{\hbar\omega_{\mathbf{k}}/k_{\mathrm{B}}T} - 1}. \tag{15.72}$$

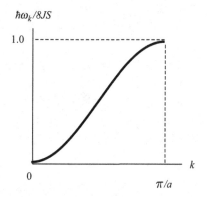

Figure 15.5   Spin wave dispersion curve for a one-dimensional ferromagnet.

For a general 3D isotropic system at low temperatures, using Eq. 15.71, $\omega_{\mathbf{k}} = Ak^2$ and A is a constant, we can write

$$E = \frac{V}{8\pi^3} \int_0^\infty \frac{4\pi Ak^4 dk}{e^{\hbar\omega_{\mathbf{k}}/k_B T} - 1} = BT^{5/2}, \tag{15.73}$$

where B is a constant. Taking a temperature derivative yields an expression for the heat capacity proportional to $T^{3/2}$, and this is consistent with experimental measurements.

A similar study for the antiferromagnetic case, where $J < 0$ and spins align in opposite directions in the ground state at zero temperature, can be described using two sublattices A and B with opposite spins. This assumes that each A site is surrounded by B sites and vice versa. Assuming a π rotation around the x-axis on the B lattice transforms the state into a ferromagnetic state with the same energy. The transformations $x \to x$, $y \to -y$, and $z \to -z$ yields spin transformations

$$S_i^{\pm} \to S_i^{\mp}; \; S_i^z \to -S_i^z. \tag{15.74}$$

The ferromagnetic Hamiltonian with nearest-neighbor interaction of Eq. (15.64) now becomes

$$H = 2|J| \sum_{\langle ij \rangle} \left[-S_i^z S_j^z + \frac{1}{2}(S_i^+ S_j^+ + S_i^- S_j^-) \right]. \tag{15.75}$$

Since neighboring pairs appear twice in the summation, a factor of 2 is inserted. Following the procedure used for the ferromagnetic lattice with appropriate changes for the antiferromagnetic state, we obtain

$$H = -|J|NZS^2 + 2|J|S \sum_{\langle ij \rangle} \left[a_i^\dagger a_j^\dagger + a_i a_j + a_i^\dagger a_i + a_j^\dagger a_j \right], \tag{15.76}$$

which is analogous to the ferromagnetic case given by Eq. (15.68). Again introducing a Fourier transform

$$a_i = \frac{1}{\sqrt{N}} \sum_{\mathbf{k}} a_{\mathbf{k}} e^{i\mathbf{k}\cdot\mathbf{r}_i}, \tag{15.77}$$

we obtain

$$H = -|J|NZS^2 + |J|S \sum_{\mathbf{k}} \sum_{\delta} \left[(a_{\mathbf{k}}^\dagger a_{-\mathbf{k}}^\dagger + a_{\mathbf{k}} a_{-\mathbf{k}}) \cos(\mathbf{k}\cdot\delta) + 2a_{\mathbf{k}}^\dagger a_{\mathbf{k}} \right], \tag{15.78}$$

which can be diagonalized using the transformation

$$a_{\mathbf{k}} = (\cosh \gamma_{\mathbf{k}}) c_{\mathbf{k}} + (\sinh \gamma_{\mathbf{k}}) c_{-\mathbf{k}}^\dagger \tag{15.79}$$

with the restriction

$$\tanh 2\gamma_{\mathbf{k}} = -\frac{1}{Z} \sum_{\delta} \cos(\mathbf{k}\cdot\delta). \tag{15.80}$$

Equation (15.78) becomes

$$H = -NZ|J|S(S+1) + 2|J|ZS \sum_{\mathbf{k}} (c_{\mathbf{k}}^{\dagger} c_{\mathbf{k}} + \frac{1}{2}) \sqrt{1 - \tanh^2 2\gamma_{\mathbf{k}}}. \qquad (15.81)$$

This yields the antiferromagnetic ground-state energy

$$E_0 = -N|J|ZS(S+1) + |J|ZS \sum_{\mathbf{k}} \left\{ 1 - \frac{1}{Z^2} \left[\sum_{\delta} \cos(\mathbf{k} \cdot \delta) \right]^2 \right\}^{1/2} \qquad (15.82)$$

and the energy of a magnon in state $\mathbf{k}$ is

$$E(\mathbf{k}) = 2|J|S \sqrt{Z^2 - \left[\sum_{\delta} \cos(\mathbf{k} \cdot \delta) \right]^2}. \qquad (15.83)$$

For a linear chain ($Z = 2$),

$$E(k) = 4|J|S \sin k\delta, \qquad (15.84)$$

which, for $k\delta \ll 1$, yields

$$E(k) = 4|J|Sk\delta \qquad (15.85)$$

in contrast to the nonlinear k-dependence for a ferromagnetic system. The quadratic dependence on wavevectors for ferromagnetic systems and the linear dependence for antiferromagnetic systems have been verified experimentally.

15.5 Magnetism in metals

Ferromagnetism in transition metals arises from the spin-polarized unpaired electrons in partially filled narrow d-bands. When electron subshells are filled, the electrons can be thought of as paired with no net magnetic moment. The classic prototype ferromagnetic metals are Fe, Co, and Ni, which have partially filled s-bands arising from the $4s^2$ atomic orbitals and partially filled d-bands from atomic orbitals $3d^6$, $3d^7$, and $3d^8$ for Fe, Co, and Ni, respectively. Under certain conditions, other metals like Pd and Mg can exhibit ferromagnetic behavior.

A common model that provides a useful view of the way in which magnetism arises from the imbalance of electronic moments comes from the Stoner model. In this model, we do not have the localized spins assumed in the Heisenberg model, but rather electrons described by Bloch states. Although the electrons are spread out, there can be a low-energy state in which the electrons have a population with an excess of one spin direction. This can be modeled by an interaction term in the electronic Hamiltonian of the form

$$H_{\text{int}} = \frac{U}{N} \sum_{\mathbf{k},\mathbf{k}'} n_{\mathbf{k}\uparrow} n_{\mathbf{k}'\downarrow},$$
(15.86)

where $n_{\mathbf{k}\uparrow}$ and $n_{\mathbf{k}\downarrow}$ represent the occupation numbers corresponding to the states $|\mathbf{k}\uparrow\rangle$ and $|\mathbf{k}\downarrow\rangle$ and N is the number of unit cells. If two electrons of opposite spin are in the same d-shell of the same atom, it has been argued that this occupancy would contribute a positive exchange-like energy $\frac{U}{N}$, while the energies of electrons of the same spin are included only in the ground state. This model is useful, but a rigorous proof of its validity remains to be formulated. Density functional theory (see Chapter 7) provides a spin-dependent potential from first principles that leads to ferromagnetism and other spin ordering from an itinerant electron point of view.

The model H_{int} in Eq. (15.86) can be viewed as producing an internal magnetic field similar to the Weiss field. Hence, if we assume $\langle n_\uparrow \rangle$ and $\langle n_\downarrow \rangle$ to be the total number of electrons per atom of spin $\uparrow$ and $\downarrow$, respectively, it can be argued that the energies of the $\uparrow$ and $\downarrow$ spins are

$$E(\mathbf{k}\uparrow) = E(\mathbf{k}) - \mu_B H + U\langle n_\downarrow \rangle$$
(15.87)

and

$$E(\mathbf{k}\downarrow) = E(\mathbf{k}) + \mu_B H + U\langle n_\uparrow \rangle.$$
(15.88)

The chemical potentials for $\uparrow$ and $\downarrow$ are the same, and

$$\langle n_\uparrow \rangle = \int_0^\infty N_\uparrow(E) f[E(\mathbf{k}\uparrow)] dE$$
(15.89)

(with a similar expression for $\langle n_\downarrow \rangle$), where $N_\uparrow(E)$ is the density of electronic states with spin $\uparrow$, and f is the Fermi–Dirac distribution function. Further, if we assume that $N_\uparrow(E)$ is approximately the same as $N_\downarrow(E)$, and denoting both as $\tilde{N}(E) = \frac{N(E)}{2}$ (this is usually not a good approximation for real materials in the ferromagnetic phase, but we do it here for simplicity to illustrate the physics near the phase transition in the paramagnetic state), then

$$M = \mu_B \int_0^\infty \{f[E(\mathbf{k}\uparrow)] - f[E(\mathbf{k}\downarrow)]\} \tilde{N}(E) dE.$$
(15.90)

Rewriting Eq. (15.89), the net magnetic moment is

$$M = \mu_B(\langle n_\uparrow \rangle - \langle n_\downarrow \rangle)$$

$$= \mu_B \int_0^\infty \{f[E - \mu_B H + U\langle n_\downarrow \rangle] - f[E + \mu_B H + U\langle n_\uparrow \rangle]\} \tilde{N}(E) dE$$
(15.91)

as $H \to 0$ and $T \to 0$,

$$\chi = \frac{M}{H} = \frac{2\mu_B^2 \tilde{N}(E_F)}{1 - U\tilde{N}(E_F)},$$
(15.92)

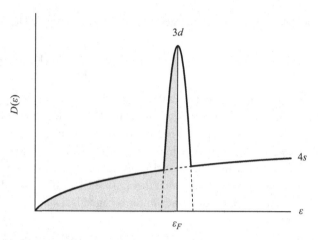

Figure 15.6   Model density of states for $3d$ and $4s$ bands in transition metals.

and the instability when $U\tilde{N}(E_F) > 1$ signals ferromagnetism. The expression for χ in Eq. (15.92) resembles the Pauli susceptibility given by Eq. (15.12) with a denominator that makes it larger. Parallel spins are favorable because of the exchange interaction. The ferromagnetic state has a higher kinetic energy than the paramagnetic state but its energy is lowered by U.

For consistency in this model of the ferromagnetic state, we require that $\langle n_\uparrow \rangle$ be larger than $\langle n_\downarrow \rangle$, and this leads to a permanent magnetic moment. When temperature effects are included, a Curie–Weiss type behavior is obtained. The Stoner model provides a starting point for exploring the origin of magnetism in transition metals. Band theory, together with models of this kind (Fig. 15.6), have given results that are consistent with experiment. For example, results including the temperature behaviors have been achieved with modern spin-dependent density functional theory.

One can still gain some insight related to the ferromagnetic properties of metals by considering the limiting case of a spin-polarized electron gas within the Hartree–Fock approximation. If the energy of each spin orientation is $E(\uparrow)$ and $E(\downarrow)$, then the total energy is the sum of these two, where (using results from Chapter 6 for a three-dimensional system)

$$E(\uparrow) = \left[\frac{3}{5} E(k_{F\uparrow}) - \frac{3e^2}{4\pi} k_{F\uparrow} \right] N_\uparrow, \tag{15.93}$$

with an identical expression for $E(\downarrow)$. The Fermi energy for spin-up is given by

$$E(\mathbf{k}_{F\uparrow}) = \frac{\hbar^2 k_{F\uparrow}^2}{2m} = \frac{\hbar^2}{2m} \left(\frac{6\pi^2 N_\uparrow}{V} \right)^{2/3}, \tag{15.94}$$

where the total number of electrons $N = N_\uparrow + N_\downarrow$, and V is the volume. For the non-aligned case, $N_\uparrow = N_\downarrow$, but if we put all the electrons into the $N_\uparrow$ state, then $N_\uparrow = N$, $N_\downarrow = 0$, and for the aligned case (A),

$$k_{F\uparrow}^{A} = \left(\frac{6\pi^2 N}{V}\right)^{1/3}. \tag{15.95}$$

Hence, for the aligned case, the energy is

$$E_A = N\left[\frac{3}{5}\frac{\hbar^2}{2m}\left(\frac{6\pi^2 N}{V}\right)^{2/3} - \frac{3e^2}{4\pi}\left(\frac{6\pi^2 N}{V}\right)^{1/3}\right] \tag{15.96}$$

and the non-aligned state (NA) has the energy (since $N_\uparrow = N_\downarrow = N/2$)

$$E_{NA} = N\left[\frac{3}{5}\frac{\hbar^2}{2m}\left(\frac{3\pi^2 N}{V}\right)^{2/3} - \frac{3e^2}{4\pi}\left(\frac{3\pi^2 N}{V}\right)^{1/3}\right]. \tag{15.97}$$

Therefore, the aligned state is lower in energy when

$$\frac{3}{5}\frac{\hbar^2}{2m}(3\pi^2 n)^{2/3}[2^{2/3} - 1] - \frac{3e^2}{4\pi}(3\pi^2 n)^{1/3}[2^{1/3} - 1] < 0. \tag{15.98}$$

The transition occurs when

$$\frac{2}{5}\pi\frac{\hbar^2}{m}(2^{1/3} + 1) = \frac{e^2}{(3\pi^2 n)^{1/3}}. \tag{15.99}$$

If we express Eq. (15.98) in terms of the electron gas parameter r_s, where

$$r_s = \frac{me^2}{\hbar^2}\left(\frac{3}{4\pi n}\right)^{1/3}, \tag{15.100}$$

then

$$r_s = \frac{2\pi}{5}(2^{1/3} + 1)\left(\frac{9\pi}{4}\right)^{1/2} = 5.45. \tag{15.101}$$

This value corresponds to a low-density metal. The element Cs has an r_s in this range; however, Cs is non-magnetic. The above free electron-like system description cannot be applied to explain the observed properties of ferromagnetic metals where d-bands are of crucial importance.

15.6 Magnetic impurities and local correlation effects

The study of magnetic impurities in metals has led to important insights, not only about local magnetic moments, but also about many other interesting physical phenomena.[5] A

[5] P. W. Anderson, "Localized magnetic states in metals," *Phys. Rev* 124(1961), 41.

successful approach for exploring this problem uses the Anderson impurity model, which has been applied to explore a number of physical properties. The starting point is to consider the case of a transition or rare earth atom incorporated in a metal. If there is no interaction between the added impurity and the spin-up or -down metal electrons, then the electronic density of states of the metal remains unchanged. However, if the impurity state interacts with the host electrons, then a modification of the impurity's local moment occurs. For example, a Co atom in Cu will exhibit a broadening of its atomic energy level into a Lorentzian peak. The $3d$ electronic states of Co can hybridize with the $4s$ states of Cu. The local moment may persist depending on the hybridization. The broadening is often large when metal sp-states are involved to hybridize with the d-states of the impurity.

The Anderson Hamiltonian has the form

$$H = \sum_{k\sigma} \epsilon_k c_{k\sigma}^\dagger c_{k\sigma} + E_d \sum_\sigma d_\sigma^\dagger d_\sigma + \sum_{k\sigma} (V_{dk} d_\sigma^\dagger c_{k\sigma} + V_{dk}^* c_{k\sigma}^\dagger d_\sigma) + U d_\uparrow^\dagger d_\uparrow d_\downarrow^\dagger d_\downarrow.$$

(15.102)

The terms in Eq. (15.102) represent the host kinetic energy, the impurity level energy E_d and associated electronic operators $d_\sigma^\dagger$ and d_σ, a hybridization term, and the onsite Coulomb "energy cost" U. If an electron is on the impurity site, its energy is E_d, but two electrons on the same site with opposite spins require energy $2E_d + U$.

If we assume the hybridization term V to be small, then for $E_F > E_d$, the impurity state is occupied; but if $E_d + U > E_F$, then double occupancy is unlikely, leading to a single-occupancy magnetic d-state. The splitting of the effective quasiparticle energy levels is U if $V = 0$. For $V \neq 0$, there is an interaction between the metal s-states (for example) and the impurity d-states. A simple approach is to consider this interaction V as a perturbation, which gives a transition rate

$$R = \pi V^2 N_s(E_d)$$

(15.103)

using the Fermi golden rule, where $N_s(E_d)$ is the metal density of states at E_d. This rate is connected to the lifetime or energy broadening of the d-level, and the sharp density of states at E_d is replaced by a Lorentzian of the form

$$D(E) = A \frac{W}{(E - E_d)^2 + W^2},$$

(15.104)

where A is a constant and W is a linewidth related to $\frac{1}{R}$ from Eq. (15.103). With this broadening, the formation of a magnetic moment is subject to the conditions that

$$D(E_F)U > 1,$$

(15.105)

(which is similar to the Stoner condition described earlier) and that the location of E_d is sufficiently deep ($E_d + U > E_F$) with respect to the Fermi level. Since $D(E_F) \sim \frac{1}{W}$, where W is the width of the resonance, the so-called Anderson criterion, i.e. $U \gtrsim W$, for developing a local moment, i.e. magnetism at an impurity, is a guide.

A related phenomenon is the Kondo[6] effect. Matthiessen's rule asserts that the temperature dependence of the resistivity of a metal can be explained by considering two components of the scattering of electrons. One arises from electron–phonon interactions and is therefore temperature dependent, and decreases with decreasing temperature, while the other contribution comes from the scattering of electrons by impurities and is assumed to be temperature independent. However, it was found that at very low temperatures, the resistivity had a minimum as a function of temperature for some impurities in metals, and this was ultimately explained in terms of scattering from magnetic impurities.

The Kondo effect is explained in the following way. At high temperatures, the electrons of the host encounter a weak and fairly isotropic scattering center around an isolated magnetic impurity as the impurity spin points in different directions in response to the temperature. As the temperature is lowered and the spin of the scattering center tends to orient, the scattering becomes more effective. If we assume that a local moment exists, then we can explore the following effective Kondo Hamiltonian:

$$H_K = \sum_{ij} t_{ij} c_i^\dagger c_j - \sum_m J\mathbf{S} \cdot \mathbf{s}_m, \tag{15.106}$$

where t_{ij} describes the hopping term for the metal electrons, and $J < 0$ implies an $s - d$ antiferromagnetic interaction between the impurity spin $\mathbf{S}$ and the host spin $\mathbf{s}_m$. This interaction causes a cloud of host electrons to surround the impurity site and the change in the dressed local moment can be large. A non-magnetic spin singlet state can be formed from the impurity and the host electrons. The temperature associated with the formation of this state is the Kondo temperature T_K. If the electronic bandwidth is W, then it can be shown that

$$k_B T_K \approx W \exp(-W/2J). \tag{15.107}$$

The signature for the Kondo effect is shown in Fig. 15.7.

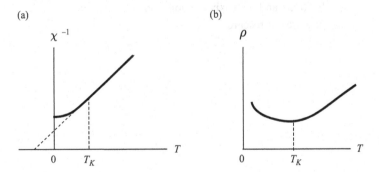

Figure 15.7 Signatures for the Kondo effect: (a) the inverse susceptibility and (b) the low-temperature resistivity.

[6] J. Kondo, "Resistance minimum in dilute magnetic alloys," *Prog. Theor. Phys.* 32(1964), 37.

Hubbard[7] introduced a model Hamiltonian which illustrates some of the properties associated with electron correlation. It uses a tight-binding-like approach and represents another path to explain magnetism from a more localized electron view compared to the band theory approach described earlier. The usual tight-binding model is augmented to provide an energy cost U for any atomic site with more than one electron. A simple form of the Hubbard Hamiltonian is

$$H_{\mathrm{H}} = -\sum_{ij} t c_i^{\dagger} c_j + U \sum_i n_{i\uparrow} n_{i\downarrow}, \tag{15.108}$$

where t represents the usual hopping term, and $n_{i\uparrow}$ and $n_{i\downarrow}$ measure the number of up and down spins at site i. Although there exists an exact solution of the basic Hamiltonian introduced by Hubbard for the one-dimensional case, complete formal extensions to three dimensions are not available; hence the mean-field version of H_{H} is commonly used. If U is greater than the bandwidth W, then the electrons should localize, since charge fluctuations will have to pay a high energy cost, and an energy gap opens up for quasielectron and quasihole excitations. Compounds with odd integral electron-to-atom ratios in a unit cell serve as the prototypical systems, and these are called Mott–Hubbard insulators. Other limits and variations have been studied extensively. For example, at half-filling, the number of electrons is equal to the number of lattice sites in the ground state. Away from half-filling, the problem becomes difficult.

A variation of the Hubbard model has also been studied extensively. It is called the T–J model and has the form

$$H_{TJ} = -\sum_{ij} t c_i^{\dagger} c_j - 2J \sum_{\langle ij \rangle} \mathbf{s}_i \cdot \mathbf{s}_j, \tag{15.109}$$

where (relating to the Hubbard model)

$$J = -\frac{2t^2}{U}. \tag{15.110}$$

Applications and approximate solutions of H_{H} and H_{TJ} are numerous. This is an active, ongoing area of research.

---

[7] J. Hubbard, "Electron correlations in narrow energy bands," *Proc. R. Soc. Lond. A* 266(1963), 238.

16 Reduced-dimensional systems and nanostructures

The electronic, transport, optical, and other properties of a material often undergo drastic changes, and new phenomena emerge, when the system is reduced in size such that quantum confinement of electrons (i.e. the wavelength of the electron is comparable to the confining structure) occurs in one or more dimensions. These reduced-dimensional systems and nanostructures have physical properties and phenomena that are of basic interest and are also potentially useful in applications. Because many of their properties are fundamentally derived from restricted geometry, the behaviors of these reduced dimensional systems and nanostructures are tunable by changing their size, and they are strongly influenced by considerations such as quantum confinement, enhanced many-electron interactions, reduced number of degrees of freedom, and symmetry effects. In this chapter, we shall discuss some basic elements of the electronic, transport, and optical properties of reduced-dimensional systems. We consider several systems, including semiconductor two-dimensional electron gas (2DEG) systems, quantum dots, graphene, carbon nanotubes, atomically thin quasi two-dimensional (2D) crystals, and molecular junctions, illustrating that small is different.

16.1 Density of states and optical properties

Electrons confined in a semiconductor quantum well or at a metal-oxide–semiconductor interface at low temperature are examples of 2DEG systems, as described in our discussion of the quantum Hall effect in Chapter 12. The electrons in such 2DEGs may be further confined, e.g. by the potential from a patterned back gate to form a one-dimensional (1D) electron gas system (a quantum wire) or a zero-dimensional system (a quantum dot). Semiconductor clusters, which are fragments of semiconductor crystals consisting of hundreds to many thousands of atoms with surface states eliminated by passivating adsorbates, or enclosure in a material that has a larger bandgap, is another class of quantum dots. Semiconductor wires of several nanometers in diameter can also be grown. Examples of other fascinating nanostructures that have been synthesized and behave like reduced-dimensional systems even at room temperature include graphene, carbon nanotubes, atomically thin quasi-2D materials consisting of mono- and few-layer Van der Waals crystals (such as hexagonal BN, transition metal dichalcogenides, etc.), and structures derived from these systems, such as nanoribbons.

The electron density of states in a reduced-dimensional system has distinct characteristics, i.e. van Hove singularities of different nature than for three dimensional systems,

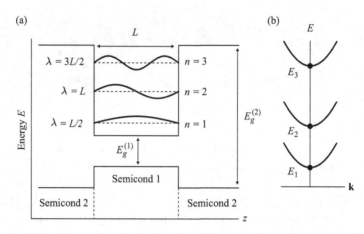

Figure 16.1 (a) Schematic of the spatial variation of the bandgap E_g of a semiconductor quantum well of width L. The lowest three **k** = 0 quantum-confined electronic states derived from the conduction band of semiconductor 1 are illustrated. (b) Schematic of the subband dispersion of the 2D electronic states. Here **k** is the in-plane wavevector.

leading to many of its characteristic properties. Let us idealize the carriers in a semiconductor quantum well, e.g. a thin layer of GaAs of thickness L sandwiched between two AlAs crystals, as free electrons with an effective mass m^*. The larger bandgap of AlAs effectively forms a potential confining the lower-energy carriers (both electrons and holes) to move within the GaAs layer, as illustrated in Fig. 16.1 for the states in the conduction band of GaAs. The low-energy states correspond to propagating planewaves along directions parallel to the layer (the x-y plane), but standing waves exist in the direction perpendicular to the layer (the z-direction). To simplify the discussion, we take the potential step in Fig. 16.1 to be infinite. The states in the conduction band of the quantum well are labeled by quantum numbers k_x, k_y, and n with wavefunctions

$$\psi_{k_x k_y n}(x, y, z) = A e^{ik_x x} e^{ik_y y} \sin\left(\frac{n\pi}{L} z\right), \tag{16.1}$$

with $n = 1, 2, 3, \ldots$, and A is a normalization constant. A similar picture holds for hole states in the valence band (not shown in Fig. 16.1). The corresponding energies are

$$E_{k_x k_y n} = \frac{\hbar^2 k_x^2}{2m^*} + \frac{\hbar^2 k_y^2}{2m^*} + \frac{\hbar^2}{2m^*} \left(\frac{n\pi}{L}\right)^2. \tag{16.2}$$

The states form separate parabolic bands labeled by n in the (k_x, k_y) space, called subbands. The separation in energy between consecutive subbands depends on n, m^*, and L. For small L, e.g. less than fifty nanometers, subband separations can be large. At low carrier density with a Fermi level above the minima of only a few bands and temperatures such that $k_B T$ is less than the subband separations, the system is considered to be in the quantum limit. In particular, if only the first subband is occupied, the system will behave as if it were a 2DEG.

The corresponding solutions for carriers confined in two or three directions are the following. For a wire with square crosssection $L_z = L_y = L$,

$$\psi_{k_x n_y n_z}(x, y, z) = A e^{ik_x x} \sin\left(\frac{n_y \pi}{L} y\right) \sin\left(\frac{n_z \pi}{L} z\right), \qquad (16.3)$$

where n_y and n_z are positive integers, A is another normalization constant, and the corresponding energies are

$$E_{k_x n_y n_z} = \frac{\hbar^2 k_x^2}{2m^*} + \frac{\hbar^2}{2m^*}\left(n_y^2 + n_z^2\right)\left(\frac{\pi}{L}\right)^2. \qquad (16.4)$$

Similarly, for a box with dimensions $L_x = L_y = L_z = L$, the solutions are

$$\psi_{n_x n_y n_z} = A \sin\left(\frac{n_x \pi}{L} x\right) \sin\left(\frac{n_y \pi}{L} y\right) \sin\left(\frac{n_z \pi}{L} z\right)$$

and

$$E_{n_x n_y n_z} = \frac{\hbar^2}{2m^*}\left(n_x^2 + n_y^2 + n_z^2\right)\left(\frac{\pi}{L}\right)^2. \qquad (16.5)$$

Because of the change in the dispersion relations, the density of states of restricted geometry systems is quite different from that of the bulk. If the density of one-particle states per unit volume in three dimensions for a parabolic band is

$$D(\varepsilon) = \frac{V}{2\pi^2}\left(\frac{2m^*}{\hbar^2}\right)^{3/2} \varepsilon^{1/2}, \qquad (16.6)$$

where V is the sample volume, then the corresponding density of states for the nth subband of a system allowed to move in only one or two directions, where the dimensionality of the system is effectively $d = 1$ and $d = 2$ respectively, is given by

$$D_n^d(\varepsilon) = C_d\left(\frac{2m^*}{\hbar^2}\right)^{\frac{d}{2}} \left(\varepsilon - E_n\right)^{\frac{d-2}{2}}, \qquad (16.7)$$

where ε is restricted to $\varepsilon > E_n$, $n = \sqrt{\sum_{i=1}^{3-d} n_i^2}$, and C_d is a constant. And, for a quantum dot, the density of states is a set of discrete δ-functions given by

$$D(\varepsilon) = 2 \sum_{n_x, n_y, n_z} \delta(\varepsilon - E_{n_x n_y n_z}). \qquad (16.8)$$

Equations (16.6)–(16.8) show that van Hove singularities and the general energy dependence of the density of states of reduced-dimensional systems are dramatically different from those of the same material in the bulk. These features are illustrated in Fig. 16.2.

The changes in the density of states lead to very different behaviors in the electronic properties of nanostructures and reduced-dimensional systems, such as their transport and optical properties, as well as thermodynamic properties such as electronic heat capacities. In particular, the positions of the van Hove singularities illustrated in Fig. 16.2 are

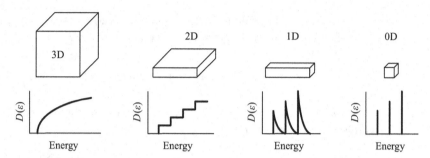

Figure 16.2   Sketch of the density of states of free particles confined to move in systems with reduced dimensions.

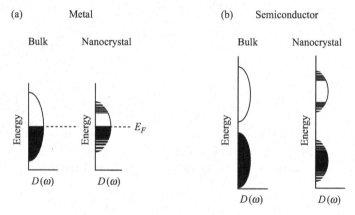

Figure 16.3   Illustration of discretized density of states in nanocrystal. (a) Metallic nanocrystal and (b) semiconductor nanocrystal.

dictated by the position of the onset of the subbands, which are explicitly and sensitively dependent on the size and dimension of the nanostructure, and hence the properties of the system may be tuned by its size and shape. A good example is the bandgap of a semiconductor quantum dot. In a quantum dot, both the electron and hole states are quantized by the finite dimensions ($L \times L \times L$ with L of the order of a few to tens of nanometers) of the semiconductor cluster or nanocrystal. (See Fig. 16.3.) To a good approximation, the energies of the lowest electron (the lowest unoccupied molecular orbital, called LUMO) and the hole (the highest occupied molecular orbital, called HOMO) states of the dot are derived from Eq. (16.5) as

$$E_{\text{LUMO}} = E_{\text{CBM}} + \frac{3\hbar^2}{2m_e^*}\left(\frac{\pi}{L}\right)^2 \tag{16.9}$$

and

$$E_{\text{HOMO}} = E_{\text{VBM}} - \frac{3\hbar^2}{2m_h^*}\left(\frac{\pi}{L}\right)^2, \tag{16.10}$$

where E_{CBM} and E_{VBM} are the energies of the conduction band minimum (CBM) and the valence band maximum (VBM), respectively. This leads to a so-called HOMO–LUMO gap or bandgap for the semiconductor quantum dot due to quantum confinement as

$$E_g^{QD} = E_g + \frac{3\hbar^2}{2\mu}\left(\frac{\pi}{L}\right)^2,$$ (16.11)

where E_g is the bulk gap of the semiconductor, L is roughly the diameter of the cluster, and μ is the reduced mass $\frac{1}{\mu} = \frac{1}{m_e^*} + \frac{1}{m_h^*}$. Hence the gap increases with decreasing dot size, and the energy change scales inversely with the reduced electron–hole mass and inversely with the diameter to the second power within this model. By changing the diameter of the quantum dots made out of the same semiconductor, one can dramatically change its transport and optical properties.

The change of the continuum of states in the bulk spectrum into a set of discrete states will not only change the absorption spectrum, but will also concentrate the nearby transitions in the bulk into individual intense transitions in a quantum dot. The concentration of the oscillator strength can be seen from the general sum rule (see Chapters 8 and 9)

$$\int \varepsilon_2(\omega)\omega \, d\omega = \frac{\pi}{2}\frac{4\pi n e^2}{m^*}.$$ (16.12)

Since the right-hand side is the same for both the nanocrystals and the bulk, Eq. (16.12) shows that the oscillator strength, which spreads out over a range of energy in the bulk, is concentrated in the quantum dot levels. The shift in absorption onset in nanometer-scale III–V and II–VI semiconductor clusters can be a large fraction of the bulk bandgap and can thus result in a tuning across a large portion of the visible spectrum. For instance, the bandgap of CdSe clusters can be tuned from 1.7 eV (deep red to the human eye) to 2.4 eV (green) by reducing its diameter from 20 nm to 2 nm. These features are seen in Fig. 16.4 for the case of CdS nanocrystal quantum dots. In Fig. 16.4, as the average size of the clusters reduce from 48 to 6.4 nm, the spectra shift from near 2.5 eV to over 4.5 eV and the absorption oscillator strength is concentrated into a small number of transitions.

Many different types of semiconductor nanocrystals have been synthesized with a variety of techniques. Figure 16.5 provides some TEM images of semiconductor quantum dot structures. Other structures, including semiconductor nanowire, nanostructures with different shapes (such as tetrapod nanocrystals), heterostructural forms from different parent materials, and superlattices of nanocrystals, are now routinely made and studied.

Equation (16.11) gives a change in the bandgap that scales as L^{-2}, where L is the diameter of the semiconductor quantum dot. This result arises from a simple model which makes use of an effective mass model with an infinitely hard and sharp potential well, and neglects many-electron effects on the quasiparticle energies. In reality, the enhancement of the gap for small clusters scales as $L^{-\alpha}$ with $\alpha < 2$. Also, excitonic effects become dominant in the optical transitions of reduced-dimensional systems. They also need to be included in any analysis of the optical properties of semiconductor nanocrystals.

In the case of metallic clusters, such as those created in atomic beam experiments of alkali elements, the electronic structure of the active valence electrons also determines the stability and structure of the clusters. Small alkali metal clusters exhibit special stability for N, the number of atoms in a cluster, at certain magic numbers. The physical origin

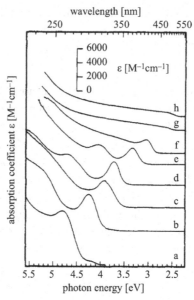

Adsorption spectra of CdS nanocrystal of different sizes, illustrating the effects of quantum confinement. As the crystal size becomes smaller (with an average size of 6.4 nm in (a) to an average size of 48 nm in (h)), the structure in the spectra move to higher energy and the absorption strength is concentrated into a small number of transitions. (After Vossmeyer *et al.*, 1994.)[1]

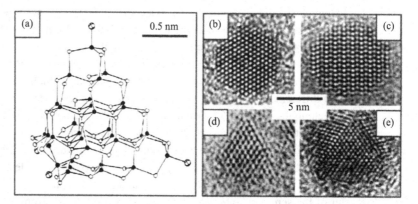

Semiconductor nanocrystal structures. (a) Structure of a $Cd_{32}S_{55}$ nanocrystal as determined by X-ray diffraction. The surface atoms are terminated by organic ligands (not shown). (b) and (c) Transmission electron micrographs (TEM) of CdSe nanocrystal with hexagonal structure, as seen viewing down two different crystallographic axes. (d) and (e) TEM of CdS/HgS/CdS quantum dot quantum wells. (After A.P. Alivisatos, 1996.)[2]

of the extra stability at these values of N is similar to the stability of closed-shell atoms in the periodic table. Each alkali metal atom contributes one electron to the binding of a cluster. The potential seen by these electrons may be modeled by a finite-size jellium model, leading to a flat-bottom potential and a set of discrete electron energy levels of different symmetry and degeneracy. As in atoms, clusters are most stable when an electron shell is completely filled. The stability of the cluster may be examined by considering the total electron energy of the system $E(N)$ and is characterized by the discontinuities in the total energy for the specific cluster size, $\Delta(N) = E(N) - E(N-1)$. The quantity

$$\Delta_2(N) = \Delta(N+1) - \Delta(N) \tag{16.13}$$

represents the relative binding energy change for clusters with N atoms compared to those with $N+1$ and $N-1$ atoms. This quantity thus measures the relative stability of clusters. Figure 16.6 illustrates this phenomenon by comparing the measured abundance of small sodium clusters with calculated $\Delta_2(N)$ from a jellium model.

16.2 Ballistic transport and quantization of conductance

As we discussed in Chapter 12, the theory for a macroscopic wire's electrical transport yields Ohm's law

$$\mathbf{j} = \sigma\mathbf{E}, \tag{16.14}$$

where σ is the conductivity tensor, a material-dependent quantity determined by the electronic structure and electron scattering properties (from impurities, defects, phonons, etc.) of the material. Consider a current through a wire of length $\mathscr{L}$ and cross-sectional area A. (See Fig. 16.7.) The current through the wire is

$$I = \int_A \sigma\mathbf{E} \cdot d\mathbf{s} = \sigma EA = \sigma A \frac{V}{\mathscr{L}} \equiv GV, \tag{16.15}$$

where V is the voltage drop between the two ends of the wire. This gives the usual expression for the conductance $G = \frac{\sigma A}{\mathscr{L}}$, which depends on the material and geometric dimensions of the wire.

Within the Drude model, $\sigma = \frac{ne^2\tau}{m^*}$, where τ is the relaxation time or the mean scattering time between scattering events. For typical metals, the mean free path ℓ ($\ell = v_F\tau$, with v_F the carrier velocity) is in the order of a micrometer. For a nano- or mesoscopic system with $\mathscr{L} < \ell$, $\tau \to \infty$, and $\sigma \to \infty$, but $A \to 0$. In this so-called ballistic limit, Eq. (16.15) is no longer valid; the conductance G is neither ∞ nor 0, but it is independent of $\mathscr{L}$ and quantized to an integer multiple of the quantum unit of conductance (for a paramagnetic

---

[1] T. Vossmeyer, L. Katsikas, M. Giersig, I. G. Popovic, K. Diesner, A. Chemseddine, A. Eychmuller, and H. Weller, "CdS nanoclusters: synthesis, characterization, size dependent oscillator strength, temperature shift of the excitonic transition energy, and reversible absorbance shift," *J. Phys. Chem.* 98(1994), 7665.

[2] A. P. Alivisatos, "Semiconductor clusters, nanocrystals, and quantum dots," *Science* 271(1996), 933.

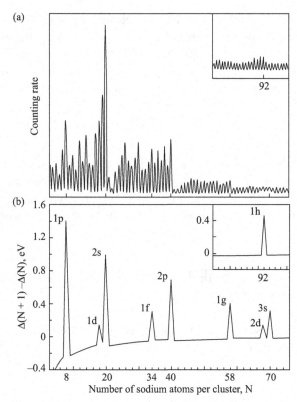

Figure 16.6 (a) Mass spectrum of sodium clusters for $N = 4 - 75$. The inset corresponds to $N = 75 - 100$. (b) Calculated change in the electronic energy difference $\Delta (N + 1) - \Delta (N)$ vs. N. The labels of the peaks correspond to the closed-shell orbitals. (After W. D. Knight *et al.*, 1984.)[3]

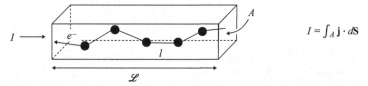

Figure 16.7 Schematic of diffusive scattering in a conductor leading to Ohm's law.

system)

$$G_0 = \frac{2e^2}{h}. \tag{16.16}$$

A system with a conductance of G_0 has a resistance of $G_0^{-1} = \frac{h}{2e^2} = 12.906 \; k\Omega$.

---

[3] W. D. Knight, K. Clemenger, W. A. de Heer, W. A. Saunders, M. Y. Chou, and M. L. Cohen, "Electronic shell structure and abundances of sodium clusters," *Phys. Rev. Lett.* 52(1984), 2141.

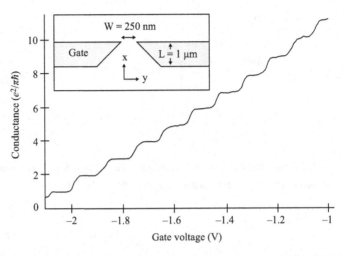

Figure 16.8 Measured point-contact conductance as a function of gate voltage for a narrow channel of 2DEG. The gate voltage is applied perpendicular to the 2DEG along the z-direction with the two top gates (indicated by the shaded areas). The conductance shows plateaus at multiples of $2e^2/h$. (After B. J. van Wees *et al.*, 1988.)[4]

The first experimental observation of quantization of conductance in the ballistic regime was made in the 1980s. The conductance of each occupied subband of a narrow channel formed by a 2DEG contributes a conductance of G_0 (in the absence of scattering) to the total conductance G. The experimental setup consists of a ballistic point contact formed in the 2DEG of a GaAs–AlGaAs heterostructure with its width controlled by a top gate. (See Fig. 16.8.) A large negative bias applied to the gate raises the electrostatic energy of the states underneath the gate to above the Fermi level, effectively driving the electrons out of the regions underneath and leaving a narrow conduction region with a width W that is tunable by varying the gate voltage.

We can understand the quantized conductance in 1D ballistic systems by using the following simple analysis. Consider a long, narrow strip defined in a 2DEG (i.e. we neglect the degree of freedom perpendicular to the x–y plane) as shown in Fig. 16.9. The electrons move under the influence of a bias voltage V so that there is a net current I through the strip. Since $G = \frac{dI}{dV}$, we can obtain the conductance by computing $I(V)$. We model the system of interest with a standard geometry for transport through nanostructures, which in our case is a narrow conductive region (C) in contact with two reservoirs of electrons, one on the left (L) with chemical potential μ_L and one on the right (R) with chemical potential μ_R, as the two contacts/leads. (See Fig. 16.9(b).) The difference in chemical potentials is the bias voltage V times the electron charge e. The wavefunctions and energies of the electronic states in the strip of width W are

$$\psi_{nk}(x,y) = C \sin\left(\frac{n\pi y}{W}\right) e^{ikx} \tag{16.17}$$

[4] B. J. van Wees, H. van Houten, C. W. J. Beenakker, J. G. Williamson, L. P. Kouwenhoven, D. van der Marel, and C. T. Foxon, "Quantized canductance of point contacts in a two-dimensional electron gas," *Phys. Rev. Lett.* 60(1988), 848.

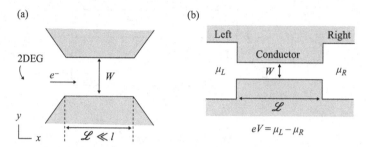

Figure 16.9   (a) Diagram of a ballistic 1D channel formed by a 2DEG. The mean free path ℓ is taken to be much larger than the channel length $\mathcal{L}$. (b) Idealized model for the conditions shown in (a).

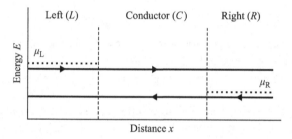

Figure 16.10   Energy diagram of the propagating electronic states of a ballistic wire in contact with a left (L) and right (R) reservoir.

and

$$E_{nk} = E_0 + \frac{\hbar^2 k^2}{2m} + \frac{\hbar^2}{2m}\left(\frac{n\pi}{W}\right)^2, \tag{16.18}$$

where C and E_0 are constants. The density of states of such a strip is schematically shown in Fig. 16.2 for the 1D case with sharp van Hove singularities at energies of the minima of the subbands. Before applying the bias, the Fermi level E_F is determined by the density of the 2DEG, and the number of occupied subbands is determined by the number of subbands that have its band minimum below E_F.

To calculate $I(V)$, we assume that steady state and quasi-equilibrium are achieved, i.e., μ_L and μ_R are well defined, and L and R are in equilibrium or nearly so. We further assume that we have reflectionless contacts so that states from reservoir L are not backscattered at the L–C interface and states from reservoir R are not backscattered at the C–R interface (see Fig. 16.10), and there is no causal relationship between left- and right-moving states. This set of assumptions is made since the true occupation function of states in the wire varies along the wire owing to the fact that this is a nonequilibrium problem. Two representative states, one left-moving ($-k$) and one right-moving ($+k$), are sketched in Fig. 16.10. At $T = 0$, all states with energy less than μ_L are occupied in reservoir L; all states with energy less than μ_R are occupied in reservoir R. At finite T, the occupation factor is given by the Fermi–Dirac distribution with the corresponding μ. Before the bias voltage V is applied, $\mu_L = \mu_R = E_F$; at steady state with bias voltage V, $\mu_L - \mu_R = eV$.

At $T = 0$, the current density $\mathbf{j}(\mathbf{r})$ at any position in conductor region C in Fig. 16.10 is given by

$$\mathbf{j}(\mathbf{r}) = e \sum_{n,k}^{occ} \mathrm{Re}\left[\psi_{nk}^*(\mathbf{r}) \mathbf{v} \psi_{nk}(\mathbf{r}) \right], \tag{16.19}$$

where

$$\mathbf{v} = \frac{\mathbf{p}}{m} = \frac{\hbar}{im}\nabla, \tag{16.20}$$

with m the carrier mass, and the sum is over all occupied states. We may compute left-moving I_L and right-moving I_R current separately and subtract to obtain $I = I_L - I_R$.

At a specific position x along the wire, I_L has a contribution from each occupied subband n of the form

$$I_{L,n}(x) = \int_A \mathbf{j}_n(\mathbf{r}) \cdot d\mathbf{s} = e \int_A \mathrm{Re}\left[\sum_{k>0}^{occ} \psi_{nk}^* v_x \psi_{nk} \right] dy\, dz. \tag{16.21}$$

In steady state, $I_{L,n}$ is independent of x, and we may rewrite ($\mathscr{L}$ is defined as in Fig. 16.9)

$$I_{L,n} = \frac{1}{\mathscr{L}} \int I_L(x) dx = \frac{e}{\mathscr{L}} \mathrm{Re}\left[\sum_{k>0}^{occ} \langle \psi_{nk}| v_x |\psi_{nk}\rangle \right] = \frac{e}{\mathscr{L}} \sum_{k>0}^{occ} \frac{1}{\hbar}\frac{d\varepsilon_{nk}}{dk}. \tag{16.22}$$

Here we make use of the relation that $\langle \psi_{nk}| v_x |\psi_{nk}\rangle$ is the band velocity and is given by $v_x = \frac{1}{\hbar}\frac{d\varepsilon_{nk}}{dk}$ for the 1D states. Since $\mathscr{L}$ is large ($\sim 1\ \mu$m), we may convert the sum over the k states into an integral over the density of states

$$\sum_{k>0}^{occ} \left(\frac{1}{\hbar}\frac{d\varepsilon_{nk}}{dk}\right) \rightarrow 2 \times \frac{\mathscr{L}}{2\pi} \int_{-\infty}^{\mu} \frac{d\varepsilon}{\left|\frac{d\varepsilon}{dk}\right|} \frac{1}{\hbar}\left(\frac{d\varepsilon_{nk}}{dk}\right), \tag{16.23}$$

where the factor 2 is for spin. Again, if the temperature is nonzero, the integral will include a Fermi factor. Equation (16.22) becomes

$$I_{L,n} = \frac{2e}{h} \int_{-\infty}^{\mu_L} d\varepsilon, \tag{16.24}$$

and similarly, we have, for each occupied subband,

$$I_{R,n} = \frac{2e}{h} \int_{-\infty}^{\mu_R} d\varepsilon. \tag{16.25}$$

The contribution of the nth occupied band to the current is

$$I_n = I_{L,n} - I_{R,n} = \frac{2e}{h} \int_{\mu_R}^{\mu_L} d\varepsilon = \frac{2e}{h}\Delta\mu = \frac{2e^2}{h}V. \tag{16.26}$$

For the last equality, we have used the fact that $\Delta\mu = eV$. The total current going through the nanowire is given by

$$I(V) = \sum_{n}^{occ} \frac{2e^2}{h}V = M\left(\frac{2e^2}{h}\right)V, \tag{16.27}$$

where M is the number of occupied subbands, and the current is independent of the details of the energy dispersion $\varepsilon(k)$ of the states and is independent of $\mathscr{L}$. Finally, for the conductance, we have

$$G = \frac{dI}{dV} = M\left(\frac{2e^2}{h}\right) = MG_0. \tag{16.28}$$

This expression explicitly shows that the conductance of the nanowires in the ballistic limit is quantized to the number of occupied subbands. In the experiment shown in Fig. 16.8, as the gate voltage becomes more negative, the effective width W of the nanowire decreases and the subband energies increase according to Eq. (16.18), resulting in fewer occupied subbands. The conductance G decreases in units of $\frac{2e^2}{h}$ until the effective width narrows to $W^2 < \frac{\hbar}{2m}\frac{\pi^2}{E_F}$.

The finite resistance of $R = G^{-1} = (MG_0)^{-1}$ of the ballistic wire originates from the contacts. It comes from the fact that the reservoirs have infinitely many current-carrying channels, and the electrons in these channels are "funneled" into a nanowire, which has far fewer channels. If the cross-sectional area is macroscopic in size, there will be many subbands below E_F and the contact resistance will be negligible, i.e. $G_c \sim MG_0$ and $R_c \sim \frac{1}{M}$, which goes to zero as $M \to \infty$. Of course, if the wire is large enough, diffusive transport takes over. The fact that G decreases in steps as the width of a nanowire decreases is a quantum confinement effect.

16.3 The Landauer formula

If the nanostructure between two metallic leads is not perfectly transmitting, Eq. (16.28) is modified. This is the case to consider for understanding electrical transport through a single-molecule junction. As seen in Fig. 16.11, the small size of the junction means that the junction conductance is dominated by the contact geometry and electronic structure, which controls the transmission of electrons through the junction. Microscopically, the energy of the molecular levels, their relative positions, how they line up with the metal Fermi level, and their level widths will control the junction transmission and resistance. Schematically, we may set up a modified, general geometry as shown in Fig. 16.12 to analyze this problem. In Fig. 16.12, C is a nanoscale resistive element, such as a molecule attached to metal leads or a nanowire with scattering centers (defects, impurities, etc.). It acts as a potential barrier to the carriers at some energy E. A state incident from the left lead is partially transmitted into any of the right-lead states with energy E and is partially reflected back to the left-lead states. Let us denote t_{ij} as the transmission amplitude from state i to state j.

Following the method used in the derivation for Eq. (16.26), one has

$$I = \frac{2e}{h} \int_{\mu_R}^{\mu_L} T(E, \mu_L, \mu_R)dE, \tag{16.29}$$

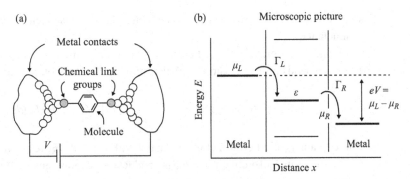

Figure 16.11 Single-molecule junction. (a) Sketch of a possible configuration. (b) Energy diagram of electronic states. Γ_L and Γ_R denote the inverse of the transition rate at the two interfaces.

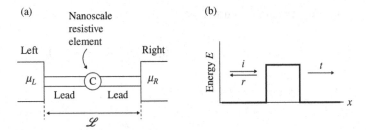

Figure 16.12 (a) Model for calculating the conductance of electron transport through a nanostructure junction. (b) The incident electron wave (i) through the nanostructure (C in (a)) represented by the potential barrier is reflected and transmitted with amplitudes r and t, respectively.

with

$$T(E, \mu_L, \mu_R) = Tr[t^\dagger t], \tag{16.30}$$

where t is the transmission matrix t_{ij}. The amount of current through the nanostructure is proportional to the probability of transmission of carriers. Equation (16.29) is commonly known as the Landauer formula.[5] The expressions are for conductance in a two-probe measurement. Extension of this approach to multiple-probe geometry is called the Büttiker–Landauer formalism.

Since $\mu_L - \mu_R = eV$, the Landauer formula (Eq. (16.29)) can be rewritten as

$$I(V) = \frac{2e}{h} \int_{\mu_R}^{\mu_R + eV} T(E, V) dE. \tag{16.31}$$

Since $T(E, V)$ is generally bias voltage dependent, the I–V characteristics of transport through nanostructures is generally nonlinear. Complex behaviors may arise from the

[5] R. Landauer, "Spatial variation of currents and fields due to localized scatterers in metallic conduction," *IBM J. Res. Dev.* 1(1957), 223.

structure of $T(E, V)$, resulting in interesting phenomena such as negative differential conductance.

At finite temperature T, the Fermi factors come into the transmission. In general, the current depends on E_F, T, and V with the following expression for the Landauer formula:

$$I(E_F, V, T) = \frac{2e}{h} \int_{-\infty}^{\infty} T(E, V)[f_L(E) - f_R(E)]dE, \tag{16.32}$$

where f_L and f_R are Fermi factors evaluated with μ_L and μ_R, respectively. For small bias, $I(V)$ is linear in V. To obtain the linear conductance at low temperatures, we consider the change in current δI as V goes from zero to δV.

$$\delta I = \frac{2e}{h} \int \left\{ \delta T(E, 0)(f_L - f_R) + T(E, 0)\delta(f_L - f_R) \right\} dE = \frac{2e}{h} \int T(E, 0)\delta(f_L - f_R)dE. \tag{16.33}$$

The first term in the curly brackets above is zero because $f_L = f_R$ at $V = 0$. For small δV,

$$\delta(f_L - f_R) \approx -\frac{\partial f}{\partial E}(e\delta V). \tag{16.34}$$

As $T \to 0$, $-\frac{\partial f}{\partial E} \to \delta(E - E_F)$, and

$$\delta I = \frac{2e^2}{h}\delta V T(E_F). \tag{16.35}$$

Equation (16.35) gives the linear response conductance as

$$G = \frac{\delta I}{\delta V} = \frac{2e^2}{h}T(E_F). \tag{16.36}$$

We note that Eq. (16.36) is a generalized form of Eq. (16.28) to the case of an imperfectly transmitting nanowire. If there are M current-carrying channels at the Fermi level and each has perfect transmission (i.e. in the ballistic limit), then $T(E_F) = M$, and Eq. (16.36) reduces to $G = MG_0$, which is Eq. (16.28).

16.4 Weak coupling and the Coulomb blockade

In the very weak coupling regime, a quantum dot or any zero-dimensional nanostructure in contact with metallic leads retains its sharp energy levels. That is, in Fig. 16.11, Γ_R and Γ_L are very small. Under these conditions, the electrical transport through the nanostructures exhibits Coulomb blockade behavior at low temperatures. The differential conductance $G = \frac{dI}{dV}$ is zero, unless the bias voltage V or a gate voltage V_g used to electrostatically shift the energy of the electrons in the dot is at a certain specific value.

The Coulomb blockade may be understood from a semiclassical analysis. In this picture, in a quantum dot with discrete energy states, the many-electron effects of adding an extra

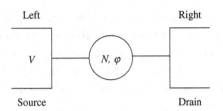

Figure 16.13 Idealized model investigating the Coulomb blockade for electron transport through a nanostructure. The electrostatic potential φ may be changed by a gate voltage.

electron to the dot may be characterized by a charging energy

$$U = \frac{e^2}{C}, \qquad (16.37)$$

where C is the capacitance of the dot. Consider a transport measurement done in the setup given in Fig. 16.13. The quantum dot is attached to a left (L) and right (R) reservoir. The contacts can be represented by an equivalent RC circuit. A bias voltage V may be applied between the source and the drain reservoirs. The dot is further capacitantly coupled to a metallic gate with capacitance C_g so that a gate voltage V_g may be applied to change the electrostatic potential on the dot.

The energy or electrochemical potential for adding an extra electron to a dot with N electrons is

$$\mu_{N+1} = \varepsilon_{N+1} + NU - \alpha e V_g, \qquad (16.38)$$

where ε_{N+1} is the single-particle orbital energy, U is the Coulomb interaction between two electrons in the dot given approximately in a classical picture by Eq. (16.37), and α is a factor which determines the electrostatic shift due to the gate voltage V_g and is given by $\alpha = \frac{C_g}{C}$. In weak coupling, the dot is filled with N electrons until $\mu(N)$ is just below μ, with the chemical potential of the leads at zero bias. An additional electron cannot hop onto the dot at small bias, as seen in Fig. 16.14(a), since there is a high energy cost (from Eq. (16.38))

$$\Delta E = \varepsilon_{N+1} - \varepsilon_N + \frac{e^2}{C}. \qquad (16.39)$$

Hence, there is no current flow at small bias and one has a Coulomb blockade. The chemical potential $\mu(N+1)$ of the dot may be changed by either a bias voltage or a gate voltage. For example, from Eq. (16.38), one more electron may be added to the dot with an additional gate voltage ΔV_g (with the new $\mu(N+1)$ equal to μ) of the value

$$\Delta V_g = \frac{1}{\alpha e}\left(\varepsilon_{N+1} - \varepsilon_N + \frac{e^2}{C}\right). \qquad (16.40)$$

In general, the Coulomb interaction energy U or the capacitance C of the dot depends on its size and shape and also on the screening properties of the material that forms the dot and the screening effects from the environment.

Charging effects and the Coulomb blockade phenomenon are important only if the energy level broadening due to tunneling and the thermal fluctuation energy $k_B T$ are small compared to $\frac{e^2}{C}$. We may estimate the level broadening using the "leaky quantum capacitor" analogy. An electron resides on a quantum dot for time $\delta t \approx RC$. Thus, from the uncertainty principle, the levels broaden in energy by

$$\delta \varepsilon \approx \frac{h}{\delta t} = \frac{h}{RC} = \left(\frac{e^2}{C} \right) \left(\frac{1}{R} \right) \left(\frac{h}{e^2} \right). \tag{16.41}$$

The level broadening would be comparable to $\frac{e^2}{C}$ if $R \approx \frac{h}{e^2}$. Thus, for observable charging effects, we need to be in the regime for the nanostructure junctions such that

$$G \ll \frac{e^2}{h} \quad \text{and} \quad \frac{e^2}{C} \gg k_B T. \tag{16.42}$$

For typical semiconductor nanocrystals of diameter $\sim$ a few nanometers, $\frac{e^2}{C} \sim$ a few hundred meV, which is significantly larger than $k_B T$ at room temperature. Also, the energy level spacing is in the order of hundreds of meV. Thus both the single-particle level spacing and charging effects determine their transport properties. On the other hand, metallic dots of the same size only have level spacing of a few meV and the energy in Eq. (16.39) is dominated by the charging term $\frac{e^2}{C}$.

In the simplest picture for transport of a metallic dot, we may neglect the $\varepsilon_{N+1} - \varepsilon_N$ term in Eq. (16.39) and focus on the charging energy $\frac{e^2}{C}$. In this model, as schematically shown in Fig. 16.14, transport is suppressed when the chemical potentials μ_L and μ_R lie between the chemical potentials $\mu(N)$ and $\mu(N+1)$ of the dot. Current flows if $\mu(N+1)$ is lowered to lie between μ_L and μ_R. This can be achieved through a gate voltage, and there is a sharp peak in the conductance for each new charge state as V_g is varied. The spacing between these Coulomb oscillations in the conductance $G(V_g)$ is given by Eq. (16.40). The chemical potential for the $N+1$ system $\mu(N+1)$ may also be pulled in to lie between μ_R and μ_L with a large bias voltage. Neglecting the $\varepsilon_{N+1} - \varepsilon_N$ term, the condition for this to occur is when $V = (2n - 1) \frac{e}{C}$ with $n = 1, 2, 3 \ldots$ For a real quantum dot, the current $I(V, V_g)$

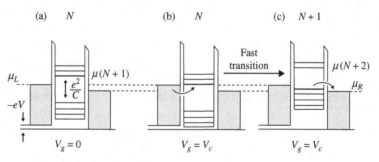

Figure 16.14 Energy level diagram of a quantum dot with different amounts of charge connected to a left (L) and right (R) reservoir with chemical potentials μ_L and μ_R, respectively. (a) Gate voltage $V_g = 0$. (b) and (c) Gate voltage $V_g = V_c$, illustrating the transfer of an electron.

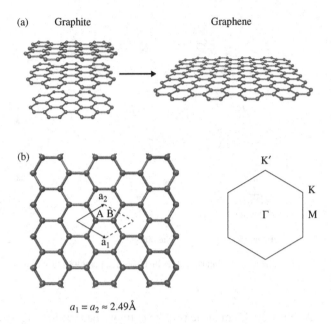

$$a_1 = a_2 \approx 2.49\text{Å}$$

Figure 16.15 (a) Atomic structure of graphite (left) and graphene (right), which is a single layer of graphite. (b) Unit cell (left) containing two carbon atoms A and B, and Brillouin zone (right) of graphene.

is a complex function of the bias and gate voltage resulting from an interplay between the charging energies and the excited-state energy levels.

16.5 Graphene, carbon nanotubes, and graphene nanostructures

Graphene is a single-atomic layer of graphite, which can be obtained by mechanical exfoliation from graphite or synthesized using different chemical or physical techniques. The carbon atoms in graphene form a honeycomb structure, as shown in Fig. 16.15. Large samples of wafer-sized ($\gtrsim$ 30" (76 cm) in diameter) single- and multi-layer graphene have been synthesized with chemical vapor deposition (CVD) and other techniques on metal substrates, which can then be transferred to other environments for study or applications. Graphene is a natural 2D material with many unusual properties because of its unique electronic structure arising from its bonding configuration and structural symmetry, and is considered a highly promising material for future electronics and other applications.[6] Andre Geim and Konstantin Novoselov were awarded the 2010 Nobel Prize in Physics for their pioneering work on graphene.

[6] A. K. Geim and K. S. Novoselov, "The rise of graphene," *Nat. Mater.* 6(2007), 183.

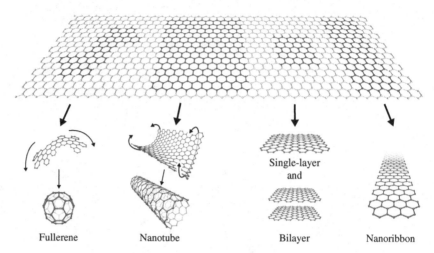

Fullerene   Nanotube   Single-layer
                                      and
                                      Bilayer   Nanoribbon

Figure 16.16 Different graphene-based nanostructures that have been synthesized and studied.

As a material, graphene is quite extraordinary. It is a highly stable 2D crystal with very few defects. Its electrical behaviors are quite exotic – the carriers have a linear E vs. $\mathbf{k}$ dispersion relation and chiral character, making them behave like 2D massless Dirac fermions, which give rise to a number of novel phenomena. Electrically, graphene has the highest carrier mobility at room temperature among known materials and a very high current-carrying capacity. Mechanically, it has a large Young's modulus, high tensile strength, and low friction. Thermally, like diamond, graphene is an extraordinarily good conductor. Moreover, graphene has excellent controllability with properties that can be tuned through electrical gating, structural patterning, etc. It is also the parent structure of many interesting nanostructures such as fullerenes, nanotubes, nanoribbons, etc., which have been synthesized and studied. (See Fig. 16.16.) Many of the novel properties of these graphene-based nanostructures are derived from the unique properties of graphene.

The crystal structure of graphene has two carbon atoms per primitive unit cell (labeled A and B in Fig. 16.15(b)) and a hexagonal Brillouin zone. Each of two atoms in the unit cell forms a triangular sublattice. An atom in sublattice A is surrounded by three atoms on sublattice B and vice versa, forming a honeycomb structure. The electronic band structure of graphene from *ab initio* density functional theory calculation is shown in Fig. 16.17. This band structure can be understood in terms of the basic interactions of the four valence electrons in the s–p shell of carbon. The σ and σ^* bands in Fig. 16.17 are basically formed by the coupling of the sp^2 hybrid orbitals in the x–y plane, giving rise to the strong bonding between C–C atoms in the plane and the stability of the honeycomb structure. The remaining p_z orbitals form the π and π^* bands, which are responsible for the transport and low-energy excitation properties of the system. A highly unusual feature of this band structure is the crossing of the π and π^* band at the Fermi level (for undoped graphene), giving rise to a set of linearly dispersing bands separated from all the other bands, and leading to a linear density of states from the charge-neutral point. This feature is a consequence of the symmetry of graphene, having electrons moving through a 2D honeycomb structure

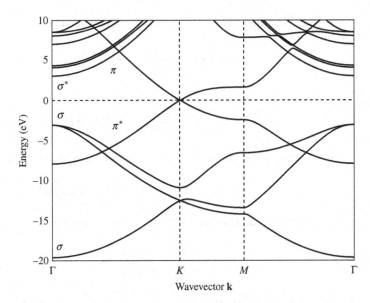

Figure 16.17 Calculated Kohn–Sham band structure (within density functional theory and the local density approximation) of graphene.

with identical potentials on both sublattices, and occurs in other 2D systems with the same symmetry. A more accurate many-body calculation for the quasiparticle band structure of graphene, including electron self-energy effects, yields a similar band structure to that shown in Fig. 16.17, except that the π bands are stretched in energy by about 30 percent.

16.5.1 The π-band electronic structure and 2D Dirac fermions in graphene

The π-band complex of graphene may be understood qualitatively by considering a tight-binding model with one p_z orbital on each atomic site. For simplicity, we assume orthogonal orbitals and nearest-neighbor interactions, but relaxation of these assumptions does not change the physics of interest here. Let us use a Cartesian coordinate system such that the positions of atom A and atom B in the unit cell are τ_A and τ_B. With d as the nearest-neighbor distance, the primitive translation vectors in Fig. 16.15(b) are

$$\mathbf{a}_1 = \left(\frac{3}{2}d, -\frac{\sqrt{3}}{2}d \right) \quad \text{and} \quad \mathbf{a}_2 = \left(\frac{3}{2}d, \frac{\sqrt{3}}{2}d \right), \tag{16.43}$$

and the reciprocal primitive lattice vectors are

$$\mathbf{b}_1 = \frac{2\pi}{d} \left(\frac{1}{3}, -\frac{1}{\sqrt{3}} \right) \quad \text{and} \quad \mathbf{b}_2 = \frac{2\pi}{d} \left(\frac{1}{3}, \frac{1}{\sqrt{3}} \right). \tag{16.44}$$

We define the Bloch sums of p_z atomic orbitals ϕ of carbon

$$\phi_{\mathbf{q}}^A(\mathbf{r}) = \frac{1}{\sqrt{N}} \sum_{\mathbf{R}} e^{i\mathbf{q}\cdot(\mathbf{R}+\boldsymbol{\tau}_A)} \phi(\mathbf{r} - \mathbf{R} - \boldsymbol{\tau}_A) \qquad (16.45)$$

and

$$\phi_{\mathbf{q}}^B(\mathbf{r}) = \frac{1}{\sqrt{N}} \sum_{\mathbf{R}} e^{i\mathbf{q}\cdot(\mathbf{R}+\boldsymbol{\tau}_B)} \phi(\mathbf{r} - \mathbf{R} - \boldsymbol{\tau}_B). \qquad (16.46)$$

The wavefunction of the electron in this tight-binding basis may be expanded as

$$\psi_{\mathbf{q}}(\mathbf{r}) = C_A \phi_{\mathbf{q}}^A(\mathbf{r}) + C_B \phi_{\mathbf{q}}^B(\mathbf{r}).$$

We solve the 2×2 Hamiltonian matrix eigenvalue problem

$$\begin{pmatrix} H_{AA}(\mathbf{q}) & H_{AB}(\mathbf{q}) \\ H_{BA}(\mathbf{q}) & H_{BB}(\mathbf{q}) \end{pmatrix} \begin{pmatrix} C_A \\ C_B \end{pmatrix} = E \begin{pmatrix} C_A \\ C_B \end{pmatrix}, \qquad (16.47)$$

where $H_{ij}(\mathbf{q}) \equiv \left\langle \phi_{\mathbf{q}}^i \middle| H \middle| \phi_{\mathbf{q}}^j \right\rangle$, and we have taken the overlap matrix to be the identity matrix. Since the environments around atoms A and B are identical, by symmetry, $H_{AA}(\mathbf{q}) = H_{BB}(\mathbf{q})$. Within the nearest-neighbor interaction assumption, $H_{AA}(\mathbf{q})$ is a constant independent of $\mathbf{q}$, which we may set to be zero, and $H_{AB}(\mathbf{q})$ may be simplified to

$$H_{AB}(\mathbf{q}) = \langle \phi(\mathbf{r} - \boldsymbol{\tau}_A) | H | \phi(\mathbf{r} - \boldsymbol{\tau}_B) \rangle \left(e^{i\mathbf{q}\cdot\boldsymbol{\ell}_1} + e^{i\mathbf{q}\cdot\boldsymbol{\ell}_2} + e^{i\mathbf{q}\cdot\boldsymbol{\ell}_3} \right), \qquad (16.48)$$

where $\boldsymbol{\ell}_1$, $\boldsymbol{\ell}_2$, and $\boldsymbol{\ell}_3$ are the vectors pointing from a given A atom to its three nearest neighbors. Defining the nearest-neighbor hopping integral as $\gamma \equiv \langle \phi(\mathbf{r} - \boldsymbol{\tau}_A) | H | \phi(\mathbf{r} - \boldsymbol{\tau}_B) \rangle$, we have

$$H_{AB}(\mathbf{q}) = H_{BA}^*(\mathbf{q}) = \gamma \left[e^{i\mathbf{q}\cdot\frac{(\mathbf{a}_1+\mathbf{a}_2)}{3}} + e^{i\mathbf{q}\cdot\left(-\frac{2}{3}\mathbf{a}_1+\frac{1}{3}\mathbf{a}_2\right)} + e^{i\mathbf{q}\cdot\left(\frac{1}{3}\mathbf{a}_1-\frac{2}{3}\mathbf{a}_2\right)} \right] = \Gamma(\mathbf{q}). \quad (16.49)$$

The Hamiltonian for graphene takes on the simple form

$$H(\mathbf{q}) = \begin{pmatrix} 0 & \Gamma(\mathbf{q}) \\ \Gamma^*(\mathbf{q}) & 0 \end{pmatrix}, \qquad (16.50)$$

and the two solutions to $H(\mathbf{q})$,

$$E_{\pm}(\mathbf{q}) = \pm |\Gamma(\mathbf{q})|, \qquad (16.51)$$

correspond to the upper ($+$) and lower ($-$) π and π^* bands of graphene in Fig. 16.17. The explicit expression for $E(\mathbf{q})$ is

$$E_{\pm}(\mathbf{q}) = \pm |\gamma| \sqrt{1 + 4\cos\left(\frac{3d}{2}q_x\right)\cos\left(\frac{\sqrt{3}d}{2}q_y\right) + 4\cos^2\left(\frac{\sqrt{3}d}{2}q_y\right)}. \qquad (16.52)$$

A value of $\gamma \approx 2.7$ eV gives a very good fit to the π-band complex of the density functional theory results in Fig. 16.17.

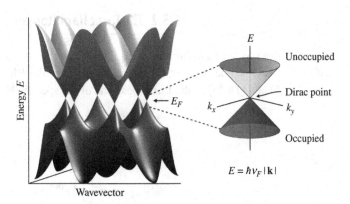

Figure 16.18 Energy surface in **k**-space for the electronic states of graphene near the Fermi energy.

For undoped graphene, there are two π electrons per unit cell. The lower π band is occupied and the upper π band is empty, thus E_F is at $E = 0$. This occurs at the $\mathbf{q} = \mathbf{K}$ or $\mathbf{q} = \mathbf{K}'$ point in the Brillouin zone. (See Fig. 16.15(b).) We note that $\mathbf{K}$ and $\mathbf{K}'$ are not the same point in general. The energy band structure corresponding to Eq. (16.52) is plotted in Fig. 16.18. It shows the bands linearly dispersing from the $\mathbf{q} = \mathbf{K}$ or $\mathbf{K}'$ point. Thus, the low-energy carriers in graphene behave like massless relativistic particles in 2D with the velocity given by the band velocity. These are known as 2D massless Dirac fermions. The energy surface for the electronic states corresponds to two cones, one inverted, which touch each other at two points (called Dirac points) in the Brillouin zone.

The **K** point corresponds to $\mathbf{q} = \frac{(\mathbf{b_1} + 2\mathbf{b_2})}{3}$. Evaluation of Eq. (16.49) gives

$$\Gamma(\mathbf{K}) = \gamma \left[e^{i\frac{2\pi}{3}} + 1 + e^{-i\frac{2\pi}{3}} \right] = 0, \qquad (16.53)$$

and hence $E(\mathbf{K}) = 0$, as discussed above. Now let us consider **q** states that are very near the Dirac point, i.e. states labeled by **k** with

$$\mathbf{q} = \mathbf{K} + \mathbf{k} \quad \text{and} \quad |\mathbf{k}| \ll |\mathbf{K}|. \qquad (16.54)$$

The function $\Gamma(\mathbf{q})$ may be expanded to first order in **k** around **K**, and we get (within an overall phase factor of $e^{i\frac{2\pi}{3}}$)

$$\Gamma(\mathbf{k}) = \left(\frac{\partial \Gamma}{\partial \mathbf{k}} \right) \cdot \mathbf{k} = \gamma \frac{3}{2} d(k_x - ik_y). \qquad (16.55)$$

The Hamiltonian of Eq. (16.50) for states near **K** takes on the form

$$H(\mathbf{k}) = \left(\frac{3}{2} \gamma d \right) \cdot \begin{pmatrix} 0 & k_x - ik_y \\ k_x + ik_y & 0 \end{pmatrix} = \hbar v_0 \begin{pmatrix} 0 & k_x - ik_y \\ k_x + ik_y & 0 \end{pmatrix}, \qquad (16.56)$$

where $v_0 = \frac{3}{2} \frac{\gamma d}{\hbar}$, which is, as we shall see, the band velocity of the π states near the Dirac point. An effective Hamiltonian of the form given by Eq. (16.56) may also be written down for states near $\mathbf{K}'$.

16.5.2 The Dirac Hamiltonian

From the form of the Hamiltonian in Eq. (16.56), for low-energy states near the $\mathbf{K}$ point with $\mathbf{q} = \mathbf{K} + \mathbf{k}$ and $|\mathbf{k}|a \ll 1$, we may represent the Bloch sum basis functions in Eqs. (16.45) and (16.46) in the form (near one of the specific $\mathbf{K}$ points)

$$\phi^A_{\mathbf{K}+\mathbf{k}}(\mathbf{r}) = \begin{pmatrix} 1 \\ 0 \end{pmatrix} e^{i\mathbf{k}\cdot\mathbf{r}} \tag{16.57}$$

and

$$\phi^B_{\mathbf{K}+\mathbf{k}}(\mathbf{r}) = \begin{pmatrix} 0 \\ 1 \end{pmatrix} e^{i\mathbf{k}\cdot\mathbf{r}}, \tag{16.58}$$

where $\begin{pmatrix} 1 \\ 0 \end{pmatrix}$ and $\begin{pmatrix} 0 \\ 1 \end{pmatrix}$ are two-component spinors representing the functions $\phi^A_K(\mathbf{r})$ and $\phi^B_K(\mathbf{r})$. The action of the Hamiltonian on the wavefunction in this sublattice basis in the long wavelength or small $\mathbf{k}$ limit is given by

$$H = \hbar v_0(-i\sigma_x \partial_x - i\sigma_y \partial_y) = v_0 \boldsymbol{\sigma} \cdot \mathbf{p}, \tag{16.59}$$

where

$$\sigma_x = \begin{pmatrix} 0 & 1 \\ 1 & 0 \end{pmatrix}, \sigma_y = \begin{pmatrix} 0 & -i \\ i & 0 \end{pmatrix}, \sigma_z = \begin{pmatrix} 1 & 0 \\ 0 & -1 \end{pmatrix} \tag{16.60}$$

are the Pauli spin matrices. Equation (16.59) is just the 2D version of the Dirac equation. The solutions of H in Eq. (16.59) are states with

$$\psi_{s\mathbf{k}}(\mathbf{r}) = \frac{1}{\sqrt{2}} \begin{pmatrix} 1 \\ se^{i\theta_\mathbf{k}} \end{pmatrix} e^{i\mathbf{k}\cdot\mathbf{r}} \tag{16.61}$$

and

$$E_{s\mathbf{k}} = s\hbar v_0 |\mathbf{k}|, \tag{16.62}$$

where $s = 1$ or -1, and $\theta_\mathbf{k}$ is the polar angle of $\mathbf{k}$ measured from the k_x direction. It is straightforward to show that $\psi_{s\mathbf{k}}(\mathbf{r})$ in Eq. (16.61) is an eigenstate of the operator

$$\sigma_\mathbf{k} = \cos\theta_\mathbf{k}\sigma_x + \sin\theta_\mathbf{k}\sigma_y \tag{16.63}$$

of eigenvalue $s = 1$ or -1. The wavefunction of the electron can then be thought of as having a pseudospin associated with it. This pseudospin is pointed either parallel or antiparallel to $\mathbf{k}$, depending on whether the state is in the $s = 1$ or $s = -1$ band. The form of $|\psi_{s\mathbf{k}}\rangle$ explicitly shows that the wavefunction is a linear combination of Bloch sums of π orbitals on the two sublattices, with a relative amplitude of 1 to $se^{i\theta_\mathbf{k}}$. The orientation of the pseudospinor thus gives the bonding character of the state with respect to the neighboring atoms. It is in this sense that the carriers in graphene have chiral character, in analogy with 2D massless neutrinos. We note here that the pseudospin has nothing to do with the real spin of the electron. The states at $E = 0$ are, in fact, eightfold degenerate since there are two valleys, $\mathbf{K}$ and $\mathbf{K}'$, and the real spin of the electron introduces two more degrees of freedom.

Equation (16.59) is the model Hamiltonian (the so-called 2D Dirac Hamiltonian) that is often used to analyze the properties associated with the low-energy electronic states of graphene. It is the characteristics of the solutions to the Dirac Hamiltonian (the linear dispersion relation and the chiral character of the wavefunctions, as given in Eqs. (16.61)–(16.62)) that give rise to many of the unusual electronic and transport properties in graphene and graphene-based nanostructures – the suppression of backscattering of carriers by slowly varying external potentials, Klein tunneling (i.e. perfect transmission of normal incident electron waves through a barrier independent of its height and width), the novel integer quantum Hall effect, the dramatic renormalization of carrier velocity, and the occurrence of new generations and branches of 2D Dirac fermions in a superlattice potential (called graphene superlattices), electron supercollimation in certain special graphene superlattices, and the rich and useful properties of carbon nanotubes, among others. We shall give a discussion of carbon nanotubes, graphene nanoribbons, and graphene superlattices in the next subsection.

16.5.3 Graphene Nanostructures

We give here a few examples of graphene-based nanostructures – graphene superlattices, carbon nanotubes, and graphene nanoribbons – to illustrate the richness of the properties of carbon nanostructures.

Graphene superlattices. A very unusual property of graphene is that if a 1D periodic scalar potential is applied to graphene, the group velocity of massless Dirac fermions in graphene is anisotropically renormalized in a counterintuitive way and sets of new massless Dirac fermions are generated.

Consider a situation where the spatial variation of the external periodic potential is much smaller than the carbon–carbon distance, so that inter-valley scattering between $\mathbf{K}$ and $\mathbf{K}'$ points in the Brillouin zone may be neglected. Under this condition, we could limit our discussion to the low-energy electronic states of graphene which have wavevector $\mathbf{k}+\mathbf{K}$ close to the $\mathbf{K}$ point, i.e. $|\mathbf{k}| \ll |\mathbf{K}|$, and we apply the Dirac Hamiltonian discussed in Section 16.5.2, that is, the Hamiltonian of the low-energy quasiparticles in the pseudospin basis $\left(\begin{smallmatrix}1\\0\end{smallmatrix}\right)e^{i\mathbf{k}\cdot\mathbf{r}}$ and $\left(\begin{smallmatrix}0\\1\end{smallmatrix}\right)e^{i\mathbf{k}\cdot\mathbf{r}}$ (where $\left(\begin{smallmatrix}1\\0\end{smallmatrix}\right)$ and $\left(\begin{smallmatrix}0\\1\end{smallmatrix}\right)$ are Bloch sums of π orbitals with wavevector $\mathbf{K}$ on the sublattices A and B, respectively) given by Eq. (16.59).

Assuming that a 1D scalar potential $V(x)$, periodic along the x-direction with periodicity L, is applied to graphene, the Hamiltonian H becomes

$$H = \hbar v_0 \left(-i\sigma_x \partial_x - i\partial_y \sigma_y + \frac{V(x)}{\hbar v_0} I \right), \tag{16.64}$$

where I is the 2×2 identity matrix. Now we perform a similarity transform $H' = U_1^{\dagger} H U_1$, using the unitary matrix

$$U_1 = \frac{1}{\sqrt{2}} \begin{pmatrix} e^{-i\alpha(x)/2} & -e^{i\alpha(x)/2} \\ e^{-i\alpha(x)/2} & e^{i\alpha(x)/2} \end{pmatrix}, \tag{16.65}$$

where $\alpha(x)$ is given by

$$\alpha(x) = 2 \int_0^x V(x') \frac{dx'}{\hbar v_0}.$$ (16.66)

Here, without losing generality, assuming an appropriate constant has been subtracted from $V(x)$ and that $V(x)$ has been shifted along the x direction so that the averages of both $V(x)$ and $\alpha(x)$ are zero, we obtain

$$H' = \hbar v_0 \begin{pmatrix} -i\partial_x & -e^{i\alpha(x)}\partial_y \\ e^{-i\alpha(x)}\partial_y & i\partial_x \end{pmatrix}.$$ (16.67)

If we are interested only in states whose wavevector $\mathbf{k} \equiv \mathbf{p} + \mathbf{G}_m/2$ (where $\mathbf{G}_m = m(2\pi/L)\hat{x} \equiv mG\hat{x}$ is a reciprocal lattice vector of the superlattice) is such that $|\mathbf{p}| \ll G$, the terms containing ∂_y in Eq. (16.67) can be treated as a perturbation. Also, to a good approximation, H' may be reduced to a 2×2 matrix using the following two states as the basis:

$$\begin{pmatrix} 1 \\ 0 \end{pmatrix}' e^{i(\mathbf{p}+\mathbf{G}_m/2)\cdot\mathbf{r}} \quad \text{and} \quad \begin{pmatrix} 0 \\ 1 \end{pmatrix}' e^{i(\mathbf{p}-\mathbf{G}_m/2)\cdot\mathbf{r}}.$$ (16.68)

Here, note that the spinors $\begin{pmatrix} 1 \\ 0 \end{pmatrix}'$ and $\begin{pmatrix} 0 \\ 1 \end{pmatrix}'$ now have a different meaning from $\begin{pmatrix} 1 \\ 0 \end{pmatrix}$ and $\begin{pmatrix} 0 \\ 1 \end{pmatrix}$.

In order to calculate these matrix elements, a Fourier transform of $e^{i\alpha(x)}$ is used:

$$e^{i\alpha(x)} = \sum_{\ell=-\infty}^{\infty} f_\ell[V] e^{i\ell Gx},$$ (16.69)

where the Fourier components $f_\ell[V]$ are determined by the periodic potential $V(x)$. In general,

$$|f_\ell| < 1,$$ (16.70)

which can be directly deduced from Eq. (16.69). The physics simplifies when the external potential $V(x)$ is an even function, and hence $\alpha(x)$ in Eq. (16.67) is an odd function. If we take the complex conjugate of Eq. (16.69) and change x to $-x$, it is evident that the $f_\ell[V]$'s are real. The analysis can also be generalized to cases other than even potentials. For states with wavevector $\mathbf{k}$ very close to $\mathbf{G}_m/2$, the 2×2 matrix M, whose elements are calculated from the Hamiltonian H' with the basis given by Eq. (16.68), can be written as

$$M = \hbar v_0 (p_x \sigma_z + f_m p_y \sigma_y) + \hbar v_0 mG/2 \cdot I.$$ (16.71)

After performing another similarity transform $M' = U_2^\dagger M U_2$ with

$$U_2 = \frac{1}{\sqrt{2}} \begin{pmatrix} 1 & 1 \\ -1 & 1 \end{pmatrix},$$ (16.72)

the final result is

$$M' = \hbar v_0 (p_x \sigma_x + f_m p_y \sigma_y) + \hbar v_0 mG/2 \cdot I.$$ (16.73)

The energy eigenvalue of the matrix M' is given by

$$E_s(\mathbf{p}) = s\hbar v_0 \sqrt{p_x^2 + |f_m|^2 p_y^2} + \hbar v_0 mG/2.$$ (16.74)

Equation (16.74) holds in general and not only for cases where the potential $V(x)$ is even. The only difference of the energy spectrum in Eq. (16.62) from that in Eq. (16.74), other than a constant energy term, is that the group velocity of quasiparticles moving along the y-direction has been changed from v_0 to $|f_m|v_0$. Thus, the electronic states near $\mathbf{k} = \frac{\mathbf{G}_m}{2}$ are also those of massless Dirac fermions but with a group velocity varying anisotropically depending on the propagation direction. The velocity along the x-direction, which is the direction of the potential variation, is totally unaffected. But the group velocity along the y-direction is always lower than v_0 (Eq. (16.70)) regardless of the form or magnitude of the periodic potential $V(x)$.

We note that other than the original Dirac points, new massless Dirac fermions are generated around the supercell Brillouin zone boundaries, i.e. the case with nonzero m values in the above formulation. It has been shown that these newly generated massless Dirac points are the only available states in a certain energy window if graphene is subjected to a 2D repulsive periodic scalar potential having triangular symmetry. Another point to note is that in graphene, under an external periodic scalar potential, a generalized pseudospin vector can be defined and used to describe the scattering properties between eigenstates, and the back-scattering processes in particular by a slowly varying impurity potential are still prohibited. On the other hand, if a 1D vector potential $\mathbf{A}(x,y) = A_y(x)\hat{y}$ is applied to graphene, it can be shown that through a transformation relation between scalar and vector potentials, the group velocity of charge carriers in graphene under a 1D periodic vector potential is renormalized isotropically. We will leave the derivation of this result as an exercise for the reader.

Carbon nanotubes. Structurally, a single-walled carbon nanotube is just a rolled-up graphene strip or ribbon whose width is of the order of a nanometer. As seen in Fig. 16.19, its geometric structure can be specified or indexed by its circumferential periodicity. Viewed in this fashion, a single-walled carbon nanotube structure may be completely specified by a pair of numbers (n, m) denoting the relative position $\mathbf{c} = n\mathbf{a}_1 + m\mathbf{a}_2$ of the pair of atoms on a graphene strip, which when rolled onto each other form the tube.

Although as discussed in Section 16.5.1, graphene is a zero-gap semiconductor with a linear dispersion near the Fermi energy, carbon nanotubes can be metals or semiconductors of different-sized gaps depending very sensitively on the diameter and helicity of the tubes, i.e. on the indices (n, m) defined in Fig. 16.19. The intimate connection between the electronic and geometric structure of the carbon nanotubes gives rise to many of the fascinating properties of various nanotube structures, in particular nanotube junctions. For

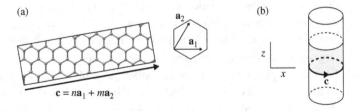

Figure 16.19   Geometric structure of an (n, m) single-walled carbon nanotube.

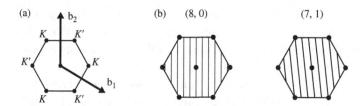

Allowed **k**-vector of the (7,1) and (8,0) tubes (solid lines) mapped onto the graphite Brillouin zone. The ideal (7,1) case essentially produces a metallic nanotube while an (8,0) tube is semiconducting.

example, one can use pairs of different nanotubes to form nanoscale metal–semiconductor junctions or semiconductor–semiconductor heterojunctions.

The physics behind this sensitivity of the electronic properties of carbon nanotubes to their structure can be understood within a band-folding picture. This is due to the unusual band structure of graphene, that has states crossing the Fermi level at only two inequivalent points in **k**-space, and to the quantization of the electron wavevector along the circumference direction. As illustrated in Fig. 16.18, an infinitely extended graphene sheet has two linearly dispersing bands that cross at the Fermi level at the **K** and **K'** points in the Brillouin zone. The Fermi surface of undoped graphene consists of these two points. When forming a tube, owing to the periodic boundary conditions imposed in the circumference direction, only a certain set of the **k**-states of graphene is allowed. The allowed set of **k**'s, indicated by the lines in Fig. 16.20, depends on the diameter and helicity of the tube. Whenever the allowed **k**'s include the point **K**, the system is a metal with nonzero density of states at the Fermi level, resulting in a 1D metal with two linear dispersing bands. When the point **K** is not included, the system is a semiconductor with different-sized gaps. We note that the states near the Fermi energy in both the metallic and the semiconducting tubes are all from states near the **K** point, and hence their transport and other properties are related to the properties of the states on the allowed lines. For example, the conduction band and valence bands of a semiconducting tube come from states along the line closest to the **K** point.

The general rules for the metallicity of the single-walled carbon nanotubes are as follows: (n, n) tubes are metals; (n, m) tubes with $n - m = 3\nu$, where ν is a nonzero integer, are very tiny-gap semiconductors; and all others are large-gap semiconductors. Strictly within the band-folding scheme, the $n - m = 3\nu$ tubes would be metals, but because of tube curvature effects, a tiny gap opens for the case where ν is nonzero. Hence, carbon nanotubes come in three varieties: large gap, tiny gap, and no gap. The (n, n) tubes, also known as armchair tubes, are always metallic within the single-electron picture, independent of curvature because of their symmetry. As the tube radius R increases, the bandgaps of the large-gap and tiny-gap varieties decrease with a $1/R$ and $1/R^2$ dependence, respectively. For most experimentally observed carbon nanotube sizes, the gap in the tiny-gap variety which arises from curvature effects would be so small that, for most practical purposes, all the $n - m = 3\nu$ tubes would be considered as metallic at room temperature. Therefore, in Fig. 16.20, a (7,1) tube would be metallic, whereas an (8,0) tube would be semiconducting.

This band-folding picture is expected to be valid for larger-diameter tubes. However, for a small-radius tube, because of its curvature, strong rehybridization among the σ and π orbitals can modify the electronic structure. Experimentally, nanotubes with radii as small as 0.35 nm have been produced.

Another very interesting aspect of carbon nanotubes is that excitonic effects (discussed in Chapter 9) are dominant in the optical response of semiconducting single-walled carbon nanotubes. Owing to enhanced electron–electron interaction and reduced screening in quasi-1D systems, theoretical studies have shown that the photophysics of semiconducting carbon nanotubes is dominated at all temperatures by excitons with large binding energies (a fraction of 1 eV), which are orders of magnitude larger than conventional semiconductors. Experimental measurements subsequently confirmed the existence and dominance of excitons. Surprisingly, it is shown that excitons exist even in metallic nanotubes. Fig. 16.21(a) compares the calculated absorption spectrum of a (8,0) single-walled carbon nanotube between the cases with and without electron–hole interactions included. With electron–hole interactions included, the spectrum is dominated by bound and resonant excitonic states. Each of the structures arising from a van Hove singularity in the non-interacting joint density of states gives rise to a series of exciton states. Fig. 16.21(b) shows the electron amplitude squared for the lowest optically active exciton plotted about a fixed hole position (solid dot) in real space; the electron is correlated to within several unit cells of the hole's position. It is a Wannier exciton in spite of the large exciton binding energy. The bottom two panels in Fig. 16.21 show the electron amplitude for two of the exciton states, one bound and one resonant, averaged over the coordinates perpendicular to the tube axis. The extent of the exciton wavefunction along the tube axis is about 20 Å for both of these states. For the (8,0) tube, the lowest-energy bound exciton has a binding energy of nearly 1 eV.

Note that the exciton binding energy for bulk semiconductors with similar bandgap sizes is in general only of the order of tens of meVs, illustrating again the dominance of many-electron Coulomb effects in reduced dimensional systems. Similar results have also been obtained for other semiconducting carbon nanotubes. In addition to the optically active (or bright) excitons shown in Fig. 16.21, there exist a number of optically inactive (or dark) excitons associated with each of the bright ones. These dark excitons also play an important role in the optical properties of the nanotubes; for example, they strongly affect the radiative lifetime of the excitons in semiconducting carbon nanotubes. Additionally, the non-Rydberg-like energy separation between the A_1', A_2', and A_3' states (Fig. 16.21) can be shown to result from a novel anti-screening effect, in which the effective electron–hole interaction is actually increased due to a strong distance-dependent variation in the electronic screening for electron–hole separations greater than the tube diameter. This unintuitive effect greatly increases the relative binding energies of the higher (A_2', A_3', etc.) states.

Graphene nanoribbons. The same mechanism of quantization of wavevectors as one goes from graphene to carbon nanotubes may be used to understand physically the electronic properties, such as the bandgaps, of graphene nanoribbons. Graphene nanoribbons have attracted considerable interest because they, unlike graphene, are systems that may provide a finite bandgap for electronic applications, which is not the case in extended

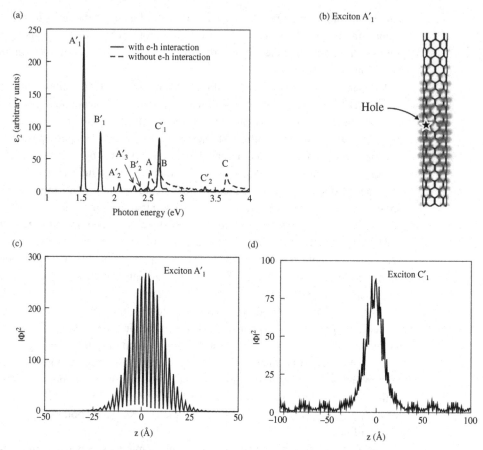

Figure 16.21 (a) Calculated optical absorption spectrum for the (8,0) single-walled carbon nanotube with (solid) and without (dashed) electron–hole interactions included. The spectrum with electron–hole interaction consists of a series of sharp peaks (broadened with a 12.5 meV Gaussian factor) corresponding to the bright bound excitons. (b) The real-space electron amplitude squared for the lowest bright state with the hole's position fixed and represented by the solid circle. (c) and (d) The electron amplitude squared averaged over the tube circumference plotted down the tube axis with the hole's position fixed at the origin. (After C.D. Spataru, S. Ismail-Beigi, L.X. Benedict, and S.G. Louie, 2004).[7]

graphene. The finite nanometer-scale width of the graphene nanoribbon restricts its electronic states to have discrete wavelengths or wavevectors along the direction perpendicular to the length of the ribbon, resulting in the possibility of opening up a bandgap, as in the case of nanotubes. Indeed, because of quantum confinement and electron interaction effects, graphene nanoribbons are shown to be semiconductors with varying size gaps depending on the width and on the geometry of the edges of the nanoribbon. In addition,

[7] C. D. Spataru, S. Ismail-Beigi, L. X. Benedict, and S. G. Louie, "Excitonic effects and optical spectra of single-walled carbon nanotubes," *Phys. Rev. Lett.* 92(2004), 077402.

some graphene nanoribbons, owing to electron exchange and correlations, have interesting magnetic edge states that can be manipulated with external probes such as an electric field. Graphene nanoribbons of different widths and edge structures have been fabricated. In particular, with the bottom-up technique of using molecular precursors to self-assemble into ribbon form, graphene nanoribbons with well-formed edges and widths down to the subnanometer scale have been synthesized.

16.6 Other quasi-2D materials

In additional to graphene, there are many other atomically thin quasi-2D materials with highly interesting properties that have been synthesized and studied. Experimental and theoretical studies of these quasi-2D materials and their nanostructures have revealed that these systems can exhibit highly unusual behaviors. Owing to their reduced dimensionality, the quasi-2D materials present opportunities for the manifestation of concepts/phenomena that may not be so prominent or may not have been seen in bulk materials. As in the case of graphene and carbon nanotubes, confinement, symmetry, many-body interactions, and environmental screening effects play a critical role in shaping qualitatively and quantitatively their electronic, transport, and optical properties, and thus their potential for applications.

Most of the quasi-2D crystals studied are mono- or few-layer forms of bulk layered materials, in which the atoms are strongly bonded within one layer but the interactions between atoms in different layers are very weak, usually van der Waals in nature. Atomically thin quasi-2D materials may be created by mechanical exfoliation and other physical and chemical means, such as molecular beam epitaxy, CVD, physical vapor transport, and others. Examples of quasi-2D materials that have been investigated, aside from graphene, include single- or few-layer crystals of hexagonal BN, transition metal dichalcogenides (such as MoS_2, $MoSe_2$, WS_2, WSe_2, etc.), metal monochalcogenides (such as GaS, GaSe, InSe, etc.), black phosphorus, and many others. Unlike graphene, these 2D materials can be semiconductors (with a range of bandgaps) or good metals, and some are superconductors.

The semiconducting transition metal dichalcogenide (TMD) monolayers are of particular interest. The structure and symmetry of the atomic orbitals in these systems, together with strong spin–orbit coupling, lead to direct-gap semiconductors with spin–orbit split band extrema at the $\mathbf{K}$ and $\mathbf{K}'$ points of the hexagonal Brillouin zone. The electronic states at the two valleys ($\mathbf{K}$ and $\mathbf{K}'$) have distinct spin textures with one valley coupled to left circular polarized light and the other to the right. This ability to couple to the valley degree of freedom of the electrons gives rise to new possibilities for manipulating the electrons in the TMD 2D crystals. Moreover, these systems exhibit dramatically strong light–exciton interactions and greatly enhanced electron–electron interactions. Exciton binding energies in monolayer TMD materials can be hundreds of meV – orders of magnitude larger than what is seen in typical bulk semiconductors. Such excitonic effects dominate the optical response in TMD mono- and few-layer systems, and create new opportunities for basic science studies and novel optoelectronic applications.

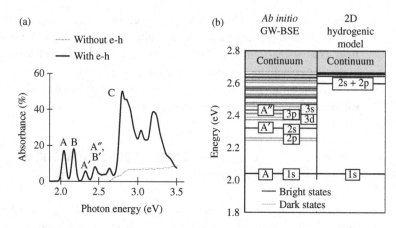

Figure 16.22 (a) Calculated absorbance of monolayer MoS_2 with (solid line) and without (dashed line) electron–hole interactions included, using *ab initio* many-body perturbation theory (the GW plus Bethe–Salpeter equation (GW-BSE) approach). The sharp exciton lines are broadened by 20 meV. There are three series of exciton states, *A*, *B*, and *C*, corresponding to transitions from the highest valence band and second highest band at **K/K′** and from the Γ point, respectively, in the Brillouin zone. (b) Energy levels of the A series excitons in monolayer MoS_2 from *ab initio* GW-BSE calculations and from a 2D hydrogenic model fit to the lowest-energy level of the *ab initio* results. (After D. Y. Qiu, F. H. da Jornada, and S. G. Louie, 2013.)[8]

Indeed, TMD monolayers have been predicted to have multiple series of excitons, arising from different regions of the Brillouin zone, with binding energies in excess of 0.6 eV and strange excitation spectra (optically bright and dark excited states of novel energy level and orbital angular momentum dependence) that cannot be explained by the usual 2D hydrogenic model. Figure 16.22 depicts the unusual excitonic features in the optical absorption spectrum of monolayer MoS_2. As in the case of carbon nanotubes, these extraordinary features originate from enhanced Coulomb interactions owing to reduced dimensionality and, importantly, an incomplete and strongly spatial-dependent screening in quasi-2D systems. Also, as for carbon nanotubes, despite their large binding energies, the TMD excitons are of the Wannier type with wavefunctions that are extended in real space and localized in **k**-space. The large exciton binding energy in the monolayer TMDs has been confirmed by combining scanning tunneling spectroscopy that measures the quasiparticle or band energies, and optical spectroscopy that measures optical transition energies.

Moreover, two-photon excitation spectroscopy and other techniques have been used to investigate the novel excitonic level structure of these systems, including the dark states. These measurements probed, for example, the optically inactive exciton $2p$ and $3p$ excited states, both in their energies and ordering with respect to other different orbital

[8] D. Y. Qiu, F. H. da Jornada, and S. G. Louie, "Optical spectrum of MoS_2: many-body effects and diversity of exciton states," *Phys. Rev. Lett.* 111(2013), 216805.

angular momentum exciton states. Another important and interesting aspect of the electronic and optical properties of TMD and other atomically thin materials is that, owing to their quasi-2D structure, they are very sensitive to environment screening. Both the electronic bandgaps and exciton binding energies can change substantially depending on the supporting substrates.

Part IV Problems

IV.1. **Green's function of a free Fermi gas**. For a free Fermi gas where $H = \sum_p \epsilon_p c_p^\dagger c_p$,

show that $c_p(\tau) = c_p(0)e^{-i\epsilon_p\tau}$. Find $G(p,t)$ and $G(p,p_0)$ for this system.

IV.2. **Momentum distribution**.

    (a) How are $G(p,0^-)$ and $G(p,0^+)$ related to the momentum distribution of a system of particles?

    (b) What is the magnitude of the discontinuity $G(p,0^+) - G(p,0^-)$?

IV.3. **Interacting Green's function**. For an interacting Green's function of the form $G(p,p_0) = \frac{Z_p}{p_0 - \epsilon_p + i\Gamma_p}$:

    (a) show that Z_p is related to the discontinuity in the momentum distribution at the Fermi surface.

    (b) show that Γ_p is proportional to $(p - p_F)^2$ near the Fermi surface.

IV.4. **Electron–phonon self energy**. Calculate the electron–phonon self energy Σ_{ph} using standard time-independent perturbation theory.

IV.5. **Random phase approximation (RPA)**. In this problem, we will dive into the Lindhard dielectric function, also called the random phase approximation for the free electron model, or equivalently, the ring approximation of polarizability for the free electron model in Green's function terminology.

    With the Green's function approach, we can get the (generalized) dielectric function as

$$\epsilon(\boldsymbol{q},\omega) = 1 - \frac{v_q}{\Omega}\sum_{k,\sigma} \frac{n_F(\epsilon_k) - n_F(\epsilon_{k+q})}{\epsilon_k - \epsilon_{k+q} + \omega + i\delta}, \quad \delta \to 0^+, \tag{IV.1}$$

    where v_q is the Fourier component of the Coulomb potential, $v_q = 4\pi e^2/q^2$, and $n_F(\epsilon)$ is the Fermi–Dirac distribution function. In the free electron model, we get the Lindhard dielectric function,

$$\epsilon_{RPA}(\boldsymbol{q},\omega) = 1 + \frac{q_{TF}^2}{2q^2}\left\{1 + \frac{m^2}{2k_Fq^3}\left[4\epsilon_F\epsilon_q - (\epsilon_q + \omega')^2\right]\ln\left[\frac{\epsilon_q + qv_F + \omega'}{\epsilon_q - qv_F + \omega'}\right]\right.$$

$$\left. + \frac{m^2}{2k_Fq^3}\left[4\epsilon_F\epsilon_q - (\epsilon_q - \omega')^2\right]\ln\left[\frac{\epsilon_q + qv_F - \omega'}{\epsilon_q - qv_F - \omega'}\right]\right\} \tag{IV.2}$$

    where $q_{TF} = 4me^2k_F/\pi$, $\omega' = \omega + i\delta, \delta \to 0^+, \hbar = 1$.

    (a) Recall the logarithmic function $\ln(z)$ defined on the complex plane and evaluate the real and imaginary part of $\epsilon_{RPA} = \epsilon_1 + i\epsilon_2$. (Hint: when evaluating ϵ_2, consider solutions in regions: (a) $q < 2k_F$: (i) $qv_F - \epsilon_q > \omega > 0$, (ii) $\epsilon_q + $

$qv_F > \omega > qv_F - \epsilon_q > 0$, (iii) $\omega > \epsilon_q + qv_F$, $qv_F - \epsilon_q > 0$; (b) $q > 2k_F$:
(i) $0 < \omega < \epsilon_q - qv_F$, (ii) $\epsilon_q + qv_F > \omega > -qv_F + \epsilon_q > 0$, (iii) $\omega > \epsilon_q + qv_F$).

(b) Prove that when $q \to \infty$, we have $\epsilon_1 \to 1$, $\epsilon_2 \to -\text{Im}(1/\epsilon)$.

(c) Prove that the real part ϵ_1 always approaches 1 at large ω.

(d) Prove that at $\omega \to 0$, we have $\epsilon_2 \to 0$, and find out the expression of ϵ_1 in this limit.

(e) Show that for values of $q < k_F$, ϵ_1 will become negative for intermediate values of ω.

(f) ϵ_2 can be written in another way:

$$\epsilon_2(q,\omega) = \frac{v_q}{2\pi} \int_0^{k_F} p^2 dp \int_{-1}^1 d\cos\theta \left[\delta(\epsilon_q + \frac{pq\cos\theta}{m} - \omega) \right. \tag{IV.3}$$
$$\left. - \delta(\epsilon_q + \frac{pq\cos\theta}{m} + \omega) \right],$$

Show that Eq. (IV.3) is equivalent to your results of ϵ_2 in (a). Also prove that ϵ_2 is antisymmetric in the frequency

$$\epsilon_2(q,\omega) = -\epsilon_2(q,-\omega). \tag{IV.4}$$

IV.6. **Dielectric function and spectral weight function.** Show in detail that the dielectric function $\epsilon(q,\omega)$ can be written in terms of a spectral weight function $f(q,\omega)$. Discuss the properties of f.

(a) Do the p integration and find the real and imaginary parts of Σ_{ph} for the jellium model.

(b) What is the effective mass correction near E_F from (a)?

IV.7. **Behavior of self energy.** Is $\Sigma_{ph}(p,\omega)$ a strong function of p or ω or both? Prove your statement.

IV.8. **Retarded Green's function and spectral function.** We know that the time-ordered Green's function is defined as

$$G_T(p, t - t') = -\frac{i}{\hbar} \langle \phi_0^N | T[c_p(t) c_p^\dagger(t')] | \phi_0^N \rangle. \tag{IV.5}$$

Here we introduce the retarded Green's function, which has a direct connection with physical quantities measured in experiments.

$$G_R(p, t - t') = -\frac{i}{\hbar} \Theta(t - t') \langle \phi_0^N | [c_p(t), c_p^\dagger(t')]_+ | \phi_0^N \rangle, \tag{IV.6}$$

which is the expectation value of the anticommutator of electron creation and destruction operators at different times.

(a) Calculate $G_T^0(p, t - t')$ and $G_R^0(p, t - t')$ for a free electron gas.

(b) Calculate $G_T^0(p, \omega)$ and $G_R^0(p, \omega)$ for a free electron gas.

(c) Define the spectral density function of an electron as

$$A(p,\omega) = -2\text{Im}[G_R(p,\omega)]. \tag{IV.7}$$

Calculate $A(\boldsymbol{p}, \omega)$ for the free electron gas.

(d) Prove the identity

$$1 = \int_{-\infty}^{\infty} \frac{d\omega}{2\pi} A(\boldsymbol{p}, \omega). \qquad (IV.8)$$

(e) Prove the identity

$$n_{\boldsymbol{p}} = \int_{-\infty}^{\infty} \frac{dE}{2\hbar\pi} f_F(E) A(\boldsymbol{p}, E/\hbar), \qquad (IV.9)$$

where $f_F(E)$ is the Fermi distribution function, with E measured from the chemical potential,

$$f_F(E) = \frac{1}{e^{\beta E} + 1}. \qquad (IV.10)$$

(f) Given the retarded Green's function for an interacting many-electron system,

$$G_R(\boldsymbol{p}, \omega) = \frac{1}{\hbar\omega - (\epsilon_p - \mu) + i\delta - \Sigma_R(\boldsymbol{p}, \omega)}, \qquad (IV.11)$$

where $\Sigma_R(\boldsymbol{p}, \omega)$ is the retarded self energy. Calculate the general form of $A(\boldsymbol{p}, \omega)$ for this retarded Green's function.

(g) If we assume the self energy takes the form

$$\Sigma_R(\boldsymbol{p}, z) = C[f(\boldsymbol{p}) - z]^{1/2}, \qquad (IV.12)$$

show that this function has a branch cut for $E = \hbar\omega > f(\boldsymbol{p})$, when $\Sigma(\boldsymbol{p}, \omega) \neq 0$, which means that

$$\Sigma(\boldsymbol{p}, \omega + i\delta) \neq \Sigma(\boldsymbol{p}, \omega - i\delta). \qquad (IV.13)$$

IV.9. **Lehmann representation.** Suppose we have two bosonic operators $\hat{A}(\boldsymbol{x}t)$ and $\hat{B}(\boldsymbol{x}t)$, and we define the time-ordered and retarded correlation functions as

$$C_T^{AB}(\boldsymbol{x}_1, \boldsymbol{x}_2; t_1 - t_2) = \left(-\frac{i}{\hbar}\right) \frac{\langle N | T[\hat{A}(\boldsymbol{x}_1 t_1) \hat{B}(\boldsymbol{x}_2 t_2)] | N \rangle}{\langle N | N \rangle}, \qquad (IV.14)$$

$$C_R^{AB}(\boldsymbol{x}_1, \boldsymbol{x}_2; t_1 - t_2) = \left(-\frac{i}{\hbar}\right) \Theta(t_1 - t_2) \frac{\langle N | [\hat{A}(\boldsymbol{x}_1 t_1), \hat{B}(\boldsymbol{x}_2 t_2)] | N \rangle}{\langle N | N \rangle}. \qquad (IV.15)$$

Use Lehmann representations of C_T^{AB} and C_R^{AB} to prove the following three properties:

(a)

$$C_R^{AB}(\boldsymbol{x}_1, \boldsymbol{x}_2; -\omega) = C_R^{AB}(\boldsymbol{x}_1, \boldsymbol{x}_2; \omega)^*, \qquad (IV.16)$$

(b)

$$C_T^{AB}(\boldsymbol{x}_1, \boldsymbol{x}_2; -\omega) = C_T^{BA}(\boldsymbol{x}_2, \boldsymbol{x}_1; \omega), \qquad (IV.17)$$

(c)

$$C_R^{AB}(\boldsymbol{x}_1, \boldsymbol{x}_2; \omega) = C_T^{AB}(\boldsymbol{x}_1, \boldsymbol{x}_2; \omega) \text{ for } \omega > 0. \qquad (IV.18)$$

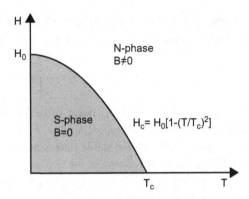

Figure IV.1   Schematic of a phase diagram of a type I superconductor.

IV.10. **Two-particle correlation function.** The non-interacting two-particle correlation function is defined as

$$L^0(x_1t_1, x_1't_1'; x_2t_2, x_2't_2') \doteq G_1(x_1t_1, x_2't_2')G_1(x_2t_2, x_1't_1'), \quad\text{(IV.19)}$$

where $G_1(x_1t_1, x_2't_2')$ and $G_1(x_2t_2, x_1't_1')$ are one-particle interacting Green's functions, and suppose we already have the quasi-particle wavefunctions $\phi_\lambda(x_1)$ and the corresponding eigenvalues ϵ_λ. For electron–hole excitation, we study the electron–hole pair excited at the same time, that is, $t_1' = t_1$ and $t_2' = t_2$. Given that the Hamiltonian is time independent, give an expression of $L^0(x_1, x_1'; x_2, x_2'; \omega)$.

IV.11. **Thermodynamics of a superconductor.** Consider a superconductor with the following characteristics. In the normal state, the N phase, the magnetization M is negligible. At a fixed temperature $T < T_c$, where T_c is the superconducting transition temperature, as the external magnetic field H is lowered below the critical field,

$$H_c(T) = H_0\left[1 - \left(\frac{T}{T_c}\right)^2\right], \quad\text{(IV.20)}$$

the normal state undergoes a phase transition to the superconducting state, the S phase. The phase diagram is shown in Figure IV.1.

(a) Show that the difference in Gibbs free energies G (in CGS units) between the two phases at temperature $T \le T_c$ is given by

$$G_S(T,H) - G_N(T,H) = \frac{1}{8\pi}\left[H^2 - H_c^2(T)\right]. \quad\text{(IV.21)}$$

(b) At $H \le H_0$, compute the latent heat of transition from the S to the N phase. (Hint: one approach is to consider a "Clausius–Clapeyron" type of analysis.)

(c) At $H = 0$, compute the discontinuity in the specific heat as the material transforms from the N to the S phase.

(d) Is the phase transition first or second order at $H = 0$?

IV.12. **Ginzburg–Landau theory of S–N phase transition**. The free energy density of the superconducting state is given relative to that of the normal state by

$$f = f_n + a|\psi|^2 + \frac{1}{2}b|\psi|^2 + c|\nabla\psi|^2, \qquad (IV.22)$$

with $a(T) = \alpha(T - T_c)/T_c$, negative below the critical temperature. It is reasonable to extend the expression, with α, $b = \beta$, and c constant above the critical temperature, although the stable state is normal, $n_s = 0$.

(a) Fill in the steps missing in the text, and derive the Ginzburg–Landau equation by minimizing the free energy. Show that the resulting equations can be written in the following form:

$$\left(\frac{i}{K}\nabla + A\right)^2 \psi = \psi - \psi^3, \qquad (IV.23)$$

$$\nabla \times \nabla \times A = -|\psi|^2 A + \frac{1}{2iK}\left(\psi^*\nabla\psi - \psi\nabla\psi^*\right). \qquad (IV.24)$$

(b) Construct the *linearized* Ginzburg–Landau equation for the material above T_c and obtain solutions for variation in one dimension.

(c) Using this description for a normal metal with $\alpha_1, \beta_1, c_1, T_{c1} < T$, find the order parameter as a function of position near a plane boundary separating the normal metal from a superconductor with $\alpha_0, \beta_0, c_0, T_{c0} > T$. (Linearize the equations for the superconductor around a constant ψ_0.) Take the order parameter everywhere continuous and obtain the boundary condition for the gradient by requiring continuous current across the boundary.

(d) Sketch the order parameter for two identical superconductors separated by a normal metal film. The film may provide a weak link between superconductors, as in the Josephson effect.

IV.13. **Numerical study of superconductivity**. A metal with electron density 10^{23}cm^{-3} and conductivity 10^{17}sec^{-1} (in esu units) becomes superconducting at 3 K. Calculate (numerically) within the Drude model the following parameters as accurately as you can:

(a) the London penetration depth;

(b) the coherence length;

(c) the Ginzburg–Landau kappa.

(d) Is this metal type I or type II?

IV.14. **Superconducting plate**. Discuss the problem of a superconducting plate of thickness d in a magnetic field using the Ginzburg–Landau equations. How does the order of the phase transition depend on the thickness of the film?

IV.15. **Superconducting cylinder**. Consider a superconducting cylinder of radius R in a magnetic field, H_0, along the axis of the cylinder. Calculate the current density, j, and the magnetic field, H, as a function of r. Plot your results.

IV.16. **Meissner effect**. Show that whenever there is a magnetic field, H, along a surface of a superconductor, there must be a current flowing through that surface at right angles to H with a surface current density, $J = c|H|$. Prove that the current is just big enough to cancel the magnetic field inside the superconductor at a depth larger than the thickness λ of the sheet.

IV.17. **Cooper pair**.

  (a) Calculate the mean square radius of a Cooper pair.

  (b) How does the energy of the bound pair vary with center of mass momentum? What does this have to do with the coherence length? Is the situation different for isolated pairs than for overlapping pairs? Why?

  (c) Does the single-pair model give a continuous spectrum above the ground state or one with an energy gap?

  (d) Which spin state (singlet or triplet) of a bound pair has lower energy for an isotropic potential? Why? Give examples of potentials in which the singlet state has lower energy and explain why. Do the same for the triplet state.

IV.18. **BCS model I**.

  (a) Calculate the Bardeen–Pines interaction at finite temperatures, i.e. put in the phonon occupation numbers.

  (b) Estimate $N(0)V$ (Coulomb) and $N(0)V$ (Bardeen–Pines) for a typical metal.

  (c) Fill in the steps in the text to find $\langle H_{BCS} \rangle$ in terms of the u's and v's for the wavefunction $\psi = \prod_k \left[u_k + v_k b_k^\dagger \right] \phi_0$, where ϕ_0 is the vacuum state.

  (d) Use (c) to derive the BCS gap equation.

  (e) What is the minimum energy required to create a superconducting quasiparticle? What is the measured "gap" in various experiments (Δ or 2Δ)? (E.g. tunneling, photoemission, heat capacity, or optical absorption measurements.)

IV.19. **BCS model II**.

  (a) Using the BCS model and assuming electron–hole symmetry, show that the energy difference between the superconducting and normal states at $T = 0$ is $-1/2N(0)\Delta^2$.

  (b) Using the BCS model, calculate the density of quasiparticle states for a BCS superconductor near the energy gap.

  (c) For the pair creation and destruction operations, calculate $[b_k, b_{k'}^\dagger]_-$, $[b_k, b_{k'}]_+$, $[b_k, b_{k'}^\dagger]_+$, b_k^2, and $b_k^{\dagger 2}$.

IV.20. **BCS model III**. Consider a metal described by a BCS reduced Hamiltonian,

$$H = \sum_k 2\varepsilon_k b_k^\dagger b_k + \sum_{k,k'} V_{k'k} b_{k'}^\dagger b_k, \qquad \text{(IV.25)}$$

and let

$$V_{kk'} = \lambda |\varepsilon_k||\varepsilon_{k'}| \qquad \text{(IV.26)}$$

for

$$|\varepsilon_k| < \hbar\omega_D, \text{and } |\varepsilon_{k'}| < \hbar\omega_D \qquad \text{(IV.27)}$$

and zero otherwise. Here ε_k is measured from the Fermi energy.

(a) By solving the energy-gap equation, find Δ_k, and also write a criterion for the existence of superconductivity. (Hint: This will be a condition on $N(0)\lambda(\hbar\omega_D)^2$ required for there to be a solution with real Δ_k.)

(b) Sketch the density of excited states as a function of energy (out to and beyond $\hbar\omega_D$). This would be called a "gapless superconductor."

IV.21. **Weakly coupled BCS superconductor**. For a weakly coupled BCS superconductor such as Al at $T = 0$, find the ratio of the radius r_p characterizing the spherical volume per pair and the coherence length. Express your answer in terms of Δ/E_F.

IV.22. **Transition temperature and frequency cutoff for electron–phonon interaction**. For the two-square-well solution of the BCS equation, maximize the transition temperature by choosing an optimum value for the first energy cutoff, ε_1.

(a) What is the optimum value for ε_1?

(b) What is the maximum transition temperature in terms of the second frequency cutoff, ε_2, K_p (phonon kernel), and K_c (Coulomb kernel)?

IV.23. **Isotope effect**. For the two-square-well model, answer the following.

(a) What is the condition for the isotope effect parameter, α ($T_c = AM^{-\alpha}$), to be zero?

(b) What are the maximum positive and negative values of the α?

(c) What is α if both cutoffs, ε_1 and ε_2, vary as $M^{-1/2}$?

IV.24. **Stoner model of itinerant magnetism**. In transition metals, the magnetic d electrons are not localized, but itinerant. For itinerant magnetism, the Heisenberg model fails, and we must consider the correlation effect among electrons. We have the Hubbard Hamiltonian in external magnetic field B,

$$H = \sum_{k\sigma} E_{k\sigma} n_{k\sigma} + U \sum_{l} n_{l\uparrow} n_{l\downarrow}, \qquad (IV.28)$$

where $E_{k\sigma} = \varepsilon_k + \sigma\mu B$, the l's are real-space lattice vectors, and the k's are reciprocal vectors.

(a) Use the mean-field method to deal with the second term in (Eq. IV.28), and prove that with a constant energy shift, we can get a Stoner model Hamiltonian as

$$H = \sum_{k\sigma} E_{k\sigma} c_{k\sigma}^{+} c_{k\sigma} + U \sum_{l\sigma} \langle n_{l\bar\sigma} \rangle c_{l\sigma}^{+} c_{l\sigma}, \qquad (IV.29)$$

where $\bar\sigma$ is $-\sigma$.

(b) In the Stoner model, we assume that $\langle n_{l\sigma} \rangle$ should be independent of the lattice position, which means that $\langle n_{l\bar\sigma} \rangle = \langle n_{\bar\sigma} \rangle$. Prove that Eq. (IV.29) can be reduced to a bilinear form

$$H = \sum_{k\sigma} \left(\tilde{E}_{k\sigma} + U\langle n_{\bar\sigma} \rangle \right) c_{k\sigma}^{+} c_{k\sigma}, \qquad (IV.30)$$

where $\tilde{E}_{k\sigma} = E_k + \sigma\mu_B B$, $\sigma = \pm 1$.

(c) Define the average density and magnetic moment as

$$n \doteq \langle n_\uparrow \rangle + \langle n_\downarrow \rangle, \tag{IV.31}$$

$$m \doteq \langle n_\downarrow \rangle - \langle n_\uparrow \rangle. \tag{IV.32}$$

Also, define the macroscopic magnetization

$$M = N\mu_B m = N\mu_B \left(\langle n_\downarrow \rangle - \langle n_\uparrow \rangle \right). \tag{IV.33}$$

Using the definition of M and the condition

$$\langle n_\sigma \rangle = \langle n_{l\sigma} \rangle = \frac{1}{N}\sum_l \langle n_{l\sigma} \rangle = \frac{1}{N}\sum_k \langle n_k \rangle = \frac{1}{N}\sum_k f(\tilde{E}_{k\sigma}), \tag{IV.34}$$

and taking the linear term with respect to external magnetic field B in the expression of M, you get the susceptibility in the Stoner model defined as

$$\chi_0(T) \doteq \frac{M(T,B)}{B}. \tag{IV.35}$$

Find the expression of $\chi_0(T)$.

(d) $\chi_0^{-1}(T) < 0$ signifies the instability of the paramagnetic phase and the emergence of the ferromagnetic phase. Prove that $\chi_0^{-1}(T) < 0$ leads to the Stoner criterion of itinerant ferromagnetism,

$$\frac{U}{2N\mu_B^2}\chi_p(T) > 1, \tag{IV.36}$$

where $\chi_p(T)$ is the Pauli susceptibility $\chi_p(T) \doteq 2\mu_B^2 \int_0^\infty dE \left(-\frac{\partial f}{\partial E} \right) \rho(E)$, where $\rho(E)$ is the density of states of the electrons.

IV.25. **Spin-1/2 ferromagnet.** For a spin-1/2 ferromagnet, within mean-field theory, show that the magnetization M as a function of temperature T in units of the Curie temperature is

$$M = \tanh\left(\frac{M}{T}\right). \tag{IV.37}$$

Show that for $T < 1$ (slightly below the Curie temperature),

$$M = \sqrt{3(1-T)}. \tag{IV.38}$$

IV.26. **Partition function of a paramagnet.** For a paramagnetic solid, with noninteracting magnetic ions having magnetic moment μ, in a weak magnetic field $\mathbf{B}$ along the z-direction:

(a) Find the average z-component of the magnetic momenta using the classical partition function.

(b) Derive the Curie law and the expression for the Curie constant.

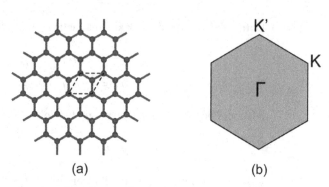

Figure IV.2 (a) Crystal structure of graphene. The primitive cell is illustrated by a dashed parallelogram. (b) Brillouin zone of graphene.

IV.27. Tight-binding model of graphene and carbon nanotubes. A single graphite sheet (called graphene) has a honeycomb structure and is a single layer of graphite, as shown in Figure IV.2(a). Assume that there is one p_z orbital (oriented perpendicularly to the sheet) on each carbon atom which forms the active valence and conduction bands of this two-dimensional crystal.

(a) Using the tight-binding method and only nearest-neighbor interactions (you may assume the overlap matrix is the unity matrix), calculate and sketch the band structure $E_n(\mathbf{k})$ for graphene. You may take the nearest-neighbor hopping integral to be $t = 2.7$ eV. Show that the bands have a linear dispersion near the $\mathbf{K}$ and $\mathbf{K}'$ points in the Brillouin zone (Figure IV.2(b)), and that if there is one electron on the p_z orbital of each of the carbon atoms, the low-energy dopant carriers in the system behave like two-dimensional massless Dirac fermions.

(b) Carbon nanotubes are structures which are graphene strips rolled into tubules with diameters of several nanometers. Because of the geometric structure of graphene, there are different ways to form a strip of a certain diameter. Show that, within the above tight-binding band picture and using periodic boundary conditions, the carbon nanotubes can be semiconductors or metals, depending sensitively on the diameter and chirality of the tube.

IV.28. Carrier properties of graphene. The unique electronic structure of graphene gives rise to some fascinating properties for the low-energy carriers in this system. We consider some of them below.

(a) Let us consider the electronic states of graphene near the corner point $\mathbf{K}$ of the hexagonal Brillouin zone and let us ignore the spin degree of freedom of the electrons. Setting the Fermi energy to be zero ($\varepsilon_F = 0$), fill in the missing steps in the text to show that these states can be described by the following Hamiltonian,

$$\hat{H} = -i\hbar v_0 \left(\sigma_x \frac{\partial}{\partial x} + \sigma_y \frac{\partial}{\partial y} \right), \tag{IV.39}$$

where $\sigma_x = \begin{pmatrix} 0 & 1 \\ 1 & 0 \end{pmatrix}$ and $\sigma_y = \begin{pmatrix} 0 & -i \\ i & 0 \end{pmatrix}$ are the Pauli matrices. (Note: it is also acceptable to find a different combination of the Pauli matrices and derivatives. This is related to the gauge freedom that you have in the Hamiltonian.)

(b) Show that the following wavefunction,

$$|sk\rangle = \frac{1}{\sqrt{2}} \begin{pmatrix} 1 \\ se^{i\theta_k} \end{pmatrix} e^{ik \cdot r} \tag{IV.40}$$

(where k is the wavevector as measured from the K point, $s = \pm 1$, and θ_k is the polar angle between k and the k_x axis), is an eigenstate of the Hamiltonian $\hat{H}$, and that the energy eigenvalue corresponding to this eigenstate is given by

$$\varepsilon_{sk} = s\hbar v_0 |k|. \tag{IV.41}$$

(c) Given the results above for the energy dispersion of the carriers, derive the expression for the density of states per unit area in real space, $D(\varepsilon)$, as a function of energy for these particles near $\varepsilon = 0$.

(d) Consider the **elastic scattering** of such a particle from an initial state $|sk\rangle$ to a final state $|sk'\rangle$ by a weak short-ranged potential $V(\mathbf{r}) = V_0 f(\mathbf{r})$, where $f(\mathbf{r})$ is a constant in a unit cell, and zero elsewhere. Determine to first order in V_0 the angular distribution of the scattering amplitude A. That is, show that $A(sk \to sk') \propto f(\theta_k, \theta_{k'})$, and find the functional dependence of $f(\theta_k, \theta_{k'})$. How is this different from normal free particles?

References

Selected books for further reading

General texts

Ashcroft, N. W. and Mermin, N. D., *Solid State Physics*. Philadelphia, PA: Saunders College, 1976.

Callaway, J., *Quantum Theory of the Solid State*. 2nd ed., Boston, MA: Academic Press, 1991.

Kittel, C., *Introduction to Solid State Physics*. 8th ed., New York, NY: Wiley, 2005.

Kittel, C., *Quantum Theory of Solids*. New York, NY: Wiley, 1963.

Madelung, O., *Introduction to Solid-State Theory*. Berlin: Springer-Verlag, 1978.

Marder, M. P., *Condensed Matter Physics*. New York, NY: Wiley, 2000.

Patterson, J. D. and Bailey, B. C., *Solid-State Physics: Introduction to the Theory*. 2nd ed., Berlin: Springer, 2010.

Ziman, J. M., *Principles of the Theory of Solids*. 2nd ed., Cambridge: Cambridge University Press, 1999.

Monographs on specific topics

Balkanski, M. and Wallis, R. F., *Semiconductor Physics and Applications*. Oxford: Oxford University Press, 2000.

Chaikin, P. M. and Lubensky, T. C., *Principles of Condensed Matter Physics*. Cambridge: Cambridge University Press, 1995.

Cohen, M. L. and Chelikowsky, J. R., *Electronic Structure and Optical Properties of Semiconductors*. Berlin: Springer-Verlag, 1989.

de Gennes, P., *Superconductivity of Metals and Alloys*. New York, NY: W. A. Benjamin, 1966.

Di Ventra, M., *Electrical Transport in Nanoscale Systems*. Cambridge: Cambridge University Press, 2008.

Kaxiras, E., *Atomic and Electronic Structure of Solids*. Cambridge: Cambridge University Press, 2003.

Mahan, G. D., *Many-Particle Physics*. 3rd ed., New York, NY: Plenum, 2000.

Martin, R. M., *Electronic Structure: Basic Theory and Practical Methods*. Cambridge: Cambridge University Press, 2004.

Schrieffer, J. R., *Theory of Superconductivity*. New York, NY: Westview Press, 1983.

Tinkham, M., *Introduction to Superconductivity*. 2nd ed., New York, NY: McGraw-Hill, 1996.

Yu, P. Y. and Cardona, M., *Fundamentals of Semiconductors: Physics and Materials Properties*. 4th ed., Berlin: Springer-Verlag, 2010.

Chapter 1

Books and reviews

Anderson, P. W., *Concepts in Solids*. New York, NY: W. A. Benjamin, 1963.
Ashcroft, N. W. and Mermin, N. D., *Solid State Physics*. Philadelphia, PA: Saunders College, 1976.
Kittel, C., *Introduction to Solid State Physics*. 8th ed., New York, NY: Wiley 2005.
Mahan, G. D., *Condensed Matter in a Nutshell*. Princeton, NJ: Princeton University Press, 2011.
Patterson, J. D and Bailey, B. C., *Solid-State Physics*. Berlin: Springer, 2007.
Pines, D., *Elementary Excitations in Solids*. New York, NY: W. A. Benjamin, 1963.
Ziman, J. M., *Principles of the Theory of Solids*. Cambridge: Cambridge University Press, 1989.

Chapter 2

Books and reviews

Falicov, L. M., *Group Theory and its Physical Applications*. Chicago, IL: University of Chicago Press, 1966.
Jones, H., *The Theory of Brillouin Zones and Electronic States in Crystals*. Amsterdam: North-Holland, 1960.
Sommerfeld, A. and Bethe, H., "Electronentheorie der metalle," in *Handbuch der Physik*. Berlin: Springer, 1933, vol. 24, pt. 2, ch. 3.
Tinkham, M., *Group Theory and Quantum Mechanics*. New York, NY: McGraw-Hill, 1964.

Chapter 3

Books and reviews

Cohen, M. L and Chelikowsky, J. R., *Electronic Structure and Optical Properties of Semiconductors*. 2nd ed., Berlin: Springer, 1989.
Fletcher, G. C., *The Electronic Band Theory of Solids*. Amsterdam: North-Holland, 1971.
Kaxiras, E., *Atomic and Electronic Structure of Solids*. Cambridge: Cambridge University Press, 2003.
Martin, R. M., *Electronic Structure: Basic Theory and Practical Methods*. Cambridge: Cambridge University Press, 2004.
Papaconstantopoulos, D. A., *Handbook of the Band Structure of Elemental Solids*. New York, NY: Plenum, 1986.
Phillips, J. C., *Bonds and Bands in Semiconductors*. New York, NY: Academic Press, 1973.
Yu, P. Y. and Cardona, M., *Fundamentals of Semiconductors: Physics and Materials Properties*. 4th ed., Berlin: Springer, 2010.

Chapter 4

Books and reviews

Bilz, H. and Kress, W., *Phonon Dispersion Relations in Insulators*. Berlin: Springer, 1979.
Born, M. and Huang, K., *Dynamical Theory of Crystal Lattices*. Oxford: Clarendon, 1954.
Wallis, R. F. (ed.), *Lattice Dynamics*. Oxford: Pergamon, 1965.
Ziman, J., *Electrons and Phonons*. Oxford: Oxford University Press, 1960.

Chapter 5

Books and reviews

Blount, E. I., "Formalisms of band theory," *Solid State Physics*. 13(1962), 305.
Marder, M. P., *Condensed Matter Physics*. New York, NY: Wiley, 2000.
Wannier, G. H., *Elements of Solid State Theory*. Cambridge: Cambridge University Press, 1959.
Xiao, D., Chang, M. C., and Niu, Q., "Berry phase effects on electronic properties," *Rev. Mod. Phys*. 82(2010), 1959.

Chapter 6

Books and reviews

Fetter, A. L. and Walecka, J. D., *Quantum Theory of Many-Particle Systems*. New York, NY: McGraw-Hill, 1971.
Kittel, C., *Quantum Theory of Solids*. New York, NY: Wiley, 1964.
Mahan, G. D., *Many-Particle Physics*. 3rd ed., New York, NY: Plenum, 2000.
Martin, R. M., *Electronic Structure: Basic Theory and Practical Methods*. Cambridge: Cambridge University Press, 2004.
Pines, D. *Elementary Excitations in Solids*. New York, NY: W. A. Benjamin, 1963.

Chapter 7

Books and reviews

Dreizler, R. M. and Gross, E. K. U., *Density Functional Theory: An Approach to the Quantum Many-Body Problem*. Berlin: Springer-Verlag, 1990.
Kaxiras, E., *Atomic and Electronic Structure of Solids*. Cambridge: Cambridge University Press, 2003.
Martin, R. M., *Electronic Structure: Basic Theory and Practical Methods*. Cambridge: Cambridge University Press, 2004.
Parr, R. G. and Yang, W., *Density-Functional Theory of Atoms and Molecules*. Oxford: Oxford University Press, 1989.
Onida, G., Reining, L., and Rubio, A., "Electronic excitations: density-functional versus many-body Green's-function approaches," *Rev. Mod. Phys*. 74(2002), 601.

Chapter 8

Books and reviews

Ashcroft, N. W. and Mermin, N. D., *Solid State Physics*. Philadelphia, PA: Saunders College, 1976.

Kittel, C., *Quantum Theory of Solids*. New York, NY: Wiley, 1963.

Martin, R. M., *Electronic Structure: Basic Theory and Practical Methods*. Cambridge: Cambridge University Press, 2004.

Pines, D., *Elementary Excitations in Solids*. New York, NY: W. A. Benjamin, 1963.

Yu, P. Y. and Cardona, M., *Fundamentals of Semiconductors: Physics and Materials Properties*. 4th ed., Berlin: Springer, 2010.

Ziman, J. M., *Principles of the Theory of Solids*. Cambridge: Cambridge University Press, 1989.

Chapter 9

Books and reviews

Abeles, F., *Optical Properties of Solids*. Amsterdam: North-Holland, 1972.

Ashcroft, N. W. and Mermin, N. D., *Solid State Physics*. Philadelphia, PA: Saunders College, 1976.

Bassani, G. F. and Parravicini, G. P., *Electronic States and Optical Transitions in Solids*. Oxford: Pergamon Press, 1989.

Cohen, M. L. and Chelikowsky, J. R., *Electronic Structure and Optical Properties of Semiconductors*. 2nd ed., Berlin: Springer, 1989.

Knox, R. S., *Theory of Excitons*. New York, NY: Academic Press, 1963.

Onida, G., Reining, L., and Rubio, A., "Electronic excitations: density-functional versus many-body Green's-function approaches," *Rev. Mod. Phys.* 74(2002), 601.

Yu, P. Y. and Cardona, M., *Fundamentals of Semiconductors: Physics and Materials Properties*. 4th ed., Berlin: Springer, 2010.

Ziman, J. M., *Principles of the Theory of Solids*. Cambridge: Cambridge University Press, 1989.

Chapter 10

Books and reviews

Ziman, J., *Electrons and Phonons*. Oxford: Oxford University Press, 1960.

Chapter 11

Books and reviews

Abrikosov, A. A., *Introduction to the Theory of Normal Metals*. New York, NY: Academic Press, 1972.

Ashcroft, N. W. and Mermin, N. D., *Solid State Physics*. Philadelphia, PA: Saunders College, 1976.

Marder, M. P., *Condensed Matter Physics*. New York, NY: Wiley, 2000.

Ziman, J. M., *Principles of the Theory of Solids*. Cambridge: Cambridge University Press, 1989.

Chapter 12

Books and reviews

Abrikosov, A. A., *Introduction to the Theory of Normal Metals*. New York, NY: Academic Press, 1972.

Ashcroft, N. W. and Mermin, N. D., *Solid State Physics*. Philadelphia, PA: Saunders College, 1976.

Hasan, M. Z. and Kane, C. L., "Colloquium: topological insulators," *Rev. Mod. Phys.* 82(2010), 3045.

Kittel, C., *Quantum Theory of Solids*. New York, NY: Wiley, 1987.

Marder, M. P., *Condensed Matter Physics*. New York, NY: Wiley, 2000.

Ziman, J. M., *Principles of the Theory of Solids*. Cambridge: Cambridge University Press, 1989.

Chapter 13

Books and reviews

Abrikosov, A. A., Gorkov, L. P., and Dzyaloshinski, I. E., *Methods of Quantum Field Theory in Statistical Physics*. Upper Saddle River, NJ: Prentice Hall, 1963.

Fetter, A. L. and Walecka, J. D., *Quantum Theory of Many-Particle Systems*. New York, NY: McGraw-Hill, 1971.

Mahan, G. D., *Many-Particle Physics*. 3rd ed., New York, NY: Plenum, 2000.

Schultz, T. D., *Quantum-Field Theory and the Many-Body Problem*. New York, NY: Gordon and Breach, 1964.

Chapter 14

Books and reviews

Cooper, L. N. and Feldman, D. (eds.), *BCS: 50 Years*. Singapore: World Scientific, 2010.

de Gennes, P. G., *Superconductivity of Metals and Alloys*. New York, NY: W. A. Benjamin, 1966.

Schrieffer, J. R., *Theory of Superconductivity*. Boulder, CO: Westview Press, 1983.

Tinkham, M., *Introduction to Superconductivity*. 2nd ed., New York, NY: McGraw-Hill, 1996.

Chapter 15

Books and reviews

Herring, C., *Magnetism IV*, eds. G. Rado and H. Suhl, New York, NY: Academic Press, 1966.

Kittel, C., *Quantum Theory of Solids*. New York, NY: Wiley, 1964.

Mattis, D., *The Theory of Magnetism*. New York, NY: Harper and Row, 1965.

White, R. M., *Quantum Theory of Magnetism*. 3rd ed., Berlin: Springer, 1970.

Chapter 16

Books and reviews

Castro Neto, A. H., Guinea, F., Peres, N. M. R., Kovoselov., K. S., and Geim, A. K., "The electronic properties of graphene," *Rev. Mod. Phys.* 81(2009), 109.

Di Ventra, M., *Electrical Transport in Nanoscale Systems*. Cambridge: Cambridge University Press, 2008.

Jorio, A., Dresselhaus, M. S., and Dresselhaus, G. (eds.), *Carbon Nanotubes: Advanced Topics in the Synthesis, Structure, Properties and Applications*. Berlin: Springer, 2008.

Tartakovskii, A. (ed.), *Quantum Dots: Optics, Electron Transport and Future Applications*. Cambridge: Cambridge University Press, 2012.

Index

PERIODIC TABLE OF ELEMENTS

With outer electron configurations of neutral atoms
in ground states and common crystal structures.

Legend:

- fcc — face-centered cubic
- bcc — body-centered cubic
- sc — simple cubic
- cub — cubic
- tetra — tetragonal
- ortho — orthorhombic
- hcp — hexagonal closed packed
- dia — diamond
- rhomb — rhombohedral
- mono — monoclinic

Each cell lists: element name, symbol, atomic number, crystal structure, outer electron configuration.

Element	Symbol	Z	Structure	Configuration
hydrogen	H	1	hcp	$1s^1$
helium	He	2	hcp	$1s^2$
lithium	Li	3	bcc	$[\bullet]2s^1$
beryllium	Be	4	hcp	$[\bullet]2s^2$
boron	B	5	tetra	$[\bullet]2s^2 2p^1$
carbon	C	6	dia	$[\bullet]2s^2 2p^2$
nitrogen	N	7	hcp	$[\bullet]2s^2 2p^3$
oxygen	O	8	cub	$[\bullet]2s^2 2p^4$
fluorine	F	9	mono	$[\bullet]2s^2 2p^5$
neon	Ne	10	fcc	$[\bullet]2s^2 2p^6$
sodium	Na	11	bcc	$[\bullet]3s^1$
magnesium	Mg	12	hcp	$[\bullet]3s^2$
aluminium	Al	13	fcc	$[\bullet]3s^2 3p^1$
silicon	Si	14	dia	$[\bullet]3s^2 3p^2$
phosphorus	P	15	cub	$[\bullet]3s^2 3p^3$
sulfur	S	16	ortho	$[\bullet]3s^2 3p^4$
chlorine	Cl	17	ortho	$[\bullet]3s^2 3p^5$
argon	Ar	18	fcc	$[\bullet]3s^2 3p^6$
potassium	K	19	bcc	$[\bullet]4s^1$
calcium	Ca	20	fcc	$[\bullet]4s^2$
scandium	Sc	21	hcp	$[\bullet]3d^1 4s^2$
titanium	Ti	22	hcp	$[\bullet]3d^2 4s^2$
vanadium	V	23	bcc	$[\bullet]3d^3 4s^2$
chromium	Cr	24	bcc	$[\bullet]3d^5 4s^1$
manganese	Mn	25	cub	$[\bullet]3d^5 4s^2$
iron	Fe	26	bcc	$[\bullet]3d^6 4s^2$
cobalt	Co	27	hcp	$[\bullet]3d^7 4s^2$
nickel	Ni	28	fcc	$[\bullet]3d^8 4s^2$
copper	Cu	29	fcc	$[\bullet]3d^{10} 4s^1$
zinc	Zn	30	hcp	$[\bullet]3d^{10} 4s^2$
gallium	Ga	31	ortho	$[\bullet]4s^2 4p^1$
germanium	Ge	32	dia	$[\bullet]4s^2 4p^2$
arsenic	As	33	rhomb	$[\bullet]4s^2 4p^3$
selenium	Se	34	hcp	$[\bullet]4s^2 4p^4$
bromine	Br	35	ortho	$[\bullet]4s^2 4p^5$
krypton	Kr	36	fcc	$[\bullet]4s^2 4p^6$
rubidium	Rb	37	bcc	$[\bullet]5s^1$
strontium	Sr	38	fcc	$[\bullet]5s^2$
yttrium	Y	39	hcp	$[\bullet]4d^1 5s^2$
zirconium	Zr	40	hcp	$[\bullet]4d^2 5s^2$
niobium	Nb	41	bcc	$[\bullet]4d^4 5s^1$
molybdenum	Mo	42	bcc	$[\bullet]4d^5 5s^1$
technetium	Tc	43	hcp	$[\bullet]4d^5 5s^2$
ruthenium	Ru	44	hcp	$[\bullet]4d^7 5s^1$
rhodium	Rh	45	fcc	$[\bullet]4d^8 5s^1$
palladium	Pd	46	fcc	$[\bullet]4d^{10}$
silver	Ag	47	fcc	$[\bullet]4d^{10} 5s^1$
cadmium	Cd	48	hcp	$[\bullet]4d^{10} 5s^2$
indium	In	49	tetra	$[\bullet]5s^2 5p^1$
tin	Sn	50	tetra	$[\bullet]5s^2 5p^2$
antimony	Sb	51	rhomb	$[\bullet]5s^2 5p^3$
tellurium	Te	52	hcp	$[\bullet]5s^2 5p^4$
iodine	I	53	ortho	$[\bullet]5s^2 5p^5$
xenon	Xe	54	fcc	$[\bullet]5s^2 5p^6$
caesium	Cs	55	bcc	$[\bullet]6s^1$
barium	Ba	56	bcc	$[\bullet]6s^2$
lutetium	Lu	71	hcp	$[\bullet]4f^{14} 5d^1 6s^2$
hafnium	Hf	72	hcp	$[\bullet]5d^2 6s^2$
tantalum	Ta	73	bcc	$[\bullet]5d^3 6s^2$
tungsten	W	74	bcc	$[\bullet]5d^4 6s^2$
rhenium	Re	75	hcp	$[\bullet]5d^5 6s^2$
osmium	Os	76	hcp	$[\bullet]5d^6 6s^2$
iridium	Ir	77	fcc	$[\bullet]5d^7 6s^2$
platinum	Pt	78	fcc	$[\bullet]5d^9 6s^1$
gold	Au	79	fcc	$[\bullet]5d^{10} 6s^1$
mercury	Hg	80	rhomb	$[\bullet]5d^{10} 6s^2$
thallium	Tl	81	hcp	$[\bullet]6s^2 6p^1$
lead	Pb	82	fcc	$[\bullet]6s^2 6p^2$
bismuth	Bi	83	rhomb	$[\bullet]6s^2 6p^3$
polonium	Po	84	sc	$[\bullet]6s^2 6p^4$
astatine	At	85	–	$[\bullet]6s^2 6p^5$
radon	Rn	86	fcc	$[\bullet]6s^2 6p^6$
francium	Fr	87	bcc	$[\bullet]7s^1$
radium	Ra	88	–	$[\bullet]7s^2$
lawrencium	Lr	103	–	$[\bullet]5f^{14} 7s^2 7p^1$
rutherfordium	Rf	104	–	$[\bullet]6d^2 7s^2$
dubnium	Db	105	–	$[\bullet]6d^3 7s^2$
seaborgium	Sg	106	–	$[\bullet]6d^4 7s^2$
bohrium	Bh	107	–	$[\bullet]6d^5 7s^2$
hassium	Hs	108	–	$[\bullet]6d^6 7s^2$
meitnerium	Mt	109	–	$[\bullet]6d^7 7s^2$
darmstadtium	Ds	110	–	$[\bullet]6d^8 7s^2$
roentgenium	Rg	111	–	$[\bullet]6d^{10} 7s^1$

* Lanthanides

Element	Symbol	Z	Structure	Configuration
lanthanum	La	57	hcp	$[\bullet]5d^1 6s^2$
cerium	Ce	58	fcc	$[\bullet]4f^1 5d^1 6s^2$
praseodymium	Pr	59	hcp	$[\bullet]4f^3 6s^2$
neodymium	Nd	60	hcp	$[\bullet]4f^4 6s^2$
promethium	Pm	61	–	$[\bullet]4f^5 6s^2$
samarium	Sm	62	rhomb	$[\bullet]4f^6 6s^2$
europium	Eu	63	bcc	$[\bullet]4f^7 6s^2$
gadolinium	Gd	64	hcp	$[\bullet]4f^7 5d^1 6s^2$
terbium	Tb	65	hcp	$[\bullet]4f^9 6s^2$
dysprosium	Dy	66	hcp	$[\bullet]4f^{10} 6s^2$
holmium	Ho	67	hcp	$[\bullet]4f^{11} 6s^2$
erbium	Er	68	hcp	$[\bullet]4f^{12} 6s^2$
thulium	Tm	69	hcp	$[\bullet]4f^{13} 6s^2$
ytterbium	Yb	70	fcc	$[\bullet]4f^{14} 6s^2$

** Actinides

Element	Symbol	Z	Structure	Configuration
actinium	Ac	89	fcc	$[\bullet]6d^1 7s^2$
thorium	Th	90	fcc	$[\bullet]6d^2 7s^2$
protactinium	Pa	91	tetra	$[\bullet]5f^2 6d^1 7s^2$
uranium	U	92	ortho	$[\bullet]5f^3 6d^1 7s^2$
neptunium	Np	93	ortho	$[\bullet]5f^4 6d^1 7s^2$
plutonium	Pu	94	mono	$[\bullet]5f^6 7s^2$
americium	Am	95	–	$[\bullet]5f^7 7s^2$
curium	Cm	96	hcp	$[\bullet]5f^7 6d^1 7s^2$
berkelium	Bk	97	–	$[\bullet]5f^9 7s^2$
californium	Cf	98	–	$[\bullet]5f^{10} 7s^2$
einsteinium	Es	99	–	$[\bullet]5f^{11} 7s^2$
fermium	Fm	100	–	$[\bullet]5f^{12} 7s^2$
mendelevium	Md	101	–	$[\bullet]5f^{13} 7s^2$
nobelium	No	102	–	$[\bullet]5f^{14} 7s^2$

Fundamental physical constants

Quantity	Symbol (expressions in CGS units)	Value	CGS	SI
Speed of light in vacuum	c	2.997 924 58	10^{10} cm s^{-1}	10^{8} m s^{-1}
Elementary charge	e	1.602 176 620 8	–	10^{-19} C
	e	4.803 204 673 0	10^{-10} esu	–
Planck's constant	h	6.626 070 040	10^{-27} erg s	10^{-34} J s
	$\hbar = h/2\pi$	1.054 571 800	10^{-27} erg s	10^{-34} J s
Electron mass	m	9.109 383 56	10^{-28} g	10^{-31} kg
Proton mass	M_p	1.672 621 898	10^{-24} g	10^{-27} kg
Proton-electron mass ratio	M_p/m	1 836.152 673 89	–	–
Inverse fine structure constant	$1/\alpha = \hbar c/e^2$	137.035 999 139	–	–
Bohr radius	$a_0 = \hbar/mc\alpha$	5.29 177 210 67	10^{-9} cm	10^{-11} m
Bohr magneton	$\mu_B = e\hbar/2mc$	9.274 009 994	10^{-21} erg G^{-1}	10^{-24} J T^{-1}
Rydberg constant	$R_\infty = \alpha/4\pi a_o$	1.097 373 156 850 8	10^{9} cm^{-1}	10^{7} m^{-1}
Rydberg energy	$\mathrm{Ry} = hcR_\infty$	2.179 872 325	10^{-11} erg	10^{-18} J
	Ry/eV	13.605 693 009	–	–
Atomic unit of energy	$\mathrm{E_h} = 2\,\mathrm{Ry}$	4.359 744 650	10^{-11} erg	10^{-18} J
	$\mathrm{E_h}/\mathrm{eV}$	27.211 386 02	–	–
Magnetic flux quantum	$\phi_0 = hc/e$	4.135 667 662	10^{-7} G cm^2	10^{-15} Wb
Conductance quantum	$G_0 = 2e^2/h$	7.748 091 731 0	–	10^{-5} S
		6.963 637 568 6	10^{7} statohm^{-1}	–
Boltzmann constant	k_B	1.380 648 52	10^{-16} erg K^{-1}	10^{-23} J K^{-1}
Permittivity of free space	ε_0	–	1	$10^{7}/4\pi c^2$
Permeability of free space	μ_0	–	1	$4\pi \times 10^{-7}$
Electron volt	eV	1.602 176 620 8	10^{-12} erg	10^{-19} J
	eV/h	2.417 989 262 3	10^{14} Hz	10^{14} Hz
	eV/hc	8.065 544 004 8	10^{3} cm^{-1}	10^{5} m^{-1}
	eV/k_B	1.160 452 21	10^{4} K	10^{4} K
Thermal energy at 300 K	$k_B T_{300\mathrm{K}}/\mathrm{eV}$	0.025 852	–	–

Source: P.J. Mohr, B.N. Taylor, and D.B. Newell (2015), "The 2014 CODATA Recommended Values of the Fundamental Physical Constants" (Web Version 7.0). National Institute of Standards and Technology, Gaithersburg, MD 20899.